Science at Oxford
1914–1939

Science at Oxford
1914–1939

Transforming an Arts University

JACK MORRELL

CLARENDON PRESS · OXFORD
1997

Oxford University Press, Great Clarendon Street, Oxford OX2 6DP
Oxford New York
Athens Auckland Bangkok Bogota Bombay
Buenos Aires Calcutta Cape Town Dar es Salaam
Delhi Florence Hong Kong Istanbul Karachi
Kuala Lumpur Madras Madrid Melbourne
Mexico City Nairobi Paris Singapore
Taipei Tokyo Toronto Warsaw
and associated companies in
Berlin Ibadan

Oxford is a trade mark of Oxford University Press

Published in the United States
by Oxford University Press Inc., New York

© Jack Morrell 1997

All rights reserved. No part of this publication may be reproduced,
stored in a retrieval system, or transmitted, in any form or by any means,
without the prior permission in writing of Oxford University Press.
Within the UK, exceptions are allowed in respect of any fair dealing for the
purpose of research or private study, or criticism or review, as permitted
under the Copyright, Designs and Patents Act, 1988, or in the case of
reprographic reproduction in accordance with the terms of the licences
issued by the Copyright Licensing Agency. Enquiries concerning
reproduction outside these terms and in other countries should be
sent to the Rights Department, Oxford University Press,
at the address above

British Library Cataloguing in Publication Data
Data available

Library of Congress Cataloging in Publication Data
Morrell, Jack.
Science at Oxford, 1914–1939: transforming an arts university/Jack Morrell.
p. cm.
Includes bibliographical references and index.
1. University of Oxford—History—20th century. 2. Science—Study
and teaching (Graduate)—England—Oxford—History—20th century.
I. Title.
Q183.4.G730946 1997
507.1'1425'74—dc21 97-7976

ISBN 0-19-820657-7

1 3 5 7 9 10 8 6 4 2

Typeset by Pure-Tech India Ltd, Pondicherry
Printed in Great Britain
on acid-free paper by
Biddles Ltd., Guildford & King's Lynn

To my
mother
and
father

PREFACE

THIS book has grown out of my chapter on the non-medical sciences 1914–39 which I was invited to write in 1986 by Brian Harrison for what became *The History of the University of Oxford*, viii: *The Twentieth Century* (Oxford, 1994) which he edited. In establishing even the basic narrative I discovered so much interesting material which I had to exclude from my chapter that I decided to write a book about the physical and biomedical sciences at Oxford 1914–39 with the aim of doing full justice to this rich empirical material and the importance of the topic. In his preface to the *History* Brian Harrison pointed out that substantial research on Oxford's recent history still needs to be done. This book was well underway before he published that call but it is, I hope, a useful response to his summons.

For information about Oxonians I am indebted to the librarians, archivists, and curators of many Oxford colleges and university institutions. My research at Oxford has been particularly facilitated by Tony Simcock of the Museum of the History of Science, by Steven Tomlinson of the Bodleian Library, and by his ever-cheerful colleagues in room 132. Two successive university archivists, Margaret MacDonald and Simon Bailey, were always helpful. Serious research on the subject of this book began in 1987 when I enjoyed a grant from the Leverhulme Trust, a supernumerary fellowship at Brasenose College, Oxford, for a term, and study-leave from the University of Bradford. Subsequently I have been indebted to the Royal Society of London for a research grant, to the Leverhulme Trust for an emeritus fellowship, and to Brasenose College for continued hospitality and encouragement. In Yorkshire I have benefited from the first-rate services provided by the university libraries of Bradford and Leeds.

In writing this book I have profited from the support and advice of many people. For valuable responses to my queries I am indebted to Sir Edward Abraham, Dr M. Adam, the late Sir Peter Allen, Professor W. Baker, the late Professor R. P. Bell, Professor B. Bleaney, Professor E. F. Caldin, Dame Mary Cartwright, Sir Derman Christopherson, the late C. H. Collie, the late Dr A. H. Cooke, Mr E. H. Cooke-Yarborough, the late Professor T. G. Cowling, Dr C. J. Danby, the late C. S. Elton, the late E. B. Ford, Mr J. E. Garfitt, Dr B. Harrison, the late Professor D. C. Hodgkin, Professor R. V. Jones, the late Lord Kearton, the late Dr H. G. Kuhn, Professor N. Kurti, Professor K. J. Laidler, Dr J. D. Lambert, Professor D. A. Long, Professor S. F. Mason, Dr H. Megaw, Mr D. Murray-Rust, Mr E. M. Nicholson, the late Dr A. G. Ogston, Dr P. C. Reynell, Mr R. S. Smith, the late

Dr L. Staveley, Dr H. J. Stern, Dr F. W. Stoyle, the late Dr L. E. Sutton, Lord Todd, the late Dr M. Tomlinson, the late Professor A. R. Ubbelohde, Dr G. J. Whitrow, Professor E. N. Willmer, Sir E. M. Wright, Professor V. C. Wynne-Edwards, and Professor J. Z. Young. I have learned much from seminars I gave at Cambridge, Leeds, London, Manchester, and Oxford. Colleagues in history of science, especially those at the universities of Leeds and Manchester, have given friendship, stimulus, and useful references. Graeme Gooday, Jon Harwood, Jeff Hughes, Gerrylynn Roberts, Jim Secord, and Paul Weindling were kind enough to read parts of the book and to offer valuable comment. John Heilbron and Simon Schaffer boosted my vanity by assuring me that my writing about twentieth-century physics was not mentally negligible. John Pickstone has read much of the typescript, has helped to sharpen its arguments, and has saved me from myopic antiquarianism. Having dragged me in middle age into the history of twentieth-century science, Brian Harrison has supported that transfer with unflagging enthusiasm. He is the chief accoucheur of this book.

For expert typing I am grateful to Hilary Mason, Eleanor Cosgrove, Sally Fox, Judy Aspinall, and Jane Weightman. At Oxford University Press Tony Morris and Anna Illingworth have aided a fellow mountain-walker far beyond the call of duty. For her forbearance I am beholden to my wife. Last but not least I remain grateful to my mother and father, to whose memory this book is dedicated.

J. M.

ACKNOWLEDGEMENTS

For permission to cite materials from the university archives, I am grateful to David Vaisey, keeper of the archives, and the registrar of the University. For permission to cite manuscripts in their care or ownership, I am indebted to: the Bodleian Library, Oxford (Asquith Commission Papers, Bodleian Library Records, C. S. Elton Papers, A. C. Hardy Papers, J. L. Harley Papers, D. C. Hodgkin Papers, W. Hume-Rothery Papers, E. A. Milne Papers, K. Mendelssohn Papers, R. A. Peters Papers, H. M. Powell Papers, Radcliffe Trustees Papers, F. Soddy Papers, L. E. Sutton Papers); Brasenose College, Oxford (Jenkinson–Curzon letters); Christ Church, Oxford (A. S. Russell Papers); Churchill College, Cambridge (J. D. Cockroft Papers); Department of Experimental Psychology, Oxford (W. Brown letter); Department of Physiology, Oxford (C. S. Sherrington collection); Department of Zoology, Oxford, Alexander Library (Oxford Bird Census collection); History of Science Museum, Oxford (E. J. Bowen collection, Queen's laboratory books, minute books of chemistry subfaculty and of Alembic Club); Lincoln College, Oxford (N. V. Sidgwick Papers); Nuffield College, Oxford (F. Lindemann Papers); Radcliffe Trustees; Rice University Library (J. S. Huxley Papers); Rockefeller Foundation archives; Royal Society of London (Council minutes, H. W. Florey Papers, F. E. Simon Papers, Warren Committee minutes); Society for the Protection of Science and Learning (papers deposited in Bodleian); University College, Oxford (A. D. Gardner autobiography); University Museum, Oxford (J. A. Douglas recollections); University of Glasgow Library (F. O. Bower Papers).

For permission to draw on previously published articles I am obliged to the University of Chicago Press for my 'W. H. Perkin, Jr., at Manchester and Oxford: From Irwell to Isis', *Osiris*, 8 (1993), 104–26 (copyright 1993 by the History of Science Society, Inc. All rights reserved); and to the Regents of the University of California for my 'Research in Physics at the Clarendon Laboratory, Oxford, 1919–1939', *Historical Studies in the Physical and Biological Sciences*, 22 (1992), 263–307.

CONTENTS

List of Illustrations   xii
List of Maps   xiii
List of Tables   xiv
Abbreviations and Conventions   xv
Glossary and Note   xvi

Introduction   1
 1. The Changing Salience of Oxford Science 1914–1939   6
 2. The Colleges and the University   52
 3. Philosophy, Liberal Education, and the Industrial Spirit   82
 4. Agriculture and Forestry   112
 5. Biomedical Sciences   161
 6. Small Sciences   217
 7. Zoology   268
 8. The Big Battalions   305
 9. Refugee Scientists   369
 10. Conclusion   433

Index   453

LIST OF ILLUSTRATIONS

1. Old Chemistry department, 1937 — 43
2. Human physiology experiment, 1930s — 173
3. Spitsbergen Expedition, Norway, 1921 — 276
4. Chemistry laboratory, Trinity College, 1921–9 — 323
5. N. V. Sidgwick and B. Lambert, 1926 — 325
6. Chemistry laboratory, Trinity College, 1937 — 329
7. K. A. G. Mendelssohn, 1930s — 405
8. Clarendon Laboratory, c.1940 — 429

LIST OF MAPS

1. Map of academic Oxford showing the main university and college buildings, 1914–1990   xviii–xix
2. Site for the proposed Department of Pathology 1922   32
3. Extension of the Museum site 1924   33
4. Science site problems 1929–1930   36
5. Plan of the Science Area 1934   39
6. Plan for the Science Area 1937   44

LIST OF TABLES

2.1.	Finalists by subject	66
2.2.	Final degree subjects as a percentage of the academic year total	67
4.1.	Ministry of Agriculture grants to Oxford agriculture and student numbers 1925–1939	117
4.2.	Main external funding of school of forestry and IFI	148
8.1.	College chemistry laboratories 1922–1923	323
9.1.	Income of Clarendon Laboratory 1919–1939	388
10.1.	Student numbers in some British universities 1934–1935	434
10.2.	Income of some British universities 1934–1935	436

ABBREVIATIONS AND CONVENTIONS

References to entries in *DNB* and *DSB* are not given as a matter of course. They are given only when the entry is the best short source or is exceptionally good. All items abbreviated in footnotes are given in full on their first occurrence, except for the following which apply throughout:

BM	*Biographical Memoirs of Fellows of the Royal Society of London*
BMJ	*British Medical Journal*
BS/M	Biological Sciences Faculty Board Minutes, OUA
BS/R	Biological Sciences Faculty Board Reports, OUA
DNB	*Dictionary of National Biography*
DSB	*Dictionary of Scientific Biography*
Franks Report	*University of Oxford: Report of Commission of Enquiry* (Oxford, 1966)
HCP	*Hebdomadal Council Papers*, OUA
LP	Lindemann Papers
Med./M	Medical Sciences Faculty Board Minutes, OUA
Med./R	Medical Sciences Faculty Board Reports, OUA
MP	Mendelssohn Papers
NS/M	Natural Sciences Faculty Board Minutes, OUA
NS/R	Natural Sciences Faculty Board Reports, OUA
OM	*Oxford Magazine*
ON	*Obituary Notices of Fellows of the Royal Society of London*
OUA	Oxford University Archives, Bodleian Library, Oxford
OUG	*Oxford University Gazette*
PS/M	Physical Sciences Faculty Board Minutes, OUA
PS/R	Physical Sciences Faculty Board Reports, OUA
SP	Simon Papers
SPSL	Society for the Protection of Science and Learning Papers

GLOSSARY AND NOTE

college Oxford colleges were autonomous, with their own income, and self-governing. They admitted undergraduates to the University, provided residential accommodation for them, controlled them within their walls, and were responsible for teaching them especially via the famous tutorial hour. The following colleges or their equivalents, listed alphabetically, appear in the text: All Souls; Balliol; Brasenose; Christ Church; Corpus Christi; Exeter; Hertford; Jesus; Keble; Lady Margaret Hall; Lincoln; Magdalen; Merton; New; Oriel; Oxford Home Students (later St Anne's); Pembroke; Queen's; St Catherine's Society; St Edmund Hall; St Hilda's; St Hugh's; St John's; St Peter's; Somerville; Trinity; University; Wadham; and Worcester. In all references to a particular college, only its name is given. Thus 'Worcester' means Worcester College and not the cathedral town

Convocation made up of all graduates who had taken their MA by leaving their name on their college's books and by paying a small sum to the University. For much of the inter-war period its powers were limited compared with those of Congregation, but it did elect the chancellor and certain professors

Congregation the sovereign body of the University which accepted or rejected the legislative proposals put to it as statutes or decrees. From 1912 membership was restricted to those members of Convocation who lived for at least twenty weeks a year within a mile and a half of the centre of Oxford. From 1925 membership was further restricted to those engaged in teaching and administration in the University

General Board of the Faculties established in 1913 to co-ordinate and supervise the work of the faculty boards which at that time numbered four (theology, law, arts, natural sciences). In the 1930s it gradually attained greater power as a result of increased financial responsibilities being given to it

Greats a degree course which covered literature, history, and philosophy of ancient Greece and Rome

Hebdomadal Council the chief administrative body of the University, which met weekly. It alone could propose legislation to Congregation. The majority of its members were elected by Congregation but not more than three members could be attached to any one college

school this normally meant a course of study culminating in a final degree examination known as finals

vice-chancellor the administrative head of the University. He was appointed from among the heads of the colleges. The office was held in rotation, according to seniority as a head of a college, for three years at a time. The incumbent was appointed by the chancellor, the titular head of the University

Note on prices and incomes

A conversion factor of about thirty to thirty-five in the value of sterling gives a very rough comparison of the monetary values of 1919–39 with those of the late 1990s.

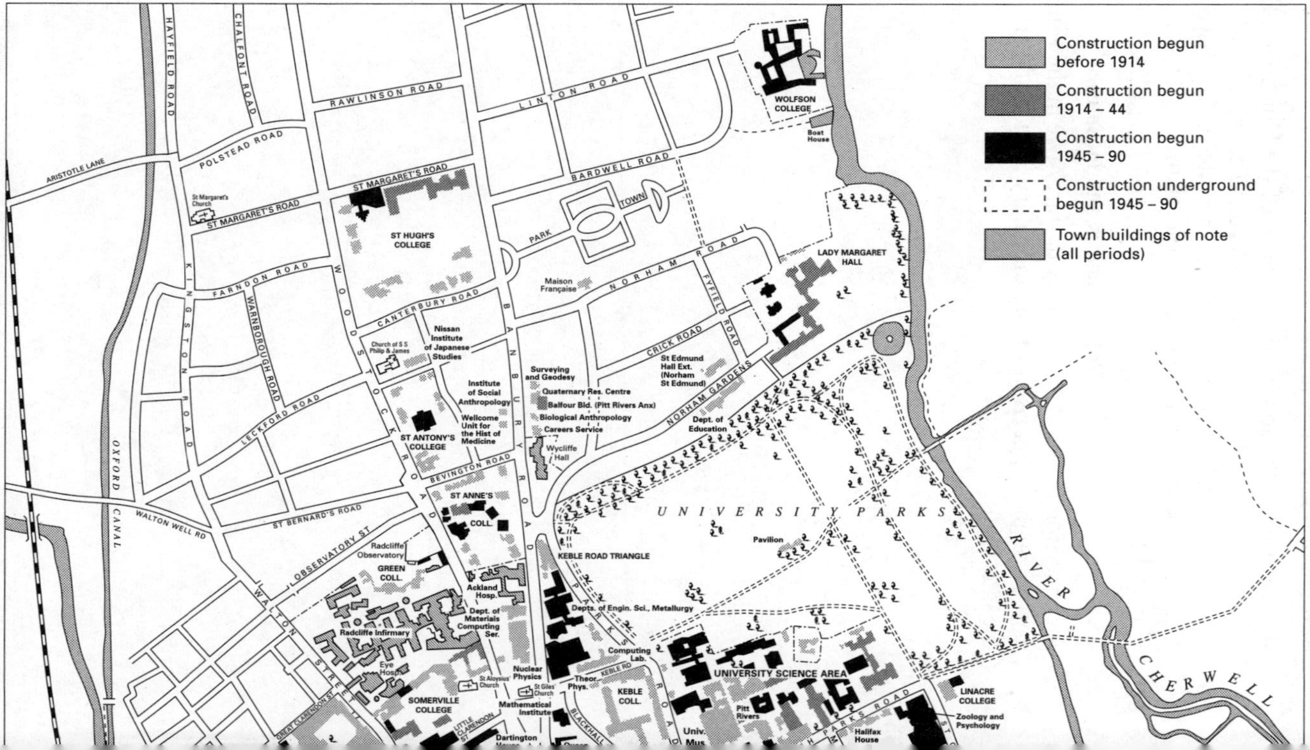

Construction begun before 1914
Construction begun 1914 – 44
Construction begun 1945 – 90
Construction underground begun 1945 – 90
Town buildings of note (all periods)
WOLFSON COLLEGE
Boat House
LADY MARGARET HALL
ST HUGH'S COLLEGE
ST ANTONY'S COLLEGE
ST ANNE'S COLL.
GREEN COLL.
SOMERVILLE COLLEGE
KEBLE COLL.
LINACRE COLLEGE
UNIVERSITY PARKS
UNIVERSITY SCIENCE AREA
KEBLE ROAD TRIANGLE
RIVER CHERWELL
OXFORD CANAL
BANBURY ROAD
WOODSTOCK ROAD
PARKS ROAD
NORHAM ROAD
NORHAM GARDENS
FYFIELD ROAD
CRICK ROAD
BARDWELL ROAD
LINTON ROAD
RAWLINSON ROAD
ST MARGARET'S ROAD
PARK TOWN
CANTERBURY ROAD
BEVINGTON ROAD
FARNDON ROAD
WARNBOROUGH ROAD
LECKFORD ROAD
KINGSTON ROAD
CHALFONT ROAD
POLSTEAD ROAD
HAYFIELD ROAD
ARISTOTLE LANE
WALTON WELL RD
ST BERNARD'S ROAD
OBSERVATORY ST
WALTON STREET
KEBLE RD
Maison Française
Nissan Institute of Japanese Studies
Institute of Social Anthropology
Wellcome Unit for the Hist of Medicine
Surveying and Geodesy
Quaternary Res. Centre
Balfour Bld. (Pitt Rivers Anx)
Biological Anthropology
Careers Service
Wycliffe Hall
St Edmund Hall Ext. (Norham St Edmund)
Dept. of Education
Pavilion
Radcliffe Observatory
Ackland Hosp.
Dept. of Materials
Computing Ser.
Radcliffe Infirmary
Eye Hosp.
Nuclear Physics
Theor. Phys.
Mathematical Institute
Depts. of Engin. Sci., Metallurgy
Computing Lab.
Pitt Rivers
Univ. Mus.
Halifax House
Zoology and Psychology
St Margaret's Church
Church of S S Philip & James
St Aloysius' Church
St Giles' Church
Dartington House

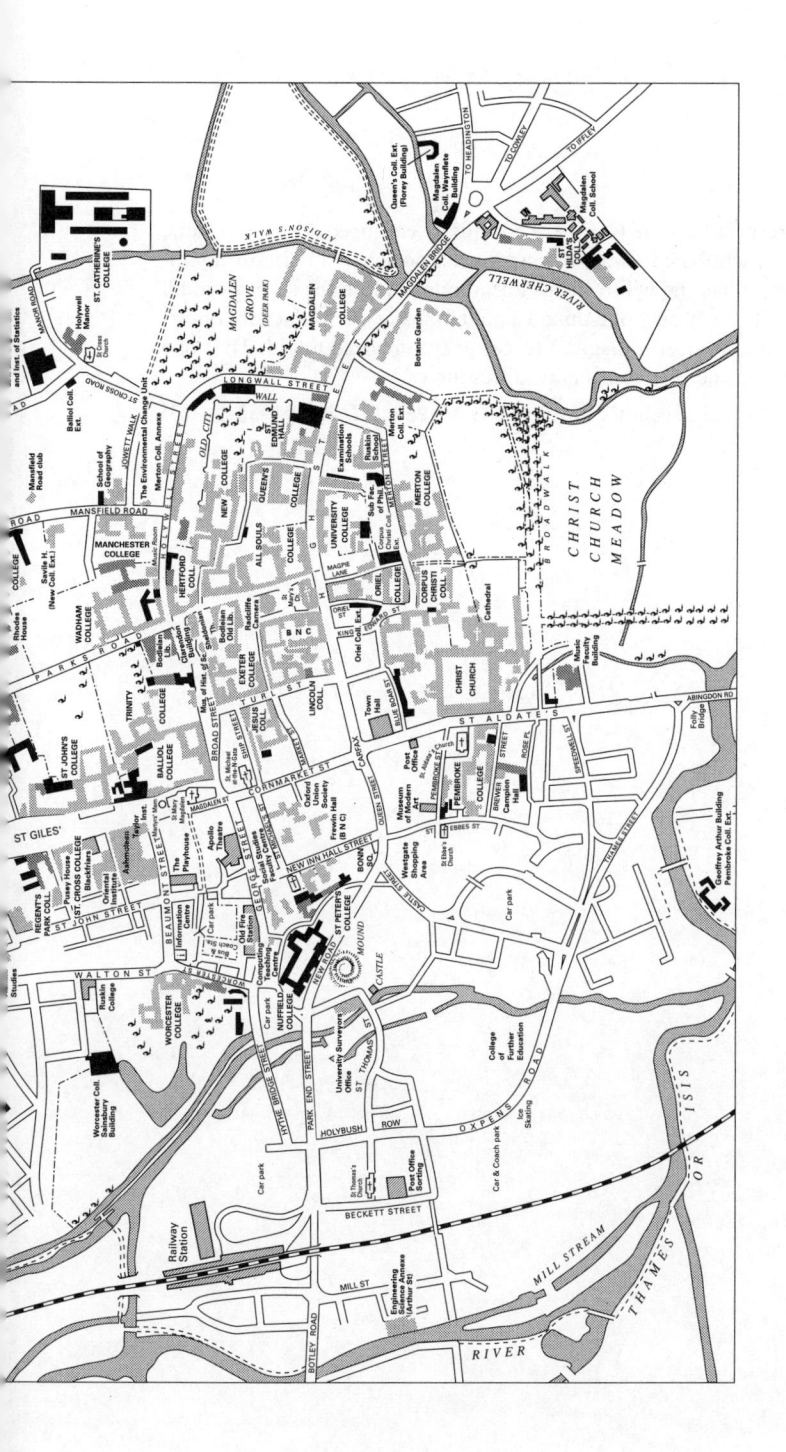

MAP 1. Map of academic Oxford showing the main university and college buildings, 1914–1990
Source: *The History of the University of Oxford*, viii. 501.

In every respect but one Oxford is splendidly equipped. It has teachers, researchers, scholars; it can command brain-power to an almost unlimited extent; it has the will to forge ahead and explore the sciences in all directions. There is only one thing wanted to make our University one of the greatest and most influential forces in the civilized world. That one thing is a large accession of financial resources.

(Vice-chancellor's oration, 5 Oct. 1932, *OUG* 63 (1932–3), 23)

INTRODUCTION

The term Oxbridge implies that the two ancient English universities had and have much in common. It is true that they are collegiate universities, that the numbers of undergraduates attending them have shown remarkable parallelism over several centuries, and that they have been disproportionately successful in producing national élites of various kinds. Yet these similarities mask an intellectual difference that became apparent in the eighteenth century and more marked later: Cambridge specialized in mathematics and science while Oxford focused on classics in particular and arts subjects in general. This difference has been emphasized by recent autobiographies and polemical works which depict Oxford as a nursery of statesmen and -women, administrators, opinion formers, and media personalities.[1] Like Evelyn Waugh's *Brideshead Revisited*, a recent volume of reminiscence has no time for laboratories and for scientists except to dismiss them as northern chemists, Adrian Moles, and stains.[2]

Yet these stereotypes ignore the remarkable fact that between 1914 and 1939 science at the University of Oxford, still renowned for its devotion to literae humaniores, was transformed. In 1914 the University saw teaching and scholarship as its prime functions. In college fellowships, undergraduate composition, college awards, university prizes, and the distribution of finalists, arts subjects were dominant. College fellows were often Oxonians who believed in liberal education and scorned vocational subjects. In some scientific fields it seemed that the miasma of the Thames valley had reduced the professoriate to perpetual torpor. Men of considerable research reputation when appointed became dodos who held on to their chairs because they had no university pension scheme and were not subject to compulsory retirement at a given age. Though there were some gifted scholars and specialist researchers in science among college fellows and the professoriate, most worked individually and research groups were few and far between. External funding from industry, individuals, and corporate philanthropists was limited. Unlike University College London, Manchester, and Cambridge,

---

[1] D. Healey, *The Time of my Life* (London, 1989); R. Jenkins, *A Life at the Centre* (London, 1991); B. Castle, *Fighting All the Way* (London, 1993); M. Thatcher, *The Path to Power* (Oxford, 1995); W. Ellis, *The Oxbridge Conspiracy* (London, 1994). Healey read classics; Jenkins and Castle philosophy, politics, and economics; Thatcher chemistry.

[2] R. Johnson (ed.), *The Oxford Myth* (London, 1988).

Oxford could not boast of a Nobel prize-winner.[3] Above all the contrast with Cambridge was dismal in mathematics, physics, engineering, and physiology, and in providing presidents of the Royal Society.[4]

Yet from 1900 some colleges and the University had been occasionally so embarrassed about Oxford's science that innovations had taken place. In order to remedy the perceived deficiencies of the university chemistry laboratory, two colleges (Queen's and Jesus) opened their own to add to the three college laboratories already run by Christ Church, Magdalen, and Balliol and Trinity in harness. Thus by 1914 chemistry was strongly supported by the colleges. Nor was the University idle in animating chemistry. In 1912 it took the then novel step of head-hunting from a provincial university a distinguished non-Oxonian researcher and research school leader to fill the Waynflete chair of chemistry. Next year it acted likewise for physiology. Thus W. H. Perkin and Sherrington became the first two established researchers in science to be poached by Oxford from northern universities. The University was also prepared to accept odd offers of external funding to launch new ventures. Forestry paid for by the India Office, agricultural economics funded by the Development Commission and the Ministry of Agriculture, and an electrical laboratory built by the Drapers' Company to remedy the failings of the Clarendon Laboratory were all established between 1905 and 1912. So too was engineering science.

By 1939 there were continuities with 1914. Oxford was still devoted to teaching and scholarship. There were still more places for undergraduates than applicants. There had been little shift in the proportion of college fellows and undergraduates in science. Inbreeding in appointments to fellowships persisted. No major new degree course in science had been introduced. No scientist had been elected head of a college. Liberal education was still the norm, applied science and vocational subjects being scorned by arts men and on occasion by scientists. There were also, however, major discontinuities. While some existing facilities were given a new research life, institutional innovations galore, often externally funded, provided new habitations for vigorous researchers. Postgraduate work flourished, partly on the basis

---

[3] William Ramsay, Nobel prize for chemistry 1904, was professor at UCL 1887–1912; Ernest Rutherford, chemistry 1908, professor of physics, University of Manchester, 1907–19; John William Strutt, Lord Rayleigh, physics 1904, professor of physics, University of Cambridge, 1879–84; J. J. Thomson, physics 1906, professor of physics, University of Cambridge, 1884–1919.

[4] There are no book-length Oxford equivalents of W. W. R. Ball, *A History of the Study of Mathematics at Cambridge* (Cambridge, 1889); J. G. Crowther, *The Cavendish Laboratory 1874–1974* (London, 1974); T. J. N. Hilken, *Engineering at Cambridge University 1783–1965* (Cambridge, 1967); G. L. Geison, *Michael Foster and the Cambridge School of Physiology: The Scientific Enterprise in Late Victorian Society* (Princeton, 1978); of presidents of the Royal Society 1900–40, J. J. Thomson 1915–20, Rutherford 1925–30, F. G. Hopkins 1930–5 occupied the office while holding a Cambridge chair, Rayleigh 1905–8 had been a Cambridge professor, and W. H. Bragg 1935–40 was a Cambridge graduate.

of the new degree of D.Phil., with dons assuming the new role of research supervisor. In some subjects the old modes of solo research or of director and pupil working as a pair were supplemented by the use of research lieutenants and by team research involving a multi-disciplinary approach. Many scientific disciplines acquired for the first time a research identity. Research in science was slowly and sometimes grudgingly accepted by the University as an institutional imperative. In 1934 the Science Area was formally designated. Generally Oxford became more visible in the national and international scientific worlds. Two professors, Soddy and Sherrington, won Nobel prizes in 1921 and 1932, Sherrington having been president of the Royal Society 1920–5. By 1939 there were five people in post who were to gain Nobel prizes later (Florey 1945, Chain 1945, Robinson 1947, Hinshelwood 1956, and Hodgkin 1964). In four cases the award was made for work done entirely in Oxford. Three of these Oxonians (Robinson 1945–50, Hinshelwood 1950–5, and Florey 1960–5) dominated the presidency of the Royal Society in the twenty years after the second war, disturbing the previous monopoly by Cantabrigians. By 1939 the contrast with Cambridge was far less embarrassing and it was no longer possible to argue that Oxford should concentrate on traditional strengths in arts subjects and leave science to Cambridge. In the 1930s Oxford's science graduates became imbued with the research spirit: 20 per cent of male science finalists but only 8 per cent of their arts equivalents went into university teaching; in the case of women science finalists, no less than 25 per cent pursued university teaching compared with 14 per cent in arts.[5]

In analysing Oxford science between 1914 and 1939 the main problem is to explain how research prospered at a time when the place of science in undergraduate education remained relatively unchanged. My main theme is therefore how scientists dealt with the structures and interests associated with the dominance of arts subjects in an academically conservative university. Their success at the research level, against considerable odds, enabled them to make a notable contribution as boffins during the second war, to provide a platform for the expansion of undergraduate teaching in science in the 1950s, and to prepare the way for Oxford's rivalling Cambridge by the 1970s as a world-famous centre for science. The subject of Oxford science between the wars is therefore not only important for the historical ecology of science but also for understanding a fundamental transformation of one of the world's major universities. As the growth of research was easily the chief element in changing the face of science in Oxford between the wars, I focus mainly on researchers and research done. Of course Oxford continued to give high priority to the teaching of undergraduates and I do not neglect that

[5] D. Greenstein, 'The Junior Members, 1900–1990: A Profile', in B. Harrison (ed.), *The History of the University of Oxford*, viii: *The Twentieth Century* (Oxford, 1994), 45–77 (68–9, 74–5).

concern. On the other hand many interesting questions about undergraduates and their studies are not systematically explored because they were not important in the controversies, both internal and external, about the status and roles of science in the University. Thus, in the account of the dire fortunes of agriculture, I note without amplification that the teaching of agriculture helped to prepare young aristocrats to manage family estates. This social fact is taken into account as a continuing element in my account of the central problems faced by agriculture. These were the vexed relations between the school and the two research institutes and the consequent disputes between the University, the Ministry of Agriculture, and the Agricultural Research Council.

Science was and is the socially organized activity of setting and solving problems concerned with the natural world, the value of the problems and the adequacy of the answers being subject to critical scrutiny.[6] One end-product of that activity was published knowledge. This book examines the habitats of various scientific enterprises in one institution at a particularly interesting time, while also trying to characterize the knowledge produced by them. Where it is feasible to do so I have argued that some features of some sciences were the results of various agents' perceiving local or national circumstances as enabling, indifferent, or hostile. In trying to relate the 'context' of science to its 'content' or 'direction', I have borne in mind E. P. Thompson's view that social and cultural formations (such as science) owe as much to agency as to circumstances.[7] Yet an awareness of the latter enables one to avoid a besetting sin of the history of universities, namely, boosterism. Too much writing on Oxford in general sees it as equivalent to England or even the world. I have tried to avoid such unpardonable exaggeration by taking cognizance of the larger themes pertaining to science in Britain and elsewhere between the wars, including the careers of science in other universities, especially Cambridge. In order to capture the peculiarities and complexities of science at Oxford I have used an exceptionally wide range of published and especially unpublished sources.

There have been some excellent recent book-length studies of the history of a science in a particular university or in several.[8] Recent biographies have illuminated the institutional history of a particular science and German

[6] J. R. Ravetz, *Scientific Knowledge and its Social Problems* (Oxford, 1971), 69–240.
[7] E. P. Thompson, *The Making of the English Working Class* (Harmondsworth, 1968), 9.
[8] Geison, *Cambridge Physiology*; K. M. Olesko, *Physics as a Calling: Discipline and Practice in the Königsberg Seminar for Physics* (Ithaca, NY, 1991); J. L. Heilbron and R. W. Seidel, *Lawrence and his Laboratory: A History of the Lawrence Berkeley Laboratory*, i (Berkeley and Los Angeles, 1989); R. E. Kohler, *From Medical Chemistry to Biochemistry: The Making of a Biomedical Discipline* (Cambridge, 1982); J. W. Servos, *Physical Chemistry from Ostwald to Pauling: The Making of a Science in America* (Princeton, 1990); J. Harwood, *Styles of Scientific Thought: The German Genetics Community 1900–1933* (Chicago, 1993).

research institutes have at last received attention.[9] But there is a shortage of book-length works which look at all the sciences in a major university in the twentieth century. Though many of them are technically daunting I have tried to write about them in a historically sensitive way which is comprehensible and interesting to historians and scientists, whether undergraduates, postgraduates, or dons, and not least to Dr Johnson's common reader.

My organizing typology is a variant on the periphery/centre model. I assume that the key to understanding Oxford was its decentralized collegiate structure which was associated with powerful interests focused on arts subjects. Relative to the aggregate wealth of the autonomous colleges, the University, which ran most of the laboratories, was impoverished. All innovators in the sciences had to cope with the dominant college structure and its associated interests. They could exploit it, accommodate to it, modify it, add to it, outflank it, or ignore it. It makes sense to analyse the careers of the various sciences according to the extent of their connection with the total life and attitudes of the colleges. This belief is reflected in the structure of the book. Chapters 1–3 examine the context of Oxford science historically, structurally, and intellectually. Chapters 4–9 deal with various sciences, usually arranged in groups. The vocational agrarian sciences are discussed in Chapter 4. Biomedical sciences, of widely varying size and relation with the colleges, are covered in Chapter 5. Various small non-vocational sciences, which struggled for survival mainly beyond the college pale, are surveyed in Chapter 6, with zoology receiving a separate chapter because it was widely recognized as an enclave of excellence. Chapter 8 is devoted to the bigger subjects of mathematics and chemistry. Physics is covered in Chapter 9 as part of an analysis of the differential reception and impact of refugee scientists from 1933. The final chapter summarizes the factors which propelled academic entrepreneurship and the disturbance of inertia.

---

[9] G. Macfarlane, *Howard Florey: The Making of a Great Scientist* (Oxford, 1979); A. J. Rocke, *The Quiet Revolution: Hermann Kolbe and the Science of Organic Chemistry* (Berkeley and Los Angeles, 1993); G. L. Geison, *The Private Science of Louis Pasteur* (Princeton, 1995); D. Cahan, *An Institute for an Empire: The Physikalisch-Technische Reichsanstalt 1871–1918* (Cambridge, 1989); J. A. Johnson, *The Kaiser's Chemists: Science and Modernisation in Imperial Germany* (Chapel Hill, NC, 1990).

1
The Changing Salience of Oxford Science 1914–1939

1.1 The War Effort

The standard view of Oxford's contribution to the First World War is based on the single though famous case of Harry Moseley, renowned as the discoverer of atomic numbers.[1] Moseley, an Oxford graduate, worked in Oxford in a private capacity from late 1913 hoping to succeed the aged Clifton in the chair of experimental philosophy. When war was declared Moseley rushed to secure a commission with the Royal Engineers who initially refused him. After several months of training he was posted in June 1915 to the Dardanelles as a signal officer. In August 1915 he was killed in action on the Gallipoli peninsula. It seemed to some important pundits, like Ernest Rutherford, that the scientist who had called the roll of the elements had been pointlessly sacrificed in the trenches: Moseley's death was a striking example of the misuse of scientific talent. He was not alone among Oxford scientists in making the supreme sacrifice. J. W. Jenkinson, lecturer in the zoology department in comparative and experimental embryology, and fellow of Exeter, was a keen member of the University Volunteer Training Corps.[2] So intense was his patriotic ardour that he applied for a commission even though he was over age. Made captain in the Worcestershire Regiment in April 1915, he too was killed at Gallipoli in June 1915. His two textbooks on embryology were reference points for his contemporaries who regarded him as a pioneer in experimental embryology. The periodical *Nature* viewed his death as epitomizing the irreparable waste to which the war had condemned Europe. One of Jenkinson's zoological colleagues, G. W. Smith, demonstrator in zoology and comparative anatomy and fellow of New College, became a captain in the Rifle Brigade and was killed in July 1916.[3] The pointless deaths of these two Oxford zoologists affected some of

---
[1] Henry Gwyn Jeffreys Moseley (1887–1915); J. L. Heilbron, *H. G. J. Moseley: The Life and Work of an English Physicist 1887–1915* (Berkeley and Los Angeles, 1974), 115–25.

[2] John Wilfred Jenkinson (1871–1915), fellow of Exeter 1909–15, *DSB*, obituary in *Nature*, 95 (1915), 456; J. W. Jenkinson, *Experimental Embryology* (Oxford, 1909), and id., *Vertebrate Embryology Comprising the Early History of the Embryo and its Foetal Membranes* (Oxford, 1913).

[3] Geoffrey Watkins Smith (1881–1916), fellow of New College 1906–16, obituary in *Nature*, 97 (1916), 502.

their contemporaries deeply.[4] Undeterred by these tragic fatalities, S. G. Scott, a histologist in the physiology department, resigned his post in July 1916 to join the Royal Army Medical Corps. He died of pneumonia in early 1918 on service in Italy.[5]

The examples of Moseley, Jenkinson, Smith, and perhaps Scott suggest that Oxford science had nothing to contribute to the war effort in the form of applied scientific expertise, but another Oxford fatality qualifies that view. Andrea Angel, a Christ Church graduate who became tutor there, showed his interest in practical chemistry before the war by drawing up a chemistry course for engineering undergraduates. During the war he was chief chemist and assistant manager at the Silvertown works of Brunner Mond which from 1915 were used for purifying trinitrotoluene. Early in 1917 the works were wrecked by huge explosions which killed sixty-nine people and demolished nearby houses. For his bravery in rescuing injured people from the works, Angel was posthumously awarded the Albert medal.[6] Angel's war-work graphically illustrates the old adage that the First World War was above all a chemists' war; but he was by no means the only Oxford chemist to contribute to the war effort. Harold Hartley of Balliol master-minded Britain's chemical warfare programme, rising to the rank of brigadier-general; and after the war he led a government mission to Germany to inspect chemical factories there.[7] Hartley's initiation into applied military research proved so agreeable to him that on his return to Oxford he styled himself General Hartley, a move which earned the caustic comment:

> When the war was over
> General Hartley
> Returned to civil life again
> partly.

Chemical warfare was also the focus of the war research of Bertram Lambert, a demonstrator in chemistry, who devised an anti-gas respirator which was being made on a large scale as the British service gas mask by 1916.[8] Two Oxford chemists, one a college fellow and the other a recent graduate, showed their versatility by working as experimental researchers for the Royal Flying Corps. R. B. Bourdillon, fellow of University, was one of the

---

[4] W. Garstang, *The Return to Oxford: A Memorial Lay* (Oxford, 1919).

[5] Samuel Geoffrey Scott (1875–1918); A. E. Boycott, 'Scott', *Journal of Pathology and Bacteriology*, 23 (1919–20), 115.

[6] Andrea Angel (1877–1917), chemical tutor at Christ Church 1912–15, obituaries in *Journal of the Chemical Society*, 111 (1917), 321–3, and *OM* 35 (1916–17), 113–14.

[7] L. F. Haber, *The Poisonous Cloud: Chemical Warfare in the First World War* (Oxford, 1986); R. MacLeod, 'The Chemists Go to War: The Mobilisation of Civilian Chemists and the British War Effort, 1914–18', *Annals of Science*, 1 (1993), 455–81; Harold Brewer Hartley (1878–1972), chemistry fellow of Balliol 1900–30; A. G. Ogston, 'Hartley', *BM* 19 (1973), 349–73 (357); E. J. Bowen, 'Hartley', *OM* 91 (1972–3), 7–8.

[8] Bertram Lambert (1881–1963), chemistry fellow of Merton 1920–47; E. J. Bowen, 'Lambert', *Nature*, 199 (1963), 1136–7.

earliest workers at the Corps' Flying School at Upavon, Wiltshire, where he worked on bomb sights and flying in clouds. H. R. Raikes, who had graduated in 1914, volunteered for military service in France where he was severely wounded in May 1915. On his recovery he joined the Corps as an experimentalist working on bombing techniques.[9]

The major thrust of Oxford's chemical contribution to the war effort came from W. H. Perkin, jun., Waynflete professor of chemistry, who introduced into Oxford industrial research, an activity which some of his colleagues regretted as a temporary necessity and others regarded as a bizarre deviation.[10] In a university which was notoriously suspicious of manufacturing, Perkin drew on his previous experience at Manchester to show that academic work and industrial research could occur side by side to mutual advantage. At Oxford during the war Perkin led small teams working on making acetone from alcohol, on a non-inflammable rubber coating for airships, and on mustard gas. But his chief contribution was to encourage industrial research in his Oxford laboratory by fine-chemical firms, especially the government-backed British Dyes formed in 1915 to strengthen and rationalize the industry. By early 1916 Perkin was chairman of British Dyes' advisory council and supervised its research department, which was composed of colonies of organic chemists working in several universities including Oxford. From 1916 Perkin had available at Oxford the newly built Dyson Perrins Laboratory and in this palace of chemistry the industrial research done by British Dyes (from 1919 to 1926 incorporated into the British Dyestuffs Corporation) persisted to 1925. As if the practice of industrial research in Oxford were not a sufficient deviation from Oxford's norms, Perkin's British Dyes' colony relied far more on young chemists recruited from outside Oxford than on Oxford graduates. After the war had ended, Oxford had its revenge: only one of these imports, Edward Hope, found permanent employment in Oxford. This made a telling contrast with the fate of one young chemist, prevented by the war from taking up his scholarship at Balliol: Cyril Hinshelwood joined the department of explosives, Queensferry Royal Ordnance Factory, where he became deputy chief chemist.[11] He worked there on the slow decomposition of solid explosives by measuring the gases evolved; this research stimulated his interest in chemical

[9] Robert Benedict Bourdillon (1889–1971), chemistry fellow of University 1913–21; E. J. Bowen, 'Bourdillon', *[University] College Record*, 6 (1971), 16–17; R. W. Clark, *Tizard* (London, 1965), 24–9, 36, 45; Humphrey Rivaz Raikes (1891–1955), chemistry fellow of Exeter 1919–27, *DNB*, obituary in *Stapeldon Magazine*, 13 (1955), 5–6.

[10] William Henry Perkin (1860–1929), Waynflete professor of chemistry 1913–29; A. J. Greenaway, J. F. Thorpe, and R. Robinson, *The Life and Work of Professor William Henry Perkin* (London, 1932); W. J. Reader, *Imperial Chemical Industries: A History*, i: *The Forerunners 1870–1926* (London, 1970), 266–75; Edward Hope (1886–1953), chemistry fellow of Magdalen 1919–53.

[11] Cyril Norman Hinshelwood (1897–1967), chemistry fellow of Trinity 1921–37, Dr Lee's professor of chemistry 1937–64; H. W. Thompson, 'Hinshelwood', *BM* 19 (1973), 375–431.

kinetics, which was to be the main focus of his subsequent research. He returned to Oxford early in 1919, took his finals in five terms, was immediately elected a research fellow of Balliol, and next year was elected tutorial fellow of Trinity. Thus the war provided a research field and an accelerated Oxford career for the future Nobel laureate.

It would be wrong to assume that Oxford's contribution to the war effort was made exclusively by its chemists. Its pathologists, physiologists, psychologists, physicists, and engineers were all heavily involved. In pathology Georges Dreyer, the professor, made three important contributions.[12] He persuaded the army to add inoculation against paratyphoid fevers to the accepted anti-typhoid inoculation. In his own department a bacteriological standards laboratory was established in 1915 under his general supervision, with A. D. Gardner providing daily direction. Its function was to prepare and to issue standard agglutinable cultures and standard agglutinating sera to army and navy hospitals for diagnosing typhoid and paratyphoid fevers. The standards laboratory was paid for by the Medical Research Committee who valued it so highly that after the war it continued as the national bacteriological standards laboratory for the United Kingdom and the dominions. Dreyer was also involved in designing a portable apparatus for supplying oxygen to men flying aeroplanes above 15,000 feet altitude. By 1917 about fifty Dreyer devices were used daily by the British air force. The war was a boon to Dreyer: it enabled him to develop via the new standards laboratory his long-standing interest in establishing a series of biological constants in human and animal physiology to act as a basis for assessing normality and hence to allow pathological states to be detected. From 1916 the war provided Dreyer with a new challenge, that of the pathology of high-altitude flying in the form of air-hunger faced by aeroplane crews. No wonder that when he returned to Oxford he found it difficult to adjust to peacetime responsibilities after the fascination and excitement of war-work.

That was not the case with Charles Sherrington, professor of physiology and renowned researcher on the central nervous system.[13] Some of his war-work was directly connected with this expertise. He served on the War Office committee on tetanus and soon devised a treatment with antiserum. For the Central Control Board Committee (Liquor Traffic) he made experimental observations on the action of alcohol on animals. But his work in occupational psychology/physiology for the War Office was a new departure. He studied industrial fatigue by experimenting on himself at a factory in

---

[12] Georges Dreyer (1873–1934), professor of pathology 1907–34; M. Dreyer, *Georges Dreyer: A Memoir by his Wife* (Oxford, 1937), 118–26, 130–68; E. W. Ainley-Walker, 'Dreyer', *Journal of Pathology and Bacteriology*, 39 (1934), 707–23; Arthur Duncan Gardner (1884–1978) ended his Oxford career as Regius professor of medicine 1948–54.

[13] Charles Scott Sherrington (1857–1952), professor of physiology 1913–36; E. G. R. Liddell, 'Sherrington', *ON* 8 (1952–3), 241–70; H. M. Sinclair, 'Sherrington and Industrial Fatigue', *Notes and Records of the Royal Society of London*, 39 (1984), 91–104.

Birmingham where for three months he worked a twelve-hour shift Monday to Friday, with one of nine-and-a-half hours per day at the weekend. He concluded that shorter hours gave greater productivity. By 1918 he had become chairman of the new Industrial Fatigue Research Board. When the war ended Sherrington returned to his work on the central nervous system, but one of his colleagues found that war-work had shifted the focus of his career: H. M. Vernon, a specialist in experimental and chemical physiology, was so fascinated by his research on fatigue for the Health of Munitions Workers' Committee, initially done with Sherrington, that he did not return to Oxford, preferring to work for the Industrial Fatigue Research Board.[14] The first to join the Board's staff, Vernon became a pioneer in industrial health and occupational psychology, publishing books on such topics as heating and ventilation, the need for a shorter working week, accidents and their prevention, and health in relation to occupation.

For two colleagues of Sherrington, C. G. Douglas and J. G. Priestley, the war provided an exciting though dangerous opportunity to show their worth as sterling disciples of J. S. Haldane by extending their work on human respiration and continuing their interest in applied as well as pure physiology. At the start of the war Douglas volunteered for the Royal Army Medical Corps and soon became the chief British authority on gas sickness; for service in the field he was awarded the military cross in 1916 and next year became physiological adviser for the Gas Directorate at GHQ.[15] Priestley, a Beit research fellow who had done classic work with Haldane on lung ventilation (1905), also joined the Medical Corps and worked with his mentor on the after-effects of gas poisoning.[16] Having contracted lung tuberculosis in 1912, Priestley did not join the intrepid Haldane in experiments on himself to test human reactions to chlorine with and without respirators. In line with his view about the importance in physiology of experimenting on the live, complete, and normal human, especially under stressful conditions, Haldane was prepared to injure himself by shutting himself for several minutes in a chamber filled with chlorine. Yet another colleague of Sherrington, W. McDougall, an experimental psychologist who worked in a laboratory in the physiology department, joined the Royal Army Medical Corps where he was a major in charge of nervously disturbed patients, especially those suffering from shell-shock.[17] In two other ways

[14] Horace Middleton Vernon (1870–1951), obituaries in *Lancet* (1951), 1: 477 and *BMJ* (1951), 1: 419.

[15] Claude Gordon Douglas (1882–1963), natural sciences fellow of St John's 1907–49, reader in metabolism 1937–42, and professor 1942–9; D. J. C. Cunningham, 'Douglas', *BM* 10 (1964), 51–74; John Scott Haldane (1860–1936), reader in physiology at Oxford 1907–13; C. G. Douglas, 'J. S. Haldane', *ON* 2 (1936), 115–39.

[16] John Gillies Priestley (1880–1941), reader in clinical physiology 1927–41; C. G. Douglas, 'Priestley', *Nature*, 147 (1941), 319–20.

[17] William McDougall (1871–1938), Wilde reader in mental philosophy 1903–20; M. Greenwood and M. Smith, 'McDougall', *ON* 3 (1939–41), 39–62.

McDougall extended his interest in fluctuation and oscillation in perceptions and actions: he supervised research on the effects of alcohol on humans and on the problems of fatigue for the Liquor Control Board, the Medical Research Committee, and the Industrial Fatigue Research Board.

The physical scientists other than chemists were also involved in war-work. F. B. Pidduck, an applied mathematician who demonstrated in the electrical laboratory of the Wykeham department of physics, spent the second half of the war researching on ballistics at the Woolwich Arsenal.[18] His colleague E. W. B. Gill was occupied with wireless intelligence, which involved intercepting German wireless messages in order to secure information about German movements on land and sea and in the air. At the end of the war Gill was in charge of the army's central wireless intelligence station at Devizes.[19] Townsend, Wykeham professor of physics and the first to measure the charge of the electron, worked as a major on wireless research for the Royal Naval Air Service, mainly at Woolwich.[20] Two other physics demonstrators, I. O. Griffith and H. Tizard, became experimental officers with the Royal Flying Corps at its Flying School, Upavon, Wiltshire, where they worked together mainly on testing bombs and aeroplanes.[21] Tizard, who had been invited to Upavon by Bourdillon, an Oxford friend and his best man at his marriage, went on to qualify in May 1916 as a test pilot and later that year was appointed scientific officer in charge of a new airfield at Martlesham, Suffolk, where he developed what was known as the Martlesham test on aircraft using in-flight measurements. By the end of the war he was controller of research and experiments at the newly created Air Ministry. Tizard thought that the war did him much good: certainly it pulled him 'out of the ruck at Oxford' and it taught him the value and interest of the application of science. It was this lesson which was partly responsible for his resignation in 1920 from his fellowship at Oriel and his university readership in thermodynamics for a post in the Department of Scientific and Industrial Research. Aeronautical research was also the focus of the war-work done by the two-man department of engineering science. Professor C. F. Jenkin researched for the Air Board on the specifications of materials used in constructing aircraft and in aircraft engines.[22] After the war Jenkin worked

[18] Frederick Bernard Pidduck (1885–1952), mathematics fellow of Corpus Christi 1921–50, reader in applied mathematics 1927–34, obituary in *Nature*, 170 (1952), 141.

[19] Ernest Walter Brudenell Gill (1883–1959), physics fellow of Merton 1909–58; E. W. B. Gill, *War, Wireless and Wangles* (Oxford, 1934).

[20] John Sealey Edward Townsend (1868–1957), Wykeham professor of physics 1900–41; H. von Engel, 'Townsend', *BM* 3 (1957), 257–72.

[21] Idwal Owen Griffith (1880–1941), fellow of St John's 1915–20, of Brasenose 1920–41, obituary in *Brazen Nose*, 7 (1931–44), 165–9; Henry Thomas Tizard (1885–1959), natural sciences fellow of Oriel 1911–20; W. Farren, 'Tizard', *BM* 7 (1961), 313–48; Clark, *Tizard*, 23–48.

[22] Charles Frewen Jenkin (1865–1940), professor of engineering science 1908–29; R. V. Southwell, 'Jenkin', *ON* 3 (1941), 575–85.

on problems that had intrigued him during the war, namely, the effects of cracks and notches on the strength of machine parts and fatigue failure in metals. Jenkin's colleague D. R. Pye joined the Flying Corps and became a close colleague of Tizard at Martlesham, testing aircraft armaments and performance. His war experience launched him into aeronautical research which was to be the chief focus of his subsequent career at the Air Ministry.[23] Oxford, therefore, made an important contribution to aeronautical research during the First World War, through the efforts of Tizard, Bourdillon, Raikes, Griffith, and Pye, all of whom worked as experimental officers for the Royal Flying Corps.

The contributions made by Oxford's senior geologists were noteworthy. Having recovered from wounds incurred in 1915 with the Gordon Highlanders, J. A. Douglas was subjected to an order compelling all officers with a knowledge of geology or mining to be used in underground warfare in France. He was wounded in 1916 when in charge of mines at St Eloi, Messenes Ridge; next year his corps of mines constructed the Grange Tunnel for the attack on Vimy Ridge.[24] Douglas's colleague Professor W. J. Sollas, aged 65 when the war broke out, made a characteristically eccentric contribution to the domestic war effort. On occasion he would lie full length on the floor in his room in the University Museum and fire a point 22 rifle at a target attached to a bookcase at the far end of the room. He missed the target but penetrated the spine of a copy of Chamberlin and Salisbury's *Geology*.[25]

Biologists made a less spectacular but more telling contribution to the war effort than Sollas. Poulton, the Hope professor of entomology who had done much to promote the cause of science at Oxford, was chiefly a propagandist. In 1915 he gave the Romanes lecture in which he denounced the British government's neglect of science before and during the war, deplored its ineptitude in permitting the export to Germany of essential raw materials for ammunition in 1915, deprecated the pusillanimous ignorance of Lord Robert Cecil (parliamentary foreign under-secretary), exposed 'the anarchy unalloyed' in the government's attitude to Germany, and called for far greater use of science by the army in the new conditions of trench warfare.[26]

---

[23] David Randall Pye (1886–1960), demonstrator in engineering science 1909–14, 1919; O. A. Saunders, 'Pye', *BM* 7 (1961), 199–205.

[24] James Archibald Douglas (1884–1978), demonstrator in geology 1905–37, professor 1937–50; J. M. Edmonds, 'Douglas', *Nature*, 274 (1978), 196; R. MacLeod, 'Kriegsgeologen and Practical Men: Military Geology and Modern Memory', *British Journal for the History of Science*, 28 (1995), 427–50.

[25] William Johnson Sollas (1849–1936), professor of geology 1897–1936; A. S. Woodward and W. W. Watts, 'Sollas', *ON* 2 (1936–8), 265–81; Douglas, dictated notes on geology at Oxford, 1902–25, Geology Department, Oxford University Museum; T. C. Chamberlin and R. D. Salisbury, *Geology: Shorter Course* (London, 1909).

[26] Edward Bagnall Poulton (1856–1943), Hope professor of zoology (entomology) 1893–1932; G. D. H. Carpenter, 'Poulton', *ON* 4 (1944), 655–80; E. B. Poulton, *Science and the Great War: The Romanes Lecture 1915* (given in Oxford, Dec. 1915) (Oxford, 1915), 33.

Though Bourne, Linacre professor of zoology and comparative anatomy, merely moved sideways from coaching Oxonian oarsmen to training Oxonian army recruits, the men who were to be his two immediate successors in the chair were both deploying their scientific expertise.[27] Goodrich, then Aldrichian demonstrator in zoology and fellow of Merton, departed from his normal research in comparative anatomy to work for the Royal Society's Grain Pests Committee: he examined the problem of weevils in wheat, using experimental breeding to determine which beetles were attacked by which parasites and at what stage of their life history. His redoubtable wife Helen organized the canning and bottling of fruit for the Board of Agriculture and then turned to bee diseases. A. C. Hardy, then an undergraduate at Exeter, volunteered for war service and, after a frustrating period in a cyclist battalion, ended by putting his deep knowledge of animal camouflage to good use as an assistant camouflage officer in home coastal defences.

Oxford's leading agriculturists and foresters did not join the armed forces but as civilians played important roles in promoting war socialism at home and in planning post-war reconstruction. In agriculture a major concern was domestic food production. W. Somerville, Sibthorpian professor of agriculture, sat on several government committees which were concerned with increasing agricultural productivity, such as the Board of Agriculture Food Production Advisory Committee, the Scottish Food Production Committee, and the Fertilizers Committee.[28] His colleague C. S. Orwin, director of the Agricultural Economics Research Institute since its foundation in 1912, studied the organization of British agriculture and in the broadest terms was concerned with the means by which maximum effect could be given to the application of scientific discoveries to the business of agriculture.[29] By 1917, when food prices and agricultural workers' wages were controlled, Orwin sat on such bodies as the Agricultural Wages Board (to which two of his staff were seconded as investigators) and the Agricultural Costings Committee. In forestry the dominant concern was increasing home production of wood in the long term as part of post-war reconstruction. In 1916 the Reconstruction Committee set up a forestry subcommittee of which William Schlich, professor of forestry, was a leading member.[30]

[27] Gilbert Charles Bourne (1861–1933), Linacre professor of zoology and comparative anatomy 1906–21; S. J. H., 'Bourne', *ON* 1 (1932–5), 126–30; Edwin Stephen Goodrich (1868–1946), Linacre professor 1921–45; G. de Beer, 'Goodrich', *ON* 5 (1945–8), 477–90; Goodrich to Huxley, 25 Aug. [1917], Huxley Papers; Alister Clavering Hardy (1896–1985), Linacre professor 1945–61; N. B. Marshall, 'A. C. Hardy', *BM* 32 (1986), 223–73.

[28] William Somerville (1860–1932), professor of agriculture 1906–25, *DNB*; *OUG*, 47 (1916–17), 324.

[29] Charles Stewart Orwin (1876–1955), director of the Agricultural Economics Research Institute 1912–41, *DNB*; *OUG* 48 (1917–18), 279 and 49 (1918–19), 260.

[30] William Schlich (1840–1925), professor of forestry 1905–19, *DNB*; N. D. G. James, *A History of English Forestry* (Oxford, 1981), 210–17.

When it reported in 1917 it recommended that extensive state afforestation be undertaken to make the United Kingdom independent of imports and to free 6 million tons of shipping devoted to importing wood. Thus Schlich helped to bring into being in 1919 the state-financed Forestry Commission.

1.2 The War and New Degree Regulations

Given the large number of Oxford's staff and undergraduates who had left the University for military service either as fighting officers or as the thinking services of the crown, the colleges and some science departments were almost empty. By spring 1916 the War Office had spotted that Oxford had much vacant accommodation and it soon requisitioned four science departments as well as eight colleges in order to establish in them a school of military aeronautics for officers of the Royal Flying Corps.[31] Though the University was far from replete with its usual complement of staff, those remaining passed some controversial legislation about three important new degrees, all of which were connected one way or another with the war. These were the extended chemistry degree approved in 1916, with the new doctorate of philosophy (D.Phil.), and the opening of the first MB to women following in 1917.

The expanded chemistry degree was mainly the brain-child of Perkin for whom research at the bench, building up a research school, and what he called output of research were top priorities.[32] He had long believed that the normal three-year B.Sc. in British universities was tediously stereotyped and glaringly deficient in that the undergraduates were not required to do any research. His early experience at Oxford quickly convinced him that in order to master organic chemistry students needed to spend more time in the laboratory. The outbreak of war in 1914 and the government's scheme of 1915 for the organization of scientific and industrial research provided Perkin with his opportunity to lead a campaign to add a fourth year of research (part two) to the existing three-year degree course (part one) in chemistry. He convinced his colleagues that research training was essential even if employers of graduate chemists were currently indifferent to it. He stressed that research experience was necessary for all honours chemistry undergraduates and that it should not be viewed as a luxury restricted to those destined for a first-class degree. His own contribution to industrial war research gave credibility to his claims that research chemists were producing

[31] *OUG* 47 (1916–17), 551; W. T. S. Stallybrass, 'Oxford in 1914–18', *Oxford*, 6 (1939), 31–45.
[32] W. H. Perkin, 'The Position of the Organic Chemical Industry', *Journal of the Chemical Society*, 107 (1915), 557–78; Report of the Subfaculty of Chemistry, n.d. [1915], NS/R/1/2; NS/M/1/3, 1 June 1915, 8 Feb. 1916; *OUG* 46 (1915–16), 328, 341–2, 350–1, 423–4, 448–50, 458–60 (458).

rapid changes in some parts of the chemical industry and that in future there would be a greater demand for research chemists. At a time of national crisis the new examination regulations in chemistry, which required each candidate to present records of experimental investigations, were quickly and unanimously approved by the University in May 1916 on 'national grounds', even though there was no compulsory research year in any other undergraduate degree course in science. Practical questions, such as the extra expense of an extra year for chemistry undergraduates and problems of extra laboratory accommodation, seem to have been swept aside.

Before the war the bulk of Americans, and many Australians and Canadians, who wished to acquire a doctorate went to German universities. Comparatively few Americans came to Oxford for this purpose because there was no postgraduate degree which was sufficiently attractive to them. As Lord Curzon had stressed in 1909, Oxford needed a larger cohort of advanced students who could be attracted by greater encouragement and rewards.[33] The Oxford D.Phil. degree was introduced to induce graduates to come to Oxford from universities in Britain, the dominions, and especially the USA, after the end of the war.[34] It was also a move to encourage the claims and practice of research at Oxford. Not surprisingly Perkin was the leading spokesman for the scientists, ably abetted by Poulton.[35] Perkin saw the D.Phil. as a means of persuading good graduates to do advanced work for three years under a supervisor and of then rewarding them, thus remedying what he regarded as a conspicuous defect of Oxford. He believed that the two existing research degrees in science at Oxford did not promote high-level supervised research: the one- or two-year B.Sc., introduced in 1895, was not sufficiently testing so it did not entice the best graduates; whereas the D.Sc., given for published solo work done over several years, was too exclusive, not a regular career option, and lacked the element of research training. Though there was a widespread feeling in the University that controversial legislation should not be introduced during the war behind the backs of soldiers, the D.Phil. was approved in May 1917 without much opposition. There was considerable debate about the relation of the D.Phil. to other research degrees available at Oxford but the long-term

[33] G. N. Curzon, *Principles and Methods of University Reform: Being a Letter Addressed to the University of Oxford* (Oxford, 1909), 187. George Nathaniel Curzon (1859–1925), chancellor of the University 1907–25.

[34] *OUG* 47 (1916–17), 184; R. Simpson, *How the PhD Came to Britain: A Century of Struggle for Postgraduate Education* (Guildford, 1983), 135–40, praises Oxford for introducing the Ph.D. into Britain.

[35] W. H. Perkin and E. B. Poulton, 'Proposed Statute for the Encouragement of Oxford Research', *OM* 35 (1916–17), 121–2; *Nature*, 98 (1916–17), 441–2; *OUG* 47 (1916–17), 251, 303–4, 352, 428–31, 448–50, 466. For Poulton's long campaign to promote Oxford as a seat of research, see [E. B. Poulton], 'The Reform of Oxford University', *Nature*, 80 (1909), 311–12; [id.], 'The Empire and University Life', *Nature*, 72 (1905), 217–18; and id., *John Viriamu Jones and Other Oxford Memories* (London, 1911), 257–79.

consequences of its introduction were neglected. It threatened to disturb the relation between the colleges and the University: postgraduate work in a collegiate university, with an elaborate and hallowed system of undergraduate teaching based in colleges, was likely to be an excrescence; and any expansion of such work was likely to fall on university professors and readers and not on hard-pressed college tutorial fellows. Large-scale expansion of post-graduation supervision could only be met properly by appointing more professors and readers, or by lightening the teaching burden of college tutors. Thus the introduction of the D.Phil. was bound to exacerbate, eventually but not immediately, the old thorny problem of the relations between the University and the colleges.

With most of the men away on war service, women became disproportionately prominent in the indigenous student body. At the same time the war had stimulated a general movement for more women to be trained in medicine. In 1917 the Oxford medical teachers took advantage of these two unusual circumstances by pressing for the first MB examination to be opened to women. They succeeded by the narrowest of margins: their proposal was approved in Convocation in June 1917 by 22 votes to 20.[36] One troublesome consequence was that henceforth women would have to be admitted to courses in human anatomy. Arthur Thomson, the professor, was prepared to let women attend lectures and demonstrations with men; but he insisted that women should do practical work in their own dissecting room and he hoped that they would study only surface anatomy down to the umbilicus.[37] The extra accommodation required was paid for, not by the University, but mainly by the Worshipful Company of Clothworkers (£1,300). The admission of women undergraduates to the anatomy department raised the awkward question of appointing a woman demonstrator to teach them. The unusual circumstances of war permitted an appointment that would have been inconceivable in peacetime: a medical graduate of the Royal University of Ireland (not a body with strong connections with Oxford) joined the anatomy staff and for the rest of the war was the sole demonstrator.[38] These innovations nourished the agitation which led to the admission of women to full membership of the University and to its degrees in 1920; and they fuelled the abolition of compulsory Greek for all undergraduates that year after forty years of controversy.

---

[36] *OUG* 47 (1916–17), 531.

[37] Arthur Thomson (1858–1935), professor of anatomy 1893–1933, obituary in *Lancet* (1935), 1: 405–6; H. M. Sinclair and A. H. T. Robb-Smith, *A Short History of Anatomical Teaching in Oxford* (Oxford, 1950), 67–8.

[38] *OUG* 48 (1917–18), 477; A. M. A. H. Rogers, *Degrees by Degrees: The Story of the Admission of Oxford Women Students to Membership of the University* (Oxford, 1938), 100–1; A. J. Engel, *From Clergyman to Don: The Rise of the Academic Profession in Nineteenth-Century Oxford* (Oxford, 1983), 223–30; Alice Bury Chance, later Mrs Carleton (1891–1979), obituary in *BMJ* 280 (1980), 1: 124.

1.3 The Asquith Commission

The war had a very important effect on Oxford's future because it occasioned the University's decision in June 1919 to apply for the first time for a grant of money from the government. From October 1919 the University received first an emergency grant and then a regular one, thus departing from the hallowed tradition of Oxford's financial and academic independence. This partial dependence of Oxford on the state was the result of Oxford's scientific contribution to the war effort, of the increased appreciation of science which it engendered, and of the associated lobbying by Oxford's science professors. Indeed it could be said that the war reduced the pervasiveness in Oxford of Dr Johnson's belief in the efficiency of ignorance: by autumn 1916 the vice-chancellor, though proud of Oxford's nine Victoria crosses, had become convinced that Britain should give up her prevalent distrust of specialist knowledge, especially science which had been badly neglected at Oxford.[39] Just before the war ended, the new vice-chancellor, H. E. D. Blakiston, was invited by H. A. L. Fisher, an Oxford graduate who was president of the Board of Education, to join a deputation in late November 1918 from British universities about financial aid to them from government after the war.[40] Though Blakiston was uneasy about Oxford receiving any money from government, he attended the deputation. At the same time he began to explore with Fisher the possibility of government assistance to meet Oxford's most pressing wants which he identified as a capital grant of £100,000 for extensions to the chemistry and engineering laboratories, and an annual grant of around £4,000 to pay for salaries of scientists. Though many Oxonians were suspicious that government aid would degenerate into government interference, Blakiston told Fisher that the establishment of regular government funding of the University would be a suitable recognition of Oxford's input into the war, especially through the acknowledged services rendered by its scientists. By March 1919 Blakiston was thinking more ambitiously. Spurred by Perkin who had contacted Fisher directly in February 1919, he sent Fisher a memorandum from the Oxford science professors who called attention to the inadequacy of funding they faced and asked for assistance from government. They stressed in particular the need to raise the salaries of teaching and research staff, and of laboratory assistants, which would cost £17,000 a year in all. They estimated that capital outlay on chemistry, geology, pathology, physiology, botany, and engineering would amount to £163,000. They hinted that the colleges should do more for science by giving more undergraduate

---

[39] *Nature*, 98 (1916–17), 123.
[40] *OUG* 49 (1918–19), 471–8 printed the Blakiston–Fisher correspondence November 1918–May 1919; Herbert Edward Douglas Blakiston (1862–1942), vice-chancellor 1917–20; Herbert Albert Laurens Fisher (1865–1940), president of the Board of Education 1916–22.

scholarships and more fellowships in science. It was this memorandum of March 1919 from the heads of science departments at Oxford which brought to a climacteric the parliamentary grant business and occasioned the setting up of a royal commission. That happened because Fisher was sympathetic to their plight: he thought that their needs were so crying that without immediate financial aid from government it would be impossible for them to carry on their current scientific work.[41] Furthermore, Fisher took an elevated view of the public role of the university, dismissing its directly vocational function. In his opinion universities existed 'not to equip students for professional posts, but to train them in disinterested intellectual habits, to give them a vision of what real learning is, to refine taste, to form judgement, to enlarge curiosity, and to substitute for a low and material outlook on life a lofty view of its resources and demands'.[42] Such views were sweet music in many Oxonian ears.

Up to this point in the negotiations it had been assumed in Oxford, and made clear to Fisher, that any government money would be channelled to individual departments, as had happened previously in a few *ad hoc* cases, and hence would not be a state subsidy in the form of a block grant to be distributed by the University itself. In April 1919, however, Fisher indicated to Blakiston a change of policy, namely, that in future there would be no grants to separate departments in any university but that each university would receive a single inclusive grant for the spending of which the university in question would be responsible. Fisher and the Chancellor of the Exchequer also insisted on a quid pro quo. They regarded Blakiston's letters as constituting or preparing the way for an application to the government for financial aid; but stressed that the government would not pay out unless the University agreed to a comprehensive inquiry being instituted by government into the total resources of the University and the colleges. Fisher made it clear that the University would be required to welcome such an inquiry as a condition of receiving even an emergency grant, and that a regular grant might be contemplated by government only after the inquiry had been completed. Blakiston was flummoxed by this response: his application was almost entirely for science departments and not for the general purposes of the University.

When Convocation debated the matter on 10 June 1919 it was agreed by a majority of 126 votes to 88 that Oxford should apply for a government grant or grants and co-operate in the inquiry which was likely to be a royal commission.[43] This narrow vote indicated that there were widespread fears that acceptance of money from government might bring in its train govern-

---

[41] H. A. L. Fisher, *An Unfinished Autobiography* (London, 1940), 115–16.

[42] H. A. L. Fisher, *The Place of the University in National Life* (Oxford, 1919), 11.

[43] *OUG* 49 (1918–19), 551. Fifteen science professors favoured government funding and an inquiry: C. F. Jenkin *et al.*, *Applications for Government Grants*, n.d. [June 1919], 3 pp. pamphlet, Bodleian Library, GA Oxon b. 141 (111). Five of them, led by Perkin and Poulton, were

ment interference, that a royal commission might recommend disturbing and unwelcome changes, that the cherished autonomy of the colleges might be violated, and that it was not expedient to lay the University open to embarrassing and maybe dangerous scrutiny merely because the science departments, and no others, were screaming for money for better laboratory accommodation and better pay for demonstrators. The opponents of the application to government for grants or a grant feared that a block grant, as opposed to grants to specific departments or for specific purposes, opened the way to control of the University by the government of the day, with its political considerations and departmental ambitions. They recalled that the great war had been *inter alia* a battle for freedom yet Oxford was proposing to sacrifice its precious independence: as Stallybrass wrote, 'we do not want Oxford to voice the views of Whitehall as Berlin voiced the views of Potsdam'.[44] Fisher had indeed given some cause for such fears: he had alluded to the block grant procedure as providing a means of exerting on universities gentle and indirect pressure which, if sympathetically applied, would be useful.

Above all the opponents of the government grant, led by A. J. Jenkinson, accused the University of negligence.[45] It had not bothered to grade the claims of the science departments forwarded to the Board of Education in March 1919; it was absurd for geology, a small department, to ask for as much as £34,000, and irresponsible of the University to accept such a figure without question. Nor had the University bothered to explore alternative sources of income to a grant which its opponents suspected would never be more than a fraction of the University's total income. These alternative sources included increased taxation of the colleges for university purposes, higher fees, giving college posts to demonstrators (i.e. teachers of undergraduates) in the science departments, paid consultancy work for government, all of which were contentious proposals, and soliciting private benefactions systematically. As Stallybrass brutally pointed out, the University as a whole was in danger of losing its financial independence for the sake of saving itself the trouble of finding non-governmental funds to pay for the science laboratories. More generally there was a considerable body of opinion which agreed with the line taken in 1909 by Curzon, chancellor of the University, namely, that a government commission could be avoided if Oxford began to put its own financial house in order.

members from 1916 of the Neglect of Science Committee: F. M. Turner, *Contesting Cultural Authority: Essays in Victorian Intellectual Life* (Cambridge, 1993), 223–4.

[44] W. T. S. Stallybrass, *OM* 37 (1918–19), 332; William Teulon Swan Stallybrass (1883–1948), *DNB*, law fellow of Brasenose 1911–36, principal 1936–48.

[45] A. J. Jenkinson, 'Applications for Government grants', *OM* 37 (1918–19), 318–19. Alfred James Jenkinson (1877–1928), philosophy fellow of Brasenose 1903–28, obituary in *Brazen Nose* (1928), 344–54.

In autumn 1919 two important events for Oxford (and also for Cambridge) took place. The first was that the University received for 1919–20 an emergency grant of £30,000, most of which was spent on the science departments.[46] This was renewed annually until in April 1922 Fisher announced that Oxford, like Cambridge, would be added to the list of universities in receipt of regular annual grants-in-aid from the Treasury, the grant to be £30,000. From 1922 to the outbreak of the Second World War, the annual grant from the government did not decline: by the late 1920s it had risen to about £85,000 and by 1936–7 had reached £100,000. The second event of autumn 1919 was that a royal commission, chaired by H. H. Asquith, an Oxonian, was appointed to inquire into Oxford and Cambridge universities. Its report of 1922 made general recommendations about the University and also revealed problems peculiar to Oxford science.

The Asquith Commission knew that the emergency grant of £30,000 p.a. since the war had saved Oxford from financial collapse.[47] Such forms of expenditure as rates, taxes, and cost of repairs had increased threefold since 1914, and simultaneously the real value of endowments had decreased *pro tanto*. Throughout the University there was a reluctance to increase the various fees charged to the increased number of undergraduates, some of whom were soldiers released from military service to graduate in two years instead of three as a result of emergency regulations. It was repugnant to charge war heroes increased fees which would have had the unacceptable effect of making Oxford even more exclusive and maintaining its unwelcome reputation as being in part a playground for the sons of the wealthier classes. In some science departments the fee charged probably did not meet the cost per undergraduate so that the rise in numbers added to the annual deficit instead of increasing net revenue. The Commission also appreciated that relative to the aggregate wealth of the colleges the University was poor: in 1920 the total college income was three times that of the total available income for the University, even though a system of graduated taxation of colleges for university purposes was in force. The Commission's solution for Oxford's dire financial problems was to suggest that Oxford follow the example of provincial universities who had increased fees by up to 50 per cent since 1920. Realizing that increases in fees would not generate enough income, the Commission recommended that government should give an annual general grant to Oxford of £100,000 and one of £10,000 per year for special purposes. In the aftermath of the falling of the Geddes axe on public expenditure early in 1922, these particular recommendations about the government's role were not immediately realizable but by the mid-1930s

[46] *OUG* 50 (1919–20), 92, 240–1, 470, 822.
[47] Herbert Henry Asquith (1852–1928) was a leading Liberal politician; *Royal Commission on Oxford and Cambridge Universities: Report*, Parliamentary Papers, x (1922), 45, 48, 52–5, 57, 94, 114, 119, 190, 192, 196, 212, 216 for its general considerations.

had been broadly implemented. Though the Commission urged that the emergency grant of £30,000 p.a. to Oxford needed to be increased to a regular one of £110,000, it was adamant that the block grant from government was a stopgap and not a solution for Oxford's relative poverty. Though it acknowledged that the existing laboratories at Oxford left much to be desired and that new laboratories for new subjects would face very great obstacles, it proclaimed that private benefaction provided the only real hope for financial viability and academic development.

The Commission recognized that academic staff were not well paid because of the effects of inflation and that they had no pension scheme to mitigate the harshness of retirement. Though it could do little about college fellowships, its recommendation that most professorial stipends be raised to £1,200 p.a. was soon implemented. So, too, were its related proposals that a compulsory retirement age and a pension scheme be introduced. The Commission was well aware that the absence of a regular pension scheme permitted declining old men to spend their dotage in important posts. At the same time the Commission was worried that some of the staff were seriously overworked as teachers so that their research suffered. That undesirable feature was inimical to the Commission's hope that Oxford and Cambridge would develop a new role by becoming centres of research and of graduate study for students from the empire and the USA. The Commission stressed that, unless both universities improved their facilities for the supervision and instruction of advanced students, their position would be imperilled. It realized that the largely autonomous colleges, dedicated to undergraduate teaching, would find it very difficult to promote advanced study and teaching. That meant that the development of postgraduate research and teaching depended on strengthening the University and its faculties at the expense of the colleges, a shift of power that at Oxford was fraught with difficulties.

The Asquith Commission took evidence from all interested parties about the state and status of science at Oxford.[48] The dominant themes which emerged were the relation of the colleges to the University, the associated question of the balance between teaching and research, the aims of Oxford science, laboratory accommodation, the layout of the science buildings, and the problem of maintaining no less than five college chemistry laboratories. The structural problem of the relation of the colleges to the University was made manifest not only financially but also in the way fellows were appointed by autonomous colleges which disregarded the needs of departments in teaching and especially in research, thus producing an imbalance between subjects. Some professors, such as Soddy and Thomson, saw the dominant

---

[48] The supporting volume of appendixes and evidence for the Asquith Commission's Report was not published; it was presumably a victim of the Geddes axe which fell on public expenditure in 1922. There are six boxes of Asquith Commission documents in the Bodleian Library, MS Top Oxon b 104–9.

college structure as damaging because it encouraged colleges to appoint as fellows inward-looking teachers who showed no loyalty to the University which in turn suffered because above all it needed men absorbed in experimental research.[49] Much concern was also expressed at heavy teaching loads, especially for those who were both college teachers and departmental demonstrators. Oxford's cherished system of the college tutorial hour, involving a college teacher and just one undergraduate, came under fire because it exhausted the college teacher of science and tended to pull him away from his departmental responsibilities. The strongest complaint was voiced about the lot of the Dr Lee's reader in chemistry, A. S. Russell, who devoted 25 hours a week to tutorials at Christ Church and 30 hours a week to demonstrating in its laboratory, a pedagogic burden which threatened his research.[50] One solution, an extreme one in the context of Oxford, was advocated, i.e. that each college needed no more than a director of science studies, the actual teaching being done by professors and demonstrators in the science departments. Apropos the aims of Oxford science there was a widespread feeling among its scientists that they spent too much time in the role of schoolmasters teaching at an elementary level, that the paramount importance of research was not sufficiently appreciated in the University, and that Oxford should not attempt to rival more modern universities in their focus on applied science. Given its situation in the University, one way forward for science was that advocated by Henry Acland years before, namely, that it should aim to be good, small, and unapplied.[51] It would function as a distinguished ornament of the University whose structure and anti-industrialism would therefore remain unchallenged. There was considerable concern expressed about the physical plant of the science departments: laboratories, whether for teaching or research, were too small, inconveniently situated, or simply non-existent.[52] If the space within laboratories was under pressure, so too was the space between the laboratories which were housed in the University Museum or near it.[53] As these laboratories had been erected piecemeal, it was widely felt that some sort of comprehensive plan for

[49] Evidence of Soddy, Thomson, and Natural Sciences Faculty Board, MS Top. Oxon. b. 107, ff. 160, 136–41, 239–45; Frederick Soddy (1877–1956), Dr Lee's professor of chemistry 1919–31 Dec. 1936.

[50] Evidence of Townsend and Tizard, MS Top. Oxon. b. 109, file 12; Alexander Smith Russell (1888–1972), Dr Lee's reader in chemistry and student, i.e. fellow, of Christ Church 1920–55.

[51] Evidence of Perkin, Bourne, Jenkin, MS Top. Oxon. b. 109, file 10; evidence of Natural Sciences Faculty Board, MS Top. Oxon. b. 107, ff. 239–45; Henry Wentworth Acland (1815–1900), Regius professor of medicine 1858–94. For Acland's vision of Oxford's science and medicine see J. B. Atlay, *Sir Henry Wentworth Acland* (London, 1903), 130–61, 197–226, and esp. 244–8, 395–7.

[52] Evidence of Sollas and Gunn, MS Top. Oxon. b. 107, ff. 156–60, 250–62; James Andrew Gunn (1882–1958), professor of pharmacology 1917–37.

[53] Evidence of Craig and Bowman, MS Top. Oxon. b. 109, file 10. Edwin Stuart Craig (1865–1939), assistant registrar 1907–24, registrar 1924–30; Herbert Lister Bowman (1874–1942), professor of mineralogy 1909–41.

science buildings, embodying provision for the future, was desirable; only then would it be possible to escape from the ad hocery which treated each scientific building as a separate problem. It was also realized that any large-scale expansion of physical plant would be problematic because, if that expansion were to be based on the Museum area, it would encroach on the University Parks which were lovingly regarded as inviolable.

On one particular issue, that of the five college chemistry laboratories, there was a vigorous difference of opinion which in Cambridge would have been regarded as tiresome.[54] The Board of Natural Sciences was at that time in favour of centralized university laboratories for chemistry; but those scientists attached to colleges which had chemistry laboratories keenly defended them against accusations that they were wasteful and unsatisfactory. They were, it was alleged, adequately equipped and they offered elasticity of management, being open in the evenings and the vacations; research prospered in them in part because a tutor was on hand in the college; and via its own laboratory a college could exert close and constant oversight over its undergraduate chemists.

The Commission's report and the subsequent legislation dodged some of these important questions pertaining to science at Oxford. Decisive power and leadership remained with the colleges, whereas at Cambridge by 1926 the University and the faculty boards were dominant. At Oxford a large part of the fees paid by students for formal teaching continued to be paid to the colleges but at Cambridge all of them went to the University. Nor did the University of Oxford benefit much from the revised version, introduced in the mid-1920s, of the graduated scheme for taxation of the colleges for university purposes. This Common University Fund in its new guise generated little more than before: the sum rose from about £17,000 p.a. in the early 1920s to around £21,000 p.a. in the late 1920s and was nearing £27,000 p.a. just before the war. Compared with the total aggregate income of the colleges, their contribution to the Common University Fund remained small beer. Even in the late 1930s only four colleges paid an annual contribution to the CUF which was greater than an annual professorial salary. Though the Commission affected only marginally the financial distribution of power between the colleges and the University, it did promote Oxford science in one direct way. It recommended a regular government grant of £100,000 p.a. to maintain laboratories, to contribute to a sites and building fund, and to endow research and advanced teaching mainly in the sciences. Though only £30,000 was immediately made available, it should be remembered that in

[54] Evidence of Natural Sciences Faculty Board, MS Top. Oxon. b. 107, ff. 239–45; evidence for college laboratories from Cronshaw and Chapman, and E. G. Hardy, MS Top. Oxon. b. 104, ff. 21–3, 89. George Bernard Cronshaw (1872–1928), fellow 1902–28, chaplain 1898–1928, lecturer and demonstrator in natural science 1900–21, bursar 1912–18, all at Queen's; David Leonard Chapman (1869–1958), fellow in chemistry at Jesus 1907–44; Ernest George Hardy (1852–1925) was vice-principal of Jesus.

1922 the biggest grant paid by government to a university was £67,500 p.a. to Imperial College, London; and that by the end of the decade the annual grant was approaching the sum recommended by the Commission.[55]

Otherwise the Commission made various recommendations which it hoped the University would eventually adopt. The Commission was aware that Oxford was a federation of colleges created on a mainly medieval pattern and it was not prepared to damage the delicate organism of a self-developed institution. It hoped to turn Oxford towards strengthening the University as compared with the colleges, and urged successfully that any government grant-in-aid should be made to the former and not the latter. The Commission's policy was to leave Oxford mainly to itself to grope for solutions to its problems, as it had been groping for years, while simultaneously giving guidelines to it. In calling for voluntary self-reformation and prudent but necessary change, the Commission echoed Lord Curzon's views of 1909 on university reform.[56]

Unlike Curzon the Commission held definite views about Oxford science.[57] It emphasized the importance of scientific research, related it to national economic advantage, but did not countenance any move to applied science or to consultancy work. The Commission was adamant that 'investigations carried on with merely technical objects and in a merely utilitarian and commercial spirit will not achieve the highest results. The disinterested pursuit of scientific investigation affords the surest means by which the nation can ultimately command the resources of nature.' At the same time the Commission agreed with the heads of science departments that 15 acres of the University Parks should be appropriated for extensions to existing laboratories and the erection of new ones. On the vexed question of the continuance of five college chemistry laboratories, the Commission showed its characteristic sane and moderate conservatism. While it admired Cambridge's centralized laboratories, it acknowledged that Oxford's college chemistry laboratories harmonized with the university ones by remedying their deficiencies. It perceived that these laboratories offered a valuable means of retaining the college-based tutorial system in science, yet it felt that they could be used as temporary homes for new subjects or exclusively for advanced work. Thus its comments offered ammunition to both the opponents and the supporters of the college laboratories. The Commission's final conclusion about Oxford science was that it was not one of Oxford's many lost causes. Of course the Commission was aware that the dominant degrees and academic experience at Oxford and Cambridge were different:

[55] Data about the Common University Fund and the government grant are taken from the University's financial accounts which were published annually as an appendix in *OUG*. See also editorials in *Nature*, 109 (1922), 428 and 110 (1922), 201–2.
[56] Curzon, *University Reform*, 211, 215.
[57] *Asquith Report*, 38–9, 45, 114, 116–18.

'Cambridge ideas are naturally coloured to a large extent by the experience of science, Oxford ideas by the experience of "Greats".' Its view of the previous record of science at Oxford was even brutal: teachers were too few in number, accommodation and apparatus were poor, the number of research students in some subjects was 'lamentably small', and generally 'the output of work' was inadequate. But the Commission did not reach a verdict often voiced in the early 1920s: that Oxford should concentrate on its strengths in arts subjects, leaving science to be pursued at Cambridge. On the contrary the Commission felt that a combination of increased fees, generous private benefactions, a regular government grant, and the postgraduate research studentships paid for by the Department of Scientific and Industrial Research would make science at Oxford financially viable and allow it to prosper especially through research. Expressions of public optimism were rare in 1922, which was dominated by the slump, unemployment, the Geddes axe on government expenditure, the crisis of party politics, and the fall of Lloyd George. Yet the Commission concluded that a great opportunity had come for Oxford science.

In a number of ways the hopes of the Asquith Commission were fulfilled against considerable odds. By 1939 it was palpably absurd to suggest that Oxford should cede science to Cambridge and concentrate on producing statesmen. Oxford science had become more visible, more distinguished, and more distinctive than it was in 1914. The increased salience of Oxford science was shown in several ways. By 1939 Oxford housed distinguished individuals with international reputations for their research; how that happened is the major theme of this book. Its scientists had received high honours and held important offices in learned societies. Research schools were established or expanded and through them science became the pioneer of postgraduate work at Oxford. In the 1930s science became more manifest architecturally and spatially. Several new buildings, funded by the University Appeal of 1937, were completed or begun by 1939. In 1934 a portion of the University Parks was set aside for new science buildings and in conjunction with the University Museum and the buildings clustered around it became known as the Science Area. At the same time the University committed itself to further forward planning of science when it bought some nearby land, known as the Keble Road triangle, for future expansion. Within the newly designated Science Area, the extended Radcliffe Science Library, in effect the University's science library, was opened in 1934. Even Oxford University Press departed from its traditional humanist concerns by showing more interest in publishing scientific journals and monographs. It is clear that there was sufficient encouragement for many innovative scientists to come to Oxford or to stay there, yet a considerable number left and their reasons for doing so reveal much about the difficulties faced by Oxford scientists between the wars.

1.4 Researchers and Postgraduates

The external recognition afforded to Oxford scientists helped to promote the cause of natural science within the University and to show that Oxford, old and beautiful, was becoming famous for its science as well as arts subjects. Between the wars two Oxford professors, Soddy and Sherrington, were awarded Nobel prizes in 1921 and 1932 respectively, part of the latter's work on the central nervous system (for which he received his prize) having been done at Oxford. Subsequently five Oxford scientists (Florey, Chain, Robinson, Hinshelwood, and Hodgkin) who were in post by 1939 received the same ultimate accolade. Another future Nobel laureate, P. B. Medawar (1960), was at Oxford for twelve years after he had graduated there in 1935 but he found its burdens and bells so intolerable that he left and did not return.[58] Some of these scientists achieved the highest honour available in the United Kingdom to a British scientist: they were elected president of the Royal Society of London. Sherrington was the first (1920–5), being followed by Robinson (1945–50), Hinshelwood (1950–5), and Florey (1960–5). Between the wars only two Oxford scientists, Sherrington (1922) and Poulton (1937), were elected president of the British Association for the Advancement of Science; but subsequently this office was occupied by a veritable cohort whose members were in post or were students there before 1939, namely, Tizard (1948), Hartley (1950), Robinson (1955), Clark (1961), Hinshelwood (1965), Lord Jackson of Burnley (1967), Medawar (1969), Lord Todd (1970), Hodgkin (1978), Lord Kearton (1979), and Dainton (1980).[59]

Oxford scientists became increasingly visible in their appropriate national disciplinary society, as officials, as councillors, and as editors. In chemistry, the biggest science department, the key representatives in the Chemical Society between the wars were N. V. Sidgwick and Robinson.[60] Sidgwick was an ordinary member of its Council for six years, a vice-president for another four years, and president 1935–7. He was also the influential first chairman of the Society's publication committee. Robinson was especially prominent after 1935, being vice-president for four years and then president. Seven other Oxford chemists sat on the Society's Council. It was also the

---

[58] Howard Walter Florey (1898–1968), professor of pathology 1935–62; Ernst Boris Chain (1906–79), a demonstrator in the department of pathology 1935–48; Robert Robinson (1886–1975), Waynflete professor of chemistry 1930–55; Dorothy Crowfoot Hodgkin (1910–94), research fellow of Somerville 1933–6, tutorial fellow 1936–77, Wolfson research professor 1960–77; Peter Brian Medawar (1915–87), demonstrator in zoology 1938–47, fellow of Magdalen 1938–44, 1945–7, fellow of St John's 1944–5.

[59] Wilfrid Edward Le Gros Clark (1895–1971), professor of anatomy 1934–61; Willis Jackson (1904–70), Oxford D.Phil. 1936; Alexander Robertus Todd (b. 1907), Oxford D.Phil. 1933; Christopher Frank Kearton (1911–92), graduated in chemistry 1933; Frederick Sydney Dainton (b. 1914), graduated in chemistry 1937.

[60] Nevil Vincent Sidgwick (1873–1952), fellow of Lincoln 1901–48; T. S. Moore and J. C. Philip, *The Chemical Society 1841–1941: An Historical Review* (London, 1947), 158–9.

case that a very small department could not only contribute to its disciplinary society but even go some way towards dominating it. Though entomology was a one-man band at Oxford, the Hope professor of zoology, Poulton, gathered round him and his collections of specimens a group of unpaid devotees who together formed an important coterie in the Entomological Society of London.[61] Poulton, who had been its president once before the war, was elected president again in 1925–6 and 1933–4; in 1933, the centenary of the Society's foundation, he was made its honorary life president. He was such a dominating figure in the Society that wags jested it should be called the Entomological Society of Oxford. Poulton was abetted by H. Eltringham (president 1931–2) and J. J. Walker (president 1919–20), each of whom also put in a good stint as secretary (Walker 1905–18 and Eltringham 1922–5). Poulton's energies were not restricted to the Entomological Society: from 1912 to 1916 he was president of the Linnean Society.

Some Oxford life scientists found it useful to promote their subjects through editing a journal. Thus Charles Elton, a pioneer in animal ecology, became the first editor of the *Journal of Animal Ecology* published by the British Ecological Society. During part of his professorial career at Oxford, Tansley edited two journals, the *New Phytologist* and the *Journal of Ecology*. When he retired from the former he ensured that he was replaced by two of his colleagues, A. R. Clapham and W. O. James, in what he regarded as the public work of the staff.[62] Another professor, Goodrich, put in a stint of just over twenty-five years as editor of the *Quarterly Journal of Microscopical Science*. Other Oxford scientists made use of the University Press on their doorstep by editing series of volumes or journals published by it or by the Clarendon Press which was its academic division. For example in 1923 Huxley and D. L. Hammick launched the Clarendon Science Series; and in 1930 the Clarendon Press captured from Cambridge the *Quarterly Journal of Mathematics* and put three Oxford mathematicians (Chaundy, Ferrar, and Poole) in charge of it. Naturally they gave space to papers produced by their mathematical colleagues at Oxford and by promising postgraduates. In the 1930s the Clarendon Press extended its repertoire into a previously suspect subject by publishing the Oxford Engineering Science Series, edited by R. V. Southwell, E. B. Moullin, and Pye, at the rate of a volume a year. In addition to editing works published by the University Press, Oxford scientists wrote monographs for it which helped to make its scientific list more reputable and

---

[61] Harry Eltringham (1873–1941), a retired businessman, worked in the Hope department as a volunteer and curator 1908–37; James John Walker (1850–1939), a retired naval commander, settled in Oxford 1904; data on Oxford entomological coterie from S. A. Neave and F. J. Griffin, *The History of the Entomological Society of London, 1833–1933* (London, 1933).

[62] Charles Sutherland Elton (1900–91), director of the Bureau of Animal Population 1932–67, reader in animal ecology 1936–67; Arthur George Tansley (1871–1955), professor of botany 1927–37; Arthur Roy Clapham (1904–90), demonstrator in botany 1930–44; William Owen James (1900–78), demonstrator in botany 1927–47, reader 1947–59.

to complement its stable of dictionaries, companions, books of verse, and such new arts ventures as the Oxford History of England. Even so nobody with a training in science was employed at Oxford by the Press until after the Second World War.[63]

While Oxford scientists were making themselves nationally visible, they were also beginning to act as supervisors to advanced students who were working for two or three years for their D.Phil. In the 1920s the D.Phil. in a scientific subject was most effectively promoted by Perkin, one of its key instigators, using graduates recruited mainly from other universities. For ambitious Oxford graduates in chemistry, working for a D.Phil. tended to be seen as a fall-back position if they failed to land an attractive academic post or a lucrative industrial one. In other subjects young Oxford graduate researchers who were academically ambitious simply did not bother to take the D.Phil. In zoology, for instance, four of the leading non-professorial staff between the wars were Oxford graduates who saw taking a D.Phil. as an irrelevance: G.R. de Beer, Elton, E. B. Ford, and J. Z. Young did not serve this sort of apprenticeship but went on to acquire an FRS.[64] The same indifference to the D.Phil. was shown by O. W. Richards, an Oxford graduate who stayed on to do research in the mid-1920s, and eventually landed an important academic job and the coveted FRS. Richards was happy to spend three years as senior Hulme scholar at Brasenose and as Christopher Welch scholar in biology, exploring systematic entomology, a field not well covered in his undergraduate course.[65] The ambiguous status of the Oxford D.Phil. for its brightest postgraduates even in the late 1930s is revealed by the case of Medawar, an undergraduate pupil of Young at Magdalen. Medawar wrote his D.Phil. thesis but did not take the degree. It served no useful purpose for him and, in any event, it cost more than the appendectomy which he had just had to pay for.[66] He chose to remain plain Mr, like his distinguished tutor Young.

Yet in general the 1930s saw the D.Phil. beginning to become the norm for graduates from Oxford and elsewhere who were looking to an academic career or to a research career in industry. As the D.Phil. gradually gained in status, so the number of research students in science in the 1930s increased

[63] Dalziel Llewellyn Hammick (1887–1966), chemistry fellow of Oriel 1921–52; Theodore William Chaundy (1889–1966), student, i.e. fellow, of Christ Church 1912–56; William Leonard Ferrar (1893–1990), fellow of Hertford 1925–59, principal 1959–64; Edgar Girard Croker Poole (1891–1940), fellow of New College 1920–40; Richard Vynne Southwell (1888–1970), professor of engineering science 1929–42; Eric Balliol Moullin (1893–1963), reader in engineering science 1929–39; P. Sutcliffe, *The Oxford University Press: An Informal History* (Oxford, 1978), 227–9.

[64] Gavin Rylands de Beer (1899–1972), demonstrator in zoology 1921–38 and fellow of Merton 1923–38; Edmund Brisco Ford (1901–88), demonstrator in zoology 1927–39, reader in ecological genetics 1939–63, professor of ecological genetics 1963–9; John Zachary Young (b. 1907), demonstrator in zoology 1933–45 and fellow of Magdalen 1931–45.

[65] Owain Westmacott Richards (1901–84); R. Southwood, 'Richards', *BM* 33 (1987), 539–71.

[66] P. B. Medawar, *Memoir of a Thinking Radish: An Autobiography* (Oxford, 1986), 71.

threefold; and the proportion of research students who were scientists more than doubled in that decade. In the session before the outbreak of the Second World War, when the total undergraduate population was 4,400, there were about 300 advanced students registered with the Committee for Advanced Studies. Eighty-five were taking a B.Litt., 30 had embarked on one to two years of research for a B.Sc., and 180 were working on their D.Phil. Some 83 of these were scientists.[67] The increased importance of research students and research degrees in the 1930s is well shown by the two cases of physics and engineering science. In the former the arrival at the Clarendon Laboratory of a group of low-temperature physicists from Breslau in 1933 led to new and effective modes of supervision of postgraduate researchers and a rapid increase in their numbers. In engineering science Southwell built up from scratch a research school which produced not only D.Phil.s who went on to pursue a successful academic career, such as D. G. Christopherson, but also those who became industrial innovators, such as S. G. Hooker, the famous aeronautic engineer.[68]

1.5 The Science Area

The increase in the number of research students in science highlighted the problem of accommodation for them and their supervisors. When the First World War broke out, most of the science departments were housed in or near the University Museum and occupied a corner of land between Parks Road and South Parks Road. Their growth, in terms of acreage built on, was an exceedingly fraught matter because of two constraints. One was the often-expressed intellectual need to have teaching and research in all the sciences, and especially in related subjects, carried out in buildings which were physically near each other. The second was that room for expansion was limited. Little space was available west and south of what was called the Museum area: existing buildings and a few college playing fields saw to that. To the east and north there was the possibility of expansion into the University Parks but these were vehemently regarded as sacrosanct by various interested parties. In 1934, after considerable argument over what the registrar saw as a wretched but important question, a compromise was reached. After a number of *ad hoc* encroachments into the Parks by science buildings, an area publicly recognized by the new title Science Area was reserved for them in 1934. The very designation of a Science Area was bizarre in that other

---

[67] *OUG* 66 (1935–6), 390; 69 (1938–9), 35–42.
[68] Derman Guy Christopherson (b. 1915), D.Phil. 1941, who held chairs at Leeds and Imperial, was then vice-chancellor of Durham University 1960–78 and master of Magdalene College, Cambridge, 1979–85; Stanley George Hooker (1907–84), D.Phil. 1935, on whom see S. G. Hooker, *Not Much of an Engineer: An Autobiography* (Shrewsbury, 1984) and P. H. J. Young *et al.*, 'Hooker', *BM* 32 (1986), 275–319.

British universities did not feel it necessary or desirable to label their collective science site in this way. As the Science Area was separate from the traditional heart of the University and the core of colleges, the very term Science Area acknowledged but at the same time distanced the place of science in a collegiate university still dominated by arts subjects.

Also in 1934 the University and the City of Oxford agreed in principle, and later in practice, that the Parks should be an inviolate private open space in perpetuity, with the exception of a strip 300 feet deep along the north-western edge of the Parks, the strip being reserved for future science buildings. This option has never been taken up by the University, showing the importance generally attached to preserving the Parks. The history of the Science Area therefore revolves around two sorts of struggle. The first was between scientists who wished to put departmental buildings in the Parks, where sites cost nothing because the University owned them, and the lovers of the Parks who wished to keep them free from spoliation. The second was within the cohort of scientists as various professors and departments competed with each other for space and status on an increasingly congested site. These two related types of controversy became acute in the 1920s but were not new: they had flared up before the First World War.

Perhaps the most important pre-war dispute concerned the location of engineering science which failed in 1912 to secure a site in the Parks and the Museum area for its first purpose-built laboratory. Cricketers, footballers, hockey-players, horticulturalists, bird-lovers, strollers, aesthetes, and north Oxford families combined to defend the Parks as one of Oxford's unique and beautiful assets. The result was that a committee led by Sir William Anson, warden of All Souls, raised money to buy land just outside the Parks at the north end of the Keble Road triangle for the engineering science laboratory. This was a victory for the environmentalists over the claims of laboratories; it also showed that in the hierarchy of laboratory sciences at Oxford engineering science was lowly placed.[69]

During the war the Museum science buildings expanded south-eastwards with the opening in 1916 of the Dyson Perrins Chemistry Laboratory on the north side of South Parks Road. This was a relatively uncontentious matter: it would have been churlish to have slowed down the war work being done by Perkin and his colleagues; and, in any event, the laboratory was close to existing outbuildings of the Museum. After the war, however, the negotiations about the government grant and the establishment of the Asquith Commission provided opportunities for Oxford scientists to ventilate their concerns about the lack of any preconceived and comprehensive scheme for

---

[69] J. S. E. Townsend, *The Proposed Engineering Laboratory* (n.p., n.d.) [Oxford, 1912]; W. Anson *et al.*, *The Parks and Science* (n.p., n.d.) [Oxford, 1912], both in Bodleian Library, GA Oxon. c. 153; H. H. Henson (ed.), *A Memoir of the Rt. Hon. Sir William Anson* (Oxford, 1920), 124–30; Sir William Reynell Anson (1843–1914), *DNB*, warden of All Souls 1881–1914.

making additions to the existing science departments in or near the Museum. It was felt by some heads of science departments that each claim for a new or extended building should no longer be treated as an isolated question, that extension of the area devoted to science buildings was desirable and imminent, and forward planning was necessary to avoid rampant ad hocery. Various elements in such a planned scheme were floated in 1920: extending eastwards from the Museum on the north side of South Parks Road, a scheme favoured by the science professors; appropriating the north-west corner of the Parks; occupying some of the space south and south-west of the University Observatory. The first two options involved encroaching on the Parks, which was unpopular in the University; the third meant invading part of the area around the Observatory, which was unacceptable to Turner, the professor of astronomy. He regarded that sacrosanct area as his Naboth's vineyard but he feared that the University and some of his fellow science professors would act against the Observatory in the style of Ahab against Naboth.[70]

By early 1924 the question of what was then officially called the Museum site had become contentious. Late in 1922 the University had accepted a munificent bequest of £100,000 from the William Dunn Trustees for a new pathology building and had decided, without dire opposition, to allocate land at the east end of South Parks Road for it (Map 2).[71] Late in 1923 the University was notified that it was to receive from the Rockefeller Foundation an offer of £75,000 towards a new biochemistry building. At the same time negotiations were taking place about the creation of an imperial forestry institute at Oxford, which led to claims by Troup, professor of forestry, that the consequent enhanced status of forestry would require a new building for it on the Museum site. This prompted Keeble, the professor of botany, to campaign for a new building for botany, also on the Museum site, which would replace his isolated department near the Botanic Gardens. In the decisions reached in spring 1924, the external benefactions for pathology and biochemistry were crucial in persuading Congregation (by 97 votes to 23) and Convocation (unanimously) to approve the setting aside of no more than 9 acres of the Parks for the extension of the science departments, including the pathology site, the extension to be mainly east of the Museum along the north side of South Parks Road (Map 3).[72] This decision about the extension of the Museum site, which was less than the 15 acres recommended

---

[70] Asquith evidence, Craig and Bowman, MS Top. Oxon. b. 109, file 10; Asquith evidence, heads of science departments, MS Top. Oxon. b. 107, ff. 366–8; University Museum. Requirements of Museum Departments 1923–4, OUA, UR/SF/MU/IB; Herbert Hall Turner (1861–1930), professor of astronomy 1893–1930.

[71] *OUG* 53 (1922–3), 112, 127, 177.

[72] *OUG* 54 (1923–4), 323–4, 333, 361, 416; 'The Allocation of the Parks', *OM* 42 (1923–4), 273–4; Frederick William Keeble (1870–1952), professor of botany 1920–6; Robert Scott Troup (1874–1939), professor of forestry 1920–39.

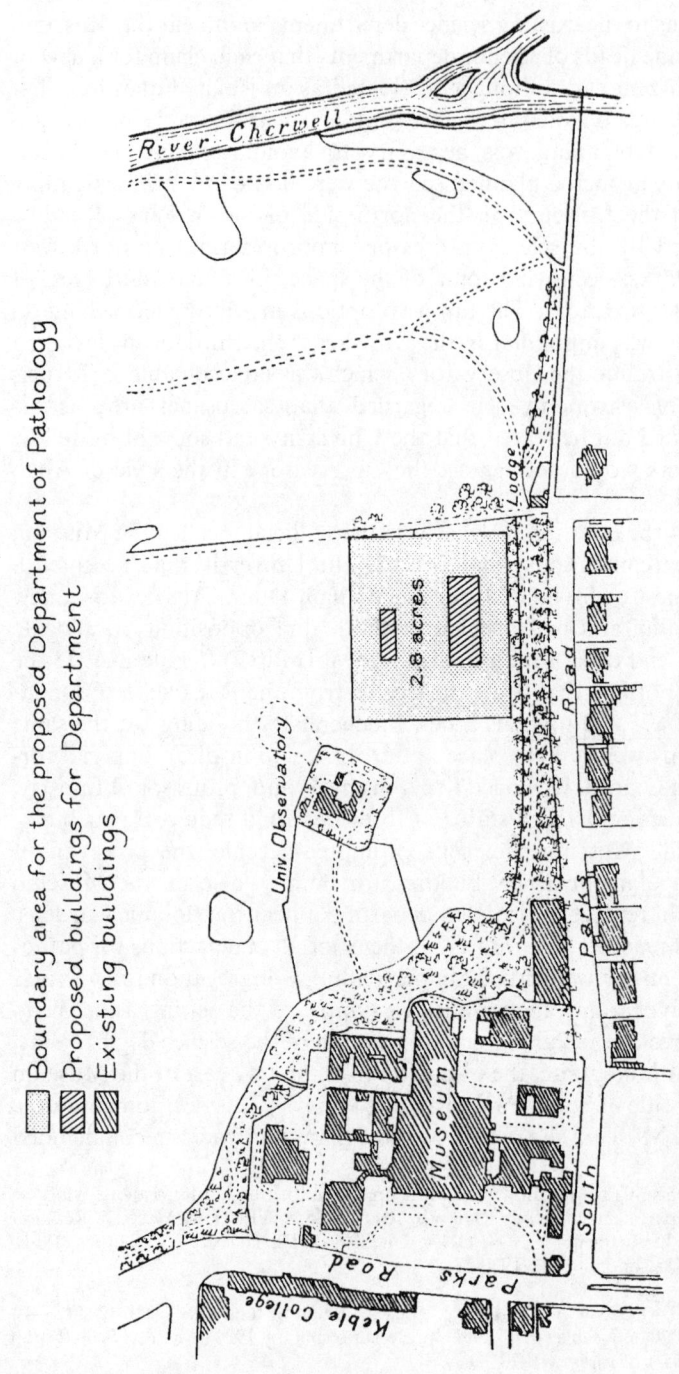

MAP 2. The site for the proposed Department of Pathology, 1922. The caption read: 'The Site for the proposed Department of Pathology, to be established by means of the donation from Sir William Dunn's Trustees under the proposed Decree No. 1 of Congregation, printed on p. 112 of the *University Gazette* of Wednesday, November 8, 1922, is indicated in red on the plan which appears below, which shows generally the position of the site (2.8 acres) in relation to the neighbouring portions of the Parks and also the proposed buildings thereon.' (The 'red' area is now shown by plain grey shading.)
Source: OUG 53 (1922–3), 127.

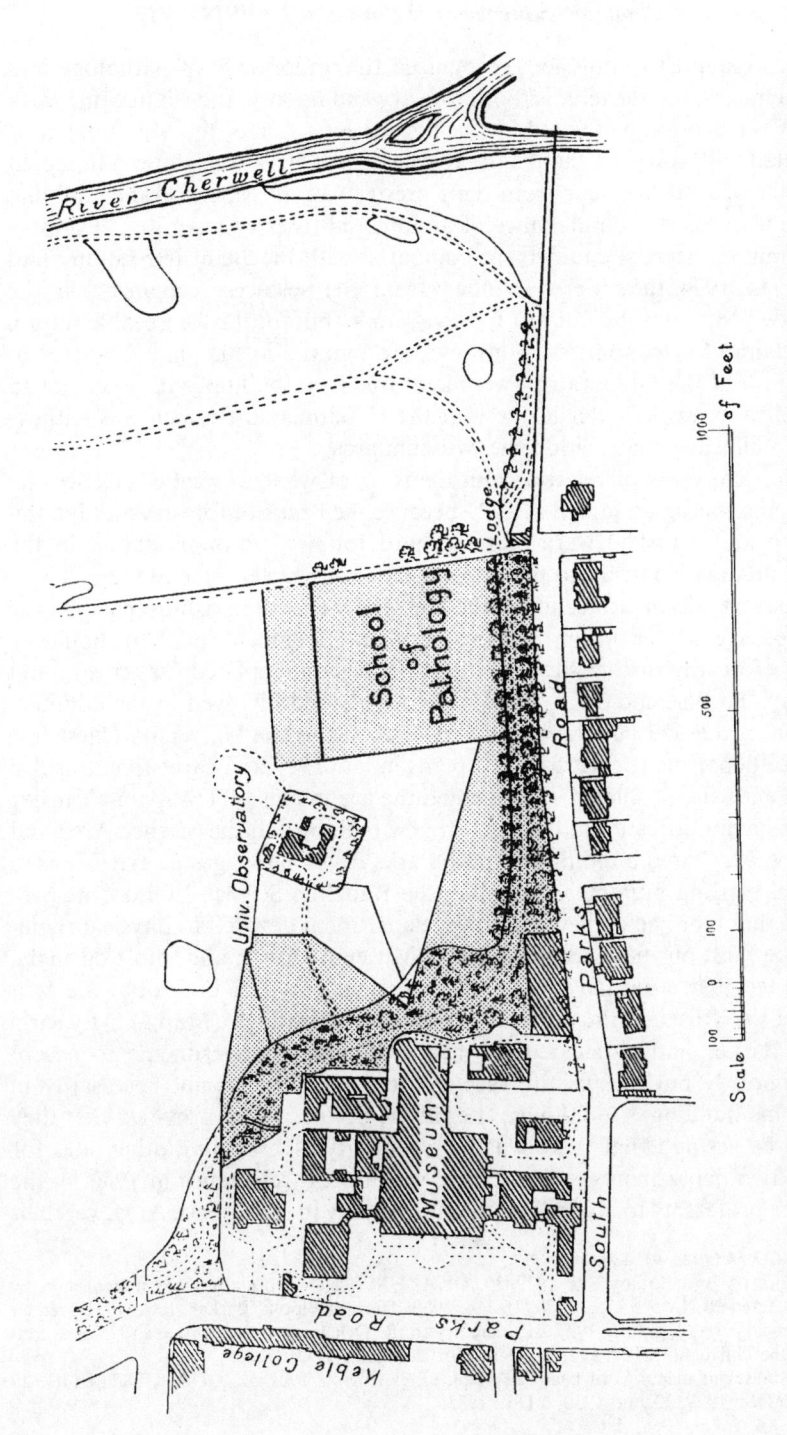

MAP 3. 'Map showing the Extension of the Museum site, proposed in Decree (2)' 1924. 'The [grey] tint indicates approximately the area which it is proposed to allocate in Decree (2) for extensions of the Scientific Departments.'
Source: *OUG* 54 (1923–4), 333.

by the Asquith Commission, recognized the importance of pathology and biochemistry, but it excluded both forestry and botany. Indeed just one week after Convocation had voted for the limited extension of the Museum site, it accepted with warmest thanks the Rockefeller offer; a week later it accepted an offer of £5,000 a year from the Forestry Commission and the Colonial Office towards the maintenance of an imperial forestry institute.[73] Forestry had limited external endowment compared with biochemistry; botany had none. In 1924 they were excluded from the sciences recognized in the extended Museum site not just for a year or so but for the foreseeable future. The claim of astronomy was, however, recognized in the plan: 2.5 acres to the south of the Observatory were kept free from buildings in order not to prejudice its work. Later in the year the Hebdomadal Council was authorized to allocate space within the Museum area.

After five years of relative calm the issue of what was called science site problems was again joined in 1929 because the heads of forestry and botany (Troup and Tansley) were pressing hard for new accommodation in the Museum area, contrary to the decision of 1924 which excluded them. There was also a widespread feeling in the University that the existing spaces in the Science Area (a term first used in a surviving document in 1929) should be used before any further slices of the Parks be contemplated for forestry and botany. To that end T. V. Barker, who had been involved in the administration of the Museum area and was secretary to the University Chest (i.e. finance department), put forward plans in late 1929 and early 1930 to solve the science site problems. He accepted the arguments of Troup and Tansley about having forestry and botany in new buildings in the Science Area and proposed a shared building on South Parks Road; he suggested extensions to several existing buildings, including the Radcliffe Science Library; he proposed that a projected new university department devoted to physical chemistry be built on the land occupied by Museum House; and, most daringly, he hinted that eventually two L-shaped wings could be built onto the west side of the Museum, thus creating a Museum quadrangle (Map 4).[74] By early 1930 Barker had persuaded the heads of science departments to accept unanimously but unofficially that forestry and botany, not necessarily in the same building, should have sites in the Science Area, provided that they could be assured that there was no possibility of obtaining other sites for those two departments.[75] Thus there was a great reluctance in 1930 by the science professors to accept forestry and botany in the Science Area, yet their

[73] *OUG* 54 (1923–4), 429, 453.
[74] Forestry and Botany Site 1929–30, OUA, UM/F/4/17, especially three memoranda by Barker dated 28 Nov., 5 Dec. 1929, 15 Jan. 1930; Thomas Vipond Barker (1881–1931), reader in chemical crystallography 1927–31, secretary to the Delegates of the Museum 1925–8, secretary to the University Chest 1928–31, fellow of Brasenose 1913–31.
[75] Minutes of meetings of heads of science departments 1901–52, OUA, UDC/M/13/1, 29 Nov., 6 Dec. 1929, 22 Jan. 1930, 6 Dec. 1933.

unanimous decision to do so, subject to a proviso, was at odds with the university policy defined in 1924. Encouraged by this apparent volte-face, in February 1930 the Hebdomadal Council decided with little further consultation that sites for forestry and botany should be found in the Science Area, provided funds for the buildings were forthcoming in a reasonable time. In its negotiations with various government departments about the Forestry Institute after February 1930, the University took the line that it had a site *in potentia* in the Science Area for a new forestry building which would house both the department and the Institute.

By 1934 a considerable sum of money for a new forestry building had been collected or promised. The Forestry Institute, then ten years old, was unsatisfactorily split between three buildings, namely, the department and two separate houses. Forestry needed a new building in order to function more effectively. The Forestry Commission and the Colonial Office agreed: in their view the University, which had promised in 1924 to provide a site for the new Institute, was dragging its feet. The University was therefore under considerable pressure from the appropriate government departments and was being harried to confirm that a site would be found in the Science Area.[76] Some science professors now realized that their decision of 1930, to accept forestry and botany in the Science Area, subject to certain conditions, was a mistake: they felt there was a serious risk that all remaining available land in that area would be spoken for if forestry were given its site, thus blocking what they regarded as the stronger claims of other and superior subjects such as chemistry. One solution to the problem was to extend the Science Area beyond its dimensions agreed in 1924, but that solution was bound to raise the hackles of the doughty defenders of the Parks who regretted the ground lost in 1924 and were determined not to lose any more. On the other hand, by 1934 the advocates of the forestry site in the Science Area had extra ammunition unavailable to them in 1930. First, weary of attrition, Tansley had withdrawn botany's claim to a site in the area, thus freeing space for some other use. Secondly, the new professor of astronomy, Plaskett, had conceded that his sort of research required only a small part of the reservation of 2.5 acres, south of the Observatory, which was guaranteed in 1924. Accordingly, Lindsay, a key figure in the Hebdomadal Council, argued forcibly early in 1934 that, as a result of these changes in the requirements of botany and astronomy, more available building land would be added to the Science Area than would be taken away from it if a new forestry building were to be built in it.[77]

---

[76] Lindsay to Lindemann, 16 Jan., 8 Feb. 1934, LP, B98/19–21, 24.

[77] Sites: Committee on Building Requirements 1934–7, OUA, UR/SF/S4, especially Tansley to Lindsay, 21 May 1934; A. D. Lindsay, 'The Forestry Statute and the Site', *OM* 52 (1933–4), 398–402; Harry Hemley Plaskett (1893–1980), professor of astronomy 1932–60; Alexander Dunlop Lindsay (1879–1952), master of Balliol 1924–49, vice-chancellor 1935–8.

ENGINEERING SCIENCE

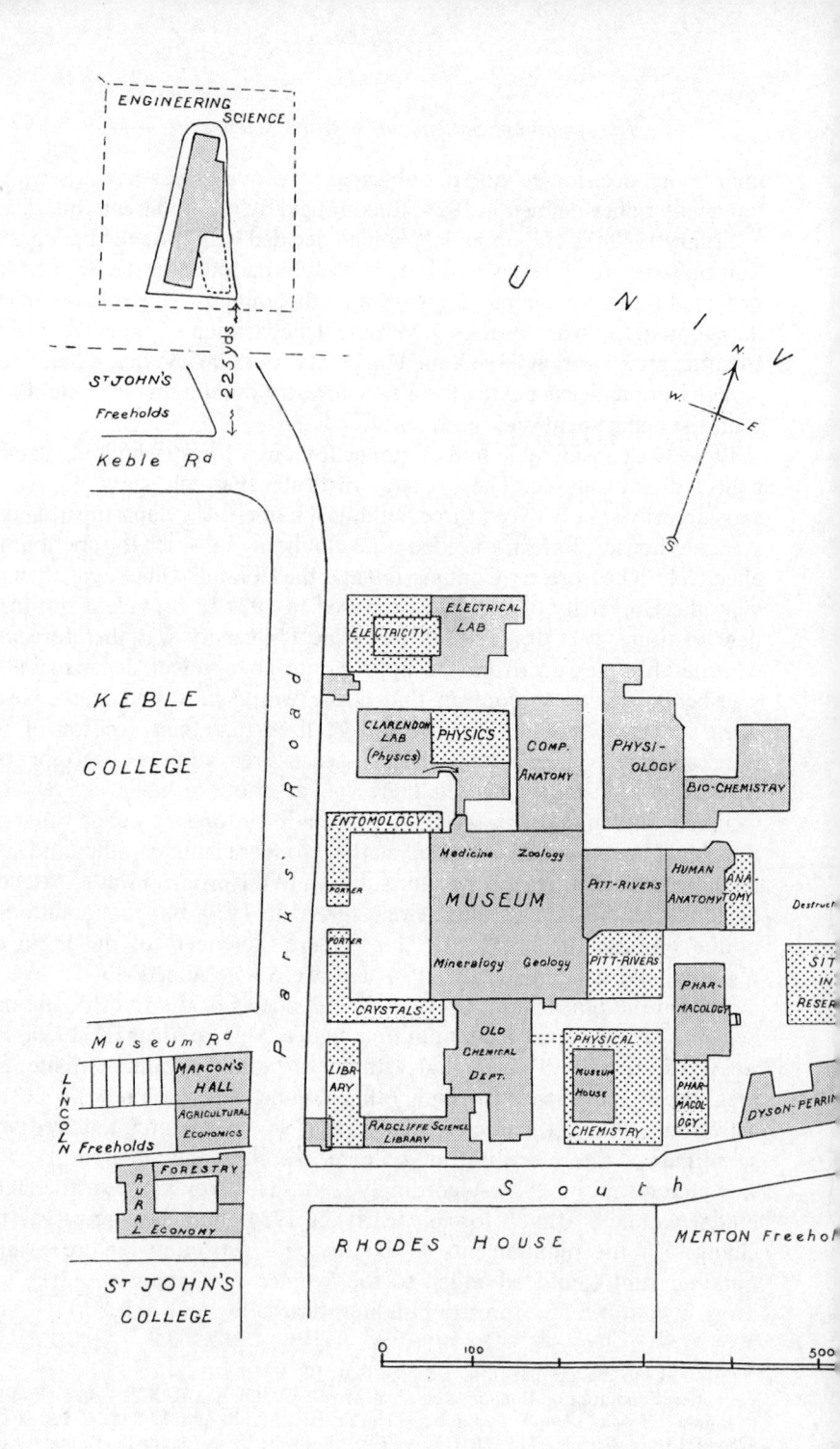

St JOHN'S
Freeholds

← 225 yds →

Keble Rᵈ

KEBLE COLLEGE

Parks' Road

ELECTRICITY | ELECTRICAL LAB

CLARENDON LAB (Physics) | PHYSICS | COMP. ANATOMY | PHYSI-OLOGY | BIO-CHEMISTRY

ENTOMOLOGY

PORTER

PORTER

Medicine Zoology
MUSEUM
Mineralogy Geology

PITT-RIVERS | HUMAN ANATOMY | ANA-TOMY

Destruct

PITT-RIVERS | PHARMACOLOGY

SITE IN RESER

CRYSTALS

Museum Rᵈ

MARCON'S HALL
AGRICULTURAL ECONOMICS

LINCOLN Freeholds

RURAL ECONOMY | FORESTRY

Sᵗ JOHN'S COLLEGE

LIBRARY | OLD CHEMICAL DEPT. | PHYSICAL | PHARMACOL-OGY | DYSON-PERRI

RADCLIFFE SCIENCE LIBRARY | MUSEUM HOUSE CHEMISTRY

South

RHODES HOUSE   MERTON Freehol

0   100   500

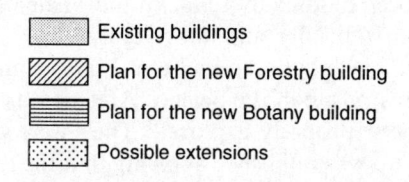

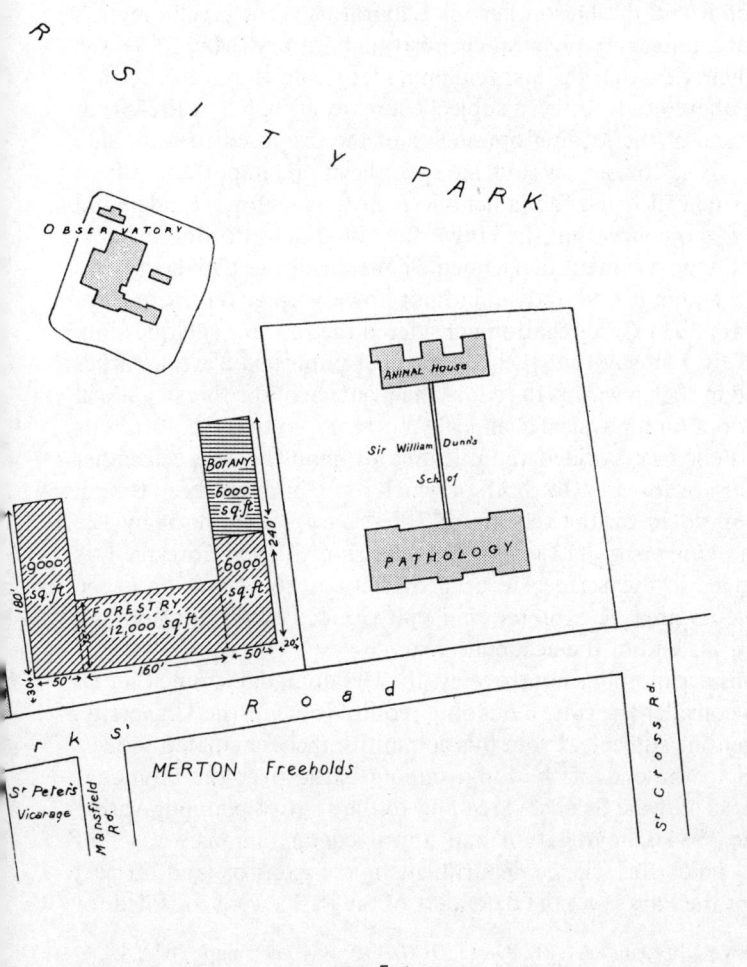

MAP 4. 'Science Site Problems', 1929–1930. The shaded buildings are the proposed extensions.
Source: *OUA*, Bodleian Library, Oxford (UM/F/4/17).

It is clear that, on the matter of the forestry site, many science professors felt that they had been diddled by Lindsay and Veale, the registrar, who had not consulted them officially and thoroughly. They were sure that their provisos of 1930 had been ignored and that Lindsay and Veale were forcing the issue of the forestry site in the Science Area: it was undeniable that sites outside had not been properly explored. They were dismayed when on 7 February 1934 a map was published showing that the location in the Science Area of the proposed forestry building would leave only a useless Polish corridor between it and the Dyson Perrins Laboratory, thus precluding the building there of a university physical chemistry laboratory (Map 5). It was outrageous in their view that the last remaining large site within the Science Area was being offered to forestry, a subject expressly excluded in 1924 from it. Lindemann, one of the leading opponents of the proposed forestry site, argued that to assist forestry would hamper the more important job of promoting what he called the fundamental sciences at Oxford. Lindemann gave short shrift to the view that the University could not afford to exhaust the patience of the government departments concerned: the University was not living in the Tudor period and should not kowtow to government.[78]

On 13 February 1934 Congregation considered the forestry site question. It was reported by Lindsay that the ground plan published a week earlier had been revised in such a way as to reduce the frontage of the forestry site in order to leave room for a physical chemical laboratory, so that the proposals debated on 13 February avoided the dilemma of choosing between either physical chemistry or forestry, for both of which room had now been found; yet Congregation voted on the scheme of 7 February, accepting it by 122 votes to 91. The University had reversed its decision of 1924: forestry was now to be included in the Science Area, provided that its expansion in the immediate Museum area be explored and that Council could not find any other site before the end of the academic year.[79]

Bruised by this acrimonious controversy, the Hebdomadal Council set up a committee to consider the future building requirements of the University, especially in scientific subjects. From this committee there emanated suggestions which led to the end of the long-running battle over the boundary between the Parks and the Science Area and to the start of planning within it. By mid-June 1934 Congregation had approved unanimously a set of proposals from which the science departments made gains of land, in part at the expense of the Parks, yet the defenders of the Parks were consoled for

[78] Lindemann to Veale, 17 Jan. 1934, LP, B98/15; *OUG* 64 (1933–4), 399 (map); OUA, UDC/M/13/1, 19 Jan. 1934; *OM* 52 (1933–4), 431–6, 459–60; Goodrich, Hinshelwood, Lindemann, Peters, Sidgwick, and Townsend, *Objections to the Proposed Grant of a Site in the Parks to a Government Forestry Institute* (n.p., n.d. [1934]), Bodleian Library, GA Oxon b. 141 (181b). Douglas Veale (1891–1973), *DNB*, registrar 1930–58; Frederick Alexander Lindemann (1886–1957), Dr Lee's professor of experimental philosophy 1919–56.

[79] *OUG* 64 (1933–4), 349, 357, 380–1, 392–3; *OM* 52 (1933–4), 459–60.

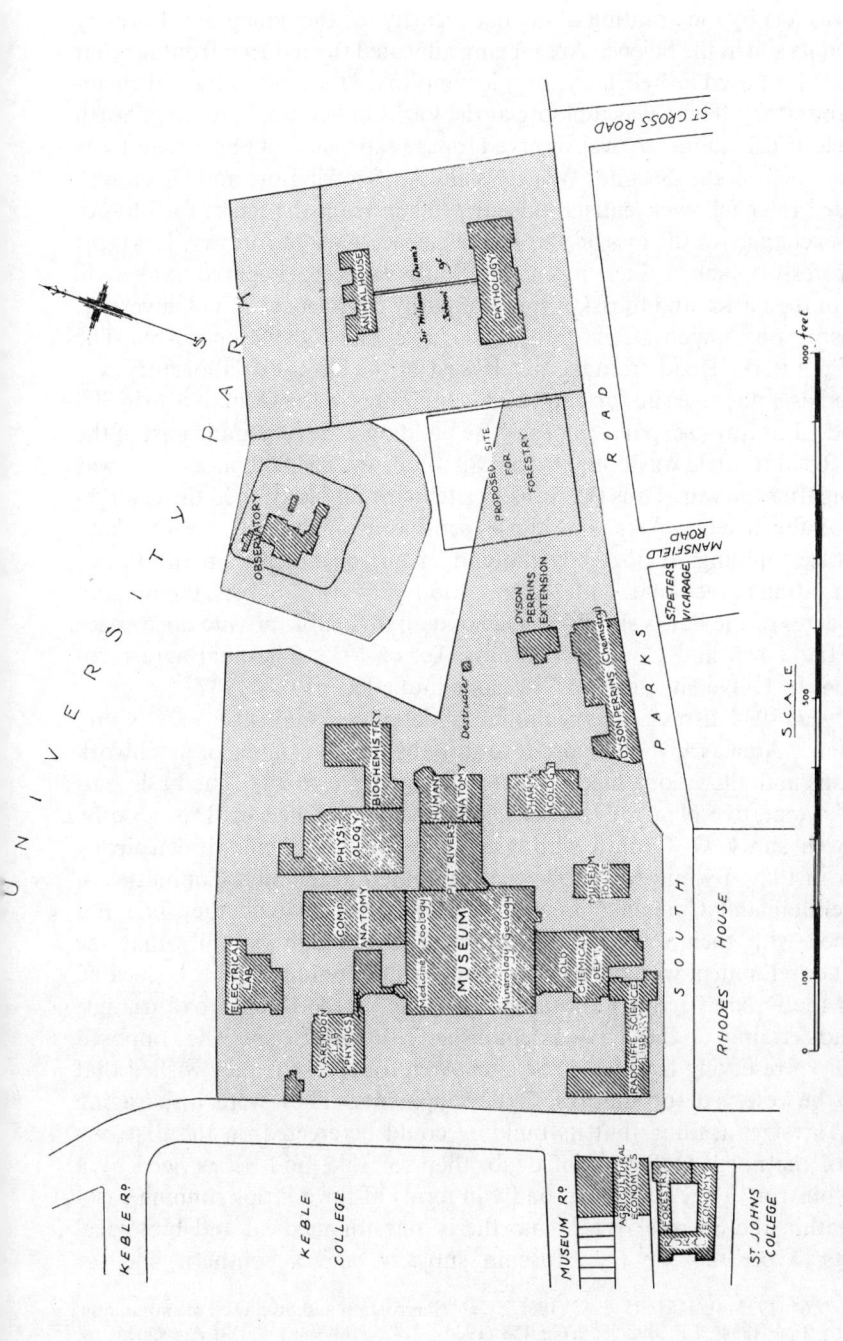

MAP 5. 'Plan of the Science Area', 1934. Note the use of the new term 'Science Area' and how the new buildings have encroached on the Parks. This map shows how the University proposed to extend the Science Area into the white area surrounding the Observatory (which had been reserved for the Observatory in 1924) in order to accommodate the proposed site for Forestry. Note also how the chemists' proposed extension to the Dyson Perrins Laboratory is represented as if it were an existing building.
Source: OUG 64 (1933–4), 389.

what was lost by the granting of greater security for the remainder. Forestry retained its site in the Science Area, being allocated the reduced frontage that had been proposed in February, thus leaving space for a new physical chemistry laboratory. There was nibbling at the Parks in two ways. An area north of the electrical laboratory was reserved for the expansion of physics, and was built on before the Second World War. As Sherrington and Southwell reminded their fellow scientists, this strip taken from the Parks for physics was in exchange for the loss of part of the Science Area to forestry. It is clear that professors such as Sherrington and Southwell were prepared to alienate lovers of the Parks, and to risk losing sympathy for science in the University, by insisting on compensation for the loss of a key site in the Science Area. The rest of the Parks Road frontage north-west of the electrical laboratory was reserved as a possible site for new science buildings in the form of a strip 300 feet wide. For future expansion of science buildings, the remaining part of the Keble Road triangle was bought from St John's for £50,000; much of it was built on after the war. Thus the forestry site question occasioned the enlargement of the Science Area and the expensive acquisition of nearby land for future building. It also saw the end of encroachments on the Parks. Congregation agreed that, with the exception of the two nibbles, the remaining 70 acres of the Parks should be declared an inviolable private open space under the Town and Country Planning Act of 1932. The final agreement between the University and the City came into effect in July 1937.[80]

The year 1934 also saw the beginning of detailed planning of space within the Science Area as a whole in order to stop the previous mode of patchwork additions and alterations made unilaterally by departments. The basic outline of a tentative plan for the Science Area was hatched in Brasenose by Southwell and I. O. Griffith who tried to revise the scheme drawn up by Barker in 1929. By mid-May 1934 they presented to the Sites Committee of the Hebdomadal Council a first draft of a scheme which they had not discussed with their scientific colleagues. They thought initially that the biggest development would be in chemistry, which could probably be located east of the Dyson Perrins Laboratory, possibly on the Keble Road triangle site, and certainly on the site of a demolished Museum House. They opposed the idea of reserving land in the Science Area for forestry: they wished that site to be reserved for unforeseen developments. They were disparaging about forestry, arguing that its building could be erected on the first two holes of the north Oxford Golf Club, then for sale, and be serviced by a 2-mile bus run. They thought broadly in terms of three strips running east–west within the Science Area, a northerly one for medical and biological subjects, a central one for Museum subjects, and a southern one for

[80] *OUG* 64 (1933–4), 416, 695–6; 65 (1934–5), 16; Sherrington and Southwell memorandum, n.d. [early June 1934], LP, B98/58; *HCP* 158 (1934), 21–2; University Parks. Agreement for Dedication as a Private Open Space 1934–49, OUA, UR/SF/CQ/IC.

chemistry. They did not contemplate a new Clarendon Laboratory in the near future. For Southwell and Griffith, a new Clarendon Laboratory could be erected in about thirty years' time on the Keble Road site. They were more concerned with congestion in the Museum and proposed the drastic solution (never implemented) of building a quadrangle west of the Museum and fronting on Parks Road.[81]

The heads of science departments agreed broadly with these proposals except that they argued that present needs were so desperate that immediate expansion of physics northwards into the Parks was required. They accepted some measure of planning perhaps because they appreciated the point made strongly by Southwell and Griffith that a planned Science Area with allocated sites would encourage benefactions. By late May 1934 Southwell and Griffith had modified their plan in response to pressure from physical scientists who saw that the Clarendon Laboratory was in an advanced state of obsolescence. They now proposed an extension of the Science Area northwards on Parks Road for physics. Though the biology faculty board had reservations about such encroachment northwards into the Parks and was concerned about the problems faced by botany, it joined the physical sciences faculty board in approving the Southwell–Griffith revised plan. This scheme, approved by Congregation in principle in June 1934, meant that the immediate beneficiary of the extended Science Area was physics, and not chemistry, whose area for expansion was reduced by the proposed forestry site, which was mischievously described by Southwell and Griffith as 'lost to science'.[82]

They were well aware that, in trying to plan for the probable requirements of science in the next thirty years, they were merely offering general ideas which they thought should govern the detailed planning of sites in the Science Area. They also knew that the time available for full discussion of the problems encountered was quite inadequate. Yet they did produce a scheme which enshrined a three-strip rationale to try to avoid the clashing of departmental interests, and it did lead to what may be regarded as ordered development within the Science Area. Their plan was broadly adopted by T. A. Lodge, the architect appointed by the University in 1935 to reconstruct the Science Area in detail.[83] Inevitably there were problems of allocation of space which were difficult to resolve. One concerned chemistry. In mid-1936 the Hebdomadal Council had decided that inorganic and physical chemistry could be combined in a new building to be erected on the site of Museum

[81] Southwell to Lindemann, 9 Mar. 1934, LP, B26/2; Sites 1934–40, OUA, UR/SF/S/1, for Southwell–Griffith draft plan 15 May 1934; Sites. Committee on Building Requirements 1934–6, OUA, UR/SF/S/1 for Veale's summary of discussion of 15 May 1934; *HCP* 158 (1934), 93–4.

[82] OUA, UDC/M/13/1, 23 May 1934; BS/M/1/1, 1 June 1934; Southwell–Griffith revised plan 28 May 1934 in *HCP* 158 (1934), 182–5 (185 quotation).

[83] Sites. Reconstruction of Science Area, 1933–8, OUA, UR/SF/S/1A; Thomas Arthur Lodge (1888–1967).

House, which would have to be demolished. This was rejected by the physical chemists who, confident of the superiority of their branch of chemistry, wanted a separate physical chemistry building east of the Dyson Perrins Laboratory, at right angles to it, and extendible northwards. Their power in the colleges was such that they secured their aims: the new physical chemistry laboratory was begun in 1938 and completed in 1941.[84] It was also agreed in 1936 that the site of Museum House would be reserved for chemistry; but it was only in 1954 that it was demolished to make way for an extended inorganic chemistry laboratory.

Territorial disagreements were not ended by the adoption of the Southwell–Griffith plan for the Science Area. Science professors continued to be touchy about their interests so that on occasion demarcation disputes occurred between them; and procedures for dealing with internal changes in the Science Area remained unformalized for several years. As late as 1944 there were two acrimonious disputes when the site of the existing physiological laboratory was marked on a planning document as that of a new entomological laboratory, and one particular area was allocated to both geology and zoology.[85] Even so the Southwell–Griffith layout plan for the Science Area permitted significant changes to be made just before the outbreak of the Second World War. In 1937 the University decided to build a new Clarendon Laboratory at a cost of nearly £80,000 on the Parks Road strip reserved for physics in 1934. This heavy expenditure was approved not only because the existing laboratory was deficient but also because a new laboratory was the key to the most urgent adjustments to be made in the Science Area, such as providing suitable accommodation for geology on the site of the current laboratory.[86] The Southwell–Griffith plan, consolidated by Lodge, was also an integral part of the university Appeal launched in 1937.

In February 1937 the University launched a well-publicized Appeal for about £500,000 and suspended it in November because most of its aims had been met. When suspended the Appeal had raised £426,912 to be spent on three main purposes, namely, the extension of the Bodleian Library; the endowment of advanced work in the humanities; and the promotion of science via new laboratories for physics, physical chemistry, and geology, extensions of some existing laboratories, funds for staffing, and a general research fund. The extension of the Bodleian Library at a total cost of just over £1 million was the top priority. In early 1932 the University had accepted an offer from the Rockefeller Foundation that it would contribute 60 per cent of the total cost of extending the Bodleian provided the

---

[84] *HCP* 161 (1935), 33–7; 162 (1935), 119–20; 164 (1936) 221–4; Development Plan 1936–7, OUA, UM/F/4/18.
[85] Sites. Reconstruction of Science Area, 1938–48, OUA, UR/SF/S/IB.
[86] 'Science Buildings', *Oxford*, 4 (1937), 50–2; 5 (1938), 48–50.

(17) A PART OF ONE OF THE CHEMISTRY LABORATORIES, HOUSED IN A BUILDING ARCHITECTURALLY AND OTHERWISE UNSUITABLE

FIG. 1 The old chemistry laboratory of 1860 modelled on the Abbot's kitchen at Glastonbury Abbey (centre) with the 1878 extension behind left and the Radcliffe Science Library (1903) right. Condemned as unsuitable in the publicity for the University Appeal (1937), it still stands. Reproduced by permission of the editor of *Oxford* from *Oxford: Special Number February 1937*.

University would raise the remaining 40 per cent by the end of 1936. By early 1937 the Foundation had promised £560,000, and the University had already raised £260,000. The main object of the Appeal was to produce £250,000 so that the University would be able to brandish £510,000 all told before the eyes of the Rockefeller Foundation. It did so: the new Bodleian building in Broad Street was begun in December 1937 and completed in 1940.[87]

The publicity associated with the Appeal stressed that the position of science within the University had reached a critical point and that Oxford could not afford to fall behind in such an important field of study. Embarrassing photographs were published in a special issue of *Oxford*, the magazine of the Oxford Society, showing science buildings that were outmoded, congested, cramped, and ramshackle (see Fig. 1). At the same time it was revealed that at last there was in existence a comprehensive general plan for

---

[87] *OUG* 63 (1932–3), 21–2; *Oxford*, 3 (1937), 5–11; 4 (1937), 45–7; 5 (1938), 45–7; 'Oxford and Present Needs in Science', *Nature*, 139 (1937), 303–4.

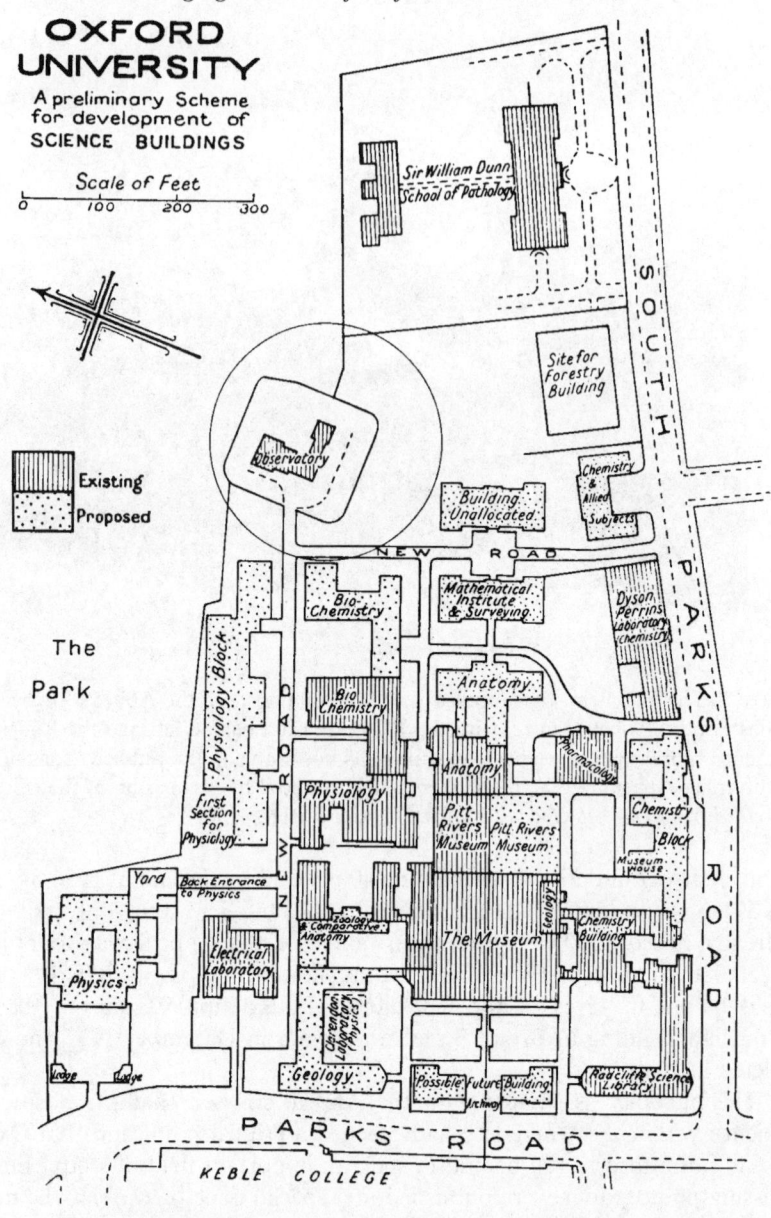

MAP 6. Plan for the Science Area, 1937. Compared with Map 5, p. 39, Forestry shrinks, new roads appear, Mathematics is at last recognized (although the plan was never implemented), and further encroachment on the Parks north of the Science Area is proposed.
Source: Oxford: Special Number February 1937, 59.

the Science Area, which would ensure its architectural dignity, produce improved accommodation, and introduce the novelty of steel-frame construction of new buildings to render them adaptable to various uses. The Appeal claimed that the scheme for the development of the science buildings provided for all possibilities over an extended period; the publication in 1937 of Lodge's detailed ground plan for the Science Area was some guarantee that the money raised for it would be well spent (Map 6).[88]

Though the Appeal stressed the importance of new laboratories for physics (Clarendon Laboratory), physical chemistry, and geology, benefactors preferred to avoid earmarking money for the new Clarendon. The new physical chemistry laboratory received £100,000 from Lord Nuffield and £10,000 from Imperial Chemical Industries, while the Shell Oil Company gave £25,000 for a new geology laboratory. It was significant that two of the three biggest earmarked benefactions, those of Nuffield and Shell, were devoted to science, the other big benefaction of £100,000 from the Rhodes Trustees being devoted to social studies. Though the Appeal was successful in its own terms, the contrast with Cambridge was telling. In 1928 the Rockefeller Foundation had agreed to contribute half of the estimated cost of £500,000 of a new university library at Cambridge, the rest to be found by the University. The Foundation also promised £450,000 towards new developments in science, provided the University raised £229,000. By 1930 this large sum had been found, with the Rockefeller conditional offer acting as a powerful lever. The Rockefeller grant to Cambridge, by providing for staff and research endowment as well as for new or extended buildings, confirmed its status as England's premier scientific university.[89] The Rockefeller Foundation never gave a big comprehensive benefaction to science at Oxford as it did at Cambridge: it preferred to discharge its largess on the Bodleian Library.

One catalyst of the creation of the Science Area was the opening in its south-west corner of the extended Radcliffe Science Library in November 1934 at a cost of £45,000.[90] This was in effect the first instalment of the Bodleian Library extension scheme. The former had, however, not always been under the superintendence of the latter. Until 1927 they were separate libraries, the University being responsible for the Bodleian and the Radcliffe Trustees being in charge of their library, mainly a scientific one. The Radcliffe Science Library had long been located in the Museum area: from 1861 it occupied premises in the Museum, and in 1903 gained its own building at

---

[88] T. A. Lodge, 'The Future of the Oxford Science Buildings', *Oxford: Special Number February 1937* (Oxford, 1937), 57–8, map p. 59.

[89] J. B. Morrell, 'Hustlers and Patrons of Science', *History of Science*, 31 (1993), 66–82.

[90] NS/M/I/3, 5 June 1925; *HCP* 130 (1925), 197–9; *OUG* 57 (1926–7), 305; 58 (1927–8), 392; 'Radcliffe Library Extension', *OM* 52 (1933–4), 760–1. The library was variously referred to as Radcliffe, Radcliffe Science, and Radcliffe (Science).

the west end of South Parks Road. In the 1920s its services to scientists were felt to be deficient: because of the lack of co-operation between the Radcliffe and Bodleian Libraries, the poor supply of periodicals by the former was a distinct handicap to researchers. Simultaneously the librarian of the Bodleian was acutely aware that his library had given little help to science. He thought that co-operation with the Radcliffe Library would save the Bodleian from being denigrated as a merely antiquarian institution, and he predicted that the study of the literature of science would become more important. Accordingly in 1927 an amicable agreement was reached between the University and the Radcliffe Trustees. The latter made a munificent benefaction to the former by donating the freehold of the Radcliffe Camera building with the adjoining land; the books (totalling about 100,000), furniture, and equipment of the Radcliffe Science Library; £1,500 p.a. towards its upkeep; and their collections already deposited in the Bodleian. Within four years, a short length of time by Oxford's standards, Congregation had decided that, as part of the scheme for improving and extending the Bodleian Library, the Radcliffe Science Library building should be extended westwards to Parks Road and northwards along it to form an L-shaped structure. Construction began in March 1933 and was complete by summer 1934. This extension of the Radcliffe Science Library meant the loss of a propitious site for future expansion of science departments, but there were two valuable gains: both the seating accommodation (64 to 188) and the shelf space (100,000 volumes to 300,000) were trebled. Moreover, the Radcliffe Library preserved an advantage over the Bodleian: so-called privileged persons had access to the book-stacks. The extension of the Radcliffe Library also met another need in that part of the first floor was partitioned temporarily to provide rooms where the mathematicians could meet, teach, and study, and maybe multiply, pending the construction of a proper mathematical institute.

1.6 Leavers

One of the most obvious features of Oxford science between the wars is that many innovative senior staff stayed and did not move elsewhere, even though on occasion they faced depressing and even gruelling circumstances. On the other hand a considerable number felt frustration with Oxford and found greater satisfaction elsewhere, especially in London and Cambridge. The colleges of the University of London attracted several established scientists from Oxford. Two applied physical scientists moved to Imperial College: Egerton, reader in thermodynamics, assumed in 1936 the chair of chemical technology, being drawn by the better facilities for research and the reduced frustrations; after thirteen hard years at Oxford, Southwell, professor of engineering science, followed him in 1942 to succeed Tizard as rector.[91] Two zoologists moved to the two oldest colleges at London. In 1925 Julian Huxley resigned his fellowship at New College and his zoology demonstratorship in order to assume the chair of zoology at King's College; while in 1938 de Beer, who liked London life, resigned his Merton fellowship and zoology demonstratorship for a readership in embryology at University College.[92] Two London colleges benefited from the frustrations felt by A. E. Jolliffe, a mathematician who was a fellow of Corpus Christi where the president was wholly unsympathetic to mathematics and to Jolliffe's financial advancement. He left Oxford in 1920 for chairs at Royal Holloway (1920–4) and King's (1924–36).[93] Only the outbreak of the Second World War prevented a sixth Oxonian, the disenchanted Clark, professor of anatomy, from assuming the chair of anatomy at University College London, which he had accepted in early 1939: the provost of UCL decided to suspend Clark's appointment for the duration of the war so his resignation from his Oxford chair was suspended—permanently, as it turned out.[94] One Oxonian, Henry Tizard, fellow of Oriel and reader in thermodynamics, felt the lure of an increased salary and of power in London via the administration of science: in 1920 he became an assistant secretary in the Department of Scientific and Industrial Research which had been established during the First World War. One Oxford scientist, R. B. Bourdillon, chemistry fellow at University, took the unusual course of resigning his post in 1921 in order to go to London to qualify in medicine at St Mary's Hospital and to pursue a research career at the National Institute for Medical Research. Of those who moved to London, no fewer than five (Egerton, Southwell, Huxley, de Beer, Tizard) gained knighthoods and pursued highly visible careers, often as administrators: Tizard and Southwell were successive rectors of Imperial College, London, while Huxley became the first director-general of UNESCO and de Beer ended his career as director of the British Museum (Natural History).

It was not unexpected that several senior men left Oxford for Cambridge. Two of them were stars of the first magnitude. G. H. Hardy, the leading English pure mathematician of his time, resigned his Savilian chair at Oxford in 1931 to return to Cambridge, partly for a more comfortable old age. In 1940 Oxford lost the world's greatest expert on Jurassic geology when W. J.

---

[91] Alfred Charles Glyn Egerton (1886–1959), reader in thermodynamics 1921–36; D. M. Newitt, 'Egerton', *BM* 6 (1960), 39–64; R. J. Egerton, *Sir Alfred Egerton: A Memoir with Papers* (n. p., 1963); D. G. Christopherson, 'Southwell', *BM* 18 (1972), 549–65.

[92] Julian Sorell Huxley (1887–1975), fellow of New College and demonstrator in zoology 1919–25; J. R. Baker, 'Huxley', *BM* 22 (1976), 207–38; E. J. W. Barrington, 'De Beer', *BM* 19 (1973), 65–93.

[93] Arthur Ernest Jolliffe (1871–1944), fellow of Corpus Christi 1891–1920, obituary in *OM* 62 (1943–4), 225; the president of Corpus Christi was Thomas Case (1844–1925).

[94] W. E. Le Gros Clark, *Chant of Pleasant Exploration* (Edinburgh, 1968), 150–1.

Arkell, a senior research fellow at New College who had failed to gain the Oxford chair of geology in 1937, left for a fellowship at Trinity, Cambridge, and never returned to Oxford.[95] At the start of the inter-war period and after it three engineers were lured by the more encouraging circumstances of their subject at Cambridge. David Pye, who had taught engineering science at Oxford before the war, returned for the summer term 1919 and was reappointed in June 1919 to take effect for five years from October 1919. He never took up this reappointment, preferring instead a fellowship at Trinity, Cambridge. Towards the end of the Second World War, A. M. Binnie and E. B. Moullin, the two senior staff under Southwell, followed Pye's example by returning to Cambridge: Binnie, a demonstrator, left Oxford in 1944 for a fellowship at Trinity; while Moullin, reader in electrical engineering and fellow of Magdalen, left in 1945 to be Cambridge's first professor in that subject.[96] Physiology and chemistry each lost one established person to Cambridge. J. B. S. Haldane, a fellow of New College, continued his father's interest in experimental human physiology, especially the role of carbon dioxide in the bloodstream. As this sort of work was not at the top of Sherrington's research agenda, Haldane moved to Cambridge in 1923 to assume a new readership in biochemistry and to gain increased emolument as well as reduced teaching. Oliver Gatty, fellow of Balliol, found chemistry teaching for just two years such a tedious diversion from original work that in 1933 he left Oxford to research at Rothamsted and then at Cambridge in zoology and colloid science.[97]

Other Oxford scientists committed what was regarded at Oxford as the ultimate crime: if leaving for Cambridge, the younger of the two old English universities, was pardonable, to go elsewhere was not. One man stepped westwards to the USA: in 1920 the frustrated William McDougall, Wilde reader in mental philosophy and fellow of Corpus Christi, left Oxford for a chair in psychology at Harvard. An English provincial university, Liverpool, managed to reverse the usual brain-drain from the north to Oxbridge by appointing two Oxonians to chairs. In 1914 W. Ramsden, fellow of Pembroke and specialist in chemical physiology, became the second occupant of the Johnston chair of biochemistry at Liverpool. After the war A. M. Carr-Saunders, a demonstrator in zoology with strong interests in popula-

[95] Geoffrey Harold Hardy (1877–1947), professor of geometry 1919–31; E. C. Titchmarsh, 'Hardy', *ON* 6 (1949), 447–61; William Joscelyn Arkell (1904–58), senior research fellow of New College 1933–40; L. R. Cox, 'Arkell', *BM* 4 (1958), 1–14.

[96] *OUG* 49 (1918–19), 572; Alfred Maurice Binnie (1901–86), demonstrator in engineering science 1924–44, FRS in 1960, reader in engineering, Cambridge, 1954–68, obituary in *The Times*, 3 Jan. 1987; Moullin, *DNB*, professor of electrical engineering, Cambridge, 1945–60.

[97] John Burdon Sanderson Haldane (1892–1964), fellow of New College 1919–22; N. W. Pirie, 'Haldane', *BM* 12 (1966), 219–49; R. W. Clark, *J. B. S.: The Life and Work of J. B. S. Haldane* (London, 1968), 55, 64; M. Weatherall and H. Kamminga, *Dynamic Science: Biochemistry in Cambridge 1898–1949* (Cambridge, 1992), 30, 40–3; Oliver Gatty (1907–40), chemistry fellow of Balliol 1931–3, obituary in *OM* 59 (1940), 24.

tion studies, eugenics, and politics, accepted the call to be the first holder of the Booth chair of social science at Liverpool.[98] J. C. Eccles, a fellow of Magdalen and demonstrator in physiology, returned in 1937 to his native Australia to become neurological king there.[99] Both Carr-Saunders and Eccles became eminent knights, the former as director of the London School of Economics and the latter as a Nobel prize-winner. One Oxford physical chemist migrated to South Africa to be a university administrator: H. R. Raikes, fellow of Exeter, went to be principal of the University of Witwatersrand, Johannesburg, in 1927. No fewer than four organic chemists who worked in the Dyson Perrins Laboratory as either demonstrators or senior researchers found it impossible to maintain or develop their careers in Oxford: G. R. Clemo, H. R. Ing, and R. D. Haworth left for provincial pastures, and J. M. Gulland went to London.[100]

Four well-known Oxford scientists migrated to industry. Harold Hartley, fellow of Balliol and knighted in 1928, was persuaded in 1930 to leave academic life to become research director for the London, Midland & Scottish Railway and to exercise for the next thirty years his superlative gifts as a chairman of many public bodies. Another chemist, M. P. Applebey, fellow of St John's, left Oxford in 1928 to become research manager at the Billingham plant of Imperial Chemical Industries where some of his former pupils were already employed. Two professors left Oxford for the world of industrial research. In late 1926 F. W. Keeble, professor of botany and knighted in 1922, was persuaded by his friend Sir Alfred Mond to join ICI as an agricultural adviser and as director from 1929 of its Jealott's Hill Agricultural Research Station at Bracknell. From 1927 his job was to sell the fertilizer produced in excess by Applebey's colleagues at Billingham. In 1929 the ageing C. F. Jenkin, professor of engineering science, was tired of

---

[98] Walter Ramsden (1868–1947), Johnston professor of biochemistry, Liverpool, 1914–31; Alexander Morris Carr-Saunders (1886–1966), demonstrator in zoology 1920–3, professor of social science, Liverpool, 1923–37, director of London School of Economics 1937–56; H. Phelps Brown, 'Carr-Saunders', *Proceedings of the British Academy*, 53 (1967), 379–89.

[99] John Carew Eccles (b. 1903), first-class honours physiology 1927, D.Phil. 1929, fellow of Exeter 1927–34, physiology fellow of Magdalen 1934–7, demonstrator in physiology 1929–37, Nobel prize-winner 1963.

[100] George Roger Clemo (1889–1983), a member of the British Dyes' research group in the Dyson Perrins Laboratory 1916–25, became professor of organic chemistry at Armstrong College, Newcastle upon Tyne, now the University, 1925–54; B. Lythgoe and G. A. Swan, 'Clemo', *BM* 31 (1985), 65–86; Harry Raymond Ing (1899–1974), demonstrator in chemistry 1920–2, 1923–6, who left Oxford in 1926 for Manchester where he did cancer research for three years, ended his career as reader in chemical pharmacology, Oxford, 1945–66; H. O. Schild and F. L. Rose, 'Ing', *BM* 22 (1976), 239–55; Robert Downs Howarth (1898–1990), researcher 1922–6, demonstrator in chemistry 1926–7, who left for Newcastle in 1927, ended as professor of chemistry at Sheffield University 1939–63; E. R. H. Jones, 'Howarth', *BM* 37 (1991), 265–76; John Masson Gulland (1898–1947), demonstrator in chemistry 1924–31, went in 1931 to the Lister Institute, London, which he left in 1936 for the chair of chemistry at Nottingham University; R. D. Howarth, 'Gulland', *ON* 6 (1948), 67–82.

teaching and wished to devote himself full-time to research on building materials which he had begun in 1924; so he resigned his chair in order to join the staff of the Building Research Station at Watford.[101]

All the established scientists who left Oxford between the wars felt the need for some emolument or increased salary; but one rich man did not. Thomas Merton, stipendless reader in spectroscopy and research fellow of Balliol, withdrew gently from research in the cramped conditions of the Clarendon Laboratory in 1924 in order to live the life of an affluent country gentleman in Herefordshire where he worked in his private laboratory, took out patents, collected Italian Renaissance pictures, and enjoyed the local salmon fishing. Like many of those scientists who left Oxford, he did not fade into obscurity. He went on to act as a long-serving treasurer for the Royal Society of London and to be knighted in 1944 for his inventions used in the Second World War.[102]

Oxford's inability to retain some of its most innovative scientists persisted into the Second World War and beyond. The most notorious cases were those of Solly Zuckerman, Ernst Chain, and Peter Medawar. The first, a South African who was a well-known researcher before he became a demonstrator in anatomy in 1934, enhanced his reputation in his Oxford period with prolific publication on the reproductive and endocrine functions of monkeys. Though he enjoyed membership of Christ Church senior common-room, no college offered him a fellowship. In 1944 Clark, his head of department who was suspicious of Zuckerman's multifarious interests, made it clear that life at Oxford would be difficult for him if he did not accept the chair of anatomy at Birmingham which had been kept open for him since 1939: in future there would be no research accommodation such as Zuckerman had enjoyed in the 1930s. Zuckerman took the hint, emigrated in 1946 to Birmingham, and stayed there twenty-two years. From his Birmingham base he was knighted and became a senior scientific adviser to government.[103] Chain, a German refugee who had arrived in Oxford in 1935, became a joint Nobel prize-winner in 1945 for his work with Howard Florey on penicillin. No college made him a fellow and the University did not

---

[101] Malcolm Percival Applebey (1884–1957), chemistry fellow of St John's 1919–28, research manager ICI, Billingham, 1928–48; J. L. S. Steel, 'Applebey', *Proceedings of the Chemical Society* (1957), 214–15; V. H. Blackman, 'Keeble', *ON* 8 (1952–3), 491–501; W. J. Reader, *Imperial Chemical Industries: A History*, ii: *The First Quarter-Century 1926–1952* (London, 1975), 81, 98–104, 158; for Keeble's role until 1932 as head of research and propaganda in fertilizers see his *Fertilisers and Food Production on Arable and Grass Land* (London, 1932); Alfred Moritz Mond (1868–1930), first chairman of ICI 1926–30.

[102] Thomas Ralph Merton (1888–1969), reader in spectroscopy 1919–23; H. Hartley and D. Gabor, 'Merton', *BM* 16 (1970), 421–40.

[103] Solly Zuckerman (1904–93), demonstrator in anatomy 1934–45, professor of anatomy at Birmingham 1943–68 (*de jure*), 1946–68 (*de facto*), chief scientific adviser to Ministry of Defence 1960–6 and to the government 1964–71; S. Zuckerman, *Monkeys, Men and Missiles: An Autobiography 1946–88* (London, 1988), 3–4.

promote him so he left Oxford for Rome in 1948 to be a research director.[104] Unlike Zuckerman and Chain, Medawar was an Oxford graduate and over a period of twelve years from 1935 held a succession of college fellowships. In his case he found college tuition increasingly onerous, time-consuming, and exhausting, so he accepted the call issued via Zuckerman to assume the chair of zoology at Birmingham in 1947; by 1960 he was a Nobel prize-winner. Medawar endured a second source of frustration at Oxford, perhaps uniquely so. Every Sunday morning the ringing of Oxford's church bells prevented him from working at his desk: repelled by their peals he was certainly not summoned by bells.[105]

[104] E. P. Abraham, 'Chain', *BM* 29 (1983), 43–91.
[105] Medawar, *Autobiography*, 53–4, 102–3.

2
The Colleges and the University

2.1 The Powers of the Colleges

Oxford was, and remains, a collegiate university in which the life of the University was broken into small units via the college system. The colleges took prime responsibility for undergraduate teaching, a goodly proportion of which was dispensed via the weekly tutorial hour, which was their pride and joy. Though some colleges were richer than others, none of them dominated Oxford in terms of numbers of fellows and undergraduates as Trinity, St John's, and King's did at Cambridge. Their distinctive feature was their status as legally independent and autonomous corporations. As such they were subject to the University only in those respects which their statutes specified. A significant example of the zeal with which colleges defended their independence arose in 1922 when Congregation rejected as illegal a proposal that the heads of the colleges be required to send to the University's registry each term a list of their undergraduates showing which subjects they were studying. The statutory position was that the University could do no more than ask any college to allow its head to dispatch the required information.[1] Thus the autonomy of colleges had a firm legal basis so that any encroachments on it could be vigorously opposed by the colleges or reluctantly accepted by them as concessions. A few college heads took the line of Farnell, rector of Exeter, who championed the interests of the University against those of the colleges or as not incompatible with them; but for every Farnell there was at least a Hazel. The principal of Jesus was an uncompromising college man who resolutely opposed any tendency towards powerful and centralized departments which he denounced as a new Leviathan.[2]

In nearly all the colleges their heads (known variously as principals, rectors, masters, provosts, wardens, and presidents) were chosen by the fellows who were responsible collectively for the government of their college. New fellows of colleges were elected by the existing fellows so that there was some danger of colleges becoming self-perpetuating oligarchies or coteries,

---

[1] *OUG* 52 (1921–2), 597; R. B. Mowat (ed.), *Letters to 'The Times' 1884–1922 Written by Thomas Case* (Oxford, 1927), 121–3.

[2] Lewis Richard Farnell (1856–1934), rector of Exeter 1913–28, vice-chancellor 1920–3; L. R. Farnell, *An Oxonian Looks Back* (London, 1934); J. N. L. Baker, *Jesus College, Oxford 1571–1971* (Oxford, 1971), 130.

especially as college fellowships were rarely advertised. Thus colleges tended to elect Oxford graduates or (other things being equal) their own graduates to fellowships, the exception being professorial fellowships where a college was required to accept as a fellow a professor appointed by the University after open public competition.

The colleges controlled the admission of undergraduates, often in an informal way by exploiting personal contacts with their graduate members who sent their sons or pupils to their old college. The University had no control over the number of undergraduates in each college or the subjects which they studied. Thus the total number of undergraduates in the University and their distribution between subjects were not determined by the University but by the colleges, each of which acted independently. On the whole the number of applicants seldom exceeded the number of undergraduate places available so that entry to Oxford was not the highly competitive affair it became after the Second World War. Colleges took almost anyone thought good enough to acquire a degree and able to pay the fees. Prospective pass-men, who had no intention of taking an honours degree, were also welcome, especially if they could 'sink a puddle' on the river as oarsmen. But even Brasenose resisted the temptation to admit brilliant sportsmen who spelled Jesus with a small g.[3] Inevitably there was discrimination in favour of applicants from the independent schools and the older grammar schools, not necessarily because Oxford was snobbish but because parents who sent their offspring to them were more likely to be able to pay the college fees. In the last academic session before the Second World War 62 per cent of first-year male undergraduates came from independent schools, 13 per cent from older grammar schools, and 19 per cent from state schools; for first-year women undergraduates (258 compared with 1,235 men) the figures were 51 per cent, 13 per cent, and 33 per cent respectively. The social exclusivity of Oxford between the wars was pointedly revealed in the distribution of scholarships and exhibitions: just 18 schools, 9 of which were boarding ones, provided 40 per cent of scholars and exhibitioners.[4] These interrelated features of informal admission of undergraduates, lack of competition for places, and social exclusiveness of the undergraduate body persisted until after the Second World War. It was only in the late 1940s that admissions tutors began to appear in the colleges; and only in the early 1960s that the Oxford Colleges Admissions Office, which administered an intercollegiate system, was established.

The colleges also controlled their own finances though they were required to make a contribution to the University through the Common University

---

[3] *Franks Report*, i. 65; B. A. Richards, 'Herbert Hart: A Brasenose View', *Brazen Nose*, 27 (1993), 6–7.

[4] *Franks Report*, i. 72–3; A. H. Halsey, *Decline of Donnish Dominion: The British Academic Professions in the Twentieth Century* (Oxford, 1992), 67.

Fund which taxed them according to their wealth by using a graduated contribution scheme with a top tax-rate of 35 per cent on the net college revenue if it exceeded £20,000. This tax system was introduced in 1882 to secure an income for the inadequately endowed University from the colleges which were relatively wealthy. The Asquith Commission regarded the CUF arrangements as so well tried that Cambridge was urged to adopt them. The scheme of 1882 persisted until 1959 when it was changed because by the early 1950s the University had at last become the richer partner mainly as a result of government grants.[5] In the inter-war period, then, most colleges had considerable freedom of choice about how to spend their income and they exercised it academically. In 1922 there were 357 academic staff in Oxford, no less than 60 per cent of these being college fellows who held no university position at all.[6] At that time and indeed until the Second World War Oxford was a confederation of independent colleges which jealously guarded their academic independence and tended to ignore the University except when they wanted it to confer degrees, to provide supplementary teaching, and to maintain such common and expensive services as big libraries, museums, and laboratories.

2.2 College Life and Loyalties

It was the autonomy of the colleges in the fundamental matters of appointments, admissions, and finance which sustained a corporate life and ethos different from that of a university department or a hall of residence. The colleges dominated the lives of fellows and undergraduates. The fellows ran the colleges, gave tutorials in them, and if they were bachelors lived in them. Similarly the social and academic lives of undergraduates revolved around the colleges: they received tuition from the fellows, or had their tuition arranged by them, and usually lived in. The colleges had, therefore, academic, residential, and social functions: they were universities in miniature in which undergraduates taking different subjects studied and lived together. Not surprisingly the colleges induced intense and persistent loyalty in both their senior and junior members. This college fetishism was perhaps more strongly developed at Oxford than Cambridge partly because the former's colleges were on average smaller than the latter's.[7] Even by the mid-1960s no Oxford college exceeded 500 in its total membership and the median size was 325. Before the Second World War, Oxford colleges were considerably smaller. In the last academic session before the Second World War no college admitted over 100 first-year undergraduates. The largest was Christ Church with 90; no fewer than twenty colleges admitted 60 or less.[8] As the

---

[5] *Franks Report*, i. 282–3.   [6] Ibid. 27.
[7] Ibid. ii. 32–3.   [8] *OUG* 69 (1938–9), 83–91.

degree courses occupied three years, with the exceptions of Greats and chemistry, the majority of Oxford colleges housed and taught less than 200 undergraduates. If the number of undergraduates was small, so too was the number of fellows in the various colleges. It seems that on a generous if approximate estimate of what constituted their academic staff only four (Christ Church, Magdalen, New College, and Queen's) had more than twenty on their roll in the 1920s. The average academic staff of the twenty-two men's colleges numbered thirteen; eleven of them had ten academic staff or less. The five women's colleges had fewer academic staff, with an average of nine per college in the 1920s.[9]

The small size of the Oxford colleges encouraged the growth or maintenance in many colleges of an ethos which was *sui generis*: Balliol prided itself on effortless superiority; New College on steady if somewhat prim moderation and responsibility. One college, Brasenose, was so renowned for its heartiness and distaste for aesthetes that an aesthete-friend of John Betjeman's always entered it limping, as he thought the hearties would be too sporting to attack one who appeared to have suffered injury on a college playing field.[10] The relative smallness of Oxford's colleges encouraged an intense social life in which it was quite possible for an undergraduate to know well a senior don in a subject different from his own: at New College Goronwy Rees, who read PPE, became friendly with G. H. Hardy, the renowned mathematician who lived on the next staircase. Like large and pleasant country houses, the colleges offered close and intimate domesticity. For college fellows, their senior common-rooms could offer civilization at its most lively and agreeable: the young Lionel Robbins found New College to be an earthly paradise with pleasures ranging from the floral magic of the college garden to the polyphony of Bach and Palestrina in the college chapel.[11] The colleges offered an ethos which resembled that of a club, a salon, or a country-house party. For some senior dons, their college and not the University claimed their total loyalty: Spooner, warden of New College, the only Oxonian of the time to bequeath an eponymous neologism to the English language—spoonerism—was uninfluential in the University, all his efforts being focused on the extended family of his college.[12] The colleges acted as protected epicurean enclaves, especially those with secluded gardens or fellows' gardens; they were refuges from the madding crowds jostling on

---

[9] *Franks Report*, ii. 32.

[10] B. Hillier, *Young Betjeman* (London, 1989), 131–2.

[11] J. Buxton and P. Williams (eds.), *New College Oxford 1379–1979* (Oxford, 1979), 112, 119, 124–5; Morgan Goronwy Rees (1909–79), principal of Aberystwyth University College, Wales, 1953–7; Lionel Charles Robbins (1898–1984), fellow of New College 1927–9, professor of economics, University of London, 1929–61.

[12] W. G. Hayter, *Spooner: A Biography* (London, 1977); William Archibald Spooner (1844–1930), warden of New College 1903–25; a spoonerism is the accidental transposition of the initial sounds of two words.

the High Street or pouring from the Cowley motor-car works.[13] Hence college fellows were often content to be good college men totally loyal to their college and its undergraduates. Hertford supplies three choice examples of dons whose lives were focused more on their college than on published research. After firsts in mathematics and natural sciences, Haselfoot was tutor in physics and responsible for science teaching for thirty-seven years, in which period he married a daughter of a Hertford fellow and occupied the offices of dean, senior tutor, and bursar. The mathematician J. E. Campbell, FRS, likewise served Hertford as a fellow for thirty-seven years, having been an undergraduate there; after the First World War he was bursar and vice-principal, the routine of which compensated for his lost research impulse. Campbell's successor was W. L. Ferrar, who was fellow for thirty-four years before being elected principal. Ferrar decided that specialist publication and college teaching plus administration could not be mixed, settled for the latter, and as principal purged his college of its Evelyn Waugh image.[14] This combination of loyal college service and reluctant publication was apparent in the science fellowship at Merton. The chemistry fellow, Bertram Lambert, researched on gas analysis and the absorption of gases by solids but was slow to publish his findings. His physics colleague Gill was domestic bursar for thirty-two years and a pioneer installer of college electric wiring and phones. Newboult, the mathematics fellow, focused his life on his college tutoring, producing many first-class honours men among his pupils, and being more interested in synthesizing known results than in seeking new ones. The only science fellow of Merton to publish heavily was de Beer.[15] That same combination of dedicated college work and indifference to output of research was apparent at New College. Though S. P. McCallum was appointed in 1927 as its first fellow in physics primarily to help the University, he turned out to be a popular tutor, an active college administrator, but not a great researcher. His chemistry colleague A. F. Walden was so committed to pedagogy but not to publication that in the college he was affectionately known as 'Teacher'. He had manuscript teaching materials, known as the Koran (on physical chemistry) and the Talmud (on organic chemistry), which undergraduates borrowed chapter by chapter, but he never published a textbook. He preferred the aesthetic pleasure of contem-

[13] M. Batey, *Oxford Gardens: The University's Influence on Garden History* (Amersham, 1982); E. S. Rohde, *Oxford's College Gardens* (London, 1932); R. T. Gunther, *Oxford Gardens: Based on Daubeny's Popular Guide to the Physic Garden of Oxford; with Notes on the Gardens of the Colleges and on the University Park* (Oxford, 1912).

[14] Charles Edward Haselfoot (1864–1936), obituary in *OM* 55 (1936–7), 183; John Edward Campbell (1862–1924), obituary in *OM* 43 (1924–5), 8; William Leonard Ferrar (1893–1990), fellow of Hertford 1925–59, principal 1959–64, obituary in *Independent*, 29 Jan. 1990.

[15] Lambert was fellow of Merton 1920–47, obituary by E. J. Bowen, *Nature*, 199 (1963), 1136–7; Gill was fellow 1909–58 and domestic bursar 1921–53; Harold Oliver Newboult (1897–1949) was fellow 1925–49, obituary in *OM* 68 (1949–50), 14; de Beer was fellow 1923–38. Merton is very well served by *The Merton College Register* (Oxford, 1964 and 1990).

plating the established edifice of knowledge to the dismaying untidiness of research.[16]

This tradition of the unselfish and long-serving college tutor, for whom published research was a secondary consideration, persisted after the Second World War. George David Parkes was for thirty years the only fellow in science at Keble. An essential but unspectacular administrator, he contributed to chemistry not papers but well-known textbooks. He was popular in Keble as a musician who founded and conducted the Iffley Choral Society, was president of the Oxford District Organists' Association, and established the Oxford Church Choirs Festival. His other interests, in archaeology and in railway history (he was a ranking expert on the Hull & Barnsley Railway), did him no harm at Keble. Witness also the case of John Griffith, classics fellow of Jesus 1938–80. He was a compulsive teacher who was convinced that his overriding responsibility was to pass on received knowledge and not to advance its frontiers; he refused to take up space in learned journals which, he alleged, should serve the interests of young scholars who had to provide evidence of research to secure appointments. These views did not prevent Griffith from being a popular university figure as public orator.[17]

This somewhat disparaging attitude to specialized research and publication was connected with several persistent features in the lives of college fellows. There was a tendency to elect as fellows recent graduates, mainly from Oxford, who had enjoyed brilliant careers as examinees and prize-winners; original research was not always regarded as an important criterion for appointment to a fellowship and it was only in the 1930s that the possession of a D.Phil. began to be seen as an advantage. Once elected, young fellows, fresh from the examination rooms, were thrust into onerous college tutorial teaching which occupied many hours a week and involved covering large areas of their subject. Inevitably this encouraged dons to pursue wide-ranging scholarship at the expense of specialized research: that was a condition of pedagogic survival as well as a source of esteem from peers. For college fellows it was difficult to keep the three balls of teaching, scholarship, and research in the air simultaneously: there was a tendency to drop research and to relinquish the aim of becoming the leading expert on a specialist topic because teaching and scholarship were required by the college tutorial system but research was not. In any event specialist publication was often seen in the colleges as a low form of academic life because it was narrow. From the same cultivated perspective, research was

[16] Stanley Powell McCallum (1895–1940), fellow of New College 1927–40, obituary in *OM* 59 (1940–1), 120; Allan Frederick Walden (1871–1956), fellow of New College 1908–39, obituary in *OM* 75 (1956–7), 132–6.

[17] George David Parkes (1899–1967), tutor at Keble 1924–30, tutorial fellow 1930–65, obituary in *OM* 86 (1967–8), 106–7; John Godfrey Griffith (1913–91), public orator 1973–8, obituary in *Independent*, 25 Apr. 1991.

viewed as an ungentlemanly and boorish German notion; and the idea of postgraduate supervision was regarded either as a Yankee device for inserting into a patrician university plebeians who could not help themselves, or as a threat to those scholars incapable of supervising D.Phil. students. Moreover Oxford's tradition of connoisseurship at high table or in the senior common-room, exercised in fellowship divine, provided an alternative or superior way of gaining or maintaining esteem in Oxford and in its outposts. This obsession with matters of polite taste and refined sensibility showed itself in conversation about food, wine, travel, literature, the arts, and railway timetables. Above all, in a university where there were endless opportunities for talking, success in verbal jousting was highly regarded: witticisms, epigrams, *le mot juste*, and argument by guffaw were the preferred forms of conversational revelry. It was perhaps not accidental that Hinshelwood and Sidgwick, two of Oxford's most successful science fellows, prospered in part as connoisseurs in the college environment. Hinshelwood was a painter who appreciated Chinese ceramics. In his college laboratory he used to read Dante in the original and in the common-room amuse his colleagues with nursery rhymes about college affairs. Sidgwick, who had acquired first-class honours at Oxford in both Greats and natural sciences, was renowned for his verbal sallies against fallacy and pretension: he was always illuminating, usually pungent, often funny, and sometimes disconcerting. As a host to visiting scientists Sidgwick persuaded his college to set aside a room permanently for his use. His generous hospitality in his college helped to make Oxford's chemistry known and respected.[18]

For their junior members the colleges offered not only tutorial teaching but an atmosphere in which friendships were easily made. For example when Stanley Hooker arrived at Brasenose he immediately welcomed its great camaraderie and *savoir-faire* after his impersonal life as an undergraduate mathematician at Imperial College. He soon met W. R. Verdon-Smith, then a law undergraduate, who in his later capacity as a managing director of the Bristol Aeroplane Company lured Hooker to join it in 1949.[19] Especially through their bachelor dons colleges offered considerable pastoral care to undergraduates who were more varied in intellectual ability than after the Second World War. Through competitive team sports on river and field undergraduates developed manliness and the ability to work with others in harmony and with zeal. Many fellows and heads of colleges did not disdain sporting prowess: on the contrary, it was widely believed that the man who

[18] T. F. H. Higham, 'Hinshelwood', *OM* 86 (1967–8), 45–6; K. Hutchison, *High Speed Gas: An Autobiography* (London, 1987), 33, 35; obituary of Sidgwick, *OM* 70 (1951–2), 284–6; L. E. Sutton, 'Sidgwick', *Proceedings of the Chemical Society* (1958), 310–19; L. E. Sutton, interview with author, 27 Nov. 1987.

[19] Stanley George Hooker (1907–84); P. H. J. Young *et al.*, 'Hooker', *BM* 32 (1986), 275–319; Hooker, *Autobiography*, 10, 15, 123; Sir William Reginald Verdon-Smith (1912–92), obituary in *Independent*, 25 June 1992.

could tackle low on the rugger field had the moral resource to tackle anything in his subsequent career. Most science departments organized their practical classes so as not to compete with afternoon sport. In these and other ways colleges provided a general education far wider than was available in a hall of residence or a specialist teaching department. That being said the formal academic teaching provided by the colleges was the tutorial in which an undergraduate, often solo, met his college tutor for an hour or so every week.

2.3 The College Tutorial

The basic aim of the tutorial between the wars was to compel undergraduates to read, think, write, and argue. The standard procedure was that the tutor set an essay title, gave a reading list, and a week later the undergraduate returned with an essay or some prepared work which was read out by its author and criticized by the tutor. The tutorial system put a premium on creativity and correction. It developed the abilities to find out from reading, to write coherent prose, and to defend one's position while accepting criticism. It encouraged the cult of oral and dialectical confidence and enlarged the powers of the mind, but it was not designed to impart factual information. In other words the tutorial could not cover the syllabus, which did include matters of fact and technique, so it was supplemented by lectures and in science subjects by laboratory work. The lectures, which were organized by the faculty boards and offered on a university-wide basis, were regarded as an optional extra in arts subjects; in the sciences, however, lectures though voluntary were more important and laboratory attendance was expected. Thus for arts undergraduates the college was the main provider of formal teaching. For undergraduate scientists there were three sources: the college which organized tutorials, the appropriate faculty board which was responsible for the lecture programme, and the appropriate department which offered laboratory work under the aegis of the appropriate faculty board. This arrangement meant that college science tutors had to dovetail their tutorials, in terms of timing and content, with the lectures and laboratory sessions to avoid overlap and to give overall coherence. This complicated organizational problem was not faced by their arts colleagues whose jobs were simpler and less onerous in that they gave tutorials and directed the reading of their charges. This difference in the teaching responsibilities of arts and science college fellows was reflected in their daily lives: the scientists split their time between the department and the college; arts tutors spent a greater proportion of their working time in college and were therefore more likely than the scientists to contribute to college life and to define its ethos.

Stephen Leacock's famous sketch of an Oxford tutorial depicted it as a social, smoky, and educative occasion: 'what an Oxford tutor does is to get a little group of students together and smoke at them. Men who have been systematically smoked at for four years turn into ripe scholars.... A well-smoked man speaks and writes English with a grace that can be acquired in no other way.' At St John's undergraduate chemists were well smoked by Applebey who required them to expound their essays at the blackboard and to read original research papers not just in English but also in French and German.[20] In practice, of course, tutorials and the nature of pastoral care varied according to the tutor. Chemistry sported some well-known tutors as well as eccentrics. Harold Hartley took a great deal of trouble with his Balliol pupils: he never missed a tutorial though he was a busy man. He steered them into jobs, laid great emphasis on exact expression on paper, and guided but did not force his pupils in their part two research year to work on his favourite topics. In his tutorials he avoided formal instruction: using penetrating questions and remarks, he gained the confidence of his pupils while making them do all the work. He challenged his pupils from the start by making them read and write about original papers and not textbooks.[21] At Trinity Hinshelwood was held in awe as an all-round tutor: though he was a physical chemist he covered inorganic and organic chemistry with his pupils. He also set demanding vacation reading on which pupils were questioned. At the end of K. J. Laidler's first term at Oxford, Hinshelwood asked him to read Sommerfeld's *Atombau und Spektrallinien* in the original, though there was an English translation of earlier editions available. Hinshelwood also gave gastronomic and bibulous training to his pupils through lavish and formal dinners with a different wine for each course. Sometimes he took his pupils to the theatre, especially to see magicians. Although Hinshelwood was a reserved person, he clearly compensated for this by exercising great courtesy and considerable hospitality to his pupils.[22] At Lincoln Sidgwick made no attempt to be a forcing tutor. He read essays casually and uncritically, but with odd illuminating comments. He discussed topics which seemed irrelevant or non-chemical. He did not require his students to attend his own lectures. He offered a liberal scientific education on which the pupil could build as he chose: Sidgwick's trademark was wide-ranging and discursive discussion in an informal

[20] S. Leacock, *My Discovery of England* (Toronto, 1922; repr. 1961), 72–96 (80); Autobiographical fragment, H. M. Powell Papers, A3.
[21] R. P. Bell, interview with author, 16 July 1987; D. M. Murray-Rust, 'Some Personal Comments on the Teaching and the Development of Research in the Balliol/Trinity Labs during the Years 1923–1931', sent to author, 8 Sept. 1987; E. J. Bowen, 'Hartley', *OM* 91 (1972–3), 7–8.
[22] K. J. Laidler, transcript of interview with Christine King (University of Ottawa, 1983), 5, 7, 35–7 (courtesy Professor Keith J. Laidler (b. 1916), first in chemistry 1938, professor of chemistry, University of Ottawa, 1955–81).

atmosphere.[23] At Magdalen Sutton, a pupil of Sidgwick's, continued this non-forcing approach. He set essays on very broad topics and asked his pupils to take them as far as they could. During the tutorial he would read the essay, sometimes asking for clarification, disputing a point, or expressing surprise at a novelty. Sutton gave a good training in working on a subject and getting a grip on it. He encouraged his pupils by giving the impression that they could illuminate him on important questions. Consequently his pupils did not leave his tutorials exhausted through having been told what they should have known: they left with some sense of achievement.[24] This idea that the tutorial involved a creative partnership between tutor and tutee was developed further by Hammick who covered all branches of chemistry at Oriel: several of his published papers developed from queries raised at tutorials. As a tutor he put the onus of study onto his pupils, frequently quoting Gibbon's view that the powers of instruction are rarely effective save in those fortunate cases where they are almost superfluous.[25]

Older tutors in chemistry seem to have had some difficulties in maintaining the vaunted advantages of the tutorial system. At Queen's Chattaway did not spoon-feed his pupils and made no attempt to transform second-class men into examination firsts. But his bibliographies were out of date; and, as a preparative organic chemist, he could not cope with physical chemistry, which his pupils studied through lectures and private reading. Yet they liked him, affectionately called him 'Poisoner', and were interested in his accounts of great chemists he had known.[26] At Jesus Chapman was an Oxford character, whose eccentricities, but not his tutorial arrangements, impressed his pupils. Chapman did not set essays regularly and gave no individual tuition so he knew little about the academic progress his charges were making. He seems to have given tutorials to four or five students together. Sometimes these were really classes on topics not covered in lectures; occasionally an individual student addressed the others on the contents of his essay. Though Chapman directed his pupils to original research papers, it is clear that at Jesus there was not the regular essay-writing on a wide range of topics and the educative discussion which were the norm in other colleges. Other rituals appealed more to Chapman. The most choice, witnessed inadvertently by his pupils, involved golf. In his college laboratory Chapman would rise from his desk, take up a golfing stance, grasp an imaginary club, practise a few swings, and drive

[23] L. E. Sutton, interview; H. Tizard, 'Sidgwick', *ON* 9 (1954), 237–58 (241).
[24] Leslie Ernest Sutton (1906–92), fellow of Magdalen 1936–73; E. J. W. Whittaker, 'Chemistry Teaching in Oxford in 1939–43 through the Eyes of an Undergraduate' (courtesy Dr B. Harrison).
[25] Dalziel Llewellyn Hammick (1887–1966), fellow of Oriel 1921–52; E. J. Bowen, 'Hammick', *BM* 13 (1967), 107–24 (108–10); information from Professor S. F. Mason.
[26] E. P. Abraham to author, 13 July 1987.

an imaginary ball through a window. Finally, with one hand over his eyes, he would peer into the distance to locate his ball.[27]

Tutorials in the life sciences seem to have maintained the key features of intensive private study and essay-writing by undergraduates, full cover of the syllabus by one tutor, and extensive discussion between the tutor and the pupil; the days of undergraduates providing answers to problems set by more than one specialist tutor had not yet dawned. At New College Julian Huxley tutored his zoology pupils, sometimes in trios, in his study in his own home in the evenings. He did most of the talking, was widely read in foreign languages, and up to the minute with the latest research. He provided heady standards for his pupils to attain.[28] Domestic tutorials were also given by de Beer who showed off his wide knowledge. With at least one musical tutee, tutorials often ended with the playing of piano duets.[29] One of de Beer's pupils was Young who tutored at Magdalen where one of his protégés was Medawar. Young's tutoring of Medawar shows how important it could be for an aspiring scientist. Young had gone to Magdalen from Marlborough School and Medawar did likewise. Young covered all zoology in his tutorials which were concerned with dialogue designed to promote, exercise, and enlarge the mind's capacities. Spotting Medawar's unusual interests, Young arranged for him to attend T. D. Weldon's philosophy tutorials at Magdalen. It is clear that the young Medawar was encouraged and exhilarated by all his tutorials. When he himself began to tutor, he followed Young in covering the whole of his subject. As tutorial fellow at Queen's, Carter thought nothing of covering all of pre-clinical medicine and all of physiology. To have sent pupils elsewhere for specialized teaching by experts would have been deemed 'dreadfully provincial' by tutors who were concerned with education and not instruction.[30] This focus on intellectual challenge was the hallmark of the tutoring of Ogston at Balliol in the late 1930s. He put emphasis on unaided individual research even with first-year undergraduates: preparation for essays involved reading research papers on topics that had not usually been covered in lectures. Nor were the topics covered always or often directed towards success in examinations.[31]

Tutorials in mathematics were inevitably different in that they focused on the solving of specified problems, working through assigned sections of books, and not on the writing of essays. Though some tutors were formal

---

[27] D. A. Long, 'The Sir Leoline Jenkins Laboratories', *Jesus College Record* (1989), 17–20; L. A. Moignard, 'A Memoir of Life in College 1932-8' (courtesy Dr B. Harrison).

[28] V. C. Wynne-Edwards to author, 3 Aug. 1987.

[29] E. J. W. Barrington, 'De Beer', *BM* 19 (1973), 65–93 (67).

[30] Medawar, *Autobiography*, 44, 51–4 (51). Thomas Dewar Weldon (1896–1958), a specialist on Kant, fellow of Magdalen 1923–58; Cyril William Carter (1898–1974), fellow of Queen's 1927–65, obituary in *Lancet* (1974), 1: 1298.

[31] Alexander George Ogston (1911–96), fellow of Balliol 1937–59; P. C. Reynell, interview with author, 1987.

and others free and easy in attitude, they seem to have tried to do more than produce technical expertise. Generally they ensured that the tutorial was the core of the course, partly by telling or advising their pupils about which lectures to attend and partly by covering the whole syllabus themselves. They also set vacation reading and helped their pupils with the ensuing difficulties. In these ways mathematics tutors encouraged their tutees to *read* for a degree. That was particularly true of Balliol's Henry Whitehead, the most accomplished mathematician at Oxford in the 1930s. Soon on mutual Christian-name terms with his pupils, he was friendly and good company. In fine weather he conducted tutorials on the college lawn, quenching his thirst with stout shandy. He did not call for written work and tended to avoid focusing the tutorial on just solving the problems which were giving difficulty to his pupils. His mode was inspirational: he talked generally about the issues involved, generalized the problems, or discussed his own research. Only final-year special subjects induced Whitehead to send his pupils for tutorials in other colleges.[32]

2.4 Academic Conservatism

There is little doubt that between the wars the colleges were arts dominated in terms of the distribution of fellows and of undergraduates. H. A. L. Fisher put the matter well when he averred that the humanities are 'so firmly rooted and the atmosphere of classical antiquity so generally diffused in Oxford that the pure and applied sciences, being comparative latecomers, are somewhat overshadowed by the older muses'.[33] The system of college scholarships and exhibitions, for which entrants competed, underlined Fisher's point. These awards were coveted because of their eleemosynary function of paying for subsistence of the award-winner and for the praise and prestige which they brought to the holder. As over half the awards were in classics, with about 20 per cent in science and 10 per cent in maths, the superiority of classics was proclaimed to serious applicants to Oxford, their schoolmasters, and their parents, though it was ignored by the wealthy playboys who entered the University as commoners. Between the wars a quarter of the undergraduates were award-holders; that meant that slightly more than one in eight of all undergraduates was a scholar or exhibitioner in classics.[34] No wonder that in 1933 the *Oxford Magazine* was in no doubt that 'literae humaniores is still the premier school; Greats men are still our natural aristocracy'.[35] This soi-disant 'natural aristocracy' was, of course, not at all natural but a dominant

---

[32] John Henry Constantine Whitehead (1904–60), fellow of Balliol 1933–47; K. C. Bowen, recollections of Balliol 1937–40, Dec. 1988 (courtesy Dr B. Harrison).
[33] H. A. L. Fisher, *The Place of the University in National Life* (Oxford, 1919), 12.
[34] *Franks Report*, ii. 46.
[35] *OM* 50 (1932–3), 293.

élite which carefully maintained and nurtured its interests and their associated structures; but such was the dominance of the humanities in the colleges that it was not uncommon in Oxford up to 1939 for the growth of the science departments to be regarded as encroachment and for scientists to be viewed as intruders. Such feelings were exacerbated by the existence of departments in science subjects but not in arts ones, and were associated with the built-in academic conservatism of the colleges. From their point of view it was sensible to appoint tutorial fellows to teach existing subjects examined in existing degree courses; and it was foolish to try to foster new subjects by appointing fellows who would have no undergraduates to teach in their college. There was thus no incentive for a college to do other than appoint fellows to teach established degree subjects. At the same time colleges often hesitated before admitting undergraduates to read subjects in which they had no fellows. Colleges were not enthusiastic recruiters of undergraduates who had minority interests which they wished to pursue in a small final honours school. Thus colleges tended to admit as undergraduates those who wished to study mainstream degree subjects. In a collegiate university pervaded by the ideals of the humanities, it was therefore difficult to disturb the related dominance of arts fellows, arts undergraduates, and arts degree courses. Unless there was severe pressure from within or without, the academic inertia of the colleges remained undisturbed.

This conservatism was exacerbated at Oxford by the colleges being more numerous and smaller than at Cambridge. No Oxford college was wealthy or big enough to be in a position to act as a locus for a coterie of scientists as Trinity, Cambridge, had done for decades.[36] From a college perspective it was easy to see science as a source of nuisance and intrusion; and, especially in the smaller impoverished colleges, to do little or nothing for it. Few colleges had a tutorial fellow in each of chemistry, mathematics, physics, and physiology, which were the four largest undergraduate degree courses in science. Some had a mathematician, a life scientist, and a physical scientist as fellows; but the smaller colleges had only one science fellow (usually a chemist or physicist) or none at all. Most colleges had a chemistry fellow, but even by 1939 almost half the colleges had no mathematician. About a third of the colleges had fellows in physics and in life sciences. Subjects such as botany and engineering were unrepresented in the colleges' fellowship. The small power-base of science in the colleges is confirmed by data from particular colleges and subjects: Corpus Christi had no chemistry or physics fellow until 1959, University no fellow in physics until 1954 and in mathematics until 1962, Balliol no physics fellow until 1950, Trinity no mathematics fellow until 1961, and no college a fellow in engineering until St Edmund Hall took the initiative in 1959. At All Souls, sometimes described

[36] Geison, *Cambridge Physiology*.

as a bastion of progressive conservatism, there had been no science fellow since the seventeenth century until 1958 when E. B. Ford was elected.[37]

Scientists were not represented among the heads of colleges. The first scientist, as opposed to a mathematician, to be head of a college was Tizard who occupied the presidency of Magdalen for four years from 1942. His reign was not trouble-free: he was resented by some fellows for being a scientist and shortly after the end of the war preferred to return to government service.[38] As vice-chancellors at Oxford were elected from the heads of colleges by seniority, it was inevitable that they were all arts men until well after 1945. Not all were sympathetic to science. Even in the late 1940s Stallybrass, principal of Brasenose and a lawyer, argued as vice-chancellor against appointing a star candidate to a science chair precisely because he would put Oxford on the map in his subject; and, by so doing, would be disruptive. Such an innovative professor would make large financial demands; he would affect the delicate balance between colleges and the University and between the humanities and science; and in a congested science area would provoke resentment from those science departments which were already on the map.[39] The arts dominance of the heads of colleges at Oxford contrasted starkly with the practice at Cambridge where in the early decades of the century leading scientists such as Hugh Anderson and J. J. Thomson were elected heads of such powerful colleges as Gonville and Caius and Trinity.[40]

The arts dominance of the colleges, shown in their fellows and heads, was also shown in the composition of the undergraduate body (Tables 2.1 and 2.2). Between the wars 80 per cent of the undergraduates studied arts and social studies, the most popular subjects at the end of the 1930s being history (21 per cent), Greats (11 per cent), law (11 per cent), philosophy, politics and economics (PPE) (11 per cent), English (10 per cent), and modern languages (10 per cent). The proportion of undergraduates reading science was 17 per cent, the popular subjects being physiology (5.1 per cent), chemistry (3.5 per cent), mathematics (2.5 per cent), and physics (1.8 per cent). Thus the total number studying these four sciences was not much bigger than that taking Greats, or law, or PPE; and the total for all the sciences put together was less than that for history alone. Indeed in the two decades between the wars physiology was the only science to make a considerable gain. The courses which grew steadily or spectacularly were not in the sciences but in arts and social studies: witness the cases of modern languages and of PPE (established in 1920). A telling contrast can be made with Cambridge: the

[37] B. Harrison (ed.), *Corpuscles: A History of Corpus Christi College, Oxford, in the Twentieth Century, written by its members* (Oxford, 1994).
[38] Medawar, *Autobiography*, 87; Clark, *Tizard*, 332–67.
[39] V. Wigglesworth, 'Pringle', *BM* 29 (1983), 530.
[40] Hugh Kerr Anderson (1865–1928), master of Gonville and Caius College 1912–28; Joseph John Thomson (1856–1940), master of Trinity 1918–40.

TABLE 2.1. *Finalists by subject*

Subject	1923–4	1928–9	1938–9
Greats	126	123	132
Theology	41	49	64
History	248	261	251
English	91	106	117
Modern Languages	58	78	122
Oriental Studies	5	2	—
Geography	—	—	39
ARTS TOTAL	569	619	725
Law	107	150	135
PPE	43	83	136
SOCIAL STUDIES TOTAL	150	233	271
Mathematics	18	24	30
Physics	20	19	22
Chemistry	37	36	42
Biochemistry	—	—	4
Physiology	33	45	61
Zoology	9	5	7
Botany	7	6	4
Geology	2	3	3
PURE SCIENCES TOTAL	126	138	173
Engineering	7	10	14
Agriculture	—	13	12
Forestry	—	6	10
APPLIED SCIENCES TOTAL	7	29	36

Source: Lists of finalists in *OUG*.

percentage of Oxford undergraduates studying the four most popular sciences was 12.9, whereas at Cambridge the figure for engineering alone was about 10 per cent.[41] Thus the old adage 'Cambridge for science, Oxford for arts' applied to the inter-war period whether one considers relative or absolute numbers of undergraduates. Though Oxford science was being transformed in these years, it was not shown in general growth at the undergraduate level. Oxford was unusual, perhaps, in that its scientists and their research schools had acquired a considerable reputation for research by 1939, some two decades before the rapid expansion

---

[41] T. J. N. Hilken, *Engineering at Cambridge University 1783–1965* (Cambridge, 1967), 164, 172; T. Tapper and B. Salter, *Oxford, Cambridge and the Changing Idea of the University* (Buckingham, 1992), 114–19.

TABLE 2.2. *Final degree subjects as a percentage of the academic year total*

Subject	1923–4	1928–9	1938–9
Greats	14.8	12.1	11.0
Theology	4.8	4.8	5.3
History	29.1	25.6	20.9
English	10.7	10.4	9.7
Modern Languages	6.8	7.7	10.2
Oriental Studies	0.6	0.2	—
Geography	—	—	3.2
ARTS TOTAL	66.8	60.7	60.4
Law	12.6	14.7	11.2
PPE	5.0	8.1	11.3
SOCIAL STUDIES TOTAL	17.6	22.9	22.6
Mathematics	2.1	2.4	2.5
Physics	2.3	1.9	1.8
Chemistry	4.3	3.5	3.5
Biochemistry	—	—	0.3
Physiology	3.9	4.4	5.1
Zoology	1.1	0.5	0.6
Botany	0.8	0.6	0.3
Geology	0.2	0.3	0.2
PURE SCIENCES TOTAL	14.8	13.5	14.4
Engineering	0.8	1.0	1.2
Agriculture	—	1.3	1.0
Forestry	—	0.6	0.8
APPLIED SCIENCES TOTAL	0.8	2.9	3.0

in the proportion and number of undergraduates reading science subjects. That meant that between the wars the cause of science could not be easily promoted via the argument from increasing or overwhelming undergraduate numbers.

2.5 The Administration of the University

Oxford's intellectual conservatism involved the reinforcing elements of dominance by the humanities in college fellowships and in the undergraduate body which was taught in the colleges. As well as academic inertia there was also a statutory element of paralysis which prevented change. One effect of the Oxford University Act (1923), which was the legal result of the report by the Asquith Commission, was to strengthen the safeguards against abuse of power by either the University or the colleges. When a deeply controversial matter arose, it was therefore difficult for any decision to be reached. Financially the University played second fiddle to the colleges, though there was some shift in the balance of power towards the former: the regular government grant, established in 1922, increased steadily and was paid to the University to be spent on its facilities. In the late 1930s there was a further shift in the balance of financial power towards the University, and away from the colleges, not by internal rearrangement but as a result of the Appeal which attracted £464,000 from alumni and well-wishers. Moreover the major part of the cost of the extension of the Bodleian Library was not met by the colleges but by the Rockefeller Foundation. In short the colleges were reluctant to lose autonomy and money, so that the system of taxing the colleges for university purposes via the Common University Fund did not lead to a deluge of money flowing into the University Chest.

All these features were reinforced by Oxford's pre-bureaucratic condition until the mid-1930s. Bureaucratic order was the achievement of Veale, who became registrar in 1930, and of his favourite vice-chancellor, Lindsay, who assumed office in 1935. The vice-chancellor was limited in his powers because the post was temporary and part-time. He held office for only three years and was appointed from the heads of colleges not according to his appropriateness but to his seniority as a college head, a post he did not relinquish while acting as vice-chancellor. Therefore the registrar, whose post was permanent and full-time, had greater opportunity than any vice-chancellor to put Oxford's administration into good working order. That was the single achievement of Veale, a master of the administrative machinery of 'veales within veales', who was the principal architect of the modern University of Oxford. He could be overbearing and at times exceeded his powers so that his enemies stigmatized the Registry as l'hôtel de Veale. As a former senior civil servant, he enjoyed many useful contacts in government. Above all he appreciated the importance of long-term planning, of overall coherence, and of anticipating snags. In pursuit of these aims he was wily: when one door shut in his face he soon found another to knock on. He was pro-active; his predecessors, two mathematicians who lacked civil service experience, tended to be re-active.[42]

The nature of the administrative machinery that Veale inherited favoured the status quo in which scientists and the interests of science were poorly represented. Though he could do little about the composition of Congregation, Oxford's sovereign body, and about the limited powers of the University's professors, he introduced financial accountability and co-ordinated

[42] *DNB* is splendid on Veale whose predecessors were: Charles Leudesdorf (1853–1924), registrar 1906–24; Edwin Stuart Craig (1865–1939), registrar 1924–30.

approaches to problems facing the University. His chief monument of the 1930s is the Science Area. Though he failed to make great changes in the composition of the General Board of the Faculties, he enhanced its position from 1937–8 by assigning to it a block grant to be doled out to competing claimants. Above all Veale saw to it that from the early 1930s the Hebdomadal Council became an active, initiating, and planning body, which reduced the power of the Chest. These achievements become clearer if we contrast the 1920s with the Veale years.

When the Asquith Commission reported in 1922 university business was small: most of it was carried out in the colleges in a clubbable way. There were in existence various university bodies, such as the Hebdomadal Council, the Chest, the General Board of the Faculties, and Congregation; on the whole each operated within its own sphere and tended to be independent of the others. There was, however, some co-ordination because Council passed proposals for Congregation to judge and when money was required the Chest had to be approached. Generally no one body in the University felt the responsibility of having a synoptic view of the University as a whole.[43] The Council in the 1920s was ostensibly in charge of administration, property, and finances but in practice had limited powers and on occasion was negligent. In 1924, for example, it had no view about the town-planning scheme produced by the City of Oxford. As the University as a corporate body thought the scheme could not possibly affect it, the appropriate government minister (Neville Chamberlain) approved the plan because the University had not objected. The scheme included the momentous development of an industrial zone at Cowley which was to make Oxford's academic square mile into the Latin quarter of Cowley. In the 1920s the Council, which was mainly elected by Congregation from its own members, did not try systematically to formulate and co-ordinate University policies. Responsibility for finance lay with the Chest which was also in the main composed of members of Congregation elected by Congregation. The Chest acted independently of Council, sometimes turning down Council decisions on policy as well as financial matters. If it did not like a proposal it said: 'Nasty new reform—out!'[44] In the 1920s it employed no trained accountant. Its accounts were so obscure that in the early 1930s Veale's intrusive eye discovered large and hidden financial reserves of about £120,000, a sum slightly greater than the annual grant then paid by government to the University. It was also the case that some heads of science departments banked departmental funds where they liked; as the Chest had no access to these accounts it had no effective financial control of such departments. When science professors wanted money they often applied directly to the Chest which dealt with each application in an *ad hoc* and sometimes

---

[43] *Franks Report*, i. 197.   [44] Ibid. 203.

condescending way. In the 1920s the General Board of the Faculties was unimpressive: it did not make academic policy and was a minnow compared with the Triton of the Chest. Congregation was composed of all teachers and senior administrators in the University and was the University's sovereign body. It was an arm of the colleges and represented direct democracy: voting was by head, irrespective of the reputation or status of the member. Congregation provided a wonderful arena in which blockers and procrastinators could exercise their skills. Congregation did not innovate: its roles were to accept or reject what was put to it. As only 10 per cent of the possible membership usually attended its meetings, it was possible for skilled obstructors of change to mount effective campaigns which curbed or destroyed initiatives.

Scientists were poorly represented on these key bodies. As late as 1935–6, of the eighteen elected members of Council only three were scientists. There were also only three scientists among the nine elected members of the Chest; this same proportion also applied to the elected representatives on the General Board. The scientific element in the Council roughly reflected the proportion of scientists in Congregation. The scientific element in the Chest and General Board corresponded approximately to the proportion of science professors in the University (23 out of 69 all told); it was more than the proportion of undergraduates reading science and slightly less than that of research students studying science (154 out of a total of 412). From a financial viewpoint, however, the representation of scientists in the Council, Chest, and General Board was inadequate because at that time the University was spending annually 50 per cent more on scientific studies than on non-scientific ones.

Veale knew that there was no one body which reviewed the aggregate expenditure on each science department as a whole: Council, the Chest, and the General Board each did something. On each of these science was not strongly represented, even though Council and the Chest tried to decide between the relative strengths of different claims made by the science departments. In addition any proposals made by Council to Congregation about expensive developments in science lacked conviction because scientists had little voice in framing them. Veale's solution was a clever one. Each year the Chest should decide how much money was available. A reconstituted General Board, having considered applications from departments and faculty boards, would define the University's priorities, carve up the block grant authorized by the Chest, and advise Council what annual grant should be voted by Congregation to each department. Veale was particularly keen that the General Board should warn Council about changes which would become more and more expensive, without compensating economies arising.[45] These

[45] *HCP*, 163 (1936), 149–52, which reproduces Veale to secretary of faculties, 28 Apr. 1936, PS/R/1/5.

procedures, in which the General Board, the Chest, and Council worked together, began to be operated in academic year 1937–8.[46] On the matter of a reconstructed General Board, Veale made only slight progress because strong vested interests were involved. Of its members elected by the various faculties, three out of nine were scientists. Veale proposed that out of eleven members elected by the faculties, five should represent science. When Congregation considered the matter in June 1937, it decided that, out of fourteen members elected by the faculties, five would be scientists. Thus that component of the General Board which was elected by the faculties was increased from 9 to 14 members in order to represent their concerns more effectively and give them more responsibility, with science gaining only marginally (5 out of 14 seats compared with 3 out of 9).[47] Veale created review mechanisms by encouraging the three science faculties (physical, biological, and medical) to add to their existing duties of supervising examinations, prescribing curricula, and framing lists of lectures, by giving informed advice to Council and the General Board. But he was powerless to remedy their weak financial position.

The growth in the 1930s of administrative centralization and control was not just the result of the ideas of leading individuals such as Veale and Lindsay. It was also prompted by the Owen affair which erupted in 1931.[48] The effects of this major embarrassment on the work done in agricultural engineering at Oxford and on the University's relations with government will be discussed elsewhere. Here we are concerned with the effects of the Owen scandal on the University's administration. In 1924 the University succumbed to the pressure exerted by the Ministry of Agriculture by appointing as director of the newly established Institute of Agricultural Engineering Brynar James Owen, who was an engineer employed by the Ministry which funded the Institute. It soon became apparent that Owen was devious on financial matters. By the end of 1924 his books were so badly kept that the university auditors warned the vice-chancellor that the finances were difficult to fathom and that the University should watch him closely. In the next four years he continued his financial obfuscation, blurring the distinction between his own private income and that of the Institute. From 1928 the Committee for Rural

[46] *OUG* 69 (1938–9), 22–3.
[47] *OUG* 67 (1936–7), 800.
[48] The account of the Owen affair draws on: Agricultural Engineering Institute. Owen Case. University Solicitors Papers, U Sol/44/1, OUA; Institute for Research in Agricultural Engineering 1923–36, UR/SF/RE/4, OUA; Committee for Rural Economy. Reports of subcommittees 1927–44, UDC/R/8/1, OUA; Owen Case Papers, MR/7/2/10, OUA, including memo 396 pp. long on behalf of Sugar Beet & Crop Driers about Owen's frauds, forgeries, and false qualifications; *HCP* 134 (1926), 135–8, 155–8, 185. Arthur Cecil McWatters (1880–1965), knighted 1929, secretary to Chest 1932–46, obituary in *The Times*, 27 Sep. 1965, followed two college fellows: John Frederick Stenning (1868–1959), fellow of Wadham 1898–1927 and warden 1927–38, secretary to Chest 1919–27; and T. V. Barker, secretary 1928–31.

Economy, which oversaw the University's activities in agriculture, was well aware of the Institute's big overdraft. By 1929 it spotted that £22,000 was missing from the Institute's accounts but was present in the balance sheet of Sugar Beet & Crop Driers, which in 1926 had bought from Owen some patents for extracting sugar from sugar beet. Owen's implausible explanation was that this money would be placed in a trust which he was forming on behalf of the Institute to receive money due to him for patents. Not suspicious of this blurring of Owen's private interests and those of the University, a Rural Economy subcommittee permitted Owen to be technical adviser to Sugar Beet & Crop Driers at £3,000 p.a. +2.5 per cent of net profits (to be never less than £1,000) for five years. By this time Owen was describing one of his Institute accounts, kept at Coutts in London, as his own, but it was secretly kept in the University's name and by 1930 was £16,000 overdrawn. In 1930 the senior proctor of the University and the secretary of the Rural Economy [i.e. Agriculture] Committee were worried by what they perceived as the peculiarities and improprieties of Owen's financial arrangements as director of the Agricultural Engineering Research Institute; but they did nothing. The Chest seems to have played no vetting role at all. It was responsible for the published annual university accounts which for 1925–6 showed the Institute as having an overdraft of £17,670 about which it did nothing. The Council knew that Owen was acting peculiarly. His report to Council for 1925 made it clear that already he had erected a sugar-beet factory at Eynsham, 6 miles from Oxford. In 1925 and 1927 Council and the Ministry of Agriculture had protracted negotiations about the patenting of inventions by researchers in university institutes financed by the Ministry, so the vice-chancellor and Council were aware in principle of the question whether patents should be vested in the University or the researcher *qua* individual. In summer 1926 Owen apologized in writing to the registrar for a financial 'irregularity' and his ignorance of university procedure. Even so Council did not vet him. Thus neither the Chest nor Council took steps to find out whether Owen's irregularities amounted to malversation. The person who unmasked Owen's financial deceits, involving forgery and the obfuscation of the difference between patents and financial accounts belonging to himself, the Institute, and the University, was Veale who in mid-March 1931 heard from a friend in the Ministry of Agriculture that Owen had forged letters from that Ministry and from the Treasury. On 20 March 1931 Owen was suspended from his university post on suspicion of forgery. By 15 May Owen had been sentenced to four years' penal servitude and twelve months' hard labour for fraudulent transactions with International Harvesters and Fords. During the remainder of 1931 his financial irregularities in the University were slowly and embarrassingly revealed; and the defects of the existing modes of control became painfully obvious.

The University responded to this scandal by tightening financial control over departments: in future all departmental accounts had to be kept with the University's banker who submitted daily records of payments to the Chest. Some fourteen heads of science departments objected at first to this infringement of their freedom: previously they had banked departmental funds where they pleased and the Chest had no access to their accounts. Another effect of the Owen scandal was that the Chest had been shown to be incompetent. Veale exploited this embarrassing fact to buttress his own position, to bring the Chest under the control of Council, and to encourage the appointment in 1932 of McWatters as the first full-time secretary to the Chest. Like Veale McWatters had read Greats, had been a civil servant, and returned to Oxford as a loyal internal modernizer. As well as these internal effects, the Owen affair had an external aspect because in 1931 Sugar Beet & Crop Driers and two other joint plaintiffs sued the University for £750,000. Their grounds were: that the patents which Owen sold to the plaintiff company in 1926 contained fraudulent statements about the superiority of a particular method of extracting sugar from beet; that some of these patents were sold by Owen acting as an agent of the University whose imprimatur gave credibility to them; that the patents turned out in practice to be worthless, which led to a heavy financial loss by the company; and lastly that the company suffered further financial losses because it followed the advice of Owen who was acting with the consent of the University and on its behalf.

This legal squabble was not quickly resolved: it hung threateningly over the University, which was liable to lose £750,000 in court, until spring 1938 when the plaintiffs at last settled out of court for £75,000.[49] For seven years, therefore, the University was prospectively liable to pay out a sum which was roughly seven times its annual grant from the government: not surprisingly this paralysing uncertainty made planning hazardous and led to great caution in spending by the University in those years. When the settlement was agreed the University successfully urged the Ministry of Agriculture to set up in May 1938 a special tribunal to advise whether and to what extent the Ministry should reimburse the University on the grounds that a measure of responsibility lay with the Ministry. At the tribunal, which met in October for four days, the University was not represented by its lawyers but by Veale whose masterly exposition took eleven hours. Early in 1939 it was announced that the tribunal had awarded £51,584 to the University, leaving it to find £24,187 as its share of the cost of the settlement.

Though the main effects of the Owen scandal were the introduction of effective financial control and caution in expenditure on capital and running costs, individuals were inevitably affected. For example in 1934–5

[49] *OUG* 69 (1938–9), 22–3; 70 (1939–40), 57; *OM* 56 (1937–8), 346–7.

G. J. Whitrow was in his second year as a university senior student and doing postgraduate research on relativity and cosmology. In that academic year his studentship was reduced in value from £250 to £215 p.a. Another person affected was Chaundy, tutor in mathematics at Christ Church. He was to have been appointed university reader in the mid-1930s initially for five years but, because of the financial uncertainty produced by the protracted Owen affair, it was decided not to spend university money that way. As he was a college fellow Chaundy was not in desperate straits financially but his promotion was delayed to his detriment and that of mathematics at Oxford.[50]

2.6 Demonstrators, Readers, and Professors

Given these evolving administrative arrangements, the prospects for scientists and their power depended on whether they were college fellows *tout court*, college fellows who demonstrated in a university department (i.e. taught undergraduates), or demonstrators who were not college fellows, readers, or professors. College fellows in science enjoyed security without any proficiency test: they could look forward either to a career for life in their college or in a few cases to moving to a senior post such as a chair in a metropolitan university. If they also demonstrated, their prospects were regularized in 1926–7 when a system of university demonstratorships was established to complement that of departmental demonstratorships which already existed. A college fellow did not have to demonstrate: if a fellow chose to be a departmental demonstrator, appointed by the professor and paid out of departmental funds, the fellow could afford to exert independence and if necessary withdraw his labour because of the financial security provided by the college. If a college fellow became a university demonstrator, appointed and paid by the University, he or she could anticipate renewal of the post and an incremental salary scale. By the mid-1930s university demonstrators who were college fellows were paid on a salary scale of £250 to £450 per year, with an increment of £50 every fourth year. Thus college fellows, whether or not they demonstrated, enjoyed decent career prospects, the many ineffable pleasures of college life, and intellectual independence. They enjoyed the freedom to launch a new line of research, to develop or to stick to a topic regarded by the professor of their subject as threatening or *passé*, and to bide their time instead of rushing into publication. In several cases they drew on their wide range of scholarship, which their college tutoring encouraged, to write large works of synthesis or several books covering different aspects of their subject.

[50] G. J. Whitrow to author, 28 May 1987; May 1934, proposed readership for Chaundy at £300 p.a., PS/R/1/4; Chaundy, reader in mathematics 1947–56, obituary in *The Times*, 16 Apr. 1966.

For demonstrators who were not college fellows the prospects were less alluring.[51] A departmental demonstrator, paid out of department funds, could not hang on indefinitely. Tenure was limited: in the early 1930s it was six years maximum. An ambitious person strove to be elevated to a university demonstratorship or, better still, to be elected to a college fellowship. If neither opening materialized, the departmental demonstrator often left Oxford. There was no regular career structure for such a person. In the case of the university demonstrator there was a paradoxical situation. The system of university demonstratorships had been introduced in 1926–7 and paid for out of the government grant to try to ensure that, in those subjects in which there were few college fellows, Oxford's reputation for teaching would be upheld. In order to give status to university demonstrators who were not college fellows, and indeed to attract and retain them, an incremental salary scale was introduced. By the mid-1930s it ran from £350 to £750 p.a., with an increment of £100 every fourth year. The incremental salary scale, in conjunction with an increased number of university demonstrators in science, created the problem of increased financial outlay per year. For example in the first full five years of the scheme, the expenditure on university demonstratorships increased from £15,750 to £21,275 p.a. Thus the introduction of regular career prospects for university demonstrators, especially those not attached to a college, accentuated the problem of finding the money for such people, especially from 1931 which saw the depression at its worst and the start of the protracted Owen affair. Sometimes matching funds were simply not available to the General Board which appointed university demonstrators. In late 1932 it had no money to appoint H. M. Powell as a university demonstrator because to have done so would have jeopardized the incremental salary scale. Powell had to wait until October 1934 when he was appointed for one year, his post being renewed for four years from October 1935. For Powell the ladder of the university demonstratorship kept him in Oxford and enabled him to become a reader in 1944 and eventually a professor in 1963.[52]

One problem facing university demonstrators without a college affiliation was that it was not easy for them to attract postgraduates either internally or externally. They had no college tutees whom they could groom for research. If, however, they were organic chemists they attracted on occasion the odd part two undergraduate from a college where the fellow was a physical chemist or where there was no fellow in chemistry; but such aspiring organic chemists usually transferred to the professor (Perkin and then Robinson) for their D.Phil. Postgraduates who came to an Oxford science department from other universities naturally registered for their D.Phil. under the professor as

---

[51] Margoliouth to Bowman, 15 Feb. 1932, notes on demonstrators, n.d. [late 1932], PS/R/1/3; Veale to secretary of faculties, 26 June 1934, PS/R/1/4.
[52] Herbert Marcus Powell (1907–91); *OUG* 64 (1933–4), 463; 65 (1934–5), 407.

the best-known person in it. The professor could hardly direct such students, who came to Oxford to work with him, to transfer to one of his demonstrators as supervisor; similarly he could not compel Oxford graduates to do that. The result was that university demonstrators without a college fellowship tended to have few D.Phil. students so it was not easy for them to build up a research school. On the other hand they had the opportunity of doing a lot of research mainly solo.

Less than 5 per cent of Oxford's academic staff were readers: their number in all subjects remained low between the wars, hovering at about two dozen. Those in science subjects, which increased from 7 in 1925 to 16 in 1939, were nearly all conferred *ad hominem*, the exceptions being the Wilde readership in mental philosophy (psychology), the Dr Lee's readership in anatomy, and the Dr Lee's readership in chemistry. Readerships were conferred on individuals distinguished in research, to promote in a university department subjects in which there was no professor or less commonly to recognize outstanding scholarship in an obscure subject. The only example of the second category was that of Fotheringham in ancient astronomy and chronology (1925).[53] Two readerships in the first category carried institutional responsibilities: Gunther's in history of science (1934) and Elton's in animal ecology in 1936. Gunther was curator of the Lewis Evans collection of scientific instruments housed in the old Ashmolean building (the basis of the Museum of History of Science formally established in 1935). Elton was in charge of the University's Bureau of Animal Population (established provisionally in 1932). In their cases their readerships were not just personal rewards for research but were associated with administrative responsibility as head of a university institution. In both cases their readerships were apparently stipendless. Gunther was appointed without salary for six years. Elton derived his emolument from his university demonstratorship (£400 p.a.) and from a fellowship at Corpus Christi (£300 p.a.).[54]

The remaining readerships in the first category in order of their establishment were: Egerton, thermodynamics, 1921 which was continued with Simon in 1936; Morison, agricultural chemistry, 1923, renamed soil science 1932; Sidgwick, chemistry, 1924; Buxton, physical anthropology, 1927; Barker, crystallography, 1927; Pidduck, applied mathematics, 1927; Dobson, meteorology, 1927; Ainley-Walker, pathology, 1927; Priestley, clinical physiology, 1927; Moullin, engineering science, 1929; Odgers, human anatomy, 1931; Gardner, bacteriology, 1937; Douglas, general metabolism, 1937; and

---

[53] The Wilde readers were William McDougall 1903–20 and William Brown (1881–1952) 1921–46; the Dr Lee's reader in anatomy was Trevor Braby Heaton (1886–1972) 1920–54; the Dr Lee's reader in chemistry was Alexander Smith Russell 1920–55; John Knight Fotheringham (1874–1936) was reader in astronomy and ancient chronology 1925–36.

[54] Robert William Theodore Gunther (1869–1940); *OUG* 64 (1933–4), 579; Charles Sutherland Elton (1900–91); *OUG* 67 (1936–7), 75.

Ford, genetics, 1939.[55] The financial terms of these readerships varied according to the individual and his circumstances. At best they carried a salary of £500 p.a., as with Simon. At their worst for the recipient, they offered nothing. Ford was appointed as stipendless reader in genetics because he was already paid as a university demonstrator.[56] None of these posts carried tenure: they were held, it seems, for a stipulated period of time, normally three to five years, but were renewable. Nearly all readerships in science indicated that the University attached special importance to a particular subject and wished to develop it in the University, but without costing the University much money. Nor did readerships affect the colleges' coffers because the posts did not carry college affiliation. Though some readers were college fellows most of them had been elected as fellows on the initiative of a college prior to the readership. Thus the University as a whole secured the services of its readers on the cheap.

There were three times as many professors as readers. During the interwar period the total number of chairs in the University rose from 66 to 74, the number in science (including pre-clinical medical subjects) increasing slightly from 19 to 22 as a result of the creation of new chairs in chemistry (1919), biochemistry (1920), and mathematics (1929).[57] Their incumbents did not enjoy the power exertable by their provincial counterparts who usually controlled the Senate, dominated the appropriate faculty boards, and ruled their department as they saw fit. Professors in provincial universities could be authoritarian and even bullying, treating their departmental staff as personal assistants or lackeys. At Oxford in contrast professors carried less power than in all metropolitan and provincial universities. The sovereign body at Oxford was Congregation, composed of all the academic staff. As there were 600–700 members of Congregation professors were in a minority. The Hebdomadal Council was elected mostly by Congregation and therefore contained few professors. In a similar way they were poorly represented by head in the faculty boards which were elected by the faculty, which was composed of the teaching staff. Thus Oxford professors were like ordinary members of a faculty, largely lacking the ex officio rights enjoyed by holders of chairs elsewhere except at Cambridge. The diffused power arrangements at Oxford meant that professors could be easily outvoted in Congregation and elsewhere by college interests and that the scientists could be vanquished without difficulty by arts people. Furthermore in Congregation and Hebdomadal Council there were no lay members to

[55] Franz Eugen Simon (1893–1956); Cecil Graham Traquair Morison (1881–1965); Leonard Halford Dudley Buxton (1889–1939); Frederick Bernard Pidduck (1885–1952); Gordon Miller Bourne Dobson (1889–1976); Ernest William Ainley-Walker (1871–1955); John Gillies Priestley (1880–1941); Paul Norman Blake Odgers (1877–1958); Arthur Duncan Gardner (1884–1978); Claude Gordon Douglas (1882–1963).
[56] *OUG* 67 (1936–7), 162; 69 (1938–9), 785.
[57] *Franks Report*, ii. 39.

whom frustrated professors could appeal. Hence the overall structure of administration and decision-making was loaded against science professors. Even in the 1950s Veale averred that an Oxford chair was not necessarily a top grade in an administrative hierarchy; and he stressed that Oxford professors occupied an official position which paradoxically deprived them of so much power that they constituted a suppressed group.[58]

Considerable parts of this book are concerned with the ways in which individual professors coped with the democratic structure, even anarchy, of the University; but here a few examples will suffice to show that the scientific professoriate was subject to limitations and could not behave as a collective barony. In the first place, general administration of the science departments as a whole was the responsibility of the Museum Delegacy which was mainly concerned with settling disputes between the science departments and with providing common services, such as heating and electricity, to the science buildings. On occasion this Delegacy, on which no head of a department could serve, ignored heads of science departments, rejected their advice, or failed to record their protests. Yet when a move was made in 1922 to replace the Delegacy by a Natural Sciences Administrative Board, composed of heads of departments, it was defeated, though narrowly, in Congregation through the efforts of Hartley and Sidgwick, two powerful college fellows in science who thought the scheme a slight on college tutors.[59] A second example concerned the financial allowances to those science professors who were in charge of laboratories. In 1922 they argued that, as laboratory work continued through most of the vacations (which *in toto* exceeded the combined length of the three eight-week terms), they should be rewarded for this extra responsibility by being paid an allowance of £250–£300 p.a. Having been frostily rejected by the Chest, the proposal took eighteen months to reach Congregation where it was rejected by 56 votes to 56. Two years later it was finally conceded that sixteen professors, each of whom was head of a department, should each receive a special allowance of £200 p.a. on top of the normal professorial salary of £1,200 p.a.[60] A third example concerns the extension of the Radcliffe Science Library in the early 1930s. The science professors protested to Council that at no stage in the discussions were they consulted collectively about the size, the location, and the timing of the extension. These questions were settled over their heads by the Hebdomadal Council, which did however consult the Museum Delegacy which the science professors regarded as an extraneous body. The Council conceded the justness of the protest by reconstituting a joint standing advisory science

[58] D. Veale, 'The University Today', in A. F. Martin and R. W. Steel (eds.), *The Oxford Region: A Scientific and Historical Survey* (London, 1954), 181–6 (183).
[59] Joint Advisory Science Committee 1934–48, UR/SF/JAS/1, OUA; *Nature* 109 (1922), 155; *HCP* 119 (1921), 215–17; *OM* 40 (1921–2), 179.
[60] *HCP* 123 (1922), 79–80, 145; *OUG* 56 (1925–6), 321; *Nature*, 113 (1924), 444; 117 (1926), 250; *OM* 42 (1923–4), 356–7.

committee, first appointed in 1923 and effectively composed of professors, to advise on matters of broad policy.[61]

Some science professors had no college fellows in their department, while others had. Either way there were grave problems. The first category of professor was the only member of his department to be part of college life because each chair was attached to a particular college via a professorial fellowship. All his colleagues, however brilliant, were excluded from the core of the University and often felt as 'unattached' demonstrators that their departmental life did not altogether compensate for being on the University's periphery. For their professor his only source of support was the University: he had no college fellows to help with teaching. If a professor belonged to the second category, he faced different problems. Those of his staff who were college fellows had more independence than those who were not; this was a distinction which sometimes created jealousy and frustration among the 'unattached'. Their professor found himself in a bizarre situation: he could control his non-fellow demonstrators but his college fellows could defy him and he could not turn to the colleges by right for help in teaching. Thus, even in the laboratories for which a professor was responsible, it was possible for a determined college fellow to create and occupy an independent fiefdom as far as research was concerned; this was a far cry from the monolithically organized laboratory in which all research was done under the leadership of a dominant professor and his research lieutenants. An Oxford professor with college fellows on his staff could not function like an autocratic laboratory director in a provincial university: his powers of controlling the research and teaching in his own laboratory were constrained by the imperatives of a collegiate university.

One further point about the science professoriate needs to be stressed. All those appointed before 1923 had tenure for life and were not required to retire at a specific age. Such men had no pension arrangements. The inevitable result was that between the wars the University was saddled with several geriatric professors whose eccentricity, incompetence, or feebleness prevented them from being successful discipline-builders, though some of them shone as Oxford 'characters'. Not all of Oxford's ageing science professors were valetudinarians who were incapable of developing their subject within the University. One of the key innovators in Oxford science, W. H. Perkin, junior, died in harness in 1929 aged 69. H. H. Turner was actively promoting seismology as a branch of astronomy when he died abroad at a conference in 1930 also aged 69. Sherrington held the chair of physiology until he resigned aged 78, while Poulton resigned from the chair of entomology when he was 77. None of these professors could be characterized as a dodo though most became set in their intellectual ways in their

---

[61] *HCP* 153 (1932), 115–21, 135; Lindemann to vice-chancellor, 23 Oct. 1932, LP, B97/2–4.

last years of service. In other cases, however, professors and their subjects languished without the University taking the drastic step of appointing a deputy professor. Mathematics was saddled with two old professors: Dixon was already 55 when in 1922 he was appointed to the Waynflete chair of pure mathematics, a post he held until 1945 when he retired aged 78; Love died in 1940 aged 77 while still the occupant of the Sedley chair of natural philosophy. In the 1930s Love's health was so frail that he always took a taxi to and from his lectures.[62] Arthur Thomson resigned from the chair of anatomy in 1933 aged 75, having taught at Oxford for no fewer than forty-eight years. It was only the uncommon exigencies of war which led to the retirement under pressure from the University of Townsend, Wykeham professor of physics, in 1941 aged 72.[63] Oxford's oldest science professor was Sollas (geology) who died in harness, *de jure* if not *de facto*, in 1937 aged 87. Up to the age of 80 he remained nimble on cliff and scree but then deafness and other ailments impaired this peculiarly appropriate solace for Sollas. He became notorious for geriatric eccentricity. In 1936 the University recognized that Sollas could no longer act responsibly as head of department so it took the rare and embarrassing step of appointing a deputy professor, J. A. Douglas, to take charge of it.[64]

The key to understanding Oxford between the wars was its decentralized collegiate structure, one that was entirely legal and associated with the dominance of certain pedagogic interests focused on arts subjects. Organizationally Oxford was a law, even an anarchy, to itself: the colleges often acted independently; and moves towards centralized control and planning occurred only when remarkable individuals and embarrassing circumstances coincided. Like the Third Reich Oxford's power structure was polycratic. It was an ancient syndicalist federation of colleges which acted as academic producer-co-operatives.[65] As such it was an academically conservative university in which new departures in teaching and research were difficult to launch and sustain. How, therefore, did any innovations in the sciences take place?

Much of the rest of this book is concerned with answering this question. In general I shall argue that two main agencies of change produced innovation. The first was internal pressure generated by instigators and activists within the University who used various resources either to transform the existing arrangements or, when frontal attack had failed or was not feasible, to adopt the less disruptive mode of trying to add to them. Such changes were often paid for by external funding of various kinds secured by academic

---

[62] Arthur Lee Dixon (1867–1955); Augustus Edward Hough Love (1863–1940); E. A. Milne, 'Love', *ON* 3 (1941), 467–82.
[63] Townsend inquiry papers, MR/7/1/7, OUA.
[64] *OUG* 67 (1936–7), 111.
[65] Halsey, *Donnish Dominion*, 173.

entrepreneurs who were dedicated discipline-builders. The second agent of change was the government, which on occasion exerted severe pressure on the University from without in exchange for funding various innovations in agriculture and in forestry. If there was neither internal nor external pressure for change, then the status quo was preserved or slow decay set in.

3
Philosophy, Liberal Education, and the Industrial Spirit

Oxford's academic conservatism was shown powerfully when threats were made to the dominance of Greats and of philosophy at Oxford. Thus in 1923 a proposal to introduce Science Greats, as a scientific analogue of the highly successful PPE, was rejected; and Oxford was slow, relative to other universities, in introducing experimental psychology. At the same time the arts dominance of the University was associated with the cherished notion that Oxford's *raison d'être* remained liberal education and not professional training. Oxford's suspicion of vocational training was made particularly manifest in its attitude to engineering. First there was no such department as engineering at Oxford: in order to stress its non-vocationality it was called engineering science from its inception in 1908. It remained a peripheral subject until the Second World War despite the clever strategies adopted in the 1930s by Southwell to increase its status. Secondly Lord Nuffield's endowment of a college for engineering named after him was appropriated by the University who saw to it that his aim of bringing together pure and applied knowledge would be implemented via social sciences and not engineering. The career of engineering at Oxford gives some substance to the generally extreme views of Wiener that British Universities turned their backs on engineering and inculated an anti-industrial spirit.[1]

3.1 Science Greats

The power of philosophy was shown in the contrasting fates of two proposals to modernize the school of literae humaniores known as Greats, a four-year degree course which focused on classical literature, history, and philosophy. It was Oxford's most famous final honours school and until 1920 was unique in that it was a combined-subject honours school in a university which otherwise offered single-subject degrees. Two proposals to

---

[1] M. Wiener, *English Culture and the Decline of the Industrial Spirit 1850–1980* (Cambridge, 1981), 22–4, 132. Wiener's views have been subjected to penetrating criticism in D. E. H. Edgerton, *Science, Technology and the British Industrial 'Decline', 1870–1970* (Cambridge, 1996) and I. Inkster, *Science and Technology in History: An Approach to Industrial Development* (London, 1991).

update Greats were brought forward as complementary both before and after the First World War. What was called Modern Greats was designed to study modern society just as Greats tried to analyse classical civilization. It was established in 1920 as the new school of Politics, Philosophy, and Economics, known as PPE; it soon became popular, attracting by the 1930s about 10 per cent of the undergraduates, which was about two-thirds of the percentage taking all science subjects.[2] In contrast with the success of PPE, Science Greats, which aimed to combine the principles, history, and philosophy of science, was rejected in 1923.

The success of the PPE scheme relied heavily on the philosophers who were looking for extra teaching which would be neither peripheral nor tiresome. Compared with Greats, PPE enabled them to teach modern philosophy, in its own right and not tied to ancient history, to final honours standard. Economics and politics were previously not catered for at that level so the economists and the historians (who taught politics) were prepared to accept a subordinate place in the PPE scheme for economics and politics because for the first time they were recognized as finals honours subjects. There was considerable opposition to PPE, ranging from those who thought the contemporary world was not a fit subject for academic study to those who alleged that it would be a shallow and sketchy rag-bag. These and other objections were defeated not only because the economists and historians found it prudent to make common cause with the philosophers, a powerful race of men who assumed intellectual jurisdiction over others; but also because PPE was in line with the general ethos of reconstruction after the war and it suited the concerns of the older students returning to the University after war service. For these men PPE was more likely than Greats to help them to create a country fit for heroes.

As early as 1912 considerable support in principle for Science Greats was being mustered among the philosophers and the scientists. Some of the latter thought that the existing honours courses were too specialized so they contemplated a school devoted to modern philosophy and natural sciences, the chief focus being the principles of philosophy pertinent to natural sciences. By 1914 the literae humaniores faculty board had worked out a scheme which involved three finals papers in philosophy, one in history of philosophy from Descartes to the present, one in history of science from Copernicus to the present, two in the principles of science (chosen from mathematics, physics and chemistry, biology, psychology), and two on special scientific topics. War prevented further discussion but early in 1919 Science Greats was resuscitated by the faculty boards of literae humaniores and natural sciences. By 1921 there was cause for optimism. In 1920 the

---

[2] D. N. Chester, *Economics, Politics and Social Studies in Oxford, 1900–85* (London, 1986), 30–40; D. Scott, *A. D. Lindsay: A Biography* (Oxford, 1971), 49–50, 90; E. L. Hargreaves, 'Combined Schools at Oxford (1914–1923)', *OM* 84 (1965–6), 344–50.

scientists had scored a noteworthy victory in securing the abolition of compulsory Greek *tout court*, after a protracted debate about whether undergraduates in honours mathematics or science should be excused this requirement. Next year book lists for courses were being drawn up by two formidable quintets: Hartley, Townsend, Lindemann, G. H. Hardy, and Sidgwick in physical and chemical sciences; and Keeble, Goodrich, J. A. Douglas, J. B. S. Haldane, and J. S. Huxley in biological and geological sciences. Moreover Science Greats was perceived as a scientific equivalent of PPE. It was designed to produce statesmen of science but not practising scientists.[3] By 1922, however, fears about staffing were becoming so prevalent that the idea was floated of launching the course on a three-year trial run using voluntary staff. There was also concern about the probable paucity of undergraduates who would embark on the new course and a widespread feeling that it would be foolish to launch two new honours schools in quick succession. These fears helped to dish the Science Greats scheme when it was defeated by 66 votes to 33 in Congregation in February 1923. The project had some eloquent and formidable defenders, such as Hartley from the science side and Professors Webb (philosophy of the Christian religion), Joachim (logic), and Myres (ancient history). They stressed the bridging function of Science Greats, its efficacy in combating narrow specialization, and its potential as a vehicle for general education. These arguments were ignored by the philosophy tutor at New College, H. W. B. Joseph, who was tough, stocky, imperturbable, and totally self-assured. Joseph stressed instead the difficulties of finding suitable people to teach Science Greats, given its ambitiously wide range, and of recruiting suitable undergraduates with appropriate interests and capacities, especially from the smaller colleges. He argued that, as there were already honours schools in various sciences, Science Greats would attract just a few candidates who were not satisfied by the existing science courses. Science Greats was also attacked philosophically. Collingwood argued that science and philosophy were different activities which required special training. Objecting to the way in which philosophy was in his view too much in thrall to scientific methods, Collingwood stressed that science makes assumptions while philosophy questions them all; and that science has fields of study whereas philosophy has none and tries to avoid the arbitrary cutting-up of reality into compartments.[4] Ironically Science Greats was revived just before the Second World

[3] NS/M/1/3, 6 May 1913; *HCP* 99 (1914), 280–9; NS/R/1/3, 26 Apr. 1921; NS/M/1/3, 10 May 1921.
[4] NS/R/1/3, 12 June 1922; NS/M/1/3, 20 June 1922; *OM* 41 (1922–3), 194, 198, 242, 301, 320, 340; *OUG* 53 (1922–3), 328; R. F. Harrod, *The Prof.: A Personal Memoir of Lord Cherwell* (London, 1959), 57–9; Harold Henry Joachim (1868–1938), Wykeham professor of logic 1919–35; Clement Charles Julian Webb (1865–1954), Oriel professor of the philosophy of the Christian religion 1920–30; John Linton Myres (1869–1954), Wykeham professor of ancient history 1910–39; Horace William Brindley Joseph (1867–1943), philosophy fellow of New College

War in the form of a two-year course for science undergraduates who had passed their first university examination. It was proposed to cover philosophy, the principles of science, and history of science as a bridging subject, and to recruit a reader in philosophy of science; but in October 1939 the proposal was postponed *sine die*.[5] The fate of Science Greats aroused the ire of Sidgwick, who had himself taken first-class honours degrees in both Greats and natural sciences. The caustic Sidgwick blamed the philosophers: 'philosophy in Oxford, on which more intellect, time, and money are spent than in any university, with less result, will remain uncontaminated... and will continue to interpret the facts of nature, not as they are but as they were believed to be a century ago.'[6]

3.2 Experimental Psychology

The dominance at Oxford of ratiocinative philosophy had much to do with the late arrival there of experimental psychology when compared with German and American universities, where it prospered in the 1880s and 1890s, and with Cambridge where it began in 1897.[7] Though there was provision for the teaching of psychology from 1898 when the Wilde readership in mental philosophy was established, experimental psychology flourished in only one decade in the next forty years. In 1936 formal recognition of research in experimental psychology was at last granted in the form of an Institute of Experimental Psychology which could confer only D.Phil.s and B.Sc.s (then a one-year research degree). In the 1940s teaching in experimental psychology was formalized, first in 1941 for postgraduates by the inauguration of a one-year diploma and then in 1947 for undergraduates through the creation of a final honours school in psychology, philosophy, and physiology. Again in 1947 psychology received the ultimate accolade of a chair. Thus it took almost fifty years for psychology in its experimental form to be fully recognized by the University.[8]

The history of psychology at Oxford began in 1898 when Henry Wilde endowed a readership in mental philosophy.[9] Wilde was a Mancunian who

1895–1932; Robin George Collingwood (1889–1943), fellow of Pembroke 1912–35, professor of philosophy 1935–41.

[5] PS/R/1/7, 5 June 1939; PS/M/1/1, 31 Oct. 1939.
[6] Scott, *Lindsay*, 50.
[7] L. S. Hearnshaw, *A Short History of British Psychology 1840–1940* (London, 1964), 120, 168–74. For a perceptive overview see R. C. Oldfield, 'Psychology in Oxford: 1898–1949. Part I', *Quarterly Bulletin of the British Psychological Society*, 9 (1950), 345–53, 'Part II', 10 (1950), 382–7.
[8] J. Passmore, *A Hundred Years of Philosophy* (London, 1957), 252–3.
[9] Henry Wilde (1833–1919); W. W. Haldane Gee, 'Wilde', *Memoirs and Proceedings of the Manchester Literary and Philosophical Society*, 63 (1920), 1–16; H. Wilde, *On Celestial Ejectamenta* (Oxford, 1910); Wilde readership in mental philosophy, 1898–1916, MR/7/2/11, OUA.

was an inventor of the electric dynamo, a successful electrical engineer, and an accomplished experimental scientist (FRS 1886). In 1884 he retired from business and devoted himself to philanthropy. A self-made man of humble origins, he fixed his attention not only on the Manchester Literary and Philosophical Society but also on Oxford, establishing the Wilde readership (£10,000 endowment), the John Locke scholarship in mental philosophy in 1898 (£3,000), the Wilde lecturership in natural and comparative religion in 1908 (£4,000), and in 1909 the annual Halley lecture on astronomy and terrestrial magnetism (£600). He himself gave the first Halley lecture in Oxford in May 1910 when he was 77 years old. In 1905 he donated £100 to the department of entomology for specimens illustrating mimicry and protective resemblance. In his will he left the residue of his estate (£10,000) to the University.

The litigious Wilde had firm views about his readership in mental philosophy. He intended that it be securely based in the faculty of literae humaniores and that it should promote Lockian methods among undergraduates. He insisted, therefore, that his reader should study the human mind by observation and reflection on experience, while excluding any form of experiment. Wilde took a derogatory view of psychology and especially experimental psychology: he thought the former a lower branch of mental philosophy and inappropriate for Oxford which educated the governing classes in their future relations with the empire and the world at large; apropos the latter he was convinced that 'the philosophy of a specialist in experimental psychology is necessarily of a low grade'.[10] The first holder of the readership, G. F. Stout, observed the terms of his appointment scrupulously: not an experimentalist but on good terms with several philosophers, Stout introduced psychology as an optional special subject in Greats. This arrangement subsequently gave a route by which a handful of Greats candidates, such as Cyril Burt, William Brown, and J. C. Flugel, could learn psychology and become experimentalists.[11]

When Stout left Oxford in 1903 for the chair of logic at St Andrews, he was succeeded by William McDougall who was well qualified in natural science, medicine, anthropology, and experimental psychology. As an avowed and energetic experimentalist, McDougall quickly established a laboratory in the department of physiology where he was encouraged by Professor Gotch and later by Sherrington who succeeded Gotch in 1913.[12] McDougall styled his enterprise the department of psycho-physics, thus firmly aligning it by name and location with the natural sciences. Two of

[10] Wilde to vice-chancellor, 22 May 1908, 11 Nov. 1908, MR/7/2/11.
[11] George Frederick Stout (1860–1944); Cyril Burt (1883–1971); L. S. Hearnshaw, *Cyril Burt, Psychologist* (London, 1979), 11–12; William Brown (1881–1952), obituary in *BMJ* (1952), 1: 1136; John Carl Flugel (1884–1955). Brown took Greats finals in 1905, Burt and Flugel in 1906.
[12] M. Greenwood and M. Smith, 'McDougall', *ON* 3 (1939–41), 39–62; Francis Gotch (1853–1913), professor of physiology 1895–1913.

his three main lines of research, on colour vision and on fluctuations in perception, were on the border with physiology. A third line, the development of mental tests using quantitative methods, was advocated by McDougall but carried out in schools by Brown, Burt, and Flugel, who were better statisticians. Though McDougall could not keep such distinguished young researchers in Oxford (Brown left in 1905, Burt and Flugel in 1909), he had the satisfaction of converting them from philosophers to experimental psychologists whose subsequent careers were not negligible: Brown succeeded his mentor as Wilde reader, Burt became famous and then in the 1970s notorious for his experimental work on intelligence tests, while Flugel taught for many years at University College London.

McDougall's penchant for experimentalism made him unpopular with Wilde who took vigorous but unsuccessful steps to remove him. In 1908 McDougall, the only candidate, was unanimously re-elected for five more years to the readership, which led to complaints by Wilde to the vice-chancellor that McDougall had deviated from the terms of the readership and that the University was shoddily evading its responsibility to enforce them. In 1913 and 1918 McDougall was re-elected for a third and fourth time, the electors recognizing his FRS (1912) and his remarkable output of four major books between 1905 and 1912, though they were deeply embarrassed by the statutes of the readership.[13] McDougall was not a materialist, but his form of experimental psychology put him in intellectual no man's land at Oxford. He was regarded as neither a scientist nor a philosopher *pur sang* and fell between two stools. Though he experimented in the physiology laboratory, some scientists suspected him of being a metaphysician; though his lectures appeared in the lecture list of the faculty of literae humaniores, some philosophers regarded him as representing an impossible and non-existent branch of science.[14] In 1912 McDougall secured a fellowship at Corpus Christi but the supply of good young researchers via the psychology paper in Greats had dried up. His leading research student immediately before the war was not an Oxford graduate: she was May Smith, a lecturer in educational psychology at a local training college for teachers, who worked part-time with him.[15] During the war, when McDougall was serving his country, the space for experimental physiology in the physiology laboratory was lost to the Royal Flying Corps and research on aviation. On his return to Oxford in 1919, after five years of important war-work, the rush of undergraduates to physiology made it impossible for him to reoccupy his

---

[13] Vice-chancellor to Wilde, 25 Feb., 20 and 21 Nov. 1908; Wilde to vice-chancellor, 22 May, 11 Nov. 1908, all in MR/7/2/11, OUA. W. McDougall, *An Introduction to Social Psychology* (London, 1908), had gone through 5 editions by 1912, 8 by 1914, and 14 by 1919.

[14] W. McDougall, 'Autobiography', in C. Murchison (ed.), *A History of Psychology in Autobiography*, i (Worcester, Mass., 1930), 191–223 (207).

[15] May Smith (1879–1968); M. Smith, 'Autobiography', *Occupational Psychology*, 23 (1949), 74–80.

former accommodation of six rooms. Lacking facilities for experiment, which he had previously enjoyed, conscious that psychology was not a central subject in curricula and examinations at Oxford, and strapped financially at a salary of £400 p.a., McDougall left Oxford in summer 1920, having accepted an invitation to occupy the chair of psychology at Harvard. The invitation was flattering, the Harvard department was well equipped for experimentation, he liked the USA as a land of romantic possibilities, and he relished its large-scale experimental approach to life.[16] Negatively he was dismayed by Oxford's antagonistic attitude to psychology. In 1926 Oxford's view of it as a 'nasty little science' still so rankled with him that he anticipated that perhaps before the end of the century 'even the University of Oxford may begin to take interest in the human mind, and may set her hallmark upon psychology by giving it a recognised place among her studies'.[17]

In 1921 William Brown was elected to the Wilde readership for five years. Unlike his two predecessors he was an Oxford graduate who had read Greats and physiology; but, given his interests, that was no royal road to success in promoting psychology. When he returned to Oxford as a psychologist who was medically qualified, he specialized in two areas. The first was mental measurement in which he employed statistical methods. The second was psychotherapy which provided for him between the wars a lucrative consultancy in London and fresh psychological material. From 1925 to 1931 he also worked as a psychotherapist at King's College Hospital, London. Not surprisingly his attendance in term time left so much to be desired that in 1926 the electors to the readership split evenly on offering it to McDougall. They then re-elected Brown by 4 votes to 2 for five years from a strong field led by J. D. Mabbott, A. C. Ewing, and A. B. Brown.[18] This narrow squeak did not deter Brown, whose attendance continued to cause dissatisfaction, from taking an initiative in 1928 about experimental psychology. He proposed to the vice-chancellor that a laboratory, of which he was prepared to be director, be established and that psychology be given greater prominence in final examination schemes. The philosophers tried to sabotage the first proposal by stressing that experimental psychology was expressly excluded from the terms of the Wilde readership. The medical faculty welcomed Brown's first proposal, seeing experimental psychology as essen-

[16] McDougall, 'Autobiography', 211–12.
[17] J. C. Flugel, *A Hundred Years of Psychology* (London, 1933), 271; W. McDougall, *An Outline of Abnormal Psychology* (London, 1926), p. xii (quotation).
[18] For Brown's interests see W. Brown and G. H. Thomson, *The Essentials of Mental Measurement* (Cambridge, 1911); W. Brown, *Psychology and Psychotherapy* (London, 1921); id., *Talks on Psychotherapy* (London, 1923); Professorial elections. Meetings miscellaneous. 1898–1938, 30 July 1926, UDC/M/41/1, OUA; John David Mabbott (1898–1988), philosophy fellow of St John's 1924–63, president 1963–9; Alfred Cyril Ewing (1899–1973), reader in philosophy, Cambridge, 1954–66, and fellow of Jesus 1962–6; Alfred Barratt Brown (1887–1947), principal of Ruskin College, Oxford, 1926–44.

tially a biological science and as sufficiently important to warrant a chair and a department. The biology faculty agreed but thought a reader in experimental psychology should direct the laboratory. By summer 1929 an agreed scheme had been developed by the three faculty boards of medicine, literae humaniores, and biology. A department of experimental psychology was to be instituted for postgraduate research, thus avoiding conflict with Greats, and run by a director at a salary of £1,000 p.a. The scheme, which placed the director but not his department in the biology faculty, envisaged a total of £1,550 p.a. being spent on various salaries, with capital outlay of £1,000, and annual upkeep of £200. The General Board added that the directorship should not be held with the Wilde readership, while the philosophers sought to sink Brown totally by insisting that the directorship be a full-time appointment. By early 1930 it had become clear that the scheme for a department of experimental psychology was impossible because the University could not finance it. Even so, the General Board and the Hebdomadal Council both approved it in principle.[19]

In June 1931 Brown was re-elected to the readership, which had been advertised in Oxford, for a third period of five years, his salary having been raised to £500 p.a. and presumably an understanding having been reached that he cease practice at King's College Hospital.[20] Sensing that he was secure and inspired by the creation in 1931 of a chair in experimental psychology at Cambridge, Brown soon turned to denouncing mentalist philosophers, especially Oxonian ones; to reviving the cause of experimental psychology; and to an unsuccessful attempt to gain a chair for himself at Oxford.[21] He complained that the pre-eminence of Oxford's philosophers had prevented psychology from achieving independence there. He alleged that undergraduates who took psychology as a special subject in Greats did so at their peril owing to the distrust shown by philosophy teachers to psychology as a growing science. He claimed that experimental psychology was a definite field of research quite separate from general philosophy; as such it employed its own methods of observation under pre-arranged conditions, of verification, of quantification in empirical work, and of the application of statistical techniques. Brown made no bones about his belief that psychology took precedence over philosophy: the former had no quarrel with the latter simply because 'philosophical questions only arise in the domain of the mind when psychology has had its say'; and as an applied subject psychology had become indispensable. Rejoicing in his own experience as a psychotherapist, he told the philosophers that mind in its essence is unconscious and that therefore the hallowed Lockian

[19] *HCP* 142 (1929), 111–14; 145 (1930), 9–10, 123, 214.
[20] Professorial elections, 4 Nov. 1930, 31 May 1931, 5 June 1931, UDC/M/41/1, OUA.
[21] 30 May 1933, UDC/M/41/1, OUA; W. Brown, 'The Importance of Psychology: I', *OM* 51 (11 May 1933), 629–30; W. Brown, 'Psychology: Part II', *OM* 51 (18 May 1933), 665–8 (668).

procedures of introspection and direct observation were inadequate for understanding the unconscious aspects of the mind. Clearly Brown was trying to subvert the dominance of the philosophers in the University and their approach to mental philosophy. He knew that arts people in Oxford thought that there was more to the soul than could be revealed by psychology and that literature and history were more profound. He also realized that psychology could not penetrate honours schools, which were, it was often claimed, already overloaded, and that an honours degree in psychology was impracticable.[22]

Late in 1935 Brown produced his trump card.[23] Mrs Hugh Watts, a friend and patient of his, made an offer to the University of £10,000 to help to found an institute of experimental psychology. She was particularly keen to promote clinical methods of prolonged psychotherapy. Her offer was made anonymously but communicated to the University by Brown. He was soon rewarded: in late November 1935 he was re-elected to the Wilde readership for five years with effect from June 1936. The University honoured its favourable view of 1929 about experimental psychology and accepted the bequest. It then faced the two related problems of staffing and finance. Brown had been the chief promoter of experimental psychology and had secured external funding for it; but the terms of his Wilde readership precluded it. Faced by this dilemma, the University rescinded its opinion of 1929 that the Wilde reader should not be director of experimental psychology and in May 1936 appointed Brown as director until he ceased to be Wilde reader. Apparently Brown was a stipendless director of the Institute of Experimental Psychology, so the University waived its previous objection to pluralism partly as a reward to Brown and partly to avoid paying a separate salary to him *qua* director. Brown secured the appointment of an assistant director at a salary of £300 p.a. in the form of his protégé William Stephenson, a non-Oxonian with doctorates in physics and psychology who was a supervisor of postgraduate research in psychology at University College London, and an expert on mental testing. The related question of finance for the Institute was solved for a time by recourse to the Social Studies Research Fund endowed in summer 1935 by the Rockefeller Foundation with £5,000 p.a. for five years: from that Fund a capital grant of £500 and an annual grant of £150 for five years were given to the Institute. Thus the Institute came into being in 1936 because external funding from an individual (Mrs Watts) and from a philanthropic foundation (Rockefeller) enabled the University to break the impasse it had created for itself in 1930. Up in the Elysian fields Wilde was no doubt enraged by the way in which the

---

[22] 'Psychology in Oxford', *OM* 53 (1934–5), 181–2.
[23] *OUG* 66 (1935–6), 163, 479–80, 612, 639; *Nature*, 138 (1936), 14–15; *HCP* 162 (1935), 67–8, 202; 163 (1936), 51–2, 89–90; Brown to Oldfield, 17 Dec. 1949, Library, Department of Experimental Psychology; William Stephenson (b. 1902).

University violated his benefaction by appointing his reader as director of an institute devoted to experimental psychology.

In October 1936 the Institute of Experimental Psychology, housed outside the Science Area in a former school on Banbury Road, began its work which was limited by the University to research and postgraduate supervision. Though its activities were confined, its existence showed that in Oxford psychology was no longer a subordinate handmaiden to philosophy. Brown was therefore not as extravagant as before in the claims he now made for experimental psychology. He admitted that introspection remained undeniably central as a procedure, and that experimentation involved introspection under pre-arranged conditions. But he was glad that at last Oxford students could enter the realms of precision and verifiability to supplement theory. He saw great possibilities for the Institute in work on perception, fatigue, volition, mental variation, psychiatry, industrial efficiency, and vocational guidance.[24] In the event Brown's enthusiasm for experimental psychology, for which he had campaigned so long, soon waned, leaving Stephenson to administer the Institute in financial circumstances more difficult than those envisaged in 1929 and to look after all the research students (4 in 1937–8; 7 in 1938–9). Stephenson's services were recognized by Mrs Watts who did not conceal that she was the Institute's benefactor in 1936 when she gave £200 p.a. for two years from June 1938 to bring his salary up to £500 p.a.[25] Stephenson extended his interest in mental testing by setting intelligence tests for Oxford County Council, and by 1939 successfully sought permission for the Institute to set tests for other educational committees and charge them fees. Stephenson was concerned to make money for his Institute which was handicapped in attracting researchers by its meagre equipment of which he was ashamed.[26] To this end he projected a positivist view of the Institute: to the frequently made charge that psychology was so awash with contending theories that it was not a proper subject, he replied that the Institute favoured no conflicting schools of thought because 'there are no such things in experimental psychology'. At the same time Stephenson saw the founding of the Institute as only the first act of recognition of psychology: it still needed the greater security and status of being a first-degree subject, as at Cambridge.

Two significant changes took place during the Second World War. In 1941 a one-year postgraduate diploma was established, with a strong emphasis on experimental and clinical psychology on the assumption that brain and nerve systems constituted the core of psychology. Stephenson had learned that in a collegiate and teaching university it was difficult to focus exclusively on research: he realized that a 'steady undercurrent of sound teaching' was

---

[24] W. Brown, 'Psychology at Oxford', *BMJ* (1936), 1: 1121–2.
[25] *OUG* 69 (1938–9), 283–4; 70 (1939–40), 433–4.
[26] W. Stephenson, 'The Institute of Experimental Psychology', *OM* 56 (1937–8), 607–9 (608); *HCP* 173 (1939), 78, 161–2, 264–5, 293–4.

needed to complement it.[27] The diploma gave the Institute for the first time formal teaching obligations. Next year Stephenson was made reader in experimental psychology. On Brown's resignation in 1945 as director of the Institute, Stephenson replaced him. Next year Brown resigned the Wilde readership and was succeeded in 1947 by Brian Farrell, an Oxonian who had read PPE and taught philosophy in South Africa. Brown's paradoxical pluralism was thus ended and his departure provided new opportunities. Early in 1947 a new undergraduate combined degree in psychology, philosophy, and physiology was established. Later that year George Humphrey, an ageing Oxonian, who was a philosopher with strong psychological interests, was brought from Canada and elected first holder of the new chair of psychology, after two years of acrimonious debate about whether to create it.[28] Next year Stephenson, who had taken a leading part in planning the degree and had run the Institute for a decade, mollified his chagrin by migrating to the USA for a chair at the University of Chicago and subsequent prominence at the University of Missouri.

It is clear that between the wars and afterwards the experimental psychologists felt deprived and embattled. Oxford was *the* home of professional philosophers in Britain, with more of them to the square mile than elsewhere in Europe except the Vatican. Their attitude to psychologists was generally that of lions looking for prey: with their well-groomed conservatism they objected to behaviouristic animal and comparative psychology, suspected the claims made by psychotherapists such as Brown, and deemed experimental psychology to be dangerous and fraudulent. It was dangerous because it insulted Christianity by putting the human soul in a pair of scales. It was fraudulent because, it was alleged, it required the wholesale abolition of distinctions essential for the study of reason and of will: there was more to logic and ethics than mere psychology.[29] For influential Oxford philosophers such as H. A. Prichard and Gilbert Ryle psychology was a hotchpotch of loosely related inquiries and not a unitary inquiry or tree of inquiries; and in its quantitative experimental form ducked the central question of causal explanation. No wonder that in 1951 the Wilde reader complained that the philosophical climate at Oxford was still not as congenial as it might be for undergraduates reading psychology.[30]

[27] W. Stephenson, 'A Diploma in Psychology', *OM* 59 (1940–1), 341–2.

[28] G. Humphrey, 'Five Years in the Oxford Chair', *British Journal of Psychology*, 44 (1953), 381–3; George Humphrey (1889–1966), professor of philosophy, Queen's University, Kingston, Ontario, 1924–47; professor of psychology, Oxford, 1947–56; Brian Farrell (b. 1912), lecturer in philosophy, University of Witwatersrand, 1938–47, Wilde reader, Oxford, 1947–79.

[29] R. G. Collingwood, *An Autobiography* (Oxford, 1939), 94–5; G. Humphrey, *On Psychology Today: An Inaugural Lecture Delivered before the University of Oxford on 25 May 1949* (Oxford, 1949).

[30] Passmore, *Philosophy*, 252–3; G. Ryle, *The Concept of Mind* (London, 1949), 323, 327; B. Farrell, 'Psychology', *OM* 69 (1950–1), 311–12; Harold Arthur Prichard (1871–1947), philosophy fellow at Trinity 1898–1924, White's professor of moral philosophy 1928–37; Gilbert Ryle

3.3 Liberal Education versus Professional Training

The dominance of the colleges at Oxford was associated with the ideal of liberal education, that is, that the chief function of being an undergraduate was to acquire mental skills and ornamental accomplishments, preferably in a Christian framework. The chief arena for this acquisition of mental culture was the college. Through the college tutorial the undergraduate was compelled to think, to write, and to justify a point of view: the emphasis was on debate and critical evaluation and not on the imparting of techniques and information which were acquired elsewhere, for example in laboratories and libraries. This emphasis on disciplined training of the mind via the college tutorial provided education in depth. As the Franks Report stressed in the mid-1960s, this did not mean going down into a subject until it was exhausted: 'the depth to be plumbed was in the man himself' who learned, it was hoped, how to assess evidence and to reason rigorously.[31] If the tutorial was the formal vehicle for liberal education, college life provided the informal means by which cultivated gentlemen and ladies were formed. Thus Oxford colleges offered a general education, mainly through arts subjects, designed to mould character and to prepare their undergraduates for service in the higher echelons of politics, administration, business, teaching, and the learned professions.[32] Before the Second World War Oxford did not give great attention to such low aims as the production of experts and the training of undergraduates for a particular career. Rather Oxford enabled its graduates, with their habits of thought, intellectual skills, and gentlemanly culture, to occupy a range of posts in various forms of public service, preferably of high status. That was true even of chemistry, the service science *par excellence*, i.e. a resource for other sciences and for a variety of industries.[33] Though chemistry undoubtedly had its practical applications, the teaching of chemistry at Oxford was not based on utilitarian or technical training: on the contrary it aimed at as wide and as general an outlook as possible.[34]

Liberal education at Oxford prepared undergraduates for what Halsey has described as 'the style of life of the ruling class of society, which means, in the case of Oxford up to the middle of the twentieth century, the amateur gentlemanly administrator of imperial Britain'.[35] As such it had to be exclusive because it had the political function of preserving the power of the various élites in which Oxonians were powerful or dominant. The particular form of liberal education as practised at Oxford therefore fostered

(1900–76), philosophy fellow at Christ Church 1925–40, Waynflete professor of philosophy 1945–68.

[31] *Franks Report*, i. 113.
[32] Halsey, *Donnish Dominion*, 61.
[33] D. M. Knight, *Ideas in Chemistry: A History of the Science* (London, 1992), 171–9.
[34] C. N. Hinshelwood, 'The School of Chemistry', *OM* 48 (1929–30), 743–5.
[35] Halsey, *Donnish Dominion*, 171.

94 Philosophy, Liberal Education, and the Industrial Spirit

a suspicion of vocational training, especially of specific training for certain occupations. Oxford's emphasis on liberal education denigrated or ostracized vocational subjects other than those pursued as respected and ancient learned professions such as law and medicine. Oxford tended to distance itself from training which was overtly and exclusively practical and utilitarian.[36] It followed that vocational training or professional instruction, in subjects not regarded as the basis of traditional learned professions, met difficulties at Oxford. No subject illustrates this point better than engineering which was in many Oxford eyes dangerously tainted by contact with the worlds of commerce and industry.

3.4 Engineering

Compared with Cambridge, two colleges of London University, several provincial universities, and two Scottish universities, Oxford was slow to respond to the growing academic interest in scientific engineering or engineering science.[37] Its chair was established in 1907, a final honours degree became available in 1909, and a purpose-built laboratory for it was ready in late 1914. In contrast the chair at Cambridge dated from 1875, its mechanical sciences tripos from 1892, and a costly laboratory from 1894. Further north in Glasgow in the 1860s W. J. M. Rankine, Regius professor of civil engineering and mechanics, had defined, theoretically and practically, the core of academic engineering to be 'engineering science', namely, taking a quantitative and mathematical approach which tried to provide optimal solutions to practical engineering problems.[38] No wonder that when the Oxford chair was established, having been agreed in principle since 1882, the vice-chancellor was relieved that at last Oxford had joined other first-rate universities. Even so, engineering was peripheral in the University: the number of academic staff remained small until the Second World War; and no college appointed a tutorial fellow in engineering until the late 1950s. When the site of a laboratory was being discussed in 1912 its opponents thought it would be at home near the canal and railway where its chimneys could emit black smoke.[39] When built it was excluded from what

---

[36] Tapper and Salter, *Oxford, Cambridge and the Changing Idea of the University*, 10–11.

[37] P. Lundgreen, 'Engineering Education in Europe and the USA, 1750–1930: The Rise to Dominance of School Culture and the Engineering Professions', *Annals of Science*, 47 (1990), 33–75, esp. 46–51, 62–4; Hilken, *Engineering at Cambridge*; R. A. Buchanan, 'The Rise of Scientific Engineering in Britain', *British Journal for the History of Science*, 18 (1985), 218–33.

[38] B. Marsden, 'Engineering Science in Glasgow: Economy, Efficiency and Measurement as Prime Movers in the Differentiation of an Academic Discipline', *British Journal for the History of Science*, 25 (1992), 319–46.

[39] J. Howarth, '"Oxford for Arts": The Natural Sciences 1880–1914', in M. Brock (ed.), *The History of the University of Oxford*, vii (in press) (courtesy of Dr Janet Howarth); Townsend, *Engineering Laboratory*.

became known as the Science Area. Moreover, Oxford avoided higher technology: with characteristic suspicion of any subject which might sully the University by making positive contact with industry and by indulging in vocational training, Oxford named its department 'engineering science', which gave the impression of more autonomy than 'applied science'.

Well before the first war one college had taken the initiative in promoting engineering. Using a bequest from Thomas Millard, Trinity built its own Millard Engineering Laboratory in 1886 and in the same year appointed Frederick Jervis-Smith as its first Millard lecturer in experimental mechanics and engineering. Jervis-Smith retired acrimoniously in 1908 when Charles Frewen Jenkin was elected first holder of the chair of engineering science. But Jenkin had no accommodation so the University rented the Millard Laboratory which was his home until 1914 when the engineering laboratory was built.[40] The Millard Laboratory was also the chief locus of a bizarre venture begun in 1905; this was the Oxford diploma in engineering and mining which enabled the holder to take a B.Sc. at Birmingham in only two years. This enterprise seemed fraudulent to outside critics who urged successfully that a proper university chair at Oxford was needed.[41]

Though Jenkin had an impressive pedigree, he soon met difficulties. The son of Fleeming Jenkin, the second great pioneer of Scottish university engineering, he had been educated at Edinburgh and Cambridge and then had gained considerable industrial experience. By 1908 he was more interested in scientific than commercial engineering; even so he applied for the chair belatedly and on the advice of J. A. Ewing, his Cambridge mentor and devoted pupil of his father. Jenkin knew that at Oxford engineering science faced the daunting prospect of having to make its way and to find its place among older and respected subjects. He knew that he lacked tenure and would be subjected to demeaning re-election every five years. Perhaps he was privately embarrassed by his record of publication, then confined to two papers of 1908 which were not about engineering but on calcareous sponges. His personality was not that of an academic entrepreneur who could promote a peripheral subject: a fastidious, austere, shy, and other-worldly man, he was no dealer in personalities and no merchant of small talk. His abrupt unclubbability perhaps explains why he had no college affiliation until 1912 when Brasenose elected him a fellow.[42]

Jenkin's approach to engineering science was that the University should teach over a three-year degree course the scientific aspects of engineering,

---

[40] K. J. Laidler, 'Chemical Kinetics and the Oxford College Laboratories', *Archive for History of Exact Sciences*, 38 (1988), 197–283, esp. 229–30, 235; Frederick Jervis-Smith (1848–1911).

[41] M. Sanderson, *The Universities and British Industry 1850–1970* (London, 1972), 39.

[42] R. V. Southwell, 'Jenkin', *ON* 3 (1939–41), 575–85; Ewing to Jenkin, 27 Apr. 1908, MR/7/1/5, OUA; M. Holroyd, 'Jenkin', *OM* 59 (1940–1), 72–3. James Alfred Ewing (1855–1935), professor of mechanism and applied mechanics, Cambridge, 1890–1903.

and that the practical side would still have to be learned through apprenticeship (minimum two years, though three years on average). In order to avoid the horrid taint of practical utility, Jenkin stressed that at Oxford there was no place for early specialization in some branch of engineering, for technical hand work, for practical details of manufacturing, and for mere skill in draughtsmanship in engineering drawing, all of which were more appropriately located in a technical college or a provincial university. At Oxford under Jenkin the focus of teaching was the scientific equipment necessary for an engineer; in the laboratory that meant an emphasis on the measurement of physical quantities. He was not an expansionist: content with a small undergraduate cohort, he saw large and elaborate apparatus as a snare and therefore unnecessary. In these ways he portrayed engineering science as non-threatening both in its nature and size.[43]

By the time of the outbreak of the First World War engineering science was struggling. Jenkin had only one or two staff to help him. By far his best catch was David Pye, a Cambridge graduate, in 1909; but the war and then Cambridge lured Pye from Oxford. Jenkin's own research was undistinguished: in his first ten years in the chair he published only one paper solo, the other two being joint ones with Pye. He had no permanent university laboratory until Christmas 1914; even then only two-thirds was built because of shortage of funds. At £800 p.a. Jenkin's own salary was below the average for science professors at Oxford and all of it was provided not by the University but by Brasenose.[44] The colliery engineers' diploma in scientific engineering and mining subjects, set up in 1905, was becoming embarrassing: no students took it and in 1913 the Home Office wanted to scrap it.[45]

During the war Jenkin worked first in the Admiralty Air Department and then became director of the materials section of the technical department of the Aircraft Production Department of the Ministry of Munitions. His focus was the specification of materials used in aircraft, with particular reference to breakages and failures of parts.[46] When he returned to Oxford his research continued his war-work: it dealt with problems that had eluded explanation during the war, in particular fatigue in materials and the effects of cracks and notches on the strength of machine parts. Indeed much of his research seemed to follow his appointments to consultancy posts

[43] C. F. Jenkin, *Engineering Science: An Inaugural Lecture on the Training for the Engineering Profession* (Oxford, 1908); id., *University of Oxford: Engineering Science* (Oxford, 1921). For the differences between the laboratory and the workshop in engineering see G. Gooday, 'Teaching Telegraphy and Electrotechnics in the Physics Laboratory: William Ayrton and the Creation of an Academic Space for Electrical Engineering in Britain 1873–1884', *History of Technology*, 13 (1991), 73–111.
[44] Jenkin to vice-chancellor, 21 Jan. 1913, *HCP* 94 (1913), 71.
[45] *HCP* 95 (1913), 35, 239–40, 299–30; 96 (1913), 11.
[46] *Aeronautical Research Committee: Report on Materials of Construction Used in Aircraft and Aircraft Engines by C. F. Jenkin* (London, 1920).

and was therefore externally driven in content. Jenkin was chairman of the materials subcommittee of the Aeronautical Research Committee and from 1924 chairman of the structures investigation committee of the Building Research Board at Watford. In the latter capacity he worked on fatigue in cement and concrete and on the brittleness of roofing materials. In 1929 Jenkin retired early not just because he was tired of burdensome teaching but also because he wished to devote himself full-time to research on various aspects of building materials. That year he simply transferred the location of such research from Oxford to Watford where he was employed until 1933.[47]

Jenkin was clearly frustrated. Though a new electrical engineering wing of the laboratory was opened in 1927, there was no money for apparatus to put in it. His staff never exceeded two, though in 1926 he did obtain the services of two new demonstrators, Binnie and Belfield, both from Cambridge; unlike their predecessors they were not birds of passage and indeed they survived Jenkin in the department.[48] He knew that after the war there was talk of dropping engineering at Oxford and leaving the subject to Cambridge which had attracted in 1919 a private endowment of a chair in aeronautical engineering and by 1922 £34,000 from private sources for a new building costing a little over £50,000. His own empirical work on strength of materials hardly fitted in with the notion that at Oxford engineering should be distinguished by its theoretical and mathematical approach and by avoiding anything savouring of a practical character. He had attracted few research students, most of whom were non-Oxonians. The few Oxonians were not graduates in mathematics or engineering science but refugees from physics and chemistry. At the undergraduate level the low standard of recruitment remained a problem, though more ambitious ones made their mark later: Robert Wynne-Edwards became the first building contractor to be president of the Institution of Civil Engineers; while Nevil Shute Norway enjoyed a vacation job with the de Havilland aircraft company, arranged for him by Jenkin, and made a career in aircraft engineering until 1938 when he became a novelist who on occasion exploited his specialist knowledge.[49]

The arrival of Southwell as Jenkin's successor in 1929 led to the transformation of the department. By 1939 it had become an outpost on the banks of the Isis of Cambridge mathematics and engineering: Southwell, Moullin (reader in electrical engineering), and the three university demonstrators

---

[47] *OUG* 52 (1921–2), 79–80; 53 (1922–3), 676; 55 (1924–5), 727–8; 60 (1929–30), 648–9.

[48] Alfred Maurice Binnie (1901–86), demonstrator in engineering science 1925–44, FRS 1960; Victor Belfield (1897–1985) as demonstrator in engineering science became the department's administrator.

[49] Robert Mere-Dydd Wynne-Edwards (1897–1974), second-class honours 1921, *DNB*; Nevil Shute Norway (1899–1960), third-class honours 1922, *DNB*; N. Shute, *Slide Rule: The Autobiography of an Engineer* (London, 1950), 39, 47–8.

(Binnie, Belfield, and A. N. Black) were all Cambridge men.[50] The output of publications had increased dramatically, two research schools were prospering, the laboratory had been extended in 1932, external activities had proliferated, and the Oxford Engineering Science Series had been launched in 1932. These changes were partly the result of Southwell's personality which was the opposite of that of Jenkin. Southwell was amiable, friendly, clubbable, and approachable; in his small laboratory he created a family atmosphere and in his home gave a weekly sherry party for his students. He had easy relations with a variety of people and was an excellent man of business in both college and university affairs.[51] Southwell was more widely recognized than Jenkin: he was already FRS when he arrived in Oxford; Jenkin became FRS after he had left. Their departures from Oxford were correspondingly different: Southwell accepted an invitation to be rector of Imperial College, London; Jenkin retired early to side-step to the Building Research Station at Watford.

Southwell's temperament was not his only strength in the Oxford context. He also held developed views about teaching and research in engineering science which helped to make it acceptable in Oxford. Southwell was the leader of a small coterie of academics who were concerned about the stagnation of engineering education in the universities; they deplored the low status of engineering teachers in universities, and wanted to respond to the pleas of some research-minded firms such as Metropolitan-Vickers and ICI who wished to recruit scientifically adept and flexible graduates. The answers provided by this coterie were to emphasize engineering science, conceived as being systematic and theoretical knowledge of somewhat idealized engineering systems, and to downgrade the drawing office and the workshop as loci of education. These aims were clearly intended to enhance the standing of academic engineering in universities, leaving technical colleges to continue training recruits for the shop-floor and drawing office.[52]

Southwell developed these general notions to suit the Oxford context which was notorious for its anti-industrial spirit.[53] He proclaimed that engineering science was science studied with a view to application, but that lectures and laboratory sessions were not of practical value themselves:

[50] Archibald Neil Black (b. 1912), demonstrator in engineering science 1935–45, Pollock reader in engineering science 1945–50, professor of mechanical engineering, Southampton University, 1950–67, professor of engineering, Southampton, 1968–72.

[51] D. G. Christopherson, 'Southwell', *BM* 18 (1972), 549–65; Christopherson to author, 11 Apr. 1988; Christopherson talk with author, 30 Sept. 1989.

[52] C. Divall, 'A Measure of Agreement: Employers and Engineering Studies in the Universities of England and Wales, 1897–1939', *Social Studies of Science*, 20 (1990), 65–112, esp. 93–5; id., 'Fundamental Science versus Design: Employers and Engineering Studies in British Universities, 1935–1976', *Minerva*, 29 (1991), 167–94.

[53] R. V. Southwell, 'The Changing Outlook of Engineering Science', *Engineering*, 146 (1938), 260–2.

Philosophy, Liberal Education, and the Industrial Spirit 99

they were modes of teaching undergraduates to think. He argued that industrialists did not want incipient specialists or trained technicians, especially from Oxford. They wanted, on the contrary, 'men of *personality* educated to take wide views', trained to think, and qualified to negotiate and control. Therefore there was no need to lengthen or intensify the three-year degree course in engineering; in that time via lectures and laboratories an adequate basic knowledge of engineering principles could be taught to undergraduates. But their 'maturity of bearing' came from their leisure pursuits pursued most propitiously in the college environment. Southwell was adamant that undergraduate leisure activities contributed to the formation of personality. Thus his ideal engineering graduate who would enter industry had breadth of background and had gained maturity of bearing 'by intercourse with men of other training'. That meant character was bred in the colleges. Southwell's views about engineering education fitted Oxford's anti-industrialism: Southwell outlawed technical handicraft and engineering practice in the pursuit of what he deemed to be 'the essential scientific equipment of an engineer'.[54] At the same time his emphasis on the importance of personality and bearing in engineering education dovetailed well with the colleges' emphasis on the shaping of gentlemen and ladies. Moreover Southwell never tried to increase the number of engineering undergraduates, which remained small and usually no more than a dozen per year. For him expansion was not a condition of survival and he saw no point in trying to rival Cambridge engineering which in the 1930s boasted 27 academic staff, 10 per cent of the total undergraduate population, and about 100 graduates a year in the mechanical sciences tripos. Before accepting the Oxford chair Southwell made it clear to the vice-chancellor that he found no attractions in a policy of continuous expansion.[55] His aim was to provide a course of the highest quality for a small number of undergraduates. The emphasis on small but good was particularly useful in the Oxford context: it permitted engineering science to survive and in some ways prosper because entrenched interests were not disturbed.

Southwell had well-developed views about research in engineering science.[56] He appreciated that, compared with the Victorian period, engineers had become defensive or diffident: their status was unclear because they seemed to be neither plumbers nor scientists. He was well aware that the advent of aeronautics had shown that old rule-of-thumb procedures were outmoded. He knew that some engineers thought that university research in engineering would soon become extinct: they feared that it would

---

[54] Draft pamphlet on *Engineering Science* [1931] in PS/R/1/2, OUA.
[55] Southwell to vice-chancellor, 1 June 1929, UC/FF/187/2, OUA.
[56] Southwell, 'Engineering Science'; id., *The Place of Engineering Science in University Studies: An Inaugural Lecture Delivered before the University of Oxford on 7 June 1930* (Oxford, 1930).

be displaced by specialist research carried out by university physicists and chemists, by *ad hoc* research done by government departments, and by the increasing research and development work carried out by larger firms. Southwell was therefore concerned to defend university research in engineering science and to define a mode for its survival.

He refused to defend university research in terms of the utility of its results: for him discovery and creation were ends in themselves. He deprecated the notion that university research could be costed as if it were an industrial product: 'no [research] work of the kind that should be done in universities can be expected to show a financial return or indeed to be anything but a loss, judged from an accountant's standpoint'.[57] Having denied himself the utilitarian justification of research in engineering, Southwell promoted it by proclaiming the general similarity of engineering to the ancient learned and noble profession of medicine. Both had academic and practical aspects which brought together science and art. Each of them was unlike a non-applied science in that each had to provide working solutions to practical questions concerned with what to do; at the same time each aspired to be more than mere rule of thumb. Southwell was particularly concerned to define engineering science as a subject which used physics and mathematics while preserving its own identity. It differed from physics in that it faced inexorable problems which had to be answered, while taking account of constraints of construction, manufacture, and materials. Engineering science differed from mathematics in that it was uninterested in abstract and exact general conclusions; instead it used and developed approximate mathematical techniques which were more useful than exact ones in coping with the factors of uncertainty and the rough data which were the daily lot of the engineer. Southwell was certain that the future for research in engineering science was assured if it could focus on what physicists and mathematical physicists were neglecting in the 1930s, namely, the behaviour of matter in bulk and the making of visualizable mechanical models as an aid to research, following the practice of such Victorian physicists as Kelvin who were distinguished in both the pure and applied realms.[58] For Southwell research in engineering science would prosper if it approximated more and more to nineteenth-century physics. At the same time Southwell thought that engineers should develop their own mathematical techniques to solve their own problems. This aspiration was derived from Bertram Hopkinson, whom Southwell revered as a mathematical engineer.[59]

[57] Southwell to University Chest, 26 Nov. 1937, UC/FF/500/2, OUA.
[58] On Kelvin see C. Smith and M. Norton Wise, *Energy and Empire: A Biographical Study of Lord Kelvin* (Cambridge, 1989).
[59] Bertram Hopkinson (1874–1918), professor of mechanism and applied mechanics at Cambridge University 1903–18, was a patron of Southwell during the First World War.

Southwell was more successful in fulfilling his aims for research than those for teaching. He gave to his department a research identity and established a productive research school. He was well qualified to do so in that he had graduated in the mechanical sciences tripos at Cambridge, he had enjoyed considerable practical experience of aeronautical engineering, and from 1925 was a fellow in mathematics at Trinity, Cambridge. Once ensconced in his Oxford chair he became increasingly dissatisfied with the limitations of pure mathematics in dealing with engineering problems. By the mid-1930s he had developed what was called his 'relaxation method' which enabled engineers to analyse the effect of any system of loading on a structural framework.[60] This new method of stress-calculation in frameworks, which took account of the unavoidable uncertainty of physical data, was Southwell's most enduring contribution to mathematics for engineering science: he called it 'mathematics with a fringe'. By 1937 he had a 'math lab' opened for 'theoretical research', using computations by machine and graphical constructions. The existence of such a laboratory might imply that his relaxation method was an automatic procedure because, after all, it involved converging on a final solution by changing one variable at a time. The rate of convergence, however, depended on the training and insight of the calculator so that the method was not at all routine. As some sojourn at Oxford under Southwell was required to gain the appropriate tacit skills, his relaxation method was not easily developed elsewhere so that his research group focused on it retained its leadership. His method remained useful until the early 1960s when computers began to displace it, it increased the interest among engineers in numerical analysis, and it played an essential role in producing propitious conditions for the arrival of computers. It also showed that engineers could profit from a combination of physical insight and mathematical analysis. That was why some members of his relaxation group (Christopherson and Black) became professors of mechanical engineering while others (Fox and Allen who had graduated in maths) moved into applied mathematics, especially numerical analysis.[61] Southwell never had more than four or five research students or collaborators at a time. Such a small number enabled him to generate a close family atmosphere and to see each researcher at least once a day. Before 1935, when his research group on relaxation methods took shape, most of his researchers were graduates of other universities, L. Chitty, J. B. B. Owen, and S. G. Hooker being the

---

[60] Christopherson, 'Southwell', 553–7; R. V. Southwell, *Relaxation Methods in Engineering Science: A Treatise on Approximate Computation* (Oxford, 1940); *OUG* 67 (1936–7), 223–6; 68 (1937–8), 215–18 (218); Southwell, 'Engineering Science', 262.

[61] Leslie Fox (1918–92), first in maths 1939, first director of Oxford University Computing Laboratory 1957–82, professor of numerical analysis 1963–83, obituary in *Independent*, 11 Aug. 1992; Deryck Norman de Garrs Allen (b. 1918), first in maths 1940, professor of applied maths, University of Sheffield, 1955–80.

best-known examples.[62] From 1935 to the war almost half his researchers were Oxonians, all of whom had graduated in mathematics or from his own department.

Southwell also promoted research in engineering science by recruiting suitable staff. His greatest coup was to secure the appointment in 1929 of a reader in electrical engineering (Moullin) as a condition of his own moving to Oxford.[63] Southwell could not have covered the teaching of electricity and magnetism; in any event he wished to develop applied electricity by bringing with him an established figure, Moullin, a Cambridge graduate in mechanical sciences who had returned to teach at Cambridge after the war. In fact Moullin joined H. J. Gough, an expert on metal fatigue, as the two leading rival candidates for the chair which in late May was offered to Southwell, who wished to leave Cambridge in order to run his own show. Initially he declined because he felt that there was no guarantee of a college fellowship for Moullin and that vague promises were inadequate. By mid-July 1929 the Rhodes Trustees had done enough to remove Southwell's hesitancy in accepting the chair: they guaranteed to pay Moullin £500 p.a. pending the raising of the necessary permanent endowment for the readership by Hartley and other Oxford supporters of engineering. Southwell accepted this arrangement and, having been elected to the chair on 25 May 1929, was re-elected on 17 July. The permanent endowment was subsequently provided in 1935 when Southwell and Hartley persuaded Donald Pollock, the chairman of Metal Industries and of British Oxygen, to provide £500 p.a. towards the salary of a permanent readership which was named after him. A fellowship for Moullin was not gained as quickly as the Rhodes endowment: he had to wait until 1932 when Magdalen made him a fellow by special election.

Moullin was a shy man who kept himself to himself. He had to teach basic electromagnetism and had little opportunity to introduce into his teaching modern material on radio etc. Generally he felt that his efforts to promote applied physics, especially electricity, had been frustrated at Oxford. Even so he did such important research there on the properties of dielectrics, the

---

[62] Letitia Chitty (1897–1982), a Cambridge mechanical sciences graduate, was research assistant to Southwell 1926–32, and lecturer in civil engineering, Imperial College, London, 1934–62; John Benjamin Brynmor Owen (b. 1910), a Cardiff graduate, Meyricke scholar Jesus 1931–2, became professor of civil engineering, Liverpool University, 1950–77; Southwell lured S. G. Hooker, an Imperial graduate, to Oxford by bothering to interview him at the Athenaeum in London; Hooker, *Autobiography*, 10.

[63] For the appointments of Southwell and Moullin see 4 and 25 May, 29 June, 17 July 1929, Meetings miscellaneous 1898–1938, UDC/M/41/1, OUA; P. Kerr, secretary of Rhodes Trust, to Hartley, 16 July 1929 and Southwell to vice-chancellor, 1 June 1929, in UC/FF/187/2, OUA; *HCP* 144 (1929), 17, 21–3, 168–9; *OUG* 66 (1935–6), 209–12; John Donald Pollock (1868–1962), *DNB*, was also rewarded by the conferment of a D.Sc. at Oxford in 1937; Herbert John Gough (1890–1965), expert on metal fatigue who ended his career as chief engineer to Unilever, on whom see S. F. Dorey, 'Gough', *BM* 12 (1966), 181–94.

mechanisms of electrical noise, and the characteristics of aerial systems that in 1945 he became the first professor of electrical engineering at Cambridge.[64] At Oxford he soon recruited postgraduate researchers, mainly non-Oxonians, and by the mid-1930s was attracting such future stars as Willis Jackson from outside and F. C. Frank from within to work on dielectric loss at radio frequencies. Moullin had useful contacts with Metropolitan-Vickers Electrical Company, Manchester, renowned as a nursery for bright young engineers. Before Jackson came to Oxford in 1933 to work for his D.Phil. his research on dielectrics done at Manchester Technical College had been suggested by Moullin. In 1936 Jackson joined the research staff of Met-Vick who, in grateful exchange, had paid for a research scholarship tenable for one year by one of their staff at Oxford under Moullin, made generous donations and loans of equipment to the department, and provided guest lecturers for it gratis.[65] Moullin's research group on dielectric loss was usefully though temporarily augmented from 1934 to 1938 by Andreas Gemant, a refugee who had taught at the Berlin Technische Hochschule and had been a research physicist for Siemens-Schuckert. He was paid £500 p.a. by Met-Vick for four years.[66] Like Southwell, Moullin set exacting standards in his research and for his postgraduates. He offered his students an austere discipline, his chief aims being to develop a 'puritanical standard of mental integrity' which involved subjecting results to all imaginable checks.[67]

Southwell also promoted research in other less obvious ways. The Edgell Sheppee scholarship was converted in 1931 to one for postgraduate research and it helped to launch the research careers of H. B. Squire and Christopherson.[68] That same year Southwell persuaded the University to pay for a £10,000 extension to his laboratory, giving it in 1932 a new wing which was devoted mainly to general research. In 1932 Southwell and Pye launched the Oxford Engineering Series, published by

[64] Moullin to Lindemann, 20 July 1940, LP, D164; W. Jackson, 'Moullin', *Nature*, 200 (1963), 404–5; Hilken, *Engineering at Cambridge*, 200–2.

[65] Frederick Charles Frank (b. 1911), first in chemistry 1933, professor of physics, Bristol University, 1954–76, FRS 1954, knighted 1977, continued to work on dielectrics; Willis Jackson, Baron Jackson of Burnley (1904–70), professor of electrical engineering, Imperial College, 1946–53, 1961–7, FRS 1953, knighted 1958, on whom see D. Gabor and J. Brown, 'Jackson', *BM* 17 (1971), 379–98; Moullin to Margoliouth, 6 Nov. 1933, PS/R/1/4; *OUG* 61 (1930–1), 691–3; 62 (1931–2), 662; 68 (1937–8), 215–18.

[66] Southwell to vice-chancellor, n.d., LP, D81; Andreas Gemant (b. 1895) left Oxford for a research post at Wisconsin University and ended his career as physicist, Detroit Electrical Company.

[67] E. B. Moullin, 'Some Thoughts about Advanced Postgraduate Courses in Electrical Engineering Subjects, if conducted in Universities', Document 5820/19, Institution of Electrical Engineers, London (courtesy Professor Colin Divall).

[68] Engineering: Mrs Edgell Sheppee's scholarship, 1930, UDC/M/53/1, OUA; Herbert Brian Squire (1909–61), first in maths 1930, professor of aviation, Imperial College, London, 1952–61, on whom see S. B. Gates and A. D. Young, 'Squire', *BM* 8 (1962), 119–35.

the Clarendon Press. By 1939 no fewer than eleven volumes, including two by Moullin and one by Southwell, had appeared, giving visibility and credibility to the department.[69] Southwell and Moullin were fast off the mark in external work on various committees. In 1930, for example, Southwell sat on eight subcommittees of the Aeronautical Research Committee, while Moullin served on committees of the Institution of Electrical Engineers, the Radio Research Board, and the Aeronautical Research Committee.[70] Nor did they flag in these external activities as the decade passed.

At the level of undergraduate teaching Southwell made a number of innovations.[71] He quickly introduced a scheme which ensured that all undergraduates gained practical experience in the long vacations as temporary apprentices in large engineering works, including from the start Met-Vick. Southwell liked and arranged variety for each undergraduate. Thus in his long vacations Christopherson worked at the Royal Aircraft Establishment, Farnborough, in the power house at ICI Billingham, and as fourteenth engineer on a transatlantic Cunard liner. As a result of changes in mathematics moderations in June 1932, more undergraduates who had acquired a sound mathematical foundation (which Southwell believed was becoming increasingly necessary for engineers) took finals in engineering. The apparatus for teaching was considerably enhanced by donations, particularly from 1932 when the new wing of the laboratory was completed. Some of these donations were valuable in cash terms as well as pedagogically. In 1934 Southwell accepted an offer for the heat laboratory of a new oil boiler from Babcock & Wilcox and of five boiler instruments from George Kent Ltd. worth in total £1,000; at that time the annual departmental grant from the University hovered around £1,500.

In spite of these innovations the quality of undergraduate recruits to engineering science remained variable and was sometimes embarrassingly low. In 1936, for example, six out of the ten finalists were awarded fourth-class honours degrees. This was a far cry from Southwell's aim of achieving sustained high quality in a small department which was intended to avoid the mass production of big departments elsewhere and to cater for the exceptionally good undergraduate.[72] On the other hand several undergraduates who attained only a low class of honours did well subsequently.

---

[69] R. V. Southwell, *An Introduction to the Theory of Elasticity for Engineers and Physicists* (Oxford, 1936); E. B. Moullin, *The Principles of Electromagnetism* (Oxford, 1932); id., *Spontaneous Fluctuations of Voltage Due to Brownian Motions, Shot Effect, and Kindred Phenomena* (Oxford, 1938).

[70] *OUG* 61 (1930–1), 691–3.

[71] Ibid.; *OUG* 62 (1931–2), 662–4; 64 (1933–4), 201–4; 65 (1934–5), 195–8; 'Gifts to Engineering Laboratory', *Oxford*, 3 (1936), 10–13; Christopherson to author, 11 Apr. 1988; Southwell to University Chest, 11 Oct. 1934, UC/FF/187/2.

[72] *OUG* 67 (1936–7), 223–6; 70 (1939–40), 237–40.

Henchley, Hopkins, and somewhat later Mintoff all acquired third-class degrees, but pursued distinguished careers in administration, academia, and politics respectively.[73] And some undergraduates chose engineering in preference to physics: they found Lindemann absent or cold while Townsend was dreary and out of touch. In contrast Southwell was amiable, approachable, helpful, ever present in his department, and an active researcher who was known to undergraduates as a leader in his field. Southwell took trouble with his pupils, advising them at crucial points in their careers and helping them to acquire jobs. Though Moullin was not generally regarded as an inspiring teacher, he gave to a few pupils such as Rawcliffe 'the feeling that here was a message from Heaven to which one *had* to listen'.[74] In fact Moullin's teaching was sufficiently effective by 1935 for Southwell to dispense with the physics teaching previously provided by Townsend's electrical laboratory staff, a move which led to strained relations between Southwell and Townsend who felt threatened.[75]

Southwell was lumbered with one embarrassing enterprise, namely the joint Birmingham and Oxford diploma in coal-mining, which had been approved in 1921 after its bumpy career before the first war. By 1930 no regulations had been made so that year a scheme was introduced whereby Oxford graduates in engineering science, who had taken geology as a special subject and somehow experienced four months' practical work in mining, attended the University of Birmingham for one year on its diploma in mining course. This joint diploma was not popular: in its first fourteen years it was conferred just once.[76]

In promoting engineering science at Oxford Southwell fought against the views that engineering was an appropriate subject for illiterate artisans and that academic engineers were men with oily rags stuffed into the pockets of denim overalls who were ill-suited, metaphorically as well as literally, to enter the senior common-rooms of the colleges.[77] His resolve was buttressed by Tizard, a leading statesman of science, whose views were like his own. Tizard did not see the German *technische hochschule* as a viable model for Oxford. He was keen that universities in general and Oxford in particular

---

[73] Douglas Victor Henchley (b. 1911), third 1933, became bursar, Administrative Staff College, Henley (information from R. G. Stansfield); H. J. Hopkins (1912–86), third 1935, professor of civil engineering, Canterbury University, New Zealand 1951–78; Dom Mintoff (b. 1916), third 1941, Prime Minister of Malta 1971–84.

[74] Gordon Hindle Rawcliffe (1910–79), first 1932, professor of electrical engineering, Bristol University, 1944–75, on whom see A. R. Collar and A. R. W. Broadway, *BM* 27 (1981), 479–503 (483).

[75] Townsend to Lindemann, 7 Oct. 1935, LP, B100; 18 June 1935, PS/M/1/1, for onslaught by Townsend and Gill on department of engineering science.

[76] *Nature*, 126 (1930), 828; Mines department [*sic*]. Diploma of Coal Mining, UR/SF/MIN/2, OUA.

[77] Medawar, *Autobiography*, 105.

should not try to ape technical schools and should avoid 'the temptation of mass production'. Engineering departments in universities should teach principles, not practice; and they should try to produce resourceful graduates who would become leaders of the profession and not its rank-and-file members. Tizard's views echoed those of Southwell: it was neither necessary nor desirable at Oxford 'to build up large schools of applied science, which is well looked after elsewhere. Quality is better than quantity. It is far more important... that Oxford should have, as it has, a small school teaching and promoting the science of engineering rather than a large school teaching its practice.'[78] Such rhetorical support was useful but it did not disturb certain political realities of which Southwell was well aware. Though his department was highly successful in its research, there were pressing problems with teaching: in 1939 Southwell was worried about the poor quality of his undergraduate intake and the total lack of undergraduate scholarships and of tutorial fellowships in his subject. He was prescient in his fear that his staff, lacking recognition by the colleges, might be lured elsewhere.[79] During the war Southwell migrated to London for the rectorship of Imperial College, while Moullin (short-listed but rejected for the Oxford chair) and Binnie returned to Cambridge for promotion and a college fellowship respectively. The disintegration that Southwell had dreaded had indeed occurred.

The sort of engineering science that Southwell pursued at Oxford meant that there was apparently no contact between his department and the large engineering plants at Cowley devoted to making motor cars. Undergraduates did not go in their long vacations for work as temporary apprentices in Morris's works or in those of the Pressed Steel Company; Met-Vick, aircraft enterprises, and Cunard liners were the preferred locations. Nor did Morris, made Lord Nuffield in 1934, donate or loan equipment to Southwell's department. It was indeed dissatisfaction with Oxford's department of engineering science which was one element in Nuffield's proposal of 1937 to endow a new college which would focus on practical engineering and on modern business methods. His generous benefaction perpetrated the double sin of aiming to promote both industrial engineering and business studies. It is not surprising that the University responded craftily: it took Lord Nuffield's money voraciously, granted eponymous fame to him by naming a new college after him, but transformed his initial proposal into a more acceptable scheme based on research and postgraduate work in social studies.

---

[78] H. T. Tizard, 'Science at the Universities', *Nature*, 134 (1934), 405–8 (407); id., 'The Needs of Oxford Science', *Oxford: Special Number* (Feb. 1937), 52–6 (53); for those who looked favourably on the *technische hochschule* see E. P. Hennock, 'Technological Education in England, 1850–1926: The Uses of a German Model', *History of Education*, 19 (1990), 299–331.
[79] *OUG* 70 (1939–40), 240.

3.5 Nuffield College

William Morris began his business career as the owner of a bicycle shop from which he built up his great motor-car manufacturing concern situated at Cowley on the outskirts of Oxford.[80] It was widely assumed in the University that the industrialization of Oxford, of which Nuffield was the chief agent, was beginning to ruin the town: the lower orders who worked at Cowley could act as a madding crowd in the centre of Oxford, disturbing the peace and repose associated with the ancient University; and Morris's cars were already beginning to generate atmosphere pollution, noise, and vibration on the High Street, one of Europe's townscape glories. The University was indeed being punished for its own lack of prescience: in 1924 when consulted by the City it had no view on the projected industrial zone at Cowley.

It was in the 1920s that Nuffield grafted onto the medieval city (a university, cathedral, court, and market town) a big motor-car plant at Cowley to which he had moved just before the first war. At that time the biggest industry in the city was related to the University; from 1830 Oxford University Press had occupied premises in Walton Street where printing and binding took place. In the 1920s Oxford acquired an industrial suburb at Cowley where large-scale industry began from scratch in the mid-1920s. Between 1923 and 1927 Morris's own work-force expanded from 1,650 to about 5,000, remaining at that level until 1939. From 1925 to 1938 Morris was Britain's largest car producer. He was also the driving force behind the Pressed Steel Company plant which began production at Cowley in 1926.

Nuffield was not only responsible single-handed for making Oxford into an industrial town between the wars. In the same period he was the University's biggest benefactor, acting as an individual philanthropist before his Foundation was established in 1943.[81] As a loyal son of the town of Oxford, as a self-made man who sympathized with less affluent undergraduates and deplored wastrels, and as one who wished to promote in the University modern studies with practical application, he poured money into it. He knew that some dons thought his factories had so wrecked Oxford that he should compensate the University. In the 1930s he gave to the University a total nearing £4 million, most of which was donated on his own initiative in a series of pleasant shocks and not in response to entreaties made by

---

[80] J. M. Houston, 'Oxford Industries', in Martin and Steel, *Oxford Region*, 141–6; R. C. Whiting, *The View from Cowley: The Impact of Industrialization upon Oxford* (Oxford, 1983); I. D. Scargill, 'Responses to Growth in Modern Oxford', in R. C. Whiting (ed.), *Oxford: Studies in the History of a University Town since 1800* (Manchester, 1993), 110–30; P. W. S. Andrews and E. Brunner, *The Life of Lord Nuffield: A Study in Enterprise and Benevolence* (Oxford, 1959).

[81] Andrews and Brunner, *Nuffield*, 259, cartoon opposite 273; 'Mr Morris and the Spanish Chair', *OM* 45 (1926–7), 27; A. L. P. Norrington, *Blackwell's 1879–1979: The History of a Family Firm* (Oxford, 1983), 87.

supplicants. The scale of his philanthropy to the University was vital to it financially: in the 1930s its annual grant from the University Grants Committee was about £100,000, so that in this decade Nuffield provided four times as much money for the University as did government. His largess, to which there was no equivalent at Cambridge, was expended in five ways. By 1939 he had given £200,000 to three of the poorest colleges (St Peter's, Pembroke, and Worcester); £100,000 for a new university physical chemistry laboratory; £100,000 in response to the University's Appeal; and a total of about £2,700,000 for developments in medicine, especially those designed to promote co-operation between scientific research done in a laboratory and clinical practice in the wards of a hospital. Nuffield's medical benefactions were clearly intended by him to end the divorce between theory and practice.[82] In promoting this general notion he had the benefit of sustained advice from his physician and friend Sir Farquhar Buzzard, the Regius professor of medicine, and from Hugh Cairns, a pioneering brain surgeon who became the first Nuffield professor of surgery in 1937. Though he relied greatly on them, he was not merely their tool. For instance he succeeded in overriding their opposition to his view that one of the new medical chairs should be in anaesthetics, a subject in which there was no chair elsewhere; and he persuaded the University to accept his unusual proposal. As an astute business man Nuffield was quite capable, if he wished, of attaining what he had proposed to the University: when he was a determined piper he was able to call the tune effectively.

Nuffield's other major benefaction was of £900,000 in 1937 for a new college to be built on the old and derelict canal wharf, opposite the mound of Oxford castle and the grey prison.[83] The site lay on the approach from the railway station to the town centre, and contributed to a dismal scene which made a depressing contrast with the glorious views to be enjoyed on the journey to the centre from the east over Magdalen Bridge and up the curving High Street. Nuffield wanted to provide a beautiful and dignified architectural vista from the west so in late 1936 he bought the canal wharf site for about £100,000, fully conscious that it occupied the gap between Pembroke and Worcester and below St Peter's. It was the last piece of ground near the centre of Oxford on which a college could be built. Its

[82] Andrews and Brunner, *Nuffield*, 285–306; D. Veale, 'The Nuffield Benefaction and the Oxford Medical School', in K. Dewhurst (ed.), *Oxford Medicine: Essays on the Evolution of the Oxford Clinical School to Commemorate the Bicentary [sic] of the Radcliffe Infirmary 1770–1970* (Oxford, 1970), 143–53; G. Jefferson, 'Memories of Hugh Cairns', in Dewhurst, *Oxford Medicine*, 154–73; Edward Farquhar Buzzard (1871–1945), Regius professor of medicine 1928–43; Hugh William Bell Cairns (1896–1952), Nuffield professor of surgery 1937–52.

[83] My account of the early history of Nuffield College is based on: Andrews and Brunner, *Nuffield*, 7–8, 309–12; Scott, *Lindsay*, 230–8; and especially on Chester, *Economics in Oxford*, 63–82. As a former warden of Nuffield Chester had access to all pertinent archives and used them fully. Edward Frederick Lindley Wood, Lord Halifax (1881–1959), chancellor of the University 1933–59.

location offered Nuffield the possibility of completing to the west of the city an outer ring of colleges with each of which he was or would be intimately associated.

In early July 1937 Nuffield revealed his aspirations for his new college in conversation with Lord Halifax, chancellor of the University, and subsequently with Douglas Veale, the registrar. Drawing only on his own experience and apparently having taken no advice, Nuffield made the totally unexpected offer of providing money for a college of engineering and accountancy to be built on the canal wharf site. He felt that Oxford engineering was markedly inferior to that of Cambridge and wished to remedy this embarrassing situation by promoting not just engineering concerned with the practical needs of industry but also modern business studies such as accountancy. These proposals were transformed by A. D. Lindsay, the vice-chancellor, and Veale, his close colleague, into a totally different scheme for a postgraduate college devoted to social studies, in which theoretical and practical concerns were closely linked. Lord Nuffield did not object to this metamorphosis. He neither threatened nor haggled as he had done about the chair of anaesthetics. On this occasion he lacked an adviser within the University who might have maintained or strengthened his resolve to promote engineering and business studies. In any event, Lindsay and Veale reacted with amazing speed to Nuffield's oral proposal made on 8 July 1937. The very next day Veale consulted Lindsay and then saw Nuffield to whom the difficulties of his proposal were stressed. Veale argued that there was no point in competing in applied engineering with Cambridge or in commercial engineering with the provincial universities; to do so would be contrary to the tacit agreement that had been made with Cambridge in 1935 to avoid direct duplication. Veale also emphasized that Oxford colleges embraced several subjects and that there was no possibility of establishing a new residential undergraduate college devoted to just engineering and business studies. Having blocked Nuffield's proposals, Veale put to him an alternative scheme, namely, that economics and social studies, studied in a postgraduate college, would be more suitable subjects than engineering and accountancy for securing closer collaboration between theorists about present-day society and those involved practically in running it. With masterly skill, Veale did not totally neglect Nuffield's concern for engineering: he argued that physical chemistry, in which Oxford had undisputed pre-eminence, was of essential importance to engineering and other commercial sciences, so that if Nuffield wanted Oxford to rival Cambridge and to serve the cause of engineering he should endow a new university physical chemistry laboratory costing £100,000. Nuffield accepted these proposals privately and in October 1937 enshrined them in his formal public offer of £900,000 for Nuffield College. In the University's publicity and pronouncements about the new college, its history began in

October 1937 and Nuffield's initial concern with engineering was not mentioned.[84]

Lindsay and not Nuffield was responsible for the chief features of Nuffield College. Lindsay and Veale made no attempt in early July 1937 to consult informally within the University or to set up some sort of working party, with the aim of trying to work out how Nuffield's concerns with engineering and business studies could be met by the University. Instead they put to him immediately a scheme for securing some of their own cherished ambitions. Nuffield accepted their proposal because it satisfied him sufficiently but not totally: it offered the prospect of collaboration between theory and practice which he had previously supported financially; and, as the founder of an Oxford college to be named after him, he would join that select group of Balliol, Keble, Merton, Somerville, and Wadham who had secured eponymous fame in this way.[85] He was probably as concerned with his college's potential contribution to Oxford's townscape as with what was taught in it: he could easily have fostered engineering elsewhere, but he chose not to do so; and in his formal offer to the University he stressed that he wanted a college building worthy of the highest traditions of Oxford architecture. That was why his major contribution to the development of his college before the Second World War concerned the exotic design for it drawn up by Austen Harrison, an architect who wore a beard and sandals, both of which were anathema to Nuffield.[86] Harrison, who had made his career as government architect in Palestine, was appointed in June 1938. He soon produced a novel design, based on medieval Mediterranean styles, to be built in white Portland stone. Nuffield considered the design un-English: he wanted the mellow picturesqueness of pitched and not flat roofs. By mid-1939 Nuffield had induced Harrison to produce a design based on Cotswold domestic architecture in which style the college was indeed built.

From 1939 Nuffield felt uneasy about his benefaction, especially during the war when the post of subwarden was held by G. D. H. Cole, a socialist and therefore an object of suspicion for Nuffield who christened his own college 'the Kremlin'. And he felt by then that Lindsay had cheated him. It is significant that when in 1942 he opened the new physical chemistry laboratory he was generally in jovial form. E. J. Bowen had laid on a celebratory demonstration of fluorescence and explained to Nuffield that it could be used to distinguish real teeth from dentures because the former fluoresced. He tried to force Nuffield under an ultraviolet lamp but Nuffield protested that trying the experiment on him was futile as he had none of his own teeth left.

[84] 'Nuffield College', *Oxford*, 4 (1937), 10–13; A. D. Lindsay, 'Vice-Chancellor's Oration, 5 October 1938', *OUG* 69 (1938–9), 15–25 (15–16).

[85] John de Baliol (d. 1269); John Keble (1792–1866); Walter de Merton (d. 1277); Mary Somerville (1780–1872); Nicholas Wadham (1532–1609).

[86] H. Colvin, *Unbuilt Oxford* (New Haven, 1983), 166–77; Austen St Barbe Harrison (1891–1976).

When departing Nuffield said he was delighted with the laboratory to which he had contributed £100,000. It was exactly what he wanted to give to the University and 'very different from my experience with Nuffield College—cheated me he did, that man'.[87] That man was Lindsay.

[87] R. F. Barrow and C. J. Danby, *The Physical Chemistry Laboratory: The First Fifty Years* (Oxford, 1991), 82–3.

4
Agriculture and Forestry

Prima facie Oxford should have shown considerable interest in agriculture. The connection between the University and the land as a source of wealth and power was well known: much of the aggregate income of the colleges came ultimately from land and the sons of landowners and land agents were not unknown among the undergraduate body. Yet at Oxford between the wars agriculture was often seen as a mere technical craft or occupation, and as an applied subject involving a hotchpotch of sciences; at worst it was regarded as a soft option for those not interested in serious academic work. Hence its status remained low, with a pass degree instigated in 1919 and an honours degree in rural economy (the term by which agriculture was known at Oxford) not launched until 1937. The very existence of the school, which taught undergraduates, was so heavily dependent financially on the Ministry of Agriculture that it came under severe scrutiny from the Ministry and only just survived in 1931, when retrenchment became the policy of many government departments, and again in 1935. Agriculture was so poorly endowed that the University was happy for the Board of Agriculture, which became the Ministry in 1919, to set up two research institutes in the University. The first of these, established in 1912, was devoted to agricultural economics, and enjoyed a relatively undisturbed career until the Second World War; but the second, established in 1924 and concerned with agricultural engineering, was directed by the nefarious B. J. Owen who was imprisoned in 1931 for fraud. The Owen scandal led to strained relations with the Ministry which on two occasions was on the point of closing the Agricultural Engineering Institute. For the University the differing careers of the school of agriculture and its two associated research institutes indicated in two cases out of three that government funding meant government interference and control. That was one price the University paid: the other was that the institutes and not the school became the chief loci for the study of agriculture.

Forestry resembled agriculture in that it was a pass school with low status: it was not elevated to an honours school until 1945. Its institutional marginality was made most manifest in the acrimonious squabble of 1934 about the proposed site for a new forestry building in the congested Science Area. There was, however, an obvious contrast between agriculture and forestry. In the former case the department faced fierce competition between the wars not only from Cambridge but also from Reading, a university college until it became in 1926 an independent university, which was just 30 miles away down the Thames valley. In the latter case there was little competition from elsewhere in Britain and Oxford was seen by various influential external bodies as the main imperial centre for forestry research, so that in 1924 the Imperial Forestry Institute (henceforth IFI) was established there. The Forestry Commission and the Colonial Office poured money into the IFI which began to overshadow the school of forestry whose job was to teach undergraduates; at the same time they began to call the tune and insisted that in 1936 the IFI and the school be divorced and run separately. This painful period ended in 1939 with the merging of the school and Institute to form a department of forestry which was heavily funded by government bodies but controlled by the University.

4.1 The School of Agriculture

Though the careers of the school and its two associated research institutes were interrelated because each of them was funded, though to different degrees, by the Board of Agriculture and the Ministry, their histories may be treated separately because they had different functions: the school taught undergraduates in a university which many Oxonians regarded as primarily devoted to teaching, whereas the two institutes were concerned with research in two specific branches of agriculture, namely economics and engineering. These three manifestations of agriculture at Oxford will be considered in order of their appearance.

It was the inability of Oxford in the late nineteenth century to provide technical education and advisory services in agriculture that led in 1892 to the establishment of a college at Reading. Two years later its agriculture department was set up in association with Oxford and the qualification of the joint Reading–Oxford diploma was launched. In response the Hebdomadal Council proposed in 1898 an honours school of agricultural science which would be open to students from Reading whose staff would be involved in the teaching; but Congregation rejected the proposal. Only in 1909 did Oxford decide to give a qualification in agriculture when its own diploma was established. By that time St John's had taken two initiatives in providing for the school: in 1907 it not only erected a building but also supplemented the meagre endowment of the chair so that the post became permanent and full-time for William Somerville who had been elected to it in 1906.[1]

Somerville had enjoyed teaching experience as professor of agriculture at the Durham College of Science and at Cambridge; and, as assistant secretary

---

[1] J. C. Holt, *The University of Reading: The First Fifty Years* (Reading, 1977), 2, 7–8; Howarth, ' "Oxford for Arts": The Natural Sciences 1880–1914'.

to the Board of Agriculture 1902–6, he was familiar with the corridors of power in Whitehall. His reputation was based on his success as a practical farmer and not on that as a trained scientist. His discoveries were empirical and useful to farmers who wished to improve their own practice: he had shown the importance to pasture of phosphates in the form of basic slag; he had developed the technique of the experimental plot; and he had pioneered the use of animals as a direct measure of the feeding value of pasture.[2] When he assumed his chair he was aware of Oxford's suspicion of vocational training so he claimed that agriculture could escape from the 'slough of empiricism' and that it was like law and medicine in that it dealt systematically with theory and principles (of farming and estate management) which would illuminate subsequent practical work. He hoped that this programme would appeal to those undergraduates who would become landowners or agents. He also accepted that Oxford could not compete with Reading which had effectively cornered the market in giving training for rank-and-file farmers and in acting as an advisory centre for adjacent county councils. The future for Oxford agriculture, thought Somerville, was for it to be distinctive by training a small number of prospective leaders and not a large battalion of plebeians. The difficulty was that effective teaching and research depended on large initial and annual expenditure, neither of which was forthcoming from the University.[3] Until the First World War the school was funded mainly by the Board of Agriculture and the Development Commission, the latter being also responsible for the establishment in 1912 of the Institute for Research in Agricultural Economics. Thus by 1914 teaching and research in agriculture at Oxford were heavily dependent on government funding for capital and running costs, useful contributions having also been made by St John's and individual philanthropists such as Walter Morrison.[4]

The exigencies of the First World War gave some boost to applied science at Oxford so that in 1919 a pass degree in agriculture was launched to supplement the existing diploma. An honours school was out of the question because the revamped course, which included the history and economics of agriculture, was too broad and disjointed to be recognized in that way.[5] It was hoped that even a pass degree would attract undergraduates into agriculture, a subject which the war had shown to be important. That hope was

[2] E. J. Russell, *A History of Agricultural Science in Great Britain 1620–1954* (London, 1966), 243–7; on Somerville see *DNB* and C. S. Orwin, *OM* 50 (1931–2), 491–2.

[3] W. Somerville, *The Place of Rural Economy in a University Curriculum: An Inaugural Lecture Delivered at the Schools on February 1, 1907* (Oxford, 1907), 6 (quotation).

[4] Walter Morrison (1836–1921), *DNB*, a Balliol classics graduate, focused much of his philanthropy on Oxford. His biggest benefaction was £50,000 to the Bodleian Library in 1920; this was the largest monetary gift ever received by it: *OUG* 51 (1920–1), 417.

[5] Minutes of the Rural Economy Committee 1908–24, OUA, UDC/M/15/1, n.d. [autumn 1917].

amply fulfilled. The flood of post-war students led to overcrowding, understaffing, and poor teaching for about four years, thus confirming the low reputation of the school.[6] That was so in spite of the attempt made in the 1920s to give a distinct identity to the school's degree course. While Cambridge focused on the scientific aspects of agriculture and Reading (which had ended the joint diploma in 1926) looked after high farming, Oxford aimed to serve gentlemen landowners. It reduced the scientific content to the bare minimum and made no attempt to train practical farmers inside the university walls. Instead it aimed at a liberal education via the history, economics, and organization of agriculture (including farm management and agricultural law). Such a menu, which presented agriculture in the garb of an arts and social science subject, allegedly provided a good general education not only for rural landowners but also for prospective country vicars, who were both expected to exercise social leadership.[7]

In the early 1920s no comparable research identity was established. Though Morison created soil science as a viable subject through his work on the colloidal properties of soil and was rewarded with a readership in 1923, Somerville's research reverted to forestry, his first love: he did not consolidate or expand his earlier work on questions of importance to practical farmers, and, not being a trained scientist, found it difficult to diversify his research.[8] Generally the school was outpublished by the Institute for Agricultural Economics. It was the indifferent research and teaching record of the school which induced the University to take two steps which later it came to regret. Avid to boost its reputation in agriculture and to receive money from the Ministry, it lobbied successfully to become in 1924 the centre of a new advisory province (carved out of existing ones) under the agricultural advisory scheme run and paid for by the Ministry.[9] In the same year it concluded negotiations with the Ministry, which had chosen Oxford as the location for an Agricultural Engineering Institute which would be funded by the Ministry and the Development Commission. By late 1924, therefore, there were four manifestations of agriculture in Oxford, with the government in the main footing the bill. In terms of relative income, the school was clearly playing third fiddle to the two research institutes, a

[6] Evidence of Chapman and Cronshaw to Asquith Commission, 30 Sept. 1920, MS Top. Oxon. b. 109, file 10, pp. 132–4; J. A. S. Watson, notes for meeting with Ministry of Agriculture on 16 May 1940, Committee on Higher Agricultural Education, OUA, UR/SF/RE/ID.

[7] 'Agriculture at Oxford', *OM* 42 (1923–4), 282–4; S. L. Bensusan, 'The Study of Rural Economy at Oxford', *Journal of the Ministry of Agriculture*, 27 (1920–1), 272–6. From 1921 the history and economics of agriculture were covered by G. D. Amery (1890–1955), a Brasenose history graduate with strong bibliographic interests, obituary in *OM* 74 (1955–6), 92–3.

[8] C. G. T. Morison, reader in agricultural chemistry 1923–32, reader in soil science 1932–48, student (i.e. fellow) of Christ Church 1928–46; tribute in *OM* 67 (1948–9), 33; W. Somerville, *British Forestry Past and Future* (London, 1917); id., *Some Problems of Re-afforestation* (London, 1919); id., *How a Tree Grows* (London, 1927).

[9] UDC/M/15/1, 25 May 1923.

situation which Somerville disliked. Dissatisfied with having two research institutes foisted on him and suffering from illness, he cut his losses by resigning in 1925.[10] His chair was not an object of ambition. There were only three applicants, two being internal (Morison and W. R. Peel). Dissatisfied with the small field, the electors invited Keen, assistant director of Rothamsted Experimental Station, and Engledow, a rising star at Cambridge, to apply.[11] Presumably they refused so the chair was offered to Watson, a Scotsman who had spent much of his career at Edinburgh University where from 1922 he had been professor of agriculture. Though he had no connections with Oxford, Watson accepted the chair.

His early years were by no means trouble-free. There were problems of co-ordination between the school, the two research institutes, and the Advisory Centre which Watson directed gratis. Externally there was increased competition in different ways from Cambridge and Reading. While they prospered the school sputtered. Watson was research-minded and up to date (as the several editions of his textbook showed), but he lacked funds, facilities, and time for research.[12] He was handicapped by having a small staff, most of whom he had inherited from Somerville, he was swamped in term by teaching, and he failed to launch sustained research work himself with the exception of his sheep-feeding experiments. Undergraduate numbers began to drop, especially from 1927 when in four years there was a steady decline. More and more the school was overshadowed in activities and in income by the two research institutes and the Advisory Centre, even though in the late 1920s the school's grant from the Ministry was increased. By session 1929–30 each of the four manifestations of Oxford agriculture was totally or mainly dependent on the Ministry which paid out almost £29,000 to support them (see Table 4.1).

In 1931 two events occurred which affected the future of the school of agriculture. The first was the Owen scandal, which erupted in spring 1931. Among other results it proved to the Ministry that the University had been lax in its control of Owen's Institute of Agricultural Engineering into which the Ministry had poured money (*circa* £30,000 in 1925–6). Naturally the Ministry began to question all its investments in agriculture at Oxford. The Owen affair highlighted the problems of co-ordination between the school and the two research institutes and called into question the continuing

[10] J. A. S. Watson, 'The University of Oxford', *Agricultural Progress*, 14 (1937), 95–9, is curiously silent about Watson's own regime.

[11] UDC/M/41/1, 20 July, 28 Sept. 1925; William Ralph Peel had lectured on agriculture from 1923, was a research assistant in the Agricultural Economics Research Institute 1925–7, and left Oxford to join the agricultural research staff of ICI; Bernard Augustus Keen (1890–1981); Frank Leonard Engledow (1890–1985), Drapers' professor of agriculture at Cambridge 1930–57.

[12] James Anderson Scott Watson (1889–1966), obituary in *The Times*, 8 Aug. 1966, professor of agriculture, Oxford, 1925–44; J. A. S. Watson and J. A. More, *Agriculture*, went through eight editions 1924–45.

TABLE 4.1. *Ministry of Agriculture grants to Oxford agriculture and student numbers 1925–1939*

Academic year	School of agriculture	Advisory Centre	Agricultural Engineering Institute	Agricultural Economics Institute	Number of undergraduates in school	Number of research students in school
1925–6	3,300	1,676	30,231	8,812	65	4
1926–7	3,300	2,015	16,265	9,168	56	4
1927–8	3,650	2,523	12,515	10,225	55	2
1928–9	4,000	2,557	12,659	8,655	50	3
1929–30	4,000	3,028	13,002	8,972	50	3
1930–1	4,130	2,871	12,970	9,659	40	7
1931–2	3,947	2,830	12,041	9,118	41	5
1932–3	4,062	406	10,489	8,698	33	3
1933–4	4,047	0	10,321	8,110	42	1
1934–5	4,056	0	10,781	8,149	40	1
1935–6	4,270	0	10,733	8,521	48	0
1936–7	4,382	0	10,634	9,064	45	0
1937–8	4,162	0	8,250	11,582	40	0
1938–9	4,142	0	7,800	11,100	42	0

Sources: University accounts and reports of the Committee for Rural Economy, both published annually in *OUG*; UR/SF/RE/1A–1E. No differentiation between the Ministry and the Agricultural Research Council is made in this table because the Chest and even Veale did not see the difference apropos grants.

existence of the Agricultural Engineering Institute. Six months later MacDonald's first national government, dedicated to balancing its budget, had been formed and Snowden, as Chancellor of the Exchequer, had introduced an emergency budget which pruned government expenditure by £70 million, mainly by a 10 per cent cut all round. One aspect of this financial retrenchment was that all institutions providing higher agricultural education in England and Wales had their annual grants reviewed by a committee chaired by Lord De La Warr.

Oxford soon found itself in the firing line. In mid-October 1931 it was informed by the Ministry that the grant to the school for October 1931–2 would be cut from £4,000 to £3,600, i.e. by the standard 10 per cent. Three weeks later the Ministry formally told the University that as a matter of general policy institutions which attracted less than forty students were ripe for elimination as 'uneconomic units'.[13] De La Warr's committee, which had visited the University in May 1931, was uneasy: it was not satisfied with the arrangements for co-operation between the school and the two research institutes; it disapproved of the way in which teaching was carried out on a farm not managed by the School or owned by the University; and lastly and most importantly it was perturbed by the marked decline in the number of students since 1927. With forty-four students in the school in 1931–2 it was very vulnerable and the Ministry played on this weakness. It suggested that undergraduate teaching be abolished and a purely postgraduate school be established; at the same time it raised an iron fist in a velvet glove by informing the University that the future of Oxford as an advisory centre was being reviewed. In late November 1931 the Minister for Agriculture, Sir J. Gilmour, told the University that the Oxford Advisory Centre and the grant for it would be abolished from October 1932. He refrained from adding insult to injury: he did not point out that Oxford was to be the only advisory centre out of eighteen to be eliminated and that its work would be given mainly to Reading.[14] Faced with such devastating criticism and the prospect of a reduction in grants from the Ministry totalling about £3,200 (the abolition of the grant for the Advisory Centre and a 10 per cent cut from £4,000 to £3,600 in that to the school), the University decided quickly to reorganize the school of agriculture in line with the Ministry's wishes while retaining undergraduate teaching. It agreed to promote postgraduate instruction by changing the scope and character of the diploma which was thrown open to non-Oxonians; and it decided to focus research on three

[13] Committee for Rural Economy Minutes 1925–32, OUA, UR/SF/RE/1, letters from Ministry to N. Cunliffe, 13 and 14 Oct. 1931; H. E. Dale, principal assistant secretary, Ministry of Agriculture, to Cunliffe, 5 Nov. 1931.
[14] Veale memorandum, 25 Sept. 1931, UR/SF/RE/1; 9th Earl De La Warr (1900–76), parliamentary under-secretary, Ministry of Agriculture, 1930–1, parliamentary secretary and then deputy minister for fisheries 1931–5; HCP 150 (1931), 141–5, 199; John Gilmour (1876–1940), DNB, Minister of Agriculture 1931–2.

topics, namely, animal husbandry under Watson, soil science under Morison, and agricultural ecology/mycology under Cunliffe. It began to plan the reorganization of the accommodation for the school and for the two research institutes in a co-ordinated way. Foolishly in March 1932 it tried to hoodwink the Ministry by suggesting to it that the present grant for the Advisory Centre be diverted to the proposed postgraduate work of the revamped school as some sort of reward for the University providing more money for the school than hitherto. The Minister saw through this ill-advised ruse: while welcoming the University's increased financial contribution, he stressed in April 1932 that it was 'necessary that the saving consequent on the dissolution of the Advisory Institute [i.e. Centre] should be fully realised as a net saving to public funds'. There was no possibility of diverting to the school the grant previously given to the Centre. The consequence, as the Minister pointed out, was that for the time being the new scheme could not be implemented in its entirety and he asked the University to revise the revamped scheme which the University had just produced in response to criticisms of the school made by the Ministry.[15] Next month the Minister made it clear that the Oxford school had to attract 'a reasonable number of students' or its grant would once again be fundamentally questioned; and he advised the University 'to terminate the employment of some members of the Oxford staff' even though in 1929 they had received hints from the Ministry that their posts were permanent.[16]

In response the University decided to maintain the undergraduate school, not to expand research, and indeed to reduce the range of research projects. It kept animal husbandry, soil science, and agricultural ecology, while scrapping agricultural mycology. As a consequence a botanist who was a university demonstrator (H. Bancroft), a mycologist who was a university research officer (R. C. Woodward), and two laboratory assistants, one of whom had served the University for thirty-two years, were given three or six months' notice of dismissal. These moves saved £2,554 for the academic year 1932–3.[17] In response the Ministry formalized that session its scheme of an annual block grant of £3,600 (compared with £4,000 p.a. in 1928–31), which it maintained until the Second World War, plus earmarked annual research grants and researchers. By the academic year 1932–3 the Ministry was controlling the direction and pace of research in the school and was breathing down its neck apropos teaching.

There is no doubt that the University paid dearly in the early 1930s for slackness and greed in the 1920s. For instance the school leased a farm from

---

[15] Reorganization scheme 28 Nov. 1931, Veale to Cunliffe, 4 Mar. 1932, 29 Apr. 1932 all in UR/SF/RE/1; *HCP* 151 (1932), 5–10, 159–62; 152 (1932), 48–52 (48–9 quotation); Norman Cunliffe (d. 1964) was also secretary to the University's Rural Economy Committee.

[16] C. J. H. Thomas to Veale, 30 May 1932, *HCP* 152 (1932), 233–5.

[17] *HCP* 152 (1932), 235–9; *OUG* 63 (1932–3), 24.

Magdalen from 1918 to 1929 for teaching and research, and made a heavy loss in those years of £11,138. It then found it impossible to obtain money for equipping another and more suitable farm and had to resort to the generosity of Christ Church which offered one of its farms as a temporary replacement.[18] Such moves gave to the Ministry the impression of lack of commitment and bungling. Simultaneously the University was avidly accepting large annual grants from the Ministry while providing only negligible contributions to the school's running expenses; no wonder that in 1931 the Ministry insisted on a larger financial input from the University. Even in early 1932 the University had not fully appreciated that the Ministry meant business, though Veale quickly realized that a 'decent veil' ought to be thrown over the University's attempt to induce the Ministry to transfer the grant for the Advisory Centre, soon to be abolished, to the School for research.[19] By March 1932 Veale had concluded that unless the University followed the Ministry's advice it would lose grants and that on occasion the Ministry was prepared to ignore the University's views as irrelevant. Two months later Lord Hugh Cecil, who had been approached to lobby the Ministry on Oxford's behalf, told Veale that though the University had been incited by the Ministry and then let down, patriotism had to come first because the national situation was that of war in which grievous sacrifices would have to be made.[20] By summer 1932 the University and the school of agriculture had learned the hard way that the sanctified phrase 'subject to Treasury approval' was no longer a formality.

Accordingly Watson turned to St John's, of which he was a professorial fellow, and undertook to manage one of its arable farms in conjunction with his right-hand man, Skilbeck. Though this was not a farm owned by the University or the school, from 1933 it provided on a regular basis valuable research opportunities and useful teaching facilities until the outbreak of war.[21] By summer 1933 the University had spent £4,000 of its own money on providing better accommodation for the school, including a soil science laboratory.[22] Unfortunately the low reputation of the school led to further decline in the numbers of undergraduates and postgraduates. With only one research student in 1933–4 and just over forty undergraduates all told, the school was not meeting two of the Ministry's desiderata. For a second time it was examined by De La Warr's committee which visited it in June 1934. De La Warr himself had definite views. He reprobated the high cost per under-

[18] HCP 150 (1931), 141–2.
[19] Veale to Fisher, 29 Jan. 1932, UR/SF/RE/1.
[20] Veale memorandum, 31 Mar. 1932, and H. Cecil to vice-chancellor, 23 May 1932, both in UR/SF/RE/1; Lord Hugh Richard Heathcote Gascoyne Cecil (1869–1956), *DNB*, MP for the University 1910–37.
[21] *OUG* 63 (1932–3), 349; Dunstan Skilbeck (1904–89), university demonstrator in agriculture 1930–40, principal of Wye College of Agriculture 1945–68.
[22] *OUG* 64 (1933–4), 341; Veale memorandum, 22 June 1932, UR/SF/RE/IB.

graduate; he criticized the course for being insufficiently practical; he compared the school's undergraduates unfavourably with those at Cambridge and called for weeding out of inferior ones; and he asserted that those staff who were not pulling their weight should be 'displaced'.[23] When De La Warr's committee reported to the University in February 1935 its conclusions were dismal: there was no earnest of future prosperity of the school so support from the Ministry should end; the school fulfilled no national function and if abolished would not be missed; it attracted no grants from county authorities and offered no undergraduate scholarships; unlike Bangor University College and Harper Adams College, Shropshire, it was of no use to local agriculture; it attracted too few students and too often they were of poor quality. Faced for a second time with the prospect of the school being closed down, the University arranged a meeting in March 1935 of its representatives with the Minister (W. Elliot), De La Warr, and key Ministry officials. De La Warr argued strongly for the odd sacking, for shutting the undergraduate school, and for concentrating on postgraduate instruction. The Minister, however, accepted the Oxford arguments that a longer run than two years was needed to test the revamped school, that it would be unfruitful to have postgraduate work totally unconnected with undergraduate teaching, and that Watson was rigorously pursuing revision and upgrading of undergraduate teaching (including the possibility of attaching agriculture somehow to the PPE degree). Elliot decided to reprieve the school and to give it five years in which to prove itself.[24]

Some of his hopes were vain. In the next four academic years the school attracted no research students and the undergraduates never exceeded fifty all told. Research was not wide-ranging: in line with the Ministry's wishes, it was scrupulously focused on three areas. Watson, Skilbeck, and Ellis worked on animal nutrition, with Ellis (whom De La Warr wished to dismiss) summarizing some of the work in his 1937 book. Watson's forte was not publication but advising farmers through lectures, conferences, and talks on the radio. He also acted from 1931 as editor of the *Journal of the Royal Agricultural Society* as well as its centennial historian. A highly effective popularizer, he continued to revise his standard textbook on agriculture, co-authored with More, and spent much time on managing for St John's its arable mechanized farm. In 1938 he began lobbying the Ministry to give

[23] H. A. L. Fisher to De La Warr, 18 June 1934, and De La Warr to Fisher, 21 June 1934, both in UR/SF/RE/IB. For De La Warr's persistent view that small agricultural departments usually lacked purpose and should specialize see his 'Presidential Address: The Relation of Agricultural Education and Research to the Development of British Agriculture', *Journal of the Proceedings of the Agricultural Economics Society*, 5 (1938), 190–8.

[24] Veale memorandum, 27 Oct. 1934; Dale to Veale, 21 Feb. 1935; Veale memorandum, 28 Mar. 1935; Elliot to Norwood, 28 Mar. 1935, all in UR/SF/RE/IC; Walter Elliot Elliot (1888–1958), Minister of Agriculture 1932–6, obituary in *BM* 4 (1958), 73–9. On Elliot and the corporatist challenge see A. F. Cooper, *British Agricultural Policy, 1912–1936: A Study in Conservative Politics* (Manchester, 1989), 160–83.

£9,000 towards a demonstration farm which would be grafted onto a farm at Filkins, near Lechlade, which Sir Stafford Cripps had offered to the school, the value of his proposed gift being £5,000. By early 1939 the Ministry had refused to help. Soil science was pursued by Morison and his laboratory assistant Clarke, the latter and not the former being responsible for a novel book on the subject which took the line that the post-mortem approach to dead soil in the laboratory had to be supplemented by studying live soil in the field.[25] Morison amused himself by making various forays to Africa, which kept his new soil science laboratory busy working on the vast chunks of that continent he brought back. Agricultural entomology/ecology was in the hands of Cunliffe who worked for years on developing cereals which would withstand the depredations of the frit-fly.

In accord with the Ministry's persistent desire to raise the calibre of undergraduates in the school, the University raised its status in 1937 when an honours degree in agriculture (to be launched in autumn 1938) was added to the existing pass degree. At the same time a one-year postgraduate certificate of proficiency in estate management, an activity which the Ministry wished to encourage, was approved. The rationale behind these moves was that the honours school would focus on agriculture in its historical, economic, and scientific aspects, the central theme being the evolution of agriculture as revealed by studying the farmer himself. Thus no attempt was made to maintain the old standard type of agricultural curriculum which was concerned with the application of the appropriate natural sciences to the improvement and cheapening of methods of production. Instead, by focusing on the evolution of agriculture, it was hoped to produce graduates with a wide outlook and a critical mind who would become 'the future leaders of the industry' in the new types of post to be found in the Colonial Agricultural Service, the various marketing boards, county farm institutes, and not least the Ministry of Agriculture. Hitherto many Oxford graduates had become land agents, a career which had been associated with the low quality of entrants to the pass school. In future such people would take the three-year honours degree in agriculture and then do the certificate in estate management, which covered such 'purely technical subjects' as surveying, accountancy, agricultural law, and estate organization.[26] Thus the 1937

---

[25] J. C. B. Ellis, *The Feeding of Farm Livestock* (London, 1937); J. A. S. Watson, *The Cattle-Breeders' Handbook* (London, 1926); id., *The Farming Year* (London, 1938); id. and M. E. Hobbs, *Great Farmers* (London, 1937); J. A. S. Watson, *The History of the Royal Agricultural Society of England 1839–1939* (London, 1939); G. R. Clarke, *The Study of the Soil in the Field* (Oxford, 1936); Watson memoranda, 22 Nov. 1938, 24 Jan. 1939, in UR/SF/RE/IC and D; Richard Stafford Cripps (1889–1952); George Robin Clarke (1894–1979) and James Clark Bendall Ellis (1886–1963), both Oxford graduates in chemistry, were respectively departmental lecturer in soil science and university demonstrator in agricultural physics and chemistry.

[26] *OUG* 67 (1936–7), 733–4; 'Proposed Honours School of Agriculture', BS/R/1/3, n.d. [1936]; J. A. S. Watson, 'The Honours School of Agriculture', *OM* 55 (1936–7), 668–9; *Oxford*, 4 (1937), 22–3, 64.

honours degree relegated technical aspects of agriculture to a diploma, made no attempt to compete with such prosperous departments as the big battalions at Cambridge and Reading, and was in line with Oxford's persistent emphasis on a liberal education, on the production of critical minds, and on national leadership.

The creation of an honours school of agriculture did not lead to a flush of college fellowships in the subject. When war broke out none of Watson's four university demonstrators was a college fellow; only Morison, reader since 1923, had been recognized in this way. Nor did it lead to the inclusion of agriculture in the detailed plans for the Science Area drawn up in connection with the Appeal of 1937. The new honours degree did not have time to establish itself before the Second World War broke out. By autumn 1939 all but 6 per cent of registered undergraduates in the school had volunteered and been accepted for military service.[27] For Watson the war provided a welcome opportunity to develop his forte as an agricultural adviser. From 1942 to 1944 he was agricultural attaché to the British Embassy in Washington. In 1944 he resigned his Oxford chair in order to pursue the more alluring career of being the senior agricultural adviser to the Ministry of Agriculture which had twice brought the school of agriculture near to extinction and had abolished the Oxford Advisory Centre of which he had been the unpaid director.

4.2 The Agricultural Economics Research Institute

The founding in 1912 of a research institute in agricultural economics was made possible by Lloyd George's Development Fund Act of 1909. By this Act £2 million were placed in the hands of a specially created Development Commission which among other things was charged with organizing and financing agricultural research. The task of planning was given to Daniel Hall, a Balliol graduate in chemistry who was called from the directorship of the Rothamsted Experimental Station to be a commissioner. Hall saw the need for long-term agricultural research, unconstrained by the need to produce specific results of immediate or short-term economic benefit. He realized this vision by persuading the Commission to establish new research institutes in addition to supporting existing ones. His scheme involved eleven institutes being attached to existing universities, colleges, or other institutions which already had departments of agriculture.[28] Most of the institutes were based on branches of science and not on types of agricultural

---

[27] *OUG* 70 (1939–40), 220.
[28] R. Olby, 'Social Imperialism and State Support for Agricultural Research in Edwardian Britain', *Annals of Science*, 48 (1991), 509–26; H. E. Dale, *Daniel Hall, Pioneer in Scientific Agriculture* (London, 1956), 8–24, 75–105; Alfred Daniel Hall (1864–1942).

production and Hall's scheme provided for all the sciences and subjects that seemed likely to have important applications on the land. As each institute had its own specialism, e.g. animal nutrition at Cambridge and plant nutrition at Rothamsted, there was no possibility of undesirable competition or overlapping. Hall's coherent scheme of permanent and specialized institutes provided for postgraduate scholarships through which young people would be trained for the new job of agricultural researcher.

It was commonly assumed before the First World War that agricultural progress depended on the fuller application of the appropriate sciences to the techniques of production. Though Hall's research field was plant nutrition, he thought that the economic aspects of farming were sadly, even dangerously, ignored. Of all the institutes he created, that for agricultural economics was peculiarly his child. The University was keen to lay its hands on some of the money available through the Development Commission for agricultural research. The Board of Agriculture welcomed Oxford's interest in promoting agriculture: though it was aware that agriculture there was not as flourishing as at Cambridge, it thought that the economics of agriculture would be appropriate for Oxford. In 1912 the University, the Board, the Development Commission, and Hall had their interests satisfied when an Agricultural Research Institute was established at Oxford, though it did not begin work until 1913. From its inception it was heavily dependent on finance from government: in its first year its income was £950, with £600 being provided by the Board, £300 by the University, and £50 by Balliol.[29]

Hall was responsible for seeing to it that his pet subject of agricultural economics in his old University would be properly pursued: the first director of the Agricultural Economics Institute was C. S. Orwin, a pupil of Hall's at Wye College of Agriculture and then a lifelong friend. It was under Hall that Orwin began research into farm economics and costing. Moreover Orwin had enjoyed practical experience for six years as an estate manager in Lincolnshire where, on his 40 square miles, he administrated, surveyed, built, valued, accounted, and farmed. As a result of Hall's initiative Orwin and his first research assistant, Arthur Ashby, were the first British professional agricultural economists. Ashby went on to be the first professor of agricultural economics in Britain and after the Second World War succeeded Orwin as director of the Oxford Institute.[30]

Orwin directed the Institute until he retired in 1945. He gave to it not only direction but also stability. In contrast with the school of agriculture, his

---

[29] Dale, *Hall*, 96, 136; *HCP* 90 (1911), 45–6; 91 (1912), 331–2; University of Oxford, Agricultural Economics Research Institute, *Agricultural Economics 1913–1938* (Oxford, 1938), 8–11.

[30] A. B. Rodger, 'Orwin', *OM* 74 (1955), 35–6; C. S. Orwin, 'The Row I have Hoed', *Countryman*, 34 (1946), 32–9; Arthur Ashby (1886–1953), director of Agricultural Economics Research Institute, Oxford, 1946–52, elected professor of agricultural economics, Aberystwyth, 1929; E. H. Whetham, *Agricultural Economists in Britain 1900–1940* (Oxford, 1981), 31–5.

Institute was not embroiled in stormy disputes with the Ministry: there was never any question of the Institute being closed down and on the whole Orwin enjoyed good relations with Whitehall. In contrast with the Agricultural Engineering Research Institute, Orwin's was run with bureaucratic probity and never suffered an equivalent of the Owen scandal. In the 1930s, while Owen was languishing in gaol, Orwin's career prospered so much that in 1939 he acquired the first D.Litt. to be awarded by the PPE school which had previously recognized agricultural economics by making it an optional subject in finals. Whereas the Agricultural Engineering Institute in that decade was subjected to searching scrutiny by both the University and the Ministry, Orwin's was usually blessed by both. Thus in 1938 his Institute was able to issue a silver jubilee publication which showed how greatly the growth of agricultural economics was indebted to Orwin's Institute, which had trained people for work in government departments, agricultural advisory centres, the colonies, and British universities and colleges. The silver jubilee celebration was attended by W. S. Morrison, the Minister of Agriculture, who spoke enthusiastically about the value of the Institute's work to his department, to farming, and to the state.[31] Financially Orwin's Institute had become a favoured child of the Ministry: at the end of the 1930s it was receiving almost as much money per year as the Agricultural Engineering Institute and school of agriculture put together (see Table 4.1). Orwin's staff expanded greatly in his Institute's opening twenty-five years, from 2 in 1913 to 26 (including a deputy director and 9 devoted to research) at the silver jubilee. Concomitantly the accommodation for his Institute was greatly enlarged. Having begun life in a rented room at Carfax, it endured temporary accommodation until 1925 when the Ministry provided £4,000 for the alteration of Museum Cottage on Parks Road. In 1932 the University provided £1,200 for a new wing and library extension to the 1925 building.[32] There is no doubt that by 1939 the Oxford Institute for Agricultural Economics was the national centre for research into the economic problems of the land and its use.

To a large extent Orwin was responsible for his Institute's high reputation. A prolific and unignorable writer, specializing in books and not articles, he proved to the Ministry that he was the leader in his field and he showed to the University that agricultural economics was not incompatible with the University's emphasis on arts subjects, especially history. It was indeed as a historian of agriculture that Orwin was best known to the general public, with books on Wye College, the reclamation of Exmoor in the nineteenth century, the Laxton open field system, and an enduring

[31] Agricultural Economics Research Institute, *Agricultural Economics 1913–1938*, 70–9, lists members of the Institute 1913–38; 'Agricultural Economics Research in Oxford', *OM* 61 (1937–8), 732–3; William Shepherd Morrison (1893–1961), Minister of Agriculture 1936–9.
[32] 'Agricultural Economics', *Oxford*, 1 (1935), 14–16.

general work on the open field system. On the subject of agricultural costings, Orwin was the ranking expert for two decades. Drawing mainly on data obtained by his Institute, he published three major books, all of which went into second editions, between 1914 and 1926. His third line of publication concerned agricultural policy where his main contention was that land nationalization was one way in which the agricultural disadvantages of too many small farms could be mitigated.[33] Orwin also indulged in journalism, writing regularly for two north of England newspapers, the *Manchester Guardian* and the *Yorkshire Post*. Such literary output led in 1922 to his election to a fellowship at Balliol where his technical skills were subsequently put to good use when he became estates bursar in 1926. Though not an Oxonian by education, Orwin adapted easily to the collegiate university, helped in 1919 to set up the Oxford Trust (the forerunner of the Oxford Preservation Trust), and became a loyal college man. His pioneering penetration of a college was followed in 1937 when one of his research officers, Keith Murray, was elected fellow and bursar of Lincoln.[34]

It is also clear that on the whole Orwin enjoyed good relations with key figures in the Ministry of Agriculture and Development Commission. In 1917 for instance he was invited by Prothero, the president of the Board of Agriculture, to serve on the Central Agricultural Wages Board and in exchange he extracted informally a promise, which was kept, that the cost accounting work of Orwin's Institute would be placed on a permanent financial footing. Through his services to government bodies, such as advising the Board of Agriculture gratis about agricultural costings after the war, he established useful connections with Sir Henry Rew, assistant secretary in the Board, and with Sir Thomas Middleton who in 1919 vacated an assistant secretaryship of the Board to succeed Hall at the Development Commission. Above all Orwin was trusted by his close friend Hall, who was the Grand Cham of agricultural policy between the wars. It was not accidental that Hall was elected an honorary fellow of Balliol (Orwin's college) early in 1939 and that in the summer of that year he was presented in Balliol with a

---

[33] C. S. Orwin and S. Williams, *A History of Wye Church and Wye College* (Ashford, 1913); C. S. Orwin, *The Reclamation of Exmoor Forest* (London, 1929); id. and Christabel Susan Orwin, *The History of Laxton* (Oxford, 1935); id. and ead., *The Open Fields* (Oxford, 1938; 2nd edn. 1954; 3rd edn. 1967); C. S. Orwin, *Farm Accounts* (Cambridge, 1914; 2nd edn. 1924); id., *The Determination of Farming Costs* (Oxford, 1917; 2nd edn. 1921); id. and H. W. Kersey, *Estate Accounts* (Cambridge, 1926; 2nd edn. 1936); C. S. Orwin and W. R. Peel, *The Tenure of Agricultural Land* (Cambridge, 1925; 2nd edn. 1926); C. S. Orwin, *The Future of Farming* (Oxford, 1930).

[34] C. S. Orwin, 'A Countryman in Oxford', *Countryman*, 34 (1946), 241–6, esp. 242; 'The New Rector of Lincoln', *OM* 62 (1943–4), 233 stressed that Murray's agreeable personality was a passport into any society; Keith Anderson Hope Murray, Lord Murray of Newhaven (1903–93), fellow and bursar of Lincoln 1937–53, rector 1944–53, chairman of University Grants Committee 1953–63, obituary in *Independent*, 16 Oct. 1993.

volume of essays in celebration of his 75th birthday, the Festschrift being edited by Orwin.[35]

Orwin's good relations with key Ministry men helped him to overcome the truculence of Somerville, professor of agriculture, who saw Orwin and his Institute as rivals. Somerville wanted to keep his school uncontaminated so he would not co-operate in allocating permanent accommodation for the Institute in the school of agriculture. For his part Orwin was fed up with temporary and makeshift premises and wanted accommodation in the school. In 1914 Somerville was so glad to learn that Orwin's Institute would continue in temporary quarters that in celebration he gave £25 a year for two years for this accommodation. He then insisted that Orwin should dig into his own pocket for the cost of his accommodation to supplement Somerville's own donation. Orwin refused.[36] For the remainder of his tenure Somerville managed to keep Orwin's Institute at bay in temporary accommodation. In general Orwin defended his own corner purposefully without seeming to be an empire-builder. For instance in late 1928 hints were dropped that the University might reduce or discontinue its annual grant of £300 to the Institute, which in 1927 had become formally recognized as a part of the University. In reply Orwin pointed out to the registrar that the Ministry was footing much of the Institute's bills: it had paid £4,000 for the new buildings erected in 1925, its annual grant was £6,000, and it was trying to promote research of a long-term character. He argued that, as agriculture was very much a Cinderella at Oxford, it needed a university grant as a symbolic gesture of interest in an important modern subject. By March 1929 the University had decided to continue its grant of £300 p.a.: it had been warned by the Ministry not to reduce this grant to £100 because such a cut would have been 'liable to misinterpretation outside'.[37] In promoting his Institute, Orwin used succinct phrasing and dry wit. In one letter to the registrar, he pointed out that his Institute did not receive a block grant from the Ministry but sums earmarked for various purposes: 'The idea that unprofitable swings may be set off against profitable roundabouts is not recognised by the Ministry of Agriculture.' He also expressed his preference for Institute and not Institution as the correct title of his enterprise: the latter was associated with the Poor Law and the Borstal scheme.[38] As a person

[35] Minutes of Meetings of Advisory Committee to Institute for Research in Agricultural Economics, 1913–21, OUA, UDC/M/15/6, 12 May, 27 Oct. 1917, 18 May 1918, 1 Feb. 1919, 5 Feb. 1920; Rowland Edmund Prothero (1851–1937), *DNB*, president of Board of Agriculture 1916–19; Sir Henry Rew (1858–1929), assistant secretary to Board of Agriculture 1906–18; Sir Thomas Middleton (1863–1943), assistant secretary to Board of Agriculture 1906–19; Dale, *Hall*, 175–6; C. S. Orwin (ed.), *Agriculture in the Twentieth Century: Essays on Research, Practice, and Organization to be Presented to Sir Daniel Hall* (Oxford, 1939).

[36] UDC/M/15/6, 20 July, 15, 27 Oct. 1914.

[37] Agricultural Economics Research Institute 1924–42, OUA, UR/SF/RE/2, Orwin to registrar, 29 Nov. 1928, registrar to Chest, 4 Mar. 1929 (quotation).

[38] Orwin to registrar, 10 Feb. 1927, UR/SF/RE/2.

Orwin, who had a commanding physical presence, was shy and sensitive; but he was also ambitious. Just one month after Watson had been appointed to the chair of agriculture in September 1925, Orwin lobbied unsuccessfully for his own salary to be raised to Watson's (£1,200 p.a.). In 1927 he was toying with a grandiose scheme for an Imperial Bureau of Agricultural Economics to be funded by the Empire Marketing Board and in 1931 unsuccessfully sought to rename his Institute the National Institute for Agricultural Economics.[39]

Initially Orwin met suspicion of agriculture as a merely practical subject which was neither legitimate nor respectable in the University. When Orwin revealed to one college head that he was an agricultural economist, he was told, 'Yes, my wife keeps hens too.' This derogatory attitude meant that at Oxford Orwin enjoyed no sinecure in establishing agricultural economics, which could be dismissed as a minor branch of economics.[40] Orwin's response involved steady publication and high standards of scholarship which he hoped would gradually prove the worth of his Institute. Of course, the First World War helped him. In his third annual report he stressed that the war, by bringing into prominence the importance of increasing food production, had highlighted the utilities of agricultural economics. He made it clear that 'until the study ... of agricultural history, the relation of landlord and tenant, the size and arrangement of holdings, the distribution of capital, the management of labour, the influence of markets, the problem of transport—in fact, all that is meant by agricultural economics—has come to be recognised as an essential part of the education both of the landlord and the farmer, the maximum of efficiency in the agricultural industry of these islands will not be realised'.[41] Hardly a year went by without Orwin claiming that recent legislation or events confirmed the importance of agricultural economics. He was flexible in formulating the Institute's research programmes, studying new problems as they arose and discarding work that was being done elsewhere or had become redundant. For example, when the Sugar Subsidy Act was passed in 1925 the Institute quickly undertook a continuous survey of the economics of sugar-beet growing and marketing, which provided important information for the government about the effects of the Act.[42] In the inter-war period as a whole the Institute was not distracted by major educational or advisory work: it focused on research. Until 1926 it worked on agricultural costings and statistical surveys of farming practices; from 1926 it tended to concentrate on the marketing of farm produce, and from 1929 there was a definite emphasis on the study

---

[39] UDC/M/15/7, 29 Oct. 1925, 13 May 1926, 19 May 1927; UDC/M/15/8, 13 Feb. 1931; Orwin memo, May 1927, and Orwin to registrar, 17 May 1927, in UR/SF/RE/2.
[40] Rodger, 'Orwin', 35; Whetham, *Agricultural Economists*, 44.
[41] *OUG* 46 (1915–16), 315.
[42] *OUG* 57 (1926–7), 535–6.

of prices. It was indeed through their work on prices that two of Orwin's researchers, Keith Murray and Ruth Cohen, laid the basis for their subsequent distinction. In 1929 Murray began work on the prices of livestock and in 1933 Cohen started research on milk prices in the context of a peculiar situation in which liquid milk products were not subject to foreign and empire competition but butter and cheese were.[43]

Orwin's flexible and evolving research programme for his Institute enabled it to be well funded compared with the school and the Agricultural Engineering Institute. After the First World War its basic existence was ensured by the Ministry of Agriculture, the Development Commission, and the Agricultural Research Council (established in 1931). It also received temporary grants from the Empire Marketing Board and the Milk Marketing Board. Though nearly all the Institute's research was externally funded, Orwin zealously defended its intellectual independence. As a believer in academic freedom he was not afraid to publish results which were unpopular with both government and farmers. Indeed in 1913 his research on cost accounts, in which he was assisted by John Orr, was not welcomed by many farmers; and Orr was suspect because he had been dismissed for unorthodoxy by the Land Values Committee.[44] In 1925 Orwin's published views about the desirability of nationalization of land were deemed to be politically embarrassing but Orwin maintained the right of members of his Institute to the 'fullest academic liberty'. Two years later, Prewett's study of milk marketing was decidedly unwelcome to the National Farmers' Union.[45] Though the Institute was devoted primarily to research and publication in agricultural economics, it also acted as a marriage bureau: Orwin, who was widowed in 1929, married his secretary Miss C. S. Lowry, an Oxford history graduate, in 1931 and they collaborated happily on the history of the open field system. Perhaps Orwin was inspired by the example of his eventual successor, Ashby, who had married an Institute secretary in 1923.[46]

Orwin's own work and that of his Institute did not go unchallenged. His method of so-called full costings was so effectively criticized in 1927 by John King that by the early 1930s it was no longer the dominant orthodoxy. Orwin's method assumed that it was possible to determine accurately what proportions of the common and overhead costs incurred by a farm were

---

[43] Agricultural Economics Research Institute, *Agricultural Economics*, 20–60; K. A. H. Murray, *Factors Affecting the Prices of Livestock in Great Britain* (Oxford, 1931); R. L. Cohen, *The History of Milk Prices* (Oxford, 1936); *OUG* 61 (1930–1), 277; 65 (1934–5), 236–8; Ruth Louisa Cohen (1906–91) left the Institute in 1939 to become fellow in economics, Newnham College, Cambridge, of which she was principal 1954–72.

[44] Orwin report, 16 Oct. 1913, UDC/M/15/6; John Orr, research assistant 1913–17, later advisory agricultural economist, University of Manchester.

[45] UDC/M/15/6, 29 Oct. 1925, 19 May 1927; F. J. Prewett, *The Marketing of Farm Produce, Part II: Milk* (Oxford, 1927); Frank James Prewett, research officer 1925–34 and sometime editor of the *Farmers' Weekly*.

[46] Whetham, *Agricultural Economists*, 48, 67.

spent on its different products. King pointed out that such costs often amounted to 40 per cent of total costs, and that the allocation of these to separate products was arbitrary and therefore unlikely to be useful to farmers in assessing the profitability of a given product such as carrots.[47] More curiously the Institute was challenged in Oxford by a rival organization launched by John Maxton who had worked for eight years in Orwin's Institute. By 1931 Maxton was disenchanted with the Institute's work which he thought too narrow: he wanted a broader perspective such as would be available in a food research institute. The Dartington Hall Trustees, of which L. K. Elmhirst was chairman, came to his aid in 1933 when they established at Oxford the Institute of Agrarian Affairs (IAA) with Maxton as director. His job was to organize the international conferences of agricultural economists of which Elmhirst was president, to publish their proceedings, and to promote research into agricultural economics and agricultural policy. The IAA had no formal connection with the University (that was achieved in 1943) but it did challenge Orwin's Institute as an alternative source of authority. From 1933 there were in Oxford two institutes which specialized in agricultural economics. Though Maxton never published a major book, from 1934 to 1940 he edited a monthly survey of agricultural policy and opinion. It was probably no accident that in spring 1934 the Institute published the first of its annual *Agricultural Registers*, which were handbooks summarizing agricultural legislation, supplies, prices, and the general experience of the agricultural year.[48]

4.3 The Agricultural Engineering Research Institute

The career of the Agricultural Economics Institute was placid in contrast with that of the Agricultural Engineering Institute established in 1924 with B. J. Owen as its director. Much though not all of its troubled and turbulent career was the result of Owen's criminality which was unmasked in 1931. Owen turned out to be a persistent offender though a head wound he suffered in the first war may have produced mental abnormality. In 1937 he was sentenced to nine months' hard labour for financial deception, and next year sentenced to eighteen months of the same for further deception. In 1944 he was convicted of forgery and sentenced to four years' penal servitude and his wife committed suicide.[49] Such a character was uncommon even

---

[47] Whetham, *Agricultural Economists*, 52–3.
[48] Ibid. 75–8; John Purdon Maxton (1895–1952); Leonard Knight Elmhirst (1893–1974), *DNB*, bought Dartington estate, Devon, 1925, president of the International Conference of Agricultural Economists 1930–61, founder and chairman Dartington Hall Trust 1931; *Agricultural Register [1933–4]: A Record of Legislation, Organization, Supplies and Prices during the Year*, 1 (1934).
[49] Committee for Rural Economy (Owen), 1933–48, OUA, UR/SF/RE/5/2.

in the long history of Oxford eccentrics. It was not usual for a head of a research institute or department at Oxford to be tried for corruption and fraud at the Old Bailey, to be found guilty, and to be sentenced to hard labour. Certainly the extent and nature of Owen's nefarious behaviour exacerbated the somewhat stressed relations which already existed between the University and the Agricultural Engineering Institute, the Ministry and the Institute, and the University and the Ministry. On two occasions, in 1931 and 1934, the Institute was on the point of being closed down by the Ministry which financed it; and in 1937 the University subjected it to a penetrating internal inquiry. Forbidden by the University to do certain kinds of research, the Institute survived in the 1930s as a tool of the Ministry. By restoring to short-term contracts and on occasion suddenly imposed cuts in grants already being paid, the Ministry controlled the Institute and the direction and pace of its research.

The Engineering Institute was created in 1924 as a result of the Ministry's own interest in promoting the increased mechanization of agriculture at a time when horses and tradition still ruled. The age of herbicides, pesticides, combine harvesters, muck spreaders, and powered balers, grabs, and hoists had not yet dawned. In an attempt to encourage mechanization, from 1918 the Ministry conducted trials of machines on a temporary basis. Owen had worked as an engineer for the Ministry on machine-testing and on other problems, such as subsoil ploughing, haystack drying, mole-draining, and the generation of electricity by wind power. By 1923 he was regarded by key men in the Ministry as *the* man to carry out desperately needed work on agricultural engineering.[50] At the same time the Ministry decided to stop its own in-house research on agricultural machinery and to transfer it to a university. In May 1923 it took the initiative of suggesting to the University that Oxford might house Britain's first university research institute devoted to agricultural machinery. The University responded warmly to this suggestion. It was also keen to become the centre of an agricultural advisory province and was not then inclined to be backward in accepting grants from government because it had not experienced the disadvantages of so doing. Reservations were expressed about whether the proposed institute would be permanent because it was totally a government baby, being funded solely by the Ministry and the Development Commission. Questions were also asked about whether the work of the intended institute would be really scientific and not merely practical. These qualms were soon stilled: it was agreed that the term agricultural engineering would be more grateful to

---

[50] MR/7/2/10 contains the following on which I have drawn: Veale to Pantin, 28 Jan. 1958; Veale, notes for oral statement at tribunal; University of Oxford and Ministry of Agriculture, Documents; printed *The University of Oxford and the Ministry of Agriculture and Fisheries: Case for the University of Oxford. Reply by the Ministry of Agriculture and Fisheries. Rejoinder by the University of Oxford.*

Oxford ears than agricultural machinery; and the scheme had the strong backing of Sir Daniel Hall who had assured the vice-chancellor that the Ministry wanted Oxford to fill a gap by housing an institute devoted to the scientific study of agricultural engineering. Hall flattered the University by arguing that Orwin's Agricultural Economics Institute had proved to be a great success in the Ministry's eyes, that the University already had experience of accommodating an agricultural research institute, and that co-operation between the proposed institute and the schools of agriculture and engineering, and Orwin's Institute, was possible and desirable.[51]

By early 1924 the University had formally accepted the Ministry's proposal that Oxford be the location of an agricultural engineering research institute. It was to do what the Ministry wanted, i.e. to do research into the application of scientific principles to agricultural machinery and engineering, to give advice, to test machines, to survey mechanical farms, and to act as a bureau for its subject. Patenting of inventions was not envisaged. It was assumed by the University and the Ministry that the staff of the Institute would appreciate the needs of the farmer as a user of machinery and be engineers, able to remedy faults in existing machines or to devise new models for particular purposes. For the University the establishment of the Engineering Institute was yet another convenient agricultural side-show which involved the University in providing nothing more than accommodation. Though the internal status of agriculture in the University was not high, the external image of Oxford agriculture was being polished by the University's housing an advisory centre and two research institutes, all three of which were almost totally funded by the Ministry and the Development Commission. For the Ministry, however, its investments in Oxford agriculture constituted a major enterprise. The Oxford Agricultural Engineering Institute was to take over work previously done by the Ministry's research branch which had been set up in 1920, and was to be paid for doing so. The assumed advantage for the Ministry of the Oxford arrangement was that willing research experts in agriculture, agricultural economics, and engineering were to hand to aid the Agricultural Engineering Institute.

It is also clear that Owen was a protégé of the Ministry which had sent him to the USA in summer 1923 to study mechanized practices. Through the Ministry's influence he was appointed in April 1924 as director of the Agricultural Engineering Institute. He was very strongly backed by H. G. Richardson, principal secretary in charge of the Ministry's research branch, who thought Owen a genius, a view he rescinded only in the late 1920s. Having been recommended by six votes to three by a university selection committee, Owen's appointment was confirmed by the Ministry which had

[51] H. E. Dale to Somerville, 18 May 1923; H. G. Richardson to Somerville, 20 July 1923; A. D. Hall to Somerville, 16 Oct. 1923; Hall to vice-chancellor, 18 Oct. 1923, all writing from Ministry, all in University Solicitors' Papers, OUA, U Sol./44/1; *HCP* 125 (25 May 1923), 87.

the power of veto.[52] Inquiries had been made by Professor Somerville about Owen's degrees; but when he claimed to be M.Sc. (Wales) and M.Eng. (Liverpool) the University followed the Ministry in assuming that he possessed these qualifications. In 1931 it was discovered that he was not a member of either university.

Though he was an impostor Owen did pursue applied research in agricultural engineering through his Institute until he was suspended by the University. He led work on large and urgent problems which he articulated and which were related. The research on artificial drying of hay raised other questions about artificial desiccation, the most important of which were the drying and storage of sugar beet. This in turn was connected with the extraction of sugar from beet which had been desiccated using De Vecchis's process. It was the sugar-beet research which attracted the interest of the Ministry who gave the Institute a special grant of £20,000 for it in 1925–6. Its concern was in part prompted by the Sugar Subsidy Act of 1925 which gave a ten-year subsidy to British beet-growers and a guaranteed minimum price, with the result that by 1930 there had been a tenfold expansion of beet acreage and in the amount of refined sugar produced from British beet; and seventeen sugar-beet factories were at work. But there was also an element of self-reward for Owen in that in 1924 he was a member of a three-man Ministry team which had reported in early 1925 on De Vecchis's process and recommended the building in England of a complete plant, of small capacity but on a factory basis, to test and improve the process experimentally. Owen quickly exploited the Ministry's concern with stimulating the British sugar-beet industry and with increasing the domestic production of sugar. He set up in 1925 an experimental factory at Eynsham near Oxford to develop the desiccation process in order to replace the so-called diffusion process. The latter involved treating freshly harvested beet which was available for three months a year, whereas the former preserved beet by drying it and allowed the extraction period to be extended to a whole year. The desiccation process offered in principle the advantages of running the sugar factories throughout the year with a smaller plant which gave a large saving on capital costs; of stabilizing employment; and of reduced transport costs from farm to factory because drying reduced the weight of the beet to a quarter of the weight of fresh beet. At the Eynsham experimental factory Owen developed what he publicly called the Oxford process as superior to De Vecchis's mode of desiccation.[53] In October 1926 the Eynsham factory

[52] *Nature*, 113 (1924), 578; evidence of Morison, June 1933, in U Sol./44/1; Richardson to Dale, 2 Nov. 1923, and Hall to Somerville, 20 Mar. 1924, both in MR/7/2/10.

[53] E. H. Whetham, *The Agrarian History of England and Wales*, viii: *1914–39* (Cambridge, 1978), 166–9; *Report of the Commission of Enquiry on the De Vecchis Beet Sugar Process* [4 Feb. 1925], Parliamentary Papers, ix (1924–5), 623–45 (Cmd. 2343); V. E. Wilkins, *Research and the Land: An Account of Recent Progress in Agricultural and Horticultural Science in the United Kingdom* (London, 1926), 300–8; *HCP* 134 (1926), 135–8.

and patents for the new process were sold to a private company, Sugar Beet & Crop Driers Ltd., who quickly developed the process commercially through what Owen publicly described as experimental commercial work. By late 1927 the Eynsham factory had been expanded from a research station to a complete commercial factory capable of dealing with 25,000 tons of fresh beet per annum.[54] When Sugar Beet & Crop Driers discovered that the patents it had bought from Owen were worthless, it took legal action against the University in 1931 on the grounds that Owen's false representations were made in his capacity as an officer of the University and that the patents belonged to the University. Thus the Owen scandal was the unintended result of the Ministry's and government's interest in promoting the domestic sugar-beet industry by sponsoring Owen's research into the desiccation process in the hope that it could be made commercially viable.

Owen also saw that certain machines were likely to be important in the future. Though many doubted whether the American combine harvester thresher could operate effectively in the different conditions of Britain, Owen's Institute began a series of trials of this machine in 1928 and published its generally favourable findings.[55] Owen was probably the first to introduce the combine harvester into Britain and he was also quick to spot the potential importance of tractors which in the late 1920s were still in their experimental stage. At a time of falling prices it was not at all clear that tractors, which were often unreliable mechanically, were cheaper than horses. In 1930 there were about 16,000 tractors and 700,000 working horses on the farms of England and Wales. That year Owen's Institute collaborated with the Royal Agricultural Society in subjecting thirty-three tractors to comparative dynamometric tests at the World Agricultural Tractor Trials held at Wallingford and published a report which was rapidly sold out.[56] Owen's interest in tractors, like that in sugar-beet processing, had its criminal side. It was his swindling of the International Harvester Company and Fords about a farm tractor scheme that led to his arrest and conviction for forgery and fraud shortly after he was suspended by the University.

Throughout his period as director of the Agricultural Engineering Institute Owen conducted its affairs in a tempestuous way, was difficult to pin down because of his vague but self-righteous language, and created problems for the University and the Ministry before he was exposed and arrested in 1931. His irregular financial behaviour was a growing source of

[54] University of Oxford, Institute for Research in Agricultural Engineering, *A Report on the Development and Costs of the Oxford Process for the Production of Sugar from Sugar Beet* (Oxford, 1929).
[55] Agricultural Engineering Research Institute, *Report of Trials of the Combine Harvester Thresher in Wiltshire 1928. By J. E. Newman Assisted by J. H. Blackaby* (Oxford, 1929).
[56] *World Agricultural Tractor Trials, 1930: Official Report and Catalogue* (Oxford, 1930).

worry though not of action to both the Ministry and the University. At the same time Owen's patenting activities raised acute and unresolved problems for both bodies. Confusion reigned, as the University and the Ministry well knew, about the patents taken out by Owen as a result of inventions made in the Institute. Were they his private property, that of the University, or that of the Ministry which paid him? There was much correspondence, some of it acrimonious, on this matter between the University and the Ministry which had tried in late 1925 to lay down conditions relating to the patenting of inventions by persons employed in universities or other institutions in receipt of grants from the Ministry.[57] By summer 1926 the University had broadly accepted these conditions as a basis for negotiation with the Ministry; but in the upshot no agreement was reached and the University did not formally accept the Ministry's conditions. Nor did the University accept a life-line thrown to it by Sir Daniel Hall early in 1926. Deeply concerned about the difficulties that had arisen in Owen's Institute, Hall proposed to the University that agriculture at Oxford needed a secretary, whose salary would be paid by the Ministry, who would co-ordinate the activities of the school and the two institutes, take charge of all their financial and business affairs, and be the main conduit for communication between the University and the Ministry. This scheme, modelled on that already working at Cambridge, came to nothing.[58]

After Owen was suspended by the University in March 1931 the future of the Institute was closely scrutinized internally and externally. The University concluded that it could not endure the ignominy of closing the Institute; instead it should be reorganized under a new director who would focus its work on pure research and publication with some testing of machines as a subsidiary activity, thus distancing it from commercial work. In June 1931 H. J. Denham, an Oxford graduate in forestry who had some experience of engineering, was appointed director from a short list of two over the head of S. J. Wright, a Cambridge engineer with five years' experience in the engineering department of the National Physical Laboratory, who had been deputy director of the Institute from autumn 1929. Denham was appointed for three years at a salary of £1,200 p.a., then the standard professorial pay.[59]

He inherited a difficult situation including the financial chaos left by Owen. For running costs he had to work from hand to mouth, and he

---

[57] *HCP* 132 (1925), 323–6.
[58] Hall to vice-chancellor, 25 Feb. 1926, *HCP*, 133 (1926), 215–16.
[59] UDC/R/8/1, 23 June 1931; Humphrey John Denham (1893–1970) had worked for the British Cotton Industrial Research Association, Henry Simon Ltd. (a Manchester engineering firm) as director of research, and Buckmaster & Moore as a technical adviser; Samuel John Wright (1899–1975) was director of the Oxford Agricultural Engineering Research Institute 1937–42, director of the National Institute of Agricultural Engineering 1942–7, agricultural adviser to Fords 1947–64, consulting engineer to Royal Agricultural Society 1931–64.

knew that both the University and the Ministry were surveying his Institute with eagle eyes. It cannot be said that he enjoyed much security or encouragement in his six years as director. As soon as he assumed the post the University banned work on the principal problems hitherto studied by the Institute, namely, crop-drying and sugar-beet work, because Sugar Beet & Crop Driers Ltd. threatened Denham with an injunction on the ground that he was infringing its patents.[60] Thus from August 1931 Denham's Institute was forbidden to work on grain-drying, a key element in mechanized farming, the embargo being lifted only in 1940. Though ICI and the Agricultural Research Council (henceforth ARC) wanted the crippling embargo to be lifted, that did not happen in peacetime because the University could not face the prospect of further litigation arising out of the Owen affair. Denham also endured staff problems: four of them were sacked or induced to leave in autumn 1931 and those remaining suffered a 10 per cent cut in salary. It is not surprising that he found it difficult to articulate a new research programme. *Faute de mieux* his Institute undertook advisory work, which the ARC liked, and testing as its main activities, but these were widely regarded in the University as non-academic. Research tended to be uncoordinated, with the promising work on tractor dynamics being abandoned for lack of money. There was little co-operation with the school of agriculture, the Agricultural Economics Institute, and the department of engineering science. In Watson's view the Institute did not do enough fundamental research, Denham did not do much original work, the only good researcher was Wright, and generally the staff lacked people with a 'research type of mind'. For Orwin the Agricultural Engineering Research Institute had lurched into advisory work and testing as a result of the ban on crop-drying investigations and therefore belied its name: its main activity was no longer research. Southwell saw no reason for co-operation between engineering science and the Agricultural Engineering Institute: the former was unspecialized and would remain so. Furthermore in his department testing, being customer-led, was banned because it was not proper research. Southwell's conclusion was that the Institute did not do fundamental research and was too concerned with 'development' work.[61] In 1934 the Institute began to publicize its activities through its journal *Farm and Machine*, which also revealed its problems: advisory work was demanding but given gratis and it detracted from research which in any case tended to be descriptive and not constructive. The Institute could not afford to buy or rent machines and had to rely on gifts and loans from manufacturers and

[60] HCP 154 (1933), 51; Veale to Stocks, 4 July 1940, in Institute for Research in Agricultural Engineering. 1937–, OUA, UR/SF/RE/4/2.
[61] Institute for Research in Agricultural Engineering. Committee of Enquiry. 1936–7, OUA, UR/SF/RE/4A, evidence of Watson, n.d.; Southwell, 25 Jan. 1937; Watson, 28 Jan. 1937; Orwin, n.d. [late Jan. 1937]; Southwell to Veale, 8 Feb. 1937; meeting with Southwell and Orwin, 28 Jan. 1937.

farmers. By early 1937 testing work was declining and there were no resources of any kind to launch a coherent research programme.[62]

Throughout his six years as director of the Agricultural Engineering Institute Denham had to serve two different masters. The University wanted pure research and publication, uncontaminated by sordid commercial considerations, from the Institute; but the Ministry was keen for the Institute 'to get its problems from the industry', i.e. from manufacturers and farmers.[63] As the Ministry paid for the Institute it was able to call the tune more and more. The ARC, set up in 1931, also took an interest in it. In 1932 the ARC was evenly divided about the Institute's future. Half the Council's members wanted it to stay at Oxford at an annual cost of £10,000; the other half argued that it should be abolished, with its work being transferred to the National Physical Laboratory at an annual cost of £2,000.[64] Next year the Minister conducted his own inquiry into the Institute and inspected it in person. He decided that it should stay in Oxford and be reprieved until autumn 1937 when a three-year block grant from 1934 would end.[65] Overall the Ministry reduced its level of financial support after the Owen scandal. In addition it was not above cutting an annual grant half-way through an academic year. In 1932–3, for instance, the expected grant of £12,000 was cut to £10,000 in February 1933, leaving the University saddled with liabilities incurred on behalf of the Ministry. Whether the latter's grant was annual (as from 1931–4) or block (1934–7), it had the weapon of being able to cut grants when it wished and all the University could do was to protest. Thus impending cuts in grants in summer 1934 led to further reorganization of the staff of the Institute.[66] On other matters the Ministry imposed its view on the University. Though the University normally proclaimed the inviolable sanctity of the freedom of its bodies to publish their research, it acquiesced in 1933 in the Ministry's scheme for confidential testing to be done for manufacturers.[67]

In late 1936 the Institute was again in a state of crisis. First, there was the resignation in November of J. E. Newman.[68] Though the University had put an embargo on crop-drying work, Newman became more and more involved

---

[62] *Farm and Machine*, 1 (1934), 11–12; 2 (1935), 11–12; Denham evidence, 16 Feb. 1937, UR/SF/RE/4A.

[63] Wilkins's view, 17 Feb. 1937, UR/SF/RE/4A.

[64] G. W. Cooke (ed.), *Agricultural Research 1931–1981: A History of the Agricultural Research Council and a Review of Developments in Agricultural Science during the Last Fifty Years* (London, 1981), 32–3.

[65] Dale to Veale, 17 July 1933, in Institute for Research in Agricultural Engineering, 1923–36, OUA, UR/SF/RE/4.

[66] *HCP* 154 (1933), 118–19, and 155 (1933), 29–30; UDC/M/15/3, 5 June, 23 Oct. 1934.

[67] *HCP*, 154 (1933), 119–20 and 155 (1933), 21–2.

[68] Institute for Research in Agricultural Engineering, 1936 [Newman file], OUA, UR/SF/RE/4/1; *HCP* 165 (1936), 225–39 (238–9 quotations). John Ernest Newman was the Institute's expert on mechanized farming.

in designing for a local private individual, J. E. Curtis of Botley, a grass-drying machine which was tried out successfully at Hatherop. In late August *The Times* reported favourably on this machine and revealed its inventor as a member of staff of the Institute. In spite of being warned by Denham, Newman appeared at a Hampshire agricultural show, stood on the Curtis drier, and gave a salesman's address to the assembled crowd. Though Newman had received no money from Curtis, he was asked by Denham to resign or to end his work for Curtis which had begun as technical advice but had ripened imperceptibly into designing a drier for him. A university disciplinary committee judged that Newman's conduct had lacked discretion and hinted that Denham had been negligent by conniving at Newman's infraction of the embargo. Newman responded decisively. On 17 November 1936 he resigned from the Institute because he wished to continue his work with Curtis on crop-drying: 'the Hatherop drier is my baby, even if it is illegitimate, and I cannot bring myself to desert it.' Next day he was asked by telegram whether his resignation was to take effect immediately or subject to notice. He replied succinctly: 'Immediate please.'

It had been known for a few years that the current grant for the Institute would expire on 30 September 1937 when there was every possibility that it would be reorganized by the Ministry or even abolished. To prepare for these contingencies all the Institute's staff were given notice by the University in December 1936 to leave on 30 September 1937.[69] Shortly before this and perhaps prompted by the Newman affair, the University launched yet another inquiry into the work and organization of the Institute in order to prevent the Ministry, which wanted thorough reorganization of the Institute, from calling the tune totally. The university inquiry concluded that abolition of the Institute would be humiliating so it recommended a second fresh start, preferably under a new director known to farmers and manufacturers, who would see to it that only a few main lines of applied research would be pursued so that the Institute would cost the Ministry less than before. The inquiry, which reported in March 1937, gave no more than reluctant approval to the testing and development of machines for the benefit of the trade because such practices could swamp research, provoke awkward situations, and lay the University open to the charge that the development of machines was inappropriate for the world's oldest English-speaking university. No future was seen for either outside work or patenting inventions.[70]

Next month Denham resigned the directorship of the Institute. On medical advice he did not wish to be reappointed. In summer 1936 he had an operation to cure a chronic infection of his nasal sinuses, he suffered from a

---

[69] Veale to Nathan, 26 May 1937, in Institute for Research in Agricultural Engineering, 1937–41, OUA, UR/SF/RE/4B.
[70] *HCP* 166 (1937), 191–204.

rare complication following general anaesthesia, and he lost his memory and concentration for a time. Early next year he endured a duodenal ulcer brought on, according to his doctor, by worry and overwork at the Institute.[71] With Denham soon to be out of the way, a research programme acceptable to the Ministry was formulated in May 1937. It still involved some testing and advisory work, but the main focus was the research into the application of power to agricultural machinery. The programme of projected work involved just five aspects of agricultural engineering—dynamometry, tractor design, tractor-testing, tractor use, and mechanized farm surveys—some of which would be pursued at the mechanized farm, owned by St John's, which the school of agriculture was already using. This programme was presented to the Ministry as the irreducible minimum on which the Institute could continue to survive, it was flexible, and it was focused. It would require a director, paid at the professorial rate, and five permanent researchers, paid on civil service grades, to implement it; and, in deference to the Ministry, the annual expenditure would be reduced. For 1937–8 the University applied for a grant of £8,297, which was 20 per cent less than in the previous five years.[72] The Ministry and the ARC accepted the scheme but did not reveal until late September 1937 how much the grant for 1937–8 would be. Faced with this situation the University sent a further note in May 1937 to the Institute's staff saying that they were free to obtain appointments elsewhere.[73]

On 24 September 1937 the Ministry informed the registrar that the Institute would be funded for three years from 30 September 1937 almost to the amount requested by the University, and that Ministry approval of a new director would be needed.[74] Given the 20 per cent cut in funding, two members of staff, J. L. Dougan (the librarian) and C. A. C. Brown (a senior assistant), were told on 25 September that their services would probably not be required after 31 December and the rest of the staff was put on a three-month contract from 1 October. Both Dougan and Brown protested vigorously but to no avail. Dougan, who had been Oxford City librarian, a pillar of the British Red Cross, and at 62 years of age was nearing retirement, complained to the vice-chancellor that his colleagues and he were victims of negligence and mismanagement by others.[75] Brown, who had published extensively on the application of electricity to agriculture, claimed that his civil service grade of senior assistant gave him permanent tenure, which he alleged had been violated by the scheme of reorganization put to the

[71] HCP 167 (1937), 47–52, 99–101, 184–5.
[72] Ibid. 117–23.
[73] HCP 169 (1938), 62.
[74] Wilkins to Veale, 24 Sept. 1937, UR/SF/RE/4B; UDC/M/15/3, 24 Sept. 1937.
[75] HCP 169 (1938), 61–4; J. L. Dougan to vice-chancellor, 18 Dec. 1937, personal, UR/SF/RE/4B; James Lockhart Dougan (1874–1941), an Oxford graduate, had been appointed librarian by Owen in 1926 and co-edited *Farm and Machine*.

Ministry by the University in spring 1937. It seems that Brown was not aware until 25 September that he did not figure in that scheme. After thirteen years' service he felt considerable indignation and deep disappointment. The University did not relent. Brown was dismissed as a consequence of the University taking drastic measures to make the Institute, which was being given a last chance, 'efficient and productive'.[76] He had not been remiss but he suffered from a change in the policy of the Institute. In any case, the new director, S. J. Wright, appointed unanimously in late November 1937 for three years, wanted rid of Brown.[77] Thus by the end of 1937, three of the five main publishers in the Institute from 1931 to 1937 had departed (Newman, Denham, and Brown), together with its librarian.

A further staff problem erupted in spring 1938 about John Herbert Blackaby, an expert on drainage, a topic specifically excluded from the scheme of reorganization put by the University to the Ministry in May 1937 and approved by it. It was hoped that Blackaby, realizing that there was no money allocated to his expertise, would migrate or simply disappear at the end of 1937. But he did neither because, as a graded research officer, he thought his post was permanent in the eyes of the Ministry. Unlike Dougan he was not nearing retirement and unlike Brown had not been able to find another job. In early 1938 he was given three months' notice of dismissal by the University, objected to it successfully, and was retained by the Institute until the outbreak of war as a financial extra because the Ministry decided he could not be sacked but would not make available any extra money to pay him.[78]

The 1937 reorganization of the Agricultural Engineering Institute was a survival measure which was acceptable to the Ministry not least on the grounds of reduced cost (see Table 4.1). Under Wright as director, new research officers who were mainly non-Oxonians were recruited in 1938 after the clear-out of staff in 1937. Their main research before the war was on pneumatic tractor tyres, paid for by the British Rubber Producers' Association.[79] Generally they were crippled by the University's embargo on crop-drying work, a ban which limited the Institute's utility as far as the ARC and Ministry were concerned.

The Engineering Institute enjoyed no junketings in 1938 when the Economics Institute celebrated its silver jubilee. Compared with Orwin's Institute, that for Agricultural Engineering was unpopular and peripheral in the University. The contrasting careers of the two agricultural research institutes

[76] C. A. C. Brown to chairman, Committee for Rural Economy, 12 Nov. 1937, UR/SF/RE/IC; *HCP* 169 (1938), 63–4 (quotation 64). Charles Alexander Cameron Brown left the Institute for a post with the British Electrical and Allied Industries Research Association.

[77] For Wright's election see UDC/R/8/1, 23 Nov. 1937.

[78] *HCP* 169 (1938), 112–13; UDC/R/8/1, 14 Mar. 1938; Wilkins to Veale, 4 Oct. 1938, UR/SF/RE/4B.

[79] *OUG* 70 (1939–40), 222–5.

were perhaps not entirely unconnected with the low status of engineering at Oxford (despite Southwell's efforts) relative to that of economics which prospered via the PPE degree course. In his 1937 oration the vice-chancellor implied that the Agricultural Engineering Institute was irrelevant to agriculture and to the University when he argued that the problems of agriculture had become mainly economic and were no longer technical.[80] Not surprisingly the Agricultural Engineering Institute was moved from Oxford in 1942 without any of its staff having been elected to a college fellowship there.

The relation between the school of agriculture and the two agricultural research institutes remained an unsolved problem: the University wanted fruitful collaboration between all three, but the Committee for Rural Economy which oversaw their activities failed to secure it and the directors of the institutes guarded their separateness. In any case it was not until 1934 that the University provided contiguous accommodation for the school and the two institutes. Intellectually only Orwin's Institute managed to retain independence while living off government money; in contrast, the school and the Engineering Institute were in the pocket of the Ministry especially in the 1930s. In the University the reputation of agriculture became even lower as a result of the Owen scandal; and the experience of the Ministry's treatment of the school and the Engineering Institute was not happy for the University. This particular experiment in government funding led to the registrar, Veale, complaining to the Ministry that the University could not make appointments 'on the lives of particular governments' and could not 'engage staff on a monthly basis like domestic servants'.[81] From Whitehall a different view was taken, especially in the 1930s when Treasury approval of grants ceased to be a mere formality: through the Ministry, money had been poured into Oxford agriculture without commensurate results. The Ministry's hopes and disappointments were reflected in its financial support of the four manifestations of Oxford agriculture from 1925 to 1939: Orwin's Institute prospered, the school just about maintained its level of support, the Agricultural Engineering Institute lost ground, and the Advisory Centre was abolished (see Table 4.1). Clearly the Ministry felt it was receiving better value for its money from its patronage of agriculture at Cambridge, Reading, Bangor, and Aberystwyth.

Indeed Oxford agriculture was not in the same league as Cambridge's, which by the 1930s flaunted five research institutes, a flourishing graduate school, and a professor (Engledow) who gave some unity not least because he had enjoyed considerable experience as an institute researcher at Cambridge. Moreover, agriculture at Cambridge was supported by the University: in 1928 it asked for and secured £163,000 for agriculture from the Rockefeller International Education Board out of a total sum of £700,000

---

[80] *OUG* 68 (1937–8), 30.
[81] Veale to Ministry, 24 June 1932, and Veale to Dale, 1 Aug. 1932, in UR/SF/RE/1B.

requested. Part of this deal involved the University of Cambridge raising £229,000 itself. It did so, the Ministry of Agriculture contributing £50,000.[82] There was also a painful contrast with Reading. With only three faculties (letters, science, and agriculture and horticulture) it was renowned as a bucolic university which housed the British Dairy Institute and the National Institute for Research in Dairying. At Reading 'agris' enjoyed high status among the undergraduate population because as graduates they often went out to work in the empire where they became bronzed and manly in topees and shorts.[83] Thus agriculture at Reading had a major imperial function; that at Oxford, which prided itself on its connections with empire, did not.

4.4 Forestry

Forestry came to Oxford as the result of an initiative taken by the India Office in 1905. The Office ran the Royal Indian Engineering College at Cooper's Hill, Surrey, where in 1885 it had launched Britain's first school of forestry under Professor William Schlich, who had spent much of his career in the Indian Forestry Service where he had reorganized the forestry school at Dehra Dun. On the abolition of the Cooper's Hill College in 1905, the India Office took Schlich's advice and transferred him, his assistant, and the training of probationers for the India Forestry Service to Oxford, where for the third time in his career he organized a new forestry school. From 1905 to 1908 the Oxford school enjoyed a monopoly of training for the Indian Forestry Service and was maintained by the India Office. In the latter year, however, the Secretary of State for India threw open the training of Indian probationers to Cambridge and Edinburgh as well as Oxford. As Edinburgh and Cambridge were not to be assisted financially by the India Office, it decided that by 1913 its support of the Oxford school would end. As a consequence of this decision and of Schlich's reaching the civil service retirement age, in 1911 Schlich was appointed reader in forestry by the University and given the status of professor during the tenure of his readership. In that year the prospects for forestry at Oxford looked bleak: his sole assistant, who had died in 1910, had not been replaced; and his chair lacked any permanent endowment. The response of Schlich, then aged 71, was remarkable. In order to ensure the continuation of his chair and to provide security for his successor, he lobbied so successfully that by 1919 he himself had raised about £7,000, to which he gave £500, for the permanent endowment of his own chair. Having provided a salary of £900 p.a. for it, he

---

[82] *Nature*, 122 (1928), 941; R. E. Kohler, *Partners in Science: Foundations and Natural Scientists 1900–1945* (Chicago, 1991), 182–8; R. Ede, 'The University of Cambridge', *Agricultural Progress*, 15 (1938), 137–42.

[83] Holt, *Reading University*, 73.

resigned in August 1919; and R. S. Troup, a former pupil from Cooper's Hill and protégé of his, was elected to it with effect from January 1920.[84]

Troup, who had no experience of Oxford, inherited a difficult situation. In 1919 the University had established forestry as a finals school, giving pass degrees only, and shortly afterwards a postgraduate diploma, without providing either running or capital expenses. It contributed nothing to the former on a regular basis, and the forestry building in Parks Road had been erected in 1914 with £3,140 provided by St John's and £2,000 raised by Schlich. This building was placed by St John's at the disposal of the school of forestry until 1923 when its maintenance was passed by St John's to the University.[85] Apropos students an imminent change was announced by the India Office in 1920: in line with the general policy of the progressive Indianization of the public services, forest probationers (preferably Indians) were henceforth to be trained at Dehra Dun.[86] As the Oxford school of forestry existed primarily to train probationers for the Indian Forestry Service it was faced with the prospect of diversification, by serving other parts of the empire, or of extinction. At the teaching level Troup was the only permanent member of staff. Though the subject matter of the degree was wide, and vacation tours of British and Continental forests took place, Troup was assisted by three or four temporary lecturers appointed for only short periods.

Troup's response to these problems took two main forms which were interrelated. One was to try to raise forestry from a pass school to an honours school, a task accomplished in part and obliquely in 1938; the other, achieved in 1924, was to make Oxford the imperial centre of higher training in forestry by setting up a forestry institute whose relation with the school would be symbiotic. His successes owed much to his long experience of forestry in India and Burma, his reputation as a good jungle man, his straightforwardly gritty if grey personality, his role as *the* reporter to government on colonial forest problems in the 1920s and early 1930s, and his extra visibility from 1922 as editor of the Oxford Forestry Memoirs. Moreover in the 1920s he produced two impressive books which became standard works, one on the silviculture of Indian trees and the other on silvicultural systems in general.[87] His high standing as a scientific forester was confirmed by his election as FRS in 1926.

---

[84] *HCP* 122 (1922), 127–30; Endowment of Professorship of Forestry, OUA, For./SF/1/2; obituary of Schlich, *OM* 44 (1925–6), 29; on Robert Scott Troup (1874–1939) see *ON* 3 (1940), 217–19; *Nature*, 144 (1939), 699–70; *The Times*, 3 Oct. 1939; *OM* 58 (1939–40), 58.

[85] *HCP* 148 (1930), 174–7.

[86] *HCP* 116 (1920), 11–13.

[87] R. S. Troup, *Report on Forestry in Kenya Colony* (London, 1922); id., *Report on Forestry in Uganda* (London, 1922); id., *Summary of the Recommendations Contained in a Report on Forestry in Cyprus* (Nicosia, 1930); id., *The Silviculture of Indian Trees* (Oxford, 1921); id., *Silvicultural Systems* (Oxford, 1928).

Troup's campaign for an imperial centre at Oxford was launched at the British Empire Forestry Conference held in London in July 1920. He attended as a representative of the Indian Forestry Service, Schlich attending on behalf of Britain. The Conference, which was prompted by the Forestry Commission, called for the establishment in Britain of one central institution which would train forest officers for all the empire. The prime responsibility of such an institution would be to raise the standard of practice of scientific forestry by superimposing forestry training upon a general scientific education given preferably at a university. The Conference was aware of one drawback of associating an imperial forestry institute with a university, i.e. the impossibility of any outside body controlling any university, but it welcomed in principle the cheapness of having an institute as part of a university because existing buildings and staff could be used. In practice the Conference found it impossible to name a location for the proposed institute.[88]

Troup used these recommendations to argue the case for more staff for his department and the elevation of the forestry degree to an honours degree. He knew that in 1910 Convocation had rejected Congregation's proposal for an honours school on the grounds that forestry was technical and not scientific. He also realized that in future there would be competition at the undergraduate level and that Oxford forestry would be at a disadvantage if it offered no more than a pass degree. Accordingly he stressed the scientific aspects of forestry, claiming that scientific silviculture required close study of ecological factors bearing on invasion, succession, and regeneration in forests, and that practical application was a necessary adjunct to scientific groundwork. He also tilted against an assumption common at the time, namely, that open-air work was less scientific than that done in a laboratory.[89]

The Forestry Commission was so concerned about imperial forestry training that in November 1920 it set up an interdepartmental committee, chaired by Lord Clinton, to report on the problem. By late January 1921 Clinton's committee had visited Oxford, provisionally chosen it as the location for a central institution for higher training, hoped the institute could be somehow combined with the school, and assured the University that the government had no desire to interfere with or control university arrangements. In its report the Clinton committee (henceforth Clinton) recommended that Oxford should be the imperial centre for higher training in forestry and that an institute should be established in connection with the University to give it status.[90] Clinton was impressed by Oxford's successful track record

---

[88] *British Empire Forestry Conference: Resolutions Passed at the Meeting Held in London on 22 July 1920*, Parliamentary Papers, xlix (1920), 3–18 (Cmd. 865).

[89] Troup memorandum, n.d. [May 1921], NS/R/1/3.

[90] *HCP* 118 (1921), 93–5; *Report of the Interdepartmental Committee on Imperial Forestry Education*, Parliamentary Papers, xii (1921), 723–33 (Cmd. 1166); Lord Clinton (1863–1957), from 1919 member of Forestry Commission and its chairman 1927–32.

from 1908 when it began to face competition from Cambridge and Edinburgh in training Indian forest probationers: Oxford had trained 88 per cent of successful probationers. In any event Cambridge and Edinburgh were opposed to the setting up of a central training institute. The former thought that forestry training was best left to individual universities, while the latter thought that a university-based institute would be subject to undesirable external control. Troup had no such qualms. He thought he could handle any possible interference in the institute from outside because the administrative arrangements proposed by Clinton offered him day-to-day dual power: he would direct the institute in addition to running the school. Nor was there apparently any threat from the proposed constitution of the board of governors of the institute: though the board had to approve appointments of staff made by the University, half the board would represent the University and half various governments. Moreover Clinton urged that the proposed institute should recruit from forestry graduates at any approved university, thus departing from the Empire Forestry Conference (1920) view that a science degree was the best preparation for postgraduate forestry training. The Clinton Report did Troup a triple service: it proposed that a central forestry training institute be established at Oxford and be incorporated in the University; it provided a handle with which to upgrade the internal status of the undergraduate forestry school; and it enabled him to claim that Oxford was *the* imperial centre for forestry. Sadly for Troup the Treasury's parsimony saw to it that nothing was done in 1921 and 1922 about Clinton's main recommendations.[91]

These were reinforced in summer 1923 when the second Empire Forestry Conference, which Troup attended, met in Canada. Its main conclusion was that Oxford should be the location of an imperial forestry institute (henceforth IFI) devoted to postgraduate training and to research. Troup was subsequently a leading preparer of the case for an IFI which was laid before the Imperial Economic Conference which met later in 1923 in London. This Conference agreed to provide £5,000 p.a. for an IFI at Oxford.[92] Armed with this promise of money from in effect the Colonial Office and the Forestry Commission, Troup then formally proposed to the University in late 1923 that an imperial forestry training institute be set up at Oxford, stressing that it would prove of great importance to the University from an imperial point of view. By early 1924 he and the vice-chancellor had negotiated a generous deal with the Forestry Commission and the Colonial

---

[91] R. Furse, *Aucuparius: Recollections of a Recruiting Officer* (London, 1962), 77.
[92] Ibid. 108, 113, 134–5; *Imperial Economic Conference, 1923: Summary of Conclusions*, Parliamentary Papers, xii (part 1) (1923) (Cmd. 1990), 194; *Imperial Economic Conference: Record of Proceedings and Documents of the Imperial Economic Conference... Held in October and November 1923*, Parliamentary Papers, x (1924) (Cmd. 2009), 853, 869.

Office.[93] The staff, buildings, and equipment of the school and the proposed institute were to be made interchangeable as far as possible in order to reduce costs. Troup was to be in charge of both school and institute. The University was to provide temporary accommodation for the institute and an extra £300 p.a. for five years to the school to which its maximum contribution had been £300 p.a. in 1923. From October 1924 the Forestry Commission and the Colonial Office would fund the IFI at £5,000 p.a. for five years, with the Forestry Commission continuing to contribute to the school though at a reduced rate of £500 p.a. There was also a tacit agreement that before long the University would have to consider how to provide increased accommodation for forestry in a new building on a new site. In March 1924 Convocation accepted the offer from the two government departments of an IFI at Oxford.[94] No doubt it thought that the University had made a wonderful bargain: the extra expenditure on forestry as a whole involved government contributing almost seventeen times as much money for running costs as the University; the professor was to be in charge of the IFI which would, it was thought, keep it under the control of the University, even though the membership of the Board of Governors of the IFI was to be equally split between representatives of government and the University; as the School and the IFI were to share staff and facilities, it would be possible to deflect resources from the latter to the former; and, though there was still a deep suspicion of mere vocational training, the IFI was appealing as a new and prospectively powerful link between the University and the administrative, scientific, and economic aspects of the empire. Just a week before this agreement between the University and the appropriate government departments had been made, the University had decided to extend what became known as the Science Area but to exclude forestry from it. Thus the University was taking government bodies for a ride, being ready to accept external funding for the running costs of forestry, while doing nothing in return (for several years as it turned out) about a site, never mind the cost of new accommodation.

The IFI landed in Oxford, with advantageous financial arrangements for the University, because of the efforts of two Oxonians who were key men in the Forestry Commission and the Colonial Office. One was Roy Robinson, a Rhodes scholar who had graduated in natural sciences at Oxford, taken a forestry diploma under Schlich, and in 1919 joined the Forestry Commission of which he became chairman in 1932. It was Robinson's quick wit which secured £5,000, and not £3,000, for the IFI at the Imperial Economic Conference of 1923: the proposed allocation of £3,000 was, he claimed at the crucial meeting, a typist's mistake for £5,000. The chairman duly changed a

[93] Troup memorandum, 16 Nov. 1923, *HCP* 126 (1923), 137–41, 203–4; 127 (1924), 35–6, 121–3.
[94] *OUG* 54 (1923–4), 453; *Nature*, 113 (1924), 444; 114 (1924), 97–8.

'3' to '5' and the proposal was carried unanimously. Robinson was also the co-author with Troup of the document which was decisive in persuading the University to accept the IFI.[95] Even more important was Ralph Furse, an official in the Colonial Office who was in charge of recruitment to the colonial services and the Office's forestry adviser. From 1923 he was on good terms with Troup, with whom he formed a close working alliance; and from 1931 with Veale, the registrar, whom he trusted as a fellow civil servant. Furse was not only an excellent committee man but a masterly negotiator in informal circumstances: in 1930 at a lunch party he managed to persuade the private secretary of Raja Brooke of Sarawak to promise £25,000 towards a new forestry building—by the end of the fish course.[96]

The aim of the IFI was to bridge the gap between forest education, research, intelligence services, and forestry practice. It therefore carried out postgraduate training of prospective forestry officers, provided in-service courses to foresters, prosecuted research into the growth but not uses of timber, and acted as a central bureau of information. For its running costs in its first five years the IFI was totally dependent on outside funding, mainly from the Forestry Commission and the Colonial Office which between them gave every year more than the agreed sum of £5,000 p.a.: in 1928–9 they contributed £8,000 (see Table 4.2). Simultaneously the Forestry Commission contributed on average as much to the running costs of the school as did the University. By 1929 there had been no movement by the University in response to Troup's repeated pleas for better and preferably new accommodation for forestry, yet in 1928 the third Empire Forestry Conference had urged that the IFI be placed on a permanent footing. Presumably in chagrin at the way in which the University was pocketing their money and doing nothing about a new building for forestry, the Forestry Commission and the Colonial Office did not renew in 1929 the five-year agreement made in 1924 though they continued to provide finance on a yearly *ad hoc* basis in excess of that stipulated in 1924. Meanwhile the IFI lurched into debt. In 1928–9 it had an overdraft of £1,433. Next session this had grown to £2,635, even though for the first time the Empire Marketing Board had made a contribution (£2,000). The overdraft was eliminated by autumn 1931 mainly by a donation of almost £4,000 by the Empire Marketing Board. Expenditure was still rising, reaching in 1930–1 a maximum of almost £16,000, which was about seven times that of the school. From the University's point of view, here was another Owen affair in the making in that another university institute was running into financial problems, but it was run by a Board of Governors on which the University did not have a decisive voice. From the

---

[95] Roy Lister Robinson (1883–1952), *DNB*, first in geology 1907, diploma in forestry under Schlich 1908–9, vice-chairman (1929–32) and chairman (1932–52) of Forestry Commission; Furse, *Aucuparius*, 135; *HCP* 127 (1924), 121–3.

[96] Ralph Dolignon Furse (1887–1973), *DNB*, third in Greats 1909; Furse, *Aucuparius*, 237–8.

TABLE 4.2. *Main external funding of school of forestry and IFI (£)*

Academic year	Forestry Commission to school and IFI	Colonial Office to IFI	DSIR to IFI	Empire Marketing Board to IFI	Dominions to IFI	Sudan to IFI	India Office to IFI	IFI expenditure
1925–6	2,500	6,841	1,298	—	—	—	—	13,015
1926–7	2,500	5,512	1,462	—	—	—	—	12,823
1927–8	2,500	5,075	961	—	—	—	—	11,362
1928–9	3,500	4,999	511	—	—	—	—	10,685
1929–30	3,562	4,999	695	2,000	—	—	—	13,968
1930–1	3,576	4,963	796	3,918	—	—	—	15,822
1931–2	3,137	5,958	982	1,500	—	—	—	12,232
1932–3	3,525	4,339	850	—	600	50	—	10,021
1933–4	3,553	3,884	266	—	550	50	—	9,373
1934–5	3,688	4,719	150	—	550	50	—	10,411
1935–6	3,825	5,000	150	—	450	50	1,000	10,804
1936–7	3,963	5,000	150	—	700	50	1,000	11,611
1937–8	3,853	5,000	150	—	256	50	1,000	11,502
1938–9	4,111	5,042	133	—	167	50	800	11,623

Source: University accounts published annually in *OUG*.

point of view of the Forestry Commission and the Colonial Office, they were paying the piper generously and naturally they wished to call the tune more, they suspected that the IFI was subsidizing the school, and generally felt that the University did not take forestry seriously enough. Decisive action was indeed taken by the Governors of the IFI in 1931–2. They reduced expenditure by £3,000 (20 per cent) by cutting some salaries and sacking ten people, including two lecturers who were Oxonians.[97]

It was the prospective bankruptcy of the IFI which precipitated the protracted negotiations between the University, the Forestry Commission, and the Colonial Office, which began in 1930 and were concluded in 1934. In the background lurked the matter of a new building, a question irritating to all parties, and the embarrassingly small recruitment by the school (in 1926–7 as low as thirty-one all told). But the main questions were the relation of the school to the IFI, the control of the IFI, and its funding. The University wished to own any future forestry building, to avoid temporary funding for the IFI which in a financial crisis might be regarded as a university liability, to keep the professor of forestry in charge of the IFI, and above all to replace the IFI's Board of Governors with a new university board which would run both the school and the IFI. The Governors agreed about the undesirability of temporary funding and set about creating an endowment fund; they also began raising money for a new forestry building and by 1931 had secured £30,000 of the estimated £100,000. But, through the Governors, the Colonial Office and the Forestry Commission demanded such a measure of control of the IFI as was consonant with their financial investment in it.[98]

This impasse was disturbed in July 1931 when the report of the Committee on the Training of Forest Officers, set up by the Colonial Office and chaired by Sir James Irvine, was published. The Irvine Report (henceforth Irvine) concluded that the five university forestry schools in Britain were not good enough, that five schools were not needed for such a small profession, and that expanded and improved courses in all five schools would be prohibitively expensive. Accordingly it recommended a reduction in the number of undergraduate schools, and that future forest officers should do three years in forestry at a recognized university or three years as honours arts or science undergraduates, followed by at least one year of postgraduate training at the IFI which should remain in Oxford and in some way attached to the

---

[97] HCP 158 (1931), 10, 248; 159 (1931), 274; OUG 62 (1931–2), 285; 63 (1932–3), 346; Forestry Institute, 1924–32, OUA, UR/SF/F1/2, Troup memorandum, 22 Apr. 1925, 24 Oct., 7 Nov. 1930, 16 May, 9 Oct. 1931. The sacked senior staff members were: Graham Vernon Jacks (1901–77), third in chemistry 1926, a soil scientist who became deputy director of Imperial Bureau of Soil Science 1931–46, director of Commonwealth Bureau of Soils 1946–66; and Richard Frederick Nevill Aldrich-Blake (1901–72), first in botany 1923, a plant physiologist who worked on root systems, who ended his career as biology master, Stowe School, Buckingham.

[98] HCP 148 (1930), 174; 149 (1931), 9–17.

University. Though the impending closure of Cambridge's forestry school in 1933 was good news, some of the proposals made by Irvine were unwelcome to the University. Irvine wished the IFI not to be controlled by Oxford: it wanted bigger representation of appropriate governments and of other universities, with a reduced Oxford input, on a reconstituted Board of Governors. Irvine urged that, as several universities would send graduates to the IFI, it should stand in the same relation to them all and not be identified with any particular school of forestry; that meant a separate and full-time director for the IFI would be needed soon. Not surprisingly Irvine wanted an IFI independent of the University but enjoying beneficial relations with it. The dominant thrust of Irvine undercut Troup's position that the IFI was a lodger in the school: given the external funding of the IFI and the interchangeability of staff and resources between the school and the IFI, Irvine proposed to end a situation in which financially the school was surreptitiously parasitic on the IFI and its sponsors.[99]

In response to Irvine and after discussion with Lord Clinton, chairman of the IFI's Governors, the Hebdomadal Council took a strong line in late 1931. Knowing that in 1931 of all years the government would not seriously contemplate the expense of moving the IFI elsewhere, it told the Colonial Secretary in writing that if it remained a university institution the University had to have complete control over its funds and curriculum. In reply the Colonial Secretary stressed that the school and the IFI were distinct and should be self-contained and that a publicly funded IFI had to be controlled by the contributing departments of government.[100] By late 1932 little progress towards agreement had been made on what the Colonial Secretary regarded as 'a concern of high imperial importance'. The University still wanted close, immediate, and detailed control by a university body on which government departments and other universities with forestry schools (Edinburgh, Bangor, and Aberdeen) would not be represented. It still opposed the Colonial Office's scheme of a separate director of the IFI appointed by the contributing governments.[101]

By early 1934 a compromise had been reached. It gave some authority to the University while acknowledging that there had to be safeguards for the present and potential financial contributors to the IFI. The precise arrangements were that the IFI was brought under the control of the University and the IFI Governors were replaced by an IFI Committee. This advisory IFI Committee, on which outside representatives were in a majority, was made

---

[99] *The Training of Candidates and Probationers for Appointment as Forest Officers in the Government Services: Report of a Committee Appointed by the Secretary of State for the Colonies, July, 1931* (Colonial Office No. 61), esp. 35–40, 48, 51–4; Sir James Colquhoun Irvine (1877–1952), vice-chancellor of University of St Andrews, was chairman of Forest Products Research Board 1927–9.

[100] *HCP* 150 (1931), 187–90; 152 (1932), 31–2.

[101] *HCP* 153 (1932), 203–7; 155 (1933), 41–8 (quotation 46).

subordinate to a Committee for Forestry which was predominantly a university committee.

Troup was made head of the new department of forestry, which included both the school and the IFI; but the university statute did not state that the professor should be head of the IFI. Indeed provision was made for the appointment of a separate director, partly in response to the Forestry Commission's threat of November 1933 to withdraw its support unless such a post were created. At the same time the University promised to provide a site for a new forestry building, the cost of which would be paid for externally. These arrangements, passed by Congregation on 13 February 1934, were not universally popular in Oxford. Though Lindsay argued in Congregation that the University should be proud to be an imperial centre for forestry, a subject which was part of the art of statesmanship for which Oxford was famed as a nursery, his opponents remained unconvinced. Six leading Oxford scientists (Goodrich, Hinshelwood, Lindemann, Peters, Sidgwick, and Townsend) objected to handing over to the Colonial Office the largest and most valuable site in the Science Area. Others, like Southwell, argued that the new arrangements gave only the appearance of control by the University and that the funding of the IFI by external bodies, especially on a year-to-year basis, gave excessive power to them.[102]

This last objection was prescient. The Forestry Commission quickly exploited the new IFI Committee to impose its view that a separate director of the IFI was necessary: in summer 1934 it threatened to withdraw its grant of just over £3,500 p.a. to the IFI and school unless Troup relinquished the directorship in August. Though Troup objected strongly to the prospective loss of part of his empire, Veale reported to the vice-chancellor that the Forestry Commission, even if bluffing, had some telling arguments up its sleeve: though Troup was undeniably a first-rate scientific forester, it had to be conceded that he was a bad administrator; and the Commission, which suspected that the University was financing the school from the funds of the IFI, did not wish to be taken for a financial ride in the same way as the Ministry of Agriculture had been deceived by Owen, an officer of the University.[103] The Forestry Commission had a point. A university document of 1934 revealed that in 1934–5 twelve forestry staff, including secretaries, were paid all told £6,654, of which £4,604 came from the IFI. Indeed three university demonstrators (Chalk, Day, and Lloyd) received all their salary not from the University but from the IFI; and of the £3,175 spent on the

[102] Veale memorandum, 27 Nov. 1933, in Forestry Institute. Relations with University. Jan. 1933–June 1934, OUA, UR/SF/F1/2/1/1; *The Times*, 14 Feb. 1934; *OM* 52 (1933–4), 362, 396–9, 402, 432–3, 459–60, 467, 492–3; Goodrich *et al.*, printed flysheet for Congregation, 13 Feb. 1934; details of forestry statute, *OUG* 64 (1933–4), 393–5.

[103] Veale memorandum, 25 June 1934; Veale to vice-chancellor, 11 July 1934; Troup memorandum, 12 July 1934, all in Forestry Institute. Relations with University. June 1934–Sept. 1937, OUA, UR/SF/F1/2/1/2.

salaries of university demonstrators no less than £2,525 came from the IFI.[104] Faced with these facts, Veale played for time in negotiations with the Forestry Commission and by spring 1935 had secured from the Commission a promise of regular funding provided a separate director of the IFI be appointed. J. N. Oliphant, fresh from the Malayan Forest Service, was duly appointed director of the IFI from January 1936 until July 1939 on an experimental basis, with Troup continuing as professor of forestry in charge of the very small school which in the mid-1930s attracted less than ten undergraduates a year.[105]

While all this squabbling was going on, teaching and research in forestry at Oxford did not stop. The undergraduate course was very wide-ranging in subject matter and designed to enable a man to work in colonial forests often single-handed. It embraced a variety of pertinent sciences and informally it was supplemented by the sapper work of the Officer Training Corps which most forestry undergraduates joined. Field-work beginning with two afternoons a week in Bagley Wood near Oxford with the renowned Tansley was important. In the vacations there were tours of British and Continental forests. The culmination of the practical work was the forest working plan project which was required to be of such a standard that it could be put into practice and it had to cover 100 years. Though forestry was a pass school, it was probably the most comprehensive undergraduate course in pure and applied sciences in the University: from the undergraduate perspective it offered education and training for an inviting career as a teak-wallah.[106] Yet its opponents could depict it as too technical, too vocational, and not sufficiently specialized.

In research the staff were not undistinguished; this needed to be so in order to compensate for the low recruitment of the IFI (fourteen students all told 1934-5) and the school. In the 1920s Troup had seen that it was desirable to produce scientifically trained forest officers for careers in other parts of the empire besides India. He had broadened the teaching and introduced ecology, physiology, and pathology of trees. Similarly in research the focus had been broadened and a group of active researchers assembled. Sometimes they collaborated, sometimes they worked solo; many of them, even if university demonstrators, were paid mainly or wholly by the IFI, a fact ignored by Troup in his repeated claim within the University that the IFI owed its success to the school. He and his staff covered eight areas. Troup

[104] Data for meeting of Committee for Forestry, 21 Nov. 1934, UR/SF/F1/2/1/2. Laurence Chalk (b. 1896), reader in wood anatomy 1955-63; William Robert Day (1893-1967); Arthur Henry Lloyd (1893-1967), obituary in *The Times*, 30 Nov. 1967.
[105] Veale to Forestry Commission, 29 Mar. 1935, Forestry Commission to Veale, 15 Apr. 1935, UR/SF/F1/2/1/2; John Ninian Oliphant (1887-1960).
[106] For informative details about the forestry course I am indebted to notes supplied by John Edward Garfitt (b. 1914), president of University Forestry Society 1934-5, distinction BA in forestry 1935.

was a world expert on silviculture, Ray Bourne a man of European reputation on forest management, while Hiley had diversified from fungal diseases to woodland management. Day published on tree diseases and injuries, with Chalk and Burtt-Davy specializing on African trees via anatomy and botany respectively. Chrystal was a ranking expert on forest entomology, while Lloyd covered forest engineering.[107] Some of this work was published in the Oxford Forestry Memoirs series began in 1922 under Troup and in the Oxford Manuals of Forestry begun in 1928. Only half of these vigorous publishers had been undergraduates at Oxford (Bourne, Day, Hiley, Lloyd) and none was a college fellow, so that forestry for both undergraduates and staff generated its own considerable *esprit de corps* outside the colleges' pale.

The period from January 1936 to the outbreak of war was one of turbulence: the system of dual control was unworkable, with Troup and Oliphant disagreeing about everything; and problems concerning the undergraduate school and the IFI became so acute that the University was on the point of closing down forestry altogether. In the event, after protracted negotiations with the Colonial Office and acrimonious exchanges with the Forestry Commission, by autumn 1939 the degree course in forestry had been revamped, the school effectively abolished by being merged with the IFI, and the directorship and committee responsible for the IFI both abolished. At the personal level Oliphant, whose experimental period of office as director of the IFI was renewable from 1 August 1939, decided that he preferred the hot forests of Nigeria to the damp miasmas of Oxford. Troup was due to retire in 1940; but, worn out by dedicated overwork and the stress caused by the system of dual control which he opposed bitterly, he died in October 1939 just two months after the directorship of the IFI had been merged with his chair. His untimely death prevented him from enjoying his reunited empire.

The problems of Oxford forestry were exacerbated by the sustained hostility of Troup and Oliphant to each other. Troup regarded Oliphant as a wrecking usurper, did not consult him, wanted rid of him, and told Veale that Oliphant's staff hated him. For his part Oliphant thought that Troup wished to continue exploiting the IFI without giving anything in return. Their inability to agree on anything made any change in forestry teaching or

[107] A. H. Lloyd, *Engineering for Forest Rangers in Tropical Countries, with Special Reference to Burma* (Oxford, 1929); R. Bourne, *Regional Survey and its Relation to Stock-Taking of the Agricultural and Forestry Resources of the British Empire* (Oxford, 1931); W. E. Hiley, *The Fungal Diseases of the Common Larch* (Oxford, 1919); id., *The Economics of Forestry* (Oxford, 1930); id., *Improvement of Woodlands* (London, 1931); W. R. Day, *The Experimental Production and the Diagnosis of Frost Injury on Forest Trees* (Oxford, 1934); J. Burtt-Davy, *A Manual of the Flowering Plants and Ferns of the Transvaal with Swaziland, South Africa* (London, 1926, 1932); R. N. Chrystal, *Insects of the British Woodlands* (London, 1937). Ray Bourne (1889–1948), fourth in botany 1911; Wilfred Edward Hiley (1886–1961), first in botany 1908; Joseph Burtt-Davy (1870–1940), active in the Oxford Preservation Trust, president of Natural History Society of Oxfordshire; Robert Neil Chrystal (d. 1956).

research an acrimonious matter. Even during Troup's final illness in summer 1939 there was friction: the Colonial Office and Oliphant wanted Chalk to be Troup's deputy, while Troup himself wanted Day whom Oliphant opposed. In the end Troup's duties were split between Chalk and Day.[108]

Oliphant was quick off the mark in trying to show the University that the IFI under Troup was inadequate. In April 1936 he reported the results of a referendum he had organized on the IFI. He stressed that the Forestry Commission, which was empowered to promote forestry in the dominions, had felt it had been ignored by the IFI of which it took a low view. Exploiting this sort of external criticism, Oliphant argued that the IFI was narrow in outlook because dominated by staff who had worked in the Indian Forestry Service; that it was unfocused because it strayed into systematic botany; that it did not act as a clearing house for which there was great demand; and, above all, that it had little contact with the imperial territories it was supposed to serve.[109]

By early 1937 the institutional conflict between Troup and Oliphant led Bourne, Troup's longest-serving demonstrator and the backbone of the school, to resign. The particular issue on which they disagreed was the making of forestry working plans. For twelve years Bourne had dealt with half of the forestry officers who had attended the IFI from the colonies, the dominions, and India. A key feature of his teaching was his belief in comprehensive long-range planning, based on scientific and economic surveys and the preparation of working plans. He took the Indo-Continental view that forms of forest management suitable to Europe and India were *ipso facto* applicable to other parts of the world. Bourne was concerned with the supply side of forestry so the basis of his teaching was the elaborate working plan using a series of examples, based on European species, worked out in detail over a very long period of time under a variety of conditions. Therefore for Bourne general forestry education at Oxford, for pupils who mostly pursued careers in the colonies and dominions, could not be directly or exclusively concerned with forest practice outside Europe. In contrast Oliphant, a specialist on tropical forestry, regarded working plans as outmoded and out of place except for densely populated areas. He tended to focus on the demand side of forestry so he believed in pragmatic planning which constantly underwent adaptation as new forestry techniques were evolved and applied. In Bourne's view Oliphant's denigration of his beloved working plans was the leading example of the way in which education in general forestry was being subordinated to the short-term aims pursued by

[108] Veale to Furse, confidential, 25 Mar. 1938, Veale to vice-chancellor, 28 Mar. 1938, Forestry Institute. Relations with University, Oct. 1937–Apr. 1938, OUA, UR/SF/F1/2/1/3; Veale to Colonial Office, 15 Mar. 1939, Veale to Lindsay, 28 June 1939, Forestry Institute. Relations with University, Jan. 1939–Dec. 1945, OUA, UR/SF/F1/2/1/5.

[109] Oliphant, report 22 Apr. 1936, note 3 May 1936, UR/SF/F1/2/1/2.

the IFI's new director. Indeed Bourne called on the University to inquire into the difficulties posed by the system of dual control which, it should be noted, preserved the interchangeability of staff and valuable resources between the school and the IFI.[110]

Bourne played into his enemy's hands unwittingly. His resignation, announced in January 1937, was used by Oliphant, in collusion with Veale, to secure a full inquiry by a university committee into Oxford forestry. Oliphant and Veale knew that the Colonial Office was reviewing forestry education in general and that it opposed Troup's ambition to elevate forestry at Oxford to an honours school. Oliphant and Furse of the Colonial Office agreed that the desired education for the colonial forestry services was a good honours degree, not in forestry but in an allied science or sciences, followed by vocational training in forestry for two years. The Oliphant–Furse scheme effectively destroyed the undergraduate school, a fate to which Lindsay, the vice-chancellor, was not unsympathetic provided the IFI continued to be funded externally and provided that aspects of scientific forestry could be taught in appropriate departments (such as botany and entomology). Lindsay took advice from Furse about the effects of the possible closure of the school, while simultaneously asking Troup to make a case for an honours school of forestry which he (Lindsay) would then use as an occasion for discussing the abolition [sic] of the school, without irritating Troup's feelings. Troup duly obliged in April 1937 with a memorandum advocating the elevation of forestry to an honours school and attacking the Colonial Office–Oliphant scheme as not giving a proper grasp of forestry and taking too long. Next month the University set up a committee to report on the problems of forestry at Oxford.[111]

By late 1937 it was clear from confidential advice given by Furse that the Colonial Office favoured the abolition of the school, with the teaching of forestry to be provided mainly by the IFI. This Colonial Office view was adopted by the university committee of inquiry which reported in January 1938.[112] It agreed with the Colonial Secretary that the most suitable training for a forester was an honours course in science, preferably botany, followed by two years' professional training. In the long run this proposal would reduce the number of men entering forestry as a job through an undergraduate forestry course. The reviewing committee therefore concluded that the school was unlikely to be viable and should be merged with the IFI

[110] Bourne to Troup, 13 Jan. 1937; Bourne memorandum on forestry education, 30 Jan. 1937; Oliphant memorandum on working plans, 2 Feb. 1937, all in BS/R/1/4; *HCP* 166 (1937), 219–22.

[111] Veale memoranda 6 Feb., 3 Mar. 1937; Veale to Furse, 22 Feb. 1937; Oliphant to Veale, 28 Feb. 1937; Troup memorandum, 'The Future of the Oxford School of Forestry', 21 Apr. 1937, all in UR/SF/F1/2/1/2.

[112] Veale to Furse, 15 Nov. 1937; Furse to Veale, 2 Dec. 1937, in UR/SF/F1/2/1/3; *HCP*, 169 (1938), 85–97.

which would be responsible for all forestry teaching for all types of student (undergraduates, graduates in forestry from Oxford and elsewhere, honours science graduates from Oxford, and serving foresters taking refresher courses). Such amalgamation had the convenient consequence of solving the problem of dual control. The investigating committee had consulted Sir James Irvine who, contrary to his report of 1931, now agreed that under dual control the school and IFI had drifted apart and that efficient co-operation between them required undivided leadership. The committee hinted that Troup should be made head of the merged school and Institute; and it hoped that the IFI would concentrate more on research and less on acting as an information bureau.

The proposed amalgamation provoked a vigorous response from Troup. It seemed to him to be a submergence, and not a merger, of the school in the IFI; there was still uncertainty about the financing of the IFI by the Colonial Office and Forestry Commission; and, instead of being regarded as a genuine branch of university learning, forestry at Oxford would be looked on as a subject controlled by government interests and finance. In order to secure some measure of university control of the combined school and IFI, Troup wished it to be called a department, the term 'institute' being applicable only to research institutes. For some time the term 'department' had been in use to denote the school and the IFI together: the deliberations about the site for a new forestry building had referred to the department of forestry and not just to the IFI. Now in early 1938 terminology became vital: the term 'department' would preserve the close connection with the University, in fact as well as in name, by ensuring that forestry would be treated like other science departments in that the University would be responsible for its policies and academic standards.[113]

In spring 1938 the Imperial Agricultural Bureaux came as a *deus ex machina* to the University's aid by offering to pay for the bureau work which the IFI had been doing for a couple of years, thus leaving it to concentrate on research and the teaching responsibilities envisaged for it. By April 1938 the University agreed to accept an Imperial Forestry Bureau, attached to the IFI, run by its director, paid for by the Imperial Agricultural Bureaux, and controlled by it. The IAB had opened its first bureaux in 1929, funded by Commonwealth governments and usually loosely attached to a research institute, to act as clearing houses. Oxford was a natural home for a forestry bureau, thought the IAB, because the IFI was situated there. The Imperial Forestry Bureau at Oxford was an independent body which rented offices at £100 p.a. from the University.[114]

---

[113] *HCP* 169 (1938), 255–7, 270–3.
[114] Ibid. 169–71; D. Chadwick to Veale, 4 Feb. 1938, UR/SF/F1/2/1/3; Sir David Thomas Chadwick (1876–1954), secretary of Imperial Agricultural Bureaux 1928–46.

Meanwhile the report and recommendations of the investigating committee, and supporting documentation, were sent unofficially in March 1938 to the Colonial Office, but not to the Forestry Commission which in 1937-8 contributed just over £4,000 to Oxford forestry. Having been invited by the Colonial Office to give its view, the Forestry Commission did so in April 1938 with decided pique. It drew the University's attention to the great disparity between the costs of the IFI to the Commission and the results accruing to British forestry. Since 1920 the Oxford school had produced 245 graduates, of whom only 12 worked for the Forestry Commission. At the IFI the position was worse: out of a total of 306 who had taken courses in it, only 5 had found their way to the Forestry Commission. Thus the Commission, which had contributed almost £50,000 to Oxford forestry up to 1938, had recruited on average just one Oxford forester of some kind per year since 1920 to work for it. Furthermore, the chairman of the Forestry Commission, Sir Roy Robinson, disliked the university/Colonial Office proposals which involved devolving the teaching of the scientific aspects of forestry to such departments as botany and entomology to which forest botanists and a forest entomologist would be transferred. In his view forestry education should avoid teaching basic sciences such as botany and concentrate on applying these sciences to the problems of forestry. He suspected that the transferred forest botanists and the entomologist would not work full-time at forestry and would not be concerned to go into forests regularly in order to appreciate practical problems. He therefore wanted such staff to work under the director of the IFI, as part of his ideal scheme of a self-contained IFI made up of a team of twelve full-time forestry specialists. Generally Robinson thought that his own University did not take forestry seriously, and he resented the way in which the university committee of inquiry had disregarded the Commission's views. He decided, therefore, to give the University a jolt by threatening it financially. As the University seemed to wish to abolish the school, Robinson told it that the grant of £500 p.a. to it would cease automatically; he also revealed formally that the recurrent grant of around £3,500 p.a. to the IFI would end in September 1939, and that from October 1939 Oxford forestry (in the form of the merged school and IFI) would be treated like the other forestry departments in British universities and receive £500 p.a. all told.[115]

The University had drawn up its forestry scheme largely in response to the Colonial Office's wishes and then found by April 1938 that the Forestry Commission was bitterly opposed to it: the Colonial Office wanted forestry training to be preceded by the study of pure sciences such as botany, whereas the Commission thought that forestry should always be the main focus and subjects such as botany were hares that should not be chased. By threatening

[115] *HCP* 170 (1938), 248-9, 265-73; Robinson to Veale, 25 Mar. 1938, UR/SF/F1/2/1/3.

to reduce its current contribution to Oxford forestry of £4,000 p.a. all told to no more than £500 p.a. the Commission was destroying the scheme put forward by the University in harness with the Colonial Office.[116] At this juncture, Lindsay, Veale, and Oliphant decided that they would not kowtow to the Commission. They would stand up to Robinson and the Commission using some influential government personage to hold the ring, pillory the Commission's opposition to the introduction of more science into forestry training, ridicule its 'outmoded empiricism', stress that if the IFI crashed the responsibility would lie fairly and squarely with the Commission, and threaten to close the IFI by autumn 1939 and forestry altogether by summer 1940. If this approach did not work, then the University would be released from the worries of forestry because it would be abolished, not by the University but by the Commission which would incur odium on that account. Veale described this second approach as letting the Commission upset the applecart, letting the University pick up good apples (the scientific side), and leaving the rotten ones (the technical side) to be swept away.[117]

This two-pronged approach was supported by the influential Sir James Irvine who tried to soften Robinson's obduracy and by late April had succeeded to some extent. At the same time Lindsay implored his fellow Oxonian Ormsby-Gore, the Colonial Secretary responsible for the new scheme of recruitment to the colonial forest services, to control the bad behaviour of the Forestry Commission and of its chairman. The way out of the impasse was indeed provided by the Colonial Office via Furse who suggested that the Colonial Secretary, Lindsay, and Robinson should meet. At the meeting held in late May 1938 in the rooms of MacDonald, the new Colonial Secretary and a loyal Oxonian, Robinson was assured that more attention would be given to British forestry and he was persuaded to withdraw his threat of a big cut in funding. By mid-June 1938 the new financial arrangements for Oxford forestry had been agreed: for five years from August 1939 the Colonial Office would maintain its present grant of £5,000, while the Forestry Commission would pay a reduced grant of £1,750 p.a.[118]

The rest was plain sailing for the University. By May 1939 the new degree regulations had been approved. They raised the standard of the qualifica-

[116] *HCP* 170 (1938), 274–5; Lindsay note, 20 Apr. 1938, UR/SF/F1/2/1/3.
[117] Veale/Oliphant notes, n.d. [Apr. 1938] in response to Forestry Commission letter, 13 Apr. 1938, UR/SF/F1/2/1/3.
[118] Irvine to Veale, 23 Apr. 1938; Lindsay to Ormsby-Gore, private and confidential, 26 Apr. 1938; Robinson to Lindsay, 29 Apr. 1938, all in UR/SF/F1/2/1/3; Veale to Furse, 12 May 1938, Veale memorandum, 26 May 1938, in Forestry Institute. Relations with University, May 1938–Dec. 1938, OUA, UR/SF/F1/2/1/4; *HCP* 170 (1938), 250–2, 275–6; William George Arthur Ormsby-Gore (1885–1964), second in history 1907, Colonial Secretary 1936–May 1938; Malcolm John MacDonald (1901–81), second in history 1923, Colonial Secretary June–Nov. 1935, May 1938–40.

tions required to take finals in forestry: prospective finalists had to have passed science moderations at Oxford or had to have an honours degree in science. Although forestry still ranked as a pass school, which satisfied those who denigrated vocational subjects, the route to it was through honours of some sort in science.[119] Though Troup liked the honours aspect, he would have liked less science and more forestry in the new degree regulations; and, in his last public word on the subject, he reminded the University that a good forester needed not only decent academic qualifications but also 'physical energy, self-reliance, administrative ability, and capacity for handling subordinates and getting work done'.[120] The new regulations also merged the school and IFI to create a single department of forestry under the control of the professor on a day-to-day basis, with policy being decided by a new forestry committee of nine members of whom six were Oxonians, one from the Colonial Office, one from the India Office, and one from the universities of Edinburgh, Wales, and Aberdeen in turn.[121] Though Troup never enjoyed his regained power over forestry as a whole, up in the Elysian fields he rejoiced no doubt that his successor, Champion, was like himself an old Schlich pupil and an experienced Indian jungle man as well as a double Oxford first in pure sciences.[122]

By 1939 Oxford forestry was in some ways bizarre. Though the IFI was its core, the staff were all British. Though the IFI proclaimed its imperial scope, it was located in Oxford which was not at the centre of a great forest. Nor did Oxford possess a great or long tradition of forest management.[123] Yet forestry, which was widely despised at Oxford for being both vocational and merely a cluster of applied sciences, endured for several reasons. First the IFI was acceptable because it was a national and indeed imperial centre and therefore slotted easily into two established visions Oxford had of itself. Secondly, the case of forestry showed that the University was becoming somewhat embarrassed about its distancing of vocational training. As Veale noted in 1939, 'the great universities cannot cut themselves off from all kinds of vocational training.... The function of the University [of Oxford] is to see that vocational training is made as liberal as circumstances permit.'[124] Thirdly, if government bodies chose Oxford in preference to other universities to give vocational training, the University could claim that it no longer lived exclusively in an ivory tower; this was a useful point to make in negotiations with the University Grants Committee whose recurrent

[119] *OUG* 69 (1938–9), 686.
[120] R. S. Troup, 'Forestry at Oxford', *OM* 57 (1938–9), 676–8 (quotation 678).
[121] *OUG* 69 (1938–9), 181–2.
[122] Harry George Champion (1891–1979), first in chemistry 1912, first in botany 1913, professor of forestry 1940–59.
[123] R: Symonds, *Oxford and Empire: The Last Lost Cause?* (London, 1986), 132–9.
[124] Veale memorandum, 1 Oct. 1938, UR/SF/F1/2/4.

government grant was becoming central to the University's finances. Lastly, though the merged department of forestry was in the main externally funded, the control of the curriculum and of appointments had been secured by the University after protracted squabbles with both the Colonial Office and the Forestry Commission. In sharp contrast with the fate of the Agricultural Engineering Institute, forestry showed by 1939 the 'arm's-length' principle in action, namely, that a government body which wanted a job done by a university, having selected that university most likely to do the job well, paid over its grant and did not seek to interfere in how it was spent. Such a principle was agreeable to Oxford: it took the money while preserving its jealously guarded academic independence.

5
Biomedical Sciences

5.1 The Dominance of Physiology

The career of medical sciences at Oxford between the wars cannot be understood without taking into account the peculiarities of the local context. Though the medical faculty had been established in 1885, its teaching and research had little to do with the care and cure of patients in local hospitals, of which the best known was the Radcliffe Infirmary, the only general hospital in the city. As so often at Oxford, the vocational element in teaching was minimized: the aim of the medical school was to give a broad and liberal education in biomedical sciences to a relatively small number of undergraduates. When the pre-clinical period, which lasted on average about five years, was completed, Oxford undergraduates usually headed to the wards of London hospitals for their three years of clinical training where they learned the techniques of practice. On occasion in the 1920s a few undergraduates spent a year at the Radcliffe doing bugs and drugs as part of their second MB work before proceeding elsewhere for their clinical courses. Generally speaking the Oxford medical school distanced itself from clinical research and training and had little contact with the Radcliffe Infirmary, until Lord Nuffield's medical benefactions of the mid-1930s. In the 1920s the medical school freely confessed that Oxford was not a great clinical centre and that clinical research hardly existed, but it proudly proclaimed that Oxford offered exceptional opportunities for special study and research in medical science.[1]

In 1930 Ainley-Walker, Oxford's first college fellow in medicine, acknowledged that pre-clinical scientific studies pursued at Oxford tended to be divorced from actual clinical experience gained mainly in the London hospitals. Nor was there any possibility, in his view, of marrying medical sciences and practice in Oxford: the city had too small a population and an impossibly large benefaction would be needed. That largess was provided in the mid-1930s by Lord Nuffield, first in 1935 with his Nuffield Institute for Medical Research which was located adjacent to the Infirmary on

---
[1] K. J. Franklin, 'A Short Sketch of the History of the Oxford Medical School', *Annals of Science*, 1 (1936), 431–46; Dewhurst (ed.), *Oxford Medicine; University of Oxford; Information Concerning the School of Medicine, Medical Degrees and Diplomas, and Post-graduate Medical Study and Research* (Oxford, 1922, 1926).

the site of the Radcliffe Observatory which he had previously bought. This turned out to be small beer compared with the medical benefaction he made to the University in 1936 for the development of clinical medicine, especially via research; by 1939 it amounted to £2,500,000. His generosity enabled four full-time clinical professors (Cairns, surgery; Macintosh, anaesthetics; Moir, obstetrics and gynaecology; Witts, medicine) to be appointed in 1937. Inevitably much of their effort in the limited time before the Second World War was spent on providing extra beds at the Infirmary and launching research but they also oversaw appointments in morbid anatomy, clinical pathology, clinical bacteriology, and clinical biochemistry. Though the Nuffield benefaction was in part intended to bring scientific and clinical medicine into fruitful partnership, for some time it did the opposite: the separation of scientific and clinical medicine was exacerbated and not reduced.[2] The biomedical scientists, especially Howard Florey, were jealous of their Nuffield clinical colleagues mainly because by 1937 it had become clear that most of the Nuffield benefaction would be spent on clinical and not pre-clinical medicine. At the same time, grant-giving bodies outside the University tended to assume that pre-clinical medical sciences were beneficiaries of the Nuffield benefaction. Not surprisingly there were tensions between the Nuffield clinical professors and those responsible for pre-clinical medical sciences. Though Sir Farquhar Buzzard, the Regius professor of medicine, wanted clinicians and their scientific colleagues to be more closely associated in their daily work, such collaboration did not occur as a result of the Nuffield bequest: the pre-clinical professors felt their status was threatened, they were deprived of funds previously promised to them, they envied the higher salaries received by most of the Nuffield clinical professors, and they resented the way in which the Nuffield clinicians built their own departmental empires at the Radcliffe Infirmary and then, to add insult to injury, lured trained staff from the pre-clinical departments by offering higher pay. The pre-clinical laboratory sciences had what might be called credit, but the Nuffield clinical enterprises had the cash. In the last session before the war the running expenditure by the Nuffield medical benefaction amounted to £53,000, which was

[2] C. Webster, 'Medicine', in B. Harrison (ed.), *The History of the University of Oxford*, viii: *The Twentieth Century* (Oxford, 1994), 317–44; Hugh William Bell Cairns (1896–1952), professor of surgery 1937–52, on whom see G. J. Fraenkel, *Hugh Cairns: First Nuffield Professor of Surgery, University of Oxford* (Oxford, 1991); Robert Reynolds Macintosh (1897–1989), professor of anaesthetics 1937–65, on whom see J. Beinart, *A History of the Nuffield Department of Anaesthetics, Oxford, 1937–1987* (Oxford, 1987); John Chassar Moir (1900–77), professor of obstetrics and gynaecology 1937–67; Leslie John Witts (1898–1982), professor of clinical medicine 1937–65; E. W. Ainley-Walker, 'The School of Medicine', *OM* 48 (1929–30), 564–8; Ernest William Ainley-Walker (1871–1955), tutor in medicine, University College, 1903–39; A. D. Gardner and M. J. Stewart, 'Ainley-Walker', *Journal of Pathology and Bacteriology*, 71 (1956), 239–46.

almost half the University's annual income from the University Grants Committee.[3]

It was the Second World War which began to remove the suspicions that existed between the Nuffield clinical professors and the university pre-clinical medical scientists, not to mention those between the ordinary clinicians at the Radcliffe and the Nuffield professors. The London teaching hospitals became unavailable to Oxford medical undergraduates so they were taught as a *pis aller* at the Radcliffe Infirmary. But before then Oxford differed from provincial university medical schools in that its teaching in Oxford had little connection with clinical practice and clinical experience was not widely available locally. Oxford's medical school also differed from that at Cambridge which was on average three times as big. As in the case of engineering, Oxford's aim in medical sciences was not to compete with either Cambridge or the big battalions of the provincial universities: it was to remain small, distinctive, and distinguished, offering what it believed to be the highest type of education in medical sciences which would naturally produce not rank-and-file practitioners but leading figures in the higher walks of the profession. As so often at Oxford, subjects which were enabling in various ways were preferred to those which were directly vocational. In medicine that view had been adumbrated in the late 1870s by Acland, who saw clinical teaching as dangerous: it would undermine scientific research, flood the university with clinically oriented undergraduates who, obsessed with mere technique and information, would be impervious to humanistic learning, and drive Oxford undergraduates preparing for a medical career 'out of the ranks of literary, historical, or philosophical culture'.[4] The development of a complete medical school at Oxford would, it was claimed by the Aclandites, deprive the University of its unique strength in medical sciences which were taught with a thoroughness and breadth of view, not available in hospitals, which helped to produce the cultured minds so prized in Oxford. Of course large numbers would vitiate these aims so it is not surprising that on occasion between the wars the medical school tried to reduce its intake in order to enhance the already high quality of undergraduates and to preserve the cherished Oxford tradition of individual tuition. In 1936, for example, the Medical Faculty Board was so keen to reduce the intake into medicine from 60 to 45 that it contemplated the inaguration of a quota system agreed by the colleges, an increase of departmental fees, and the introduction of a preliminary examination to weed out weaker candidates.[5]

[3] Howard Walter Florey (1898–1968), professor of pathology 1935–62, Nobel prize-winner 1945, president of the Royal Society of London 1960–5; Edward Farquhar Buzzard (1871–1945), Regius professor of medicine 1928–43; for the scope of the Nuffield benefaction by summer 1939 see *OUG* 70 (1939–40), 377–92.

[4] Atlay, *Acland*, 397; Henry Wentworth Acland (1815–1900), Regius professor of medicine 1858–94.

[5] Med./R, 12 Oct. 1936, n.d. [Nov. 1936], n.d. [Jan. 1937].

The unique feature of medicine at Oxford between the wars was the dominance of physiology, a tradition that went back to Acland who thought that the subject was like the book of the gospels. Between the wars physiology at Oxford was justified for giving to its students intellectual discipline and training in research methods, both of which (it was alleged) were invaluable in the advancement of science and the art of medicine.[6] Accordingly almost all medical undergraduates took finals in physiology, the only medical subject in which honours were available. If they did well in finals physiology it gave them the BA degree which was necessary for proceeding to the BM courses and excused them from taking physiology in the BM. This feature of the examination regulations helped physiology to overhaul chemistry by the late 1920s as the biggest final honours school in pure science, with around 50–60 graduates a year in the 1930s (see Tables 2.1 and 2.2). Of the remaining four subjects in which there were chairs, biochemistry was taught as part of the physiology degree course and it also became available in 1930–1 as a supplementary subject in finals for chemistry undergraduates; but human anatomy, pathology, and pharmacology were not separate final honours schools. Even the redoubtable Florey failed to persuade the University to elevate pathology to that status in the 1930s.[7] In parallel with their relatively high rank as final degree subjects, physiology and biochemistry belonged to the biological sciences faculty as well as to the medical faculty, whereas human anatomy, pathology, and pharmacology belonged only to the latter. The lecture and laboratory courses in physiology and biochemistry available in the former faculty were almost all that was on offer in these two subjects to medical undergraduates. Thus physiology and biochemistry were studied by medical undergraduates as pure biological sciences. As Franklin noted, Sherrington's famous course in mammalian physiology started many Oxford men on the way to original research and gave nearly all medical students 'a unique insight into the normal workings of the animal body'.[8]

Of course Oxford's school of physiology under Sherrington produced well-known physicians such as Scott Bodley, famous surgeons such as Porritt, and leading administrators such as Godber. It also generated academic physiologists, including three of Sherrington's successors as professor (Liddell, Brown, and Whitteridge) and a couple of Nobel prize-winners (Granit and Eccles).[9] Oxford physiologists also permeated other biomedical

[6] *BMJ* (1936), 1: 1064–6; R. S. Creed memorandum, Med./R, 10 Feb. 1931.
[7] E. P. Abraham, 'Florey', *BM* 17 (1971), 255–302 (276).
[8] Franklin, 'Medical School', 441.
[9] Ronald Scott Bodley (1906–82), *DNB*, second in physiology 1928, physician to the Queen 1952–73; Arthur Espie Porritt (1900–94), second in physiology 1925, surgeon to the King and Queen 1946–67; George Edward Godber (b. 1908), second in physiology 1930, chief medical officer to the government 1960–73; Edward George Tandy Liddell (1895–1981), first in physiology 1918, professor of physiology 1940–60; George Lindor Brown (1903–71), professor of

departments there. Until the Second World War, and indeed for some years after it, biochemistry was dominated by those who had been trained at Oxford as physiologists. Up to 1939 most of the demonstrators and four out of five college fellows who worked in the biochemistry department (Carter, Queen's, elected 1927; Philpot, Balliol, 1934; Sinclair, Magdalen, 1937; R. H. S. Thompson, University, 1938) had read physiology as undergraduates. The other college fellow (Ogston, Balliol, 1937) had read chemistry at Oxford. The first non-Oxonian biochemist to become a college fellow was Margery Ord who was elected at Lady Margaret Hall in 1954.[10] The close association between biochemistry and physiology was reflected in the physical accommodation for biochemistry: from its birth in 1920 to 1927 it was housed in physiology; from 1927 it enjoyed it own premises which were contiguous to those of physiology, with which it shared a library, a lecture-room, and a tea-room. Contacts between the two departments were close and lasted until the early 1950s when physical separation occurred.[11] Oxford physiologists also penetrated pathology, pharmacology, human anatomy, and the Nuffield Institute, at the professorial level. Florey and William Paton read physiology as undergraduates and demonstrated in physiology before eventually returning to Oxford as professors in 1935 and 1959 respectively. In 1948 Geoffrey Dawes began a long stint as director of the Nuffield Institute for Medical Research. As late as 1975 C. G. Phillips, yet another Oxford-trained physiologist and demonstrator, moved sideways from his chair of neurophysiology at Oxford to that of human anatomy. Clinical medicine at Oxford also benefited from Oxford-trained physiologists: the key medical advisers for the 1936 Nuffield benefaction were Hugh Cairns and Sir Farquhar Buzzard, each of whom had studied physiology at Oxford. The first scientist to be head of a women's college, Janet Vaughan, was an Oxford physiologist.[12]

physiology 1960–7, a Manchester graduate, worked for six months with Eccles in Sherrington's laboratory in 1932, obituary in *BM* 20 (1974), 41–74; David Whitteridge (1912–94), first in physiology 1934, professor of physiology 1968–79; Ragnar Granit (1900–91) studied with Sherrington 1928, 1932–3, professor of neurophysiology at Royal Caroline Institute, Stockholm, 1940–67, Nobel prize 1967; John Carew Eccles, first in physiology 1927, Nobel prize 1963.

[10] Cyril William Carter, first in physiology 1921, fellow of Queen's 1927–65, reader in biochemistry 1948–65; John St Leger Philpot (1907–90), first in chemistry 1929, second in physiology 1930, fellow of Balliol 1934–45, obituary in *Balliol College Annual Record* (1990), 32–4; Hugh MacDonald Sinclair (1910–90), first in physiology 1932, fellow of Magdalen 1937–58, reader in human nutrition 1951–8; Robert Henry Stewart Thompson (b. 1912), first in physiology 1933, fellow of University College 1938–47; Ogston, first in chemistry 1933, fellow of Balliol 1937–59, reader in biochemistry 1955–9, president of Trinity 1970–8; Margery Grace Ord (b. 1927), educated at University College London, Ph.D. (London), fellow of Lady Margaret Hall 1954–88.

[11] M. G. Ord and L. A. Stocken, *The Oxford Biochemistry Department: Its History and Activities* (Oxford, 1990), 6.

[12] Florey, first in physiology 1923, demonstrator 1923–4; William Drummond Macdonald Paton (1917–93), first in physiology 1938, demonstrator 1938–9, professor of pharmacology

The dominance of physiology in the medical school was reinforced between the wars by Charles Sherrington. He had been head-hunted from Liverpool in 1913 and made his Oxford department internationally renowned for its research in neurophysiology. Honours came thick and fast to Sherrington: he was president of the Royal Society 1920–5, president of the British Association for the Advancement of Science and knighted in 1922, and a joint Nobel prize-winner in 1932. His work at Oxford indicated to philanthropic bodies and the outside world in general that world-class medical science could be done there and that any benefactions they made to it were unlikely to be wasted. The 1920s therefore witnessed large endowments of biochemistry and pathology by the Rockefeller Foundation and the William Dunn Trustees respectively. In the 1930s, however, physiology shone less, owing to Sherrington's illness and his subsequent resignation in 1935 aged 77, and there were no spectacular external endowments other than those by Lord Nuffield. The main themes of the decade concern the changes introduced by three brisk and energetic men who assumed chairs in the short space of three years: Clark in human anatomy in 1934, Florey in pathology in 1935, and J. H. Burn in pharmacology in 1937. Each was a determined department-builder with a coherent programme of research which, in the cases of Clark and Florey, soon led to the introduction of an experimental approach to their subject. In addition Florey developed team research which was to be central to the success of the penicillin project during the war; for this work Florey and his co-worker Chain joined Fleming as recipients of a Nobel prize in 1945.[13] This chapter will therefore focus on Sherrington's reign in physiology, the philanthropic benefactions of the 1920s and their effects, and the impact of the new medical triumvirate from the mid-1930s.

The death in 1913 of Francis Gotch, Waynflete professor of physiology, presented a welcome opportunity to two individuals, Charles Sherrington and John Scott Haldane, and to the University. In 1895 Sherrington had been defeated by Gotch in the election to the Oxford chair but he was not unhappy to go to Liverpool that year to occupy the chair vacated by Gotch. At Liverpool Sherrington quickly established an enviable international reputation as a neurophysiologist, his Silliman lectures at Yale University on the integrative action of the nervous system doing for the nervous system what Harvey had done for the circulation of the blood.[14] In contrast

1959–83; Geoffrey Sharman Dawes (1918–96), first in physiology 1939, director of Nuffield Institute for Medical Research 1948–85, obituary in *Independent*, 16 May 1996; Charles Garrett Phillips (1916–94), first in physiology 1938, professor of neurophysiology 1966–75, professor of anatomy 1975–83; Buzzard, fourth in physiology, 1894; Cairns studied physiology 1919–21 but did not graduate in it; Dame Janet Maria Vaughan (1899–1993), first in physiology 1922, principal of Somerville 1945–67, FRS 1979, obituary in *Independent*, 12 Jan. 1993.

[13] Wilfrid Le Gros Clark (1895–1971), professor of human anatomy 1934–62; Joshua Harold Burn (1892–1981), professor of pharmacology 1937–59.

[14] C. S. Sherrington, *The Integrative Action of the Nervous System* (New Haven, 1906).

Haldane was an Oxford resident, having been elected a demonstrator as long ago as 1887, a fellow of New College in 1897, and reader in physiology in 1907. With his colleagues Claude Douglas and John Priestley he had made the Oxford department well known for its highly unusual emphasis on experimental work on intact, normal, living humans, especially under stressful conditions as experienced in certain dangerous occupations. Sherrington and Haldane were both considered for the Oxford chair in autumn 1913, the former being elected unanimously.[15] Not for the first time the University had distanced itself from applied science by rejecting Haldane's emphasis *qua* academic physiologist on applied industrial medicine. For the University the election of Sherrington had two other advantages. He would follow the well-established Oxonian tradition of pure neurophysiology. Secondly the head-hunting of W. H. Perkin from Manchester in autumn 1912, to make Oxford chemistry distinguished via research, was already showing signs of success. A year later Sherrington was lured from the Holt chair at Liverpool to do the same for physiology, the key chair in biomedical sciences.[16] Though Perkin's devotion to industrial research was not shared by Sherrington, they were alike in that each was a middle-aged professor from a northern provincial university (Perkin 52 on appointment, Sherrington 55), each saw research as the prime academic function, and though both were totally new to Oxford their practice of various forms of connoisseurship made them acceptable and even popular in its peculiar environment.

From Burdon-Sanderson, his teacher and mentor, Gotch had derived an interest in electrophysiology. His research work was narrowly focused on one technique, the capillary electrometer used to study the action of nerves and muscle. Highly regarded until about 1900, his research reputation then declined and he was not in a position to impose his own research topic on his department or to give it an identity based on his own work. Instead various subdepartments prospered as both teaching and research units. Walter Ramsden was so successful as a biochemist that he moved to the Liverpool chair in 1914. One of the demonstrators in histology, Gustav Mann, left for a chair in New Orleans, while his successor, Samuel Scott, died in the war having tried to rescue histology from soulless micro-anatomy.[17] Experimental psychology was vigorously pursued by William McDougall. Above

[15] S. W. Sturdy, 'A Co-ordinated Whole: The Life and Work of John Scott Haldane' (University of Edinburgh, Ph.D. thesis, 1987), 271–3.

[16] For the Liverpool context of Sherrington's career there as professor 1895–1913, see T. Kelly, *For Advancement of Learning: The University of Liverpool 1881–1981* (Liverpool, 1981), esp. 96–108.

[17] John Scott Burdon-Sanderson (1828–1905), professor of physiology 1882–95, Regius professor of medicine 1895–1903; Francis Gotch (1853–1913), professor of physiology 1895–1913, was appointed *contra* Haldane and Sherrington; on Gotch, *Lancet* (1913), 2: 347–51, and on his reign W. J. O'Connor, *British Physiologists 1885–1914: A Biographical Dictionary* (Manchester, 1991), 75–119; Walter Ramsden (1868–1947), professor of biochemistry at Liverpool 1914–31; Gustav Mann (1864–1921).

all, a small research group led by Haldane, who worked chiefly on respiratory physiology, had given the Oxford department an international reputation especially through his work on the chemical control of lung ventilation. This situation, in which a variety of subspecialisms was pursued and in which the reader was better known than the professor, was not to persist under Sherrington who saw to it that biochemistry, histology, experimental psychology, and respiratory physiology were either excluded from his department or subjugated within it, and that its reputation was to be strongly based on his own neurophysiological research. During his reign at Oxford there were no departmental colloquia, presumably because he viewed them as either redundant or disruptive.

The subdepartment of respiration lost its leader, J. S. Haldane, in 1913 when, having failed to secure the chair, he resigned his readership in order to pursue research in his private laboratory in his home in north Oxford and through his various consultancies.[18] He resigned because he saw rightly that he and Sherrington could not coexist peacefully in the same department. Haldane worked chiefly on respiration, Sherrington on the nervous system. On the particular problem of the regulation of respiration, Haldane discounted control by nerves, a view which hardly endeared him to Sherrington. Haldane used normal, living, intact humans, including himself, as his experimental material; Sherrington used treated mammals, often cats. Moreover Haldane thought that human physiology had been usurped in status, wrongly and for far too long a time, by experiments on fragments of frogs and other animals. He rose late and worked late into the night, whereas Sherrington worked long daylight hours. Much of Haldane's research was carried out to promote industrial hygiene and to help those such as miners, divers, and tunnellers who endured stressful and dangerous occupations. He was therefore active as a consultant and commissioner for industry and government respectively. This emphasis on applied physiology had little appeal for Sherrington, who did not share Haldane's belief that the physiology of today would be the medicine of tomorrow. Above all, perhaps, Haldane disregarded conventional boundaries between pure and applied physiology. As Sturdy has incisively revealed, Haldane's research in the University's physiology laboratory was inseparable from his research done outside it for industry and government. He drew no distinction between pure and applied science, whereas Sherrington consciously separated them. For Sherrington there were three sorts of scientist: the investigator, the teacher,

[18] On J. S. Haldane see Sturdy, 'Haldane', esp. 262–75; C. G. Douglas, 'Haldane', *ON* 2 (1936–8), 115–39; G. E. Allen, 'J. S. Haldane: The Development of the Idea of Control Mechanisms in Respiration', *Journal of the History of Medicine and Allied Sciences*, 22 (1967), 392–412; J. G. Priestley, 'Haldane', *Nature*, 137 (1936), 566–9; J. S. Haldane, *Mechanism, Life and Personality: An Examination of the Mechanistic Theory of Life and Mind* (London, 1913), 46–7; S. W. Sturdy, 'Biology as Social Theory: John Scott Haldane and Physiological Regulation', *British Journal for the History of Science*, 21 (1988), 315–40.

and the applied scientist.[19] The first made original discoveries with little or no reference to practical ends. The second diffused knowledge won by investigators, while the third type made scientific knowledge directly serve practical ends. Whenever knowledge was successfully applied, the greatest step in Sherrington's view was that made in the laboratory of the investigator. It was the investigator who constituted the fountain head of knowledge that was taught or applied. Thus Sherrington defended a pure form of physiology unsullied by the concerns of clinical practice and his own research was focused on the nervous system as a fundamental biological problem.

Given such contrarieties between Haldane and Sherrington, it is not surprising that Haldane resigned his university post shortly after Sherrington's election to the chair. For his part Sherrington began to instigate changes in the laboratory as soon as he assumed his chair in January 1914. In a palpable hit against Haldane he complained that the apparatus available for research, though admirable for its particular purpose, did not 'represent other and more general lines of work'; and he alluded to a 'necessary shift' in the work of his department. He promised new departures in teaching and reminded the registrar that in his department research proceeded in the vacations as well as in term.[20] Clearly Sherrington was trying to demote human physiology, an aim which was not lost on Douglas and Priestley who spent much time just before the outbreak of war working in the sanctuary provided by Haldane's private laboratory.

The war brought a temporary suspension of hostilities between Sherrington and the human physiologists because all of them were involved in war-work as applied physiologists. In 1915 Sherrington was even toying with the idea that Haldane be created extraordinary professor of hygienic science at Oxford, but did not pursue it. Though much of Haldane's research on poison gases and respiration problems was conducted in his home and in military hospitals, he and Sherrington did collaborate in one piece of research: Haldane studied the effects of poison gases on decerebrated rabbits prepared by Sherrington, but few results were obtained.[21] Once the war had ended and even though he was the first chairman of the Industrial Fatigue Research Board (established 1918), Sherrington reverted to type as head of department. He ensured that neurophysiology would be the hallmark of his department not only by his own positive achievements and control of staffing but also by subjugating or excluding possible competition within his department. Histology was given to H. M. Carleton, who was deeply

---

[19] J. C. Eccles and W. C. Gibson, *Sherrington: His Life and Thought* (Berlin, 1979), 206–8; Lord Cohen of Birkenhead, *Sherrington: Physiologist, Philosopher and Poet* (Liverpool, 1958); E. G. T. Liddell, 'Sherrington', *ON* 8 (1952–3), 241–70; R. Granit, *Charles Scott Sherrington: An Appraisal* (London, 1966); J. Swazey, 'Sherrington', *DSB*.
[20] *HCP* 97 (1914), 133–5 (quotations), 293–5.
[21] Med./M, 15 May 1915; S. Sturdy to author, 20 Dec. 1991.

loyal to Sherrington and best known as a textbook reviser and author and therefore unthreatening to the department's research identity.[22] When McDougall returned from the armed forces, Sherrington told him there was no space available for experimental psychology, a deprivation which was partly responsible for McDougall's resignation. William Brown, McDougall's successor, fared no better. Though Sherrington had encouraged the growth of psychology in Liverpool before the war, he gave it short shrift at Oxford when the war had ended.[23] In contrast the department of biochemistry, which was created in 1920 (the year McDougall left Oxford), was temporarily housed in Sherrington's department until 1927 when it moved to its own premises, adjacent to Sherrington's. In each situation biochemistry posed no threat to Sherrington's neurophysiological hegemony. Human physiology, however, was a different matter. Before 1914 J. S. Haldane had enjoyed the services of three collaborators, his son J. B. S. Haldane, Priestley, and Douglas. Each of these was to survive as a respiratory physiologist after the First World War but in his own way.

J. B. S. Haldane enjoyed an unusual childhood, acting as his father's laboratory assistant and mathematical calculator at home, and aided him from 1908 in research on diving.[24] In his first year as an undergraduate at New College 1911–12 he read mathematics, studied genetic linking in guinea-pigs, and was responsible for the mathematical part of the work done by his father and Douglas on the combination of haemoglobin and oxygen in the blood. He then switched to Greats, gaining a first in 1914. During the war he joined his father and Douglas in testing the reaction of humans to chlorine, with and without respirators—thus living up to the family motto of 'suffer'. From 1919 to 1923 he was a fellow of New College where he taught physiology; at the same time he worked in his father's private laboratory and the university laboratory, where he was a demonstrator, on experimental human physiology using himself as a guinea-pig. A convinced atheist and a flamboyant *enfant terrible*, J.B.S. was easily lured by Hopkins to a new readership in biochemistry at Cambridge in 1923: he doubled his emolument, shed a heavy teaching load, and could concentrate on research and its supervision. He also moved into an encouraging milieu from one that had little sympathy with his brand of human physiology. Sherrington's notion of proper academic physiology excluded a fellow of New College eating sodium

---

[22] Harry Montgomerie Carleton (1896–1956), fellow of New College 1933–40; H. M. Carleton, *Histological Technique* (London, 1926, 1938, 1957).

[23] L. S. Hearnshaw, 'Sherrington, Burt and the Beginnings of Psychology in Liverpool', *Bulletin of the British Psychological Society*, 27 (1974), 9–14.

[24] John Burdon Sanderson Haldane (1892–1964); N. W. Pirie, 'Haldane', *BM* 12 (1966), 219–49; R. Clark, *J.B.S.: The Life and Work of J. B. S. Haldane* (London, 1968); Weatherall and Kamminga, *Dynamic Science*, 40–3; S. Sarkar, 'Haldane as Biochemist: The Cambridge Decade, 1923–1932', in S. Sarkar (ed.), *The Founders of Evolutionary Genetics: A Centenary Reappraisal* (Dordrecht, 1992), 53–82.

bicarbonate and ammonium chloride to discover for himself how the pH of the blood affected the action of carbon dioxide in it.

Priestley and Douglas proved to be a more durable fifth column in Sherrington's department. Priestley had read physiology at Oxford where he met the elder Haldane who became his mentor.[25] Their joint paper of 1905 on the regulation of lung ventilation established his reputation. After various hospital jobs and serious illness (tuberculosis), he returned to Oxford in 1913 as a Beit research fellow and began work on the way the kidneys affected the composition of the blood. After war-work with Haldane on gas poisoning, he returned to Oxford as Welch lecturer and eventually became reader in clinical physiology but never a college fellow. Though ill-health adversely affected his research output, he promoted human physiology by updating Haldane's textbook on respiration and by launching with Douglas in the early 1920s a practical course on human physiology, which in turn led to their popular textbook on the subject.

Douglas had a more secure and powerful base than his ally Priestley. An Oxford graduate and protégé of Haldane, Douglas was elected a fellow of St John's in natural science when 25 years old. For the next seven years he enjoyed a very close collaboration with the elder Haldane, working on human respiration and circulation and taking up consultancy work in which the pure and applied aspects were not separable. Like Priestley and Haldane he was involved in poison gas research in the First World War. On his return to Oxford he was soon made a demonstrator in general metabolism. Elected FRS in 1922, his policy in the 1920s was to snipe at Sherrington through his textbook, authored with Priestley, and through speeches to external bodies. The textbook summarized their practical course in human physiology. It was a paean to the elder Haldane: most of the book was directly inspired by his teaching and more than half of it covered characteristically Haldanian topics. Douglas and Priestley excluded the central nervous system because, as they mischievously explained, it was a specialist field.[26]

On occasion Douglas revealed his chagrin with Sherrington's remorseless concentration on animal physiology using prepared specimens. In his presidential address to the physiology section of the British Association for the Advancement of Science in 1927, Douglas judged direct experiment on anaesthetized animals to be artificial, coarse, and at best only capable of revealing potentialities not actualities.[27] He stressed that physiology should study normal life, especially that of humans, in such a way that investigation

[25] John Gillies Priestley (1880–1941), on whom see *DSB*, C. G. Douglas, 'Priestley', *Nature*, 147 (1941), 319–20; C. G. Douglas and J. G. Priestley, *Human Physiology: A Practical Course* (Oxford, 1924); J. S. Haldane and J. G. Priestley, *Respiration* (Oxford, 1935); J. S. Haldane, *Respiration* (New Haven, 1922).

[26] D. J. C. Cunningham, 'Douglas', *BM* 10 (1964), 51–74; Douglas and Priestley, *Human Physiology*, p. vi.

[27] C. G. Douglas, 'The Development of Human Physiology', *Nature*, 127 (1927), 845–8 (847).

did not interfere with the intactness of the organism, its power of self-maintenance, and the mutual interdependence of all the different organs. For Douglas humans were better subjects than animals, as the classic work of Haldane and Priestley on lung ventilation had shown over twenty years ago: there was better control of the type and intensity of the particular activity being studied, the subject was co-operative, and could report subjective impressions (see Fig. 2). In a palpable hit against Sherrington, Douglas argued that the focus of research should be the whole normal organism. That meant that no one method in physiological research should be relied on and that the teaching of human physiology was necessary, especially for medical students. Douglas believed that much of the physiology taught at Oxford was useless for them: 'physiology is not medicine: the physician sees a side of life which the physiologist does not meet in the cold aloofness of the laboratory'. The remedy was to focus on the intact and normal human organism, which would have the desirable result of not separating pure and applied physiology but of inosculating them.[28]

Not surprisingly Sherrington saw to it that Douglas and Priestley had little or no technical help with their practical class on human physiology. The effect on Douglas was that he poured his considerable energies into his undergraduate pupils at St John's and into mainly Haldanian activities. In the 1930s he continued his work, mainly via research students, on the respiratory and circulatory effects of exercise, to which he added the topic of the metabolic effects of exercise. The culmination of this research, and indeed of the Haldanian emphasis on suffering as a consequence of self-experimentation under extreme conditions, was the first ever 4-minute mile, run in 1954 at Oxford by Roger Bannister, an Oxford graduate who was Douglas's last research pupil. As a consultant Douglas became more and more involved in work for various government bodies, which dealt with the physiological and occupational problems of the armed forces and of industry. His merits were eventually recognized by the University which made him reader (1937) and then professor (1942) in general metabolism. Finally retiring as a demonstrator in 1953 aged 71, it was entirely appropriate that he presided at the Haldane centenary symposium held in Oxford in 1961.[29]

Once the First World War had ended, Sherrington made it quite clear that he would give short shrift to the Haldanians in his department when he published in 1919 his class book on practical mammalian physiology.[30] This

---

[28] Douglas, contribution to BMA meeting on place of human physiology in medical training, *Lancet* (1929), 2: 478–9.

[29] Roger Gilbert Bannister (b. 1929); D. J. C. Cunningham and B. B. Lloyd (eds.), *The Regulation of Human Respiration: The Proceedings of the J. S. Haldane Centenary Symposium* (Oxford, 1963).

[30] C. Sherrington, *Mammalian Physiology: A Course of Practical Exercises* (Oxford, 1919); E. G. T. Liddell and C. Sherrington, *Mammalian Physiology: A Course of Practical Exercises* (Oxford, 1929). George Newman, chief medical officer, Ministry of Health, praised

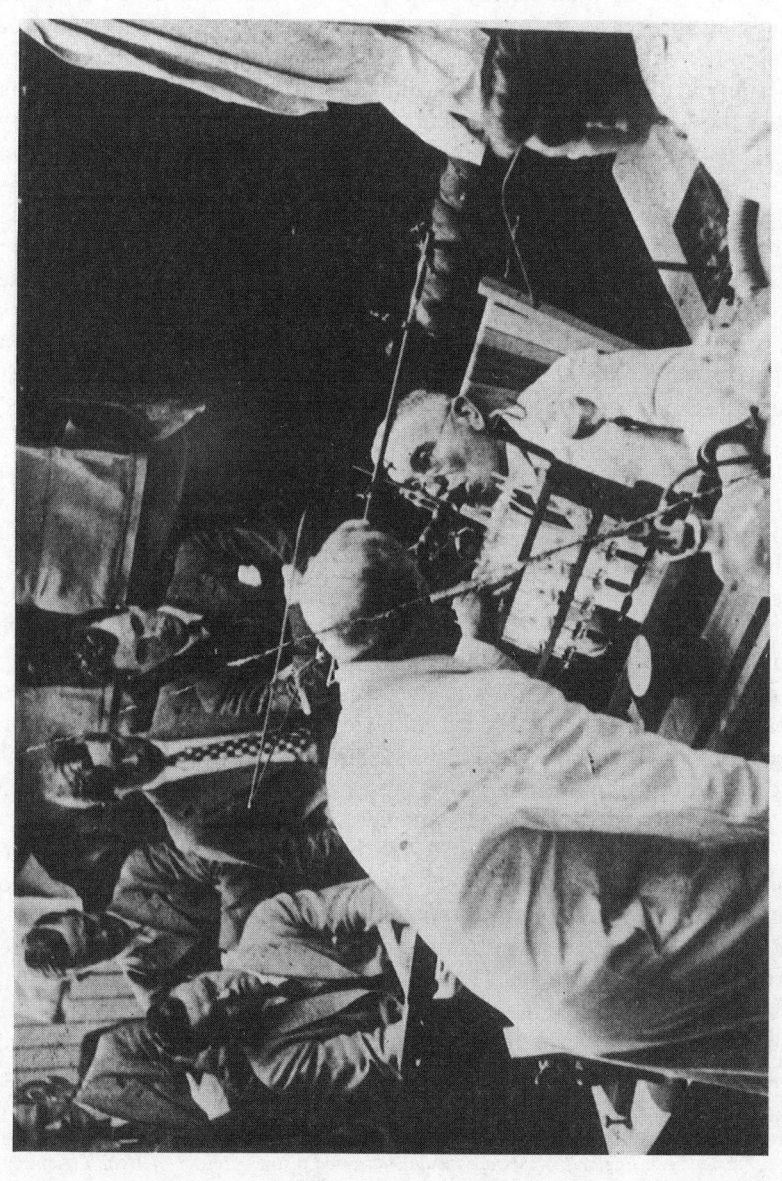

FIG. 2 J. G. Priestley (left) conducting an experiment on a live, intact, and normal fellow of St John's College, Oxford (C. G. Douglas, centre), in a human physiology class, 1930s. Reproduced by permission of the Department of Physiology, Oxford.

showed that Oxford enjoyed what was widely regarded as the most advanced course of its kind in Europe. Its origin lay in Liverpool where Sherrington tried to supplement what he regarded as the outmoded procedures based on studying the nerves and muscles of frogs: he used cats, decapitated or decerebrated, either of which allowed him to comply with the letter of the law pertaining to vivisection in the teaching of physiology. At Oxford during the war, the sparse number of students enabled Sherrington to work out by late 1915 a scheme involving animal preparations (not frogs) and twenty-two practical exercises which fitted into the Oxford academic year which was twenty-four weeks long. The plates and figures were based on his class materials, and the graphic records were those obtained by his undergraduate pupils. Characteristically Sherrington's bibliography for each exercise revealed via historical items the intellectual cost and value of observations which were repeated by the student. Ten years later a second edition, written by Sherrington and Liddell, his right-hand man, drew further on their teaching methods at Oxford and updated the original edition in scope and method.

One effect of the First World War was that death and resignations left a lacuna in the staff of Sherrington's department which he filled mainly with neurophysiologists, mostly Oxford graduates, who became his research collaborators. Sherrington's first significant move was the appointment of Liddell as demonstrator in 1921.[31] A graduate of Trinity, Liddell was elected a research fellow there in 1921. From 1921 to 1926 Liddell was the sole research collaborator of Sherrington who, as president of the Royal Society of London from 1920 to 1925, was a very busy man. As Sherrington's chief assistant in the laboratory, Liddell initiated visitors, especially those from overseas, into its characteristic research techniques. A shy man even with his own college pupils, Liddell's forte was research and its supervision. In 1923 J. F. Fulton, a Rhodes scholar at Magdalen, Sherrington's college, became a demonstrator and, while doing his D.Phil., established himself as a rapid and important publisher.[32] In 1925 Sybil Cooper, a Cambridge graduate who had just made a valuable contribution to Edgar Adrian's research on nerve impulses by helping him to instigate the use of electrical amplifiers, joined Sherrington's staff. Next year Fulton left Oxford for medical training at Harvard where he learned modern surgical techniques from Harvey Cushing who had worked in Liverpool with Sherrington in 1901. Fulton was replaced

---

Sherrington's practical course as the best applied physiology course in England and one which prepared the student for human physiology and clinical work: G. Newman, *Recent Advances in Medical Education in England: A Memorandum Addressed to the Minister of Health* (London, 1923), 36–8.

[31] C. G. Phillips, 'Liddell', *BM* 29 (1983), 333–59.

[32] On John Farquhar Fulton (1899–1960) see *DSB*; J. F. Fulton, *Muscular Contraction and the Reflex Control of Movement* (London, 1926) was dedicated to Sherrington.

by Stephen Creed, a graduate of Trinity, Oxford, who had been elected fellow in physiology at New College in 1925.[33] By 1928 Fulton had returned as a demonstrator and as a fellow of Magdalen in physiology. Next year he was joined as a demonstrator by John Eccles, another Rhodes scholar from Magdalen who, as a research fellow at Exeter from 1927, had taken his D.Phil. in 1929. Thus by 1929 of the nine staff in Sherrington's department six were neurophysiologists (Sherrington, Liddell, Cooper, Creed, Fulton, Eccles), all of whom were college fellows; the other staff were Carleton in histology, Priestley who was reader in clinical physiology, and Douglas who was in charge of respiration and general metabolism. Though Fulton left in 1930 to be professor of physiology at Yale, where he opened the first primate laboratory for experimental physiology in the USA, his loss was made good by Eccles who in the early 1930s became Sherrington's chief research collaborator. By the time of Sherrington's retirement in 1935 Eccles, who had been elected fellow in physiology at Magdalen in 1934, was by far the dominant solo publisher in the department. Like Sherrington, whom he revered as his great master, Eccles won a Nobel prize.

Sherrington's laboratory quickly attracted undergraduates who came to Oxford as Rhodes scholars from the USA and colonies. Among the finalists during the war were Wilburt Davison and Wilder Penfield, both Rhodes scholars who were to be innovators in medical education in the USA and Canada.[34] After the war Fulton arrived from Harvard, from Australia came Hugh Cairns, Howard Florey, and John Eccles, while New Zealand offered Arthur Porritt and Robert Aitken.[35] Both Fulton and Eccles, Sherrington's leading researchers in the laboratory in the 1920s and 1930s respectively, were Rhodes scholars at Magdalen (Sherrington's college), and stayed on to take their D.Phil.s and to become demonstrators. At the research level Sherrington attracted visitors who stayed months or years in what was widely regarded as the leading laboratory in the world in the field of the physiology of the central nervous system. In 1921 F. M. R. Walshe, a clinical neurologist, spent six months with Sherrington who had inspired him since 1906. Some of these migrants already held chairs elsewhere, the best-known

---

[33] Sybil Cooper (1900–70), part two natural sciences tripos (physiology) 1922, Ph.D. (Cambridge) 1927, research fellow at St Hilda's 1926–34, departmental demonstrator 1925–8, university demonstrator 1928–34, married in 1933 Richard Stephen Creed (1898–64), fellow of New College 1925–60; on Harvey Cushing (1869–1939) see J. F. Fulton, *Harvey Cushing: A Biography* (Oxford, 1946); on the importance of Cooper's research with Edgar Douglas Adrian (1889–1977), Nobel prize-winner with Sherrington 1932, see A. L. Hodgkin, 'Adrian', *BM* 25 (1979), 1–74 (29, 30), and A. Brading, 'Cooper', in L. Bindman, A. Brading, and T. Tansey (eds.), *Women Physiologists* (London, 1993), 115–17.

[34] Wilburt Cornell Davison (1892–1972), professor of paediatrics, Duke University, Durham, NC, 1927–61; Wilder Graves Penfield (1891–1976), Montreal-based physiological neuro-surgeon, on whom see J. C. Eccles and W. Feindel, *BM* 24 (1978), 473–513.

[35] Robert Stevenson Aitken (b. 1901), a Rhodes scholar at Balliol 1924–6, D.Phil. 1928, was vice-chancellor, University of Birmingham, 1953–68.

cases being Granit from Helsingfors and Olmsted from California.[36] Granit was so keen to learn the appropriate techniques for studying the retina as a nervous centre that he paid two long visits to Oxford, in 1928 and 1932–3. In 1967 he received a Nobel prize for his research on the physiology of vision, having published the previous year a warm tribute to Sherrington.

In his Liverpool period Sherrington had published mainly solo; but at Oxford he was more prepared to publish collaboratively. This was partly the result of being so busy as president of the Royal Society from 1920 to 1925 that he was forced to use the services of Liddell as his chief research collaborator and as his right-hand man in the famous mammalian laboratory class, known as the cat class. Increasing collaboration was also the result of Sherrington's advancing age which eventually forced him to give up the arduous experimental life at the laboratory bench in 1931 when he was 73 years old. Thirdly, Sherrington realized that his co-workers were capable of producing important new results and of devising instruments which enabled him not only to attack far more profitably problems on which he had been working since the 1890s but also to develop new concepts. The Oxford period was far from being a mere corroboration of the brilliant insights and results of his Liverpool days, epitomized in his famous book of 1906 on the nervous system. The Oxford equivalent to that work was a collaborative effort involving four of Sherrington's leading colleagues.[37]

The best example of the mutual interplay between a new technique and a new concept involved the myograph, which measured the tension exerted by the contraction of a muscle, and the idea of a motor unit. In 1921 Sherrington improved the myograph by amplifying the torsion of a steel wire using an optical beam that was reflected from a mirror on the rod and photographed by a camera. Optical myography, as it was called, was the technique basic to all Sherrington's research of the 1920s. In 1925 Fulton and Liddell added electromyography, using a string galvanometer, to the mechanical instrument devised by Sherrington four years before.[38] By 1930 Cooper and Eccles, under Sherrington's guidance, had improved the bearing of the mechanical myograph, so that Sherrington and Eccles could capitalize on the innovation of Fulton and Liddell by combining more effectively mechanical and electrical recording of muscle action. It was only in the early 1930s, when Sherrington had retired from the laboratory bench, that amplifiers and cathode-ray oscillographs appeared to supplement the string galvanometer.

[36] Francis Martin Rouse Walshe (1885–1973), on whom see C. G. Phillips, *BM* 20 (1974), 457–81, esp. 458, 460; James Montrose Duncan Olmsted (1886–1956), then professor of physiology, University of California, visited Oxford in the late 1920s.
[37] R. S. Creed, D. Denny-Brown, J. C. Eccles, E. G. T. Liddell, and C. S. Sherrington, *Reflex Activity of the Spinal Cord* (Oxford, 1932).
[38] The mechanical and electrical myographs are described and illustrated in Fulton, *Muscular Contraction*, 541–8; see generally D. Whitteridge, 'The Apparatus Used by Sherrington and his Pupils', *Trends in Neurosciences*, 5 (1982), 420–5.

Precision optical myography, a technique on which Sherrington set great store, was not just the pride of the laboratory in the 1920s. It also enabled Sherrington in 1925 to create the concept of a motor unit, by which he meant a spinal motor-neurone (nerve-cell) which controls and co-ordinates the actions of more than 100 muscle fibres. This idea of the motor unit was Sherrington's last theoretical contribution to the physiological anatomy of the function of spinal nerves, a topic he had worked on for forty years. At the experimental level, Denny-Brown, a researcher from New Zealand, succeeded in 1929 in recording the responses of a single motor-neurone.[39] Next year Sherrington and Eccles exploited some of Denny-Brown's techniques to study the motor unit experimentally. Sherrington's Nobel prize was in fact awarded for the isolation and functional analysis of the motor unit.

The year 1932 was the climax of Sherrington's Oxford period with the publication of the collaborative book on the reflex activity of the spinal cord and the award of a Nobel prize. The book was suggested by the Clarendon Press which asked Sherrington in 1930 to write a book on his Oxford work. He agreed, provided his leading research associates would join him. Four of them (Creed, Denny-Brown, Eccles, and Liddell) were happy to do so, with the faithful Liddell acting as editor and co-ordinator. The preface made it clear that the 1920s had been an innovative decade in Oxford neurophysiology in both concepts and techniques: the chief aim of the book was to interpret reflex response in terms of the motor unit which had recently been more effectively studied because of improved myography which involved simultaneous optico-mechanical and electrical measurements. In October 1932 Granit, a Swede, was the first of Sherrington's associates to hear that he had been awarded a Nobel prize. He organized and led the surprise celebration in Sherrington's name, calling on the rapidly assembled throng of researchers to drink to Sherrie—appropriately with sherry.[40]

As a researcher Sherrington was not an empire-builder. In the 1920s he did not campaign for improved accommodation and in the prolonged discussions pertaining to the Science Area he made no large claims for physiology. He was content with the camaraderie of a small group of researchers such as he himself had enjoyed at Cambridge in the 1880s under Gaskell and Langley.[41] As he had only a few research pupils at any one time, he did not have to approach external bodies for big grants. As a key member of the

---

[39] Derek Ernest Denny-Brown (1901–81), D.Phil. 1928, Beit research fellow attached to Magdalen (Sherrington's college), ended as professor of neurology and chief neurologist Boston City Hospital 1941–67.

[40] Creed et al., Reflex Activity, p. v; Eccles and Gibson, Sherrington, 49–52, 61–5, 70–1; J. C. Eccles, 'Life in Sherrington's Laboratory in his Last Decade at Oxford 1925–1935', typescript, Department of Physiology Library, Oxford, published in Proceedings of the Australian Physiological and Pharmacological Society, 9 (1978), 69–72.

[41] Cohen, Sherrington, 4; John Newport Langley (1852–1925), lecturer in physiology, Cambridge, 1883–1903, professor 1903–25; Walter Holbrook Gaskell (1847–1914), lecturer

Medical Research Council, he naturally directed some of its money to his own department, without reporting to the University what he was receiving.[42] His laboratory personality eschewed competitiveness: he was never an academic *homo praedatorius*.[43] He avoided jealousy and pettiness. He was courteous, considerate, and unassuming with young researchers. With ham-fisted beginners he was not deflating. He eschewed declamation, *ex cathedra* pronouncements, and aloof Olympian arrogance. Except when he was involved in sacred days dedicated to time-consuming experiments, he treated young researchers as equals. He made laboratory life into an adventure, expecting each researcher to teach him something. Though he set high intellectual standards, he never denigrated others and was unconcerned with priority of discovery. From obituaries and books about him, it is clear that he elicited intense loyalty and devotion from his associates in the laboratory, though not from those undergraduates who found his lecturing unstructured and, towards the end of his career, inaudible. There is no doubt that Sherrington was a generous adviser and helper to favourite pupils and researchers. Wilder Penfield saw Sherrington as his scientific hero so he returned to Oxford as a postgraduate and was inspired by Sherrington to be a neuro-surgeon. Even when he had achieved fame, he often felt Sherrington looking over his shoulder while performing routine brain surgery.[44] Another famous brain surgeon, Hugh Cairns, was indebted to Sherrington for securing for him a year's experience with Harvey Cushing, which enabled him to see that brain surgery was his vocation.[45] John Fulton was so upset when he finally left Oxford in 1930 that he could not bring himself to meet Sherrington, his guide, friend, and inspirer, whose cherished traditions he implanted at Yale. Homesick for Oxford, he subscribed to *The Times*.[46] When Alexander Carr-Saunders, a biologist at Oxford in the early 1920s, began to overstep his subject's boundaries, Sherrington encouraged him in his sociological work.[47] Howard Florey owed much to Sherrington and

in physiology 1883–1914. For Cambridge physiology generally see Geison, *Cambridge Physiology*.

[42] A. Landsborough Thomson, *Half a Century of Medical Research*, ii: *The Programme of the Medical Research Council (UK)* (London, 1975), 158. Sherrington sat on the MRC 1925–9, 1930–4.

[43] On Sherrington's laboratory personality see Eccles and Gibson, *Sherrington*, 184; R. S. Creed, obituary, *British Journal of Psychology*, 44 (1953), 1–4; E. G. T. Liddell, obituary, *OM* 70 (1951–2), 282–4; D. Denny-Brown, tribute, *Journal of Neurophysiology*, 20 (1957), 543–8; E. D. Adrian and R. Granit, obituaries, *Nature*, 169 (1952), 689–90. For reverence of Sherrington see D. Denny-Brown (ed.), *Selected Writings of Sir Charles Sherrington: A Testimonial Presented by the Neurologists Forming the Guarantors of the Journal Brain* (London, 1939).

[44] W. Penfield, obituary, *Nature*, 169 (1952), 688; W. Penfield, 'Sir Charles Sherrington, Poet and Philosopher', *Brain*, 80 (1957), 402–10.

[45] Cairns to Sherrington, 22 Dec. 1927, Sherrington Papers; Fraenkel, *Cairns*, 29–32, 35, 47.

[46] Eccles and Gibson, *Sherrington*, 86; Fulton to Sherrington, 2 Aug. 1930, Sherrington Papers.

the contact made through him with Fulton: during the Second World War, when Florey was working on penicillin, his children stayed with the Fultons at New Haven.[48] Sherrington was also though not intentionally a matchmaker: when Sybil Cooper became engaged to Stephen Creed, whom she had met in the physiology laboratory in 1925, she wrote a warm letter of thanks to Sherrington, explaining (without any resentment) that she would resign her post of demonstrator because of her impending marriage.[49]

Sherrington was well suited to the peculiarities of Oxford. He brought to it great intellectual distinction, but unlike some of his fellow science professors he made few institutional demands. Though his department did not eclipse the famous Cambridge school of experimental physiology, he put Oxford physiology on the international map and showed that world-class research in a biomedical science could be carried out at Oxford, but unlike W. H. Perkin and Lindemann he rarely pressed within the University for large extra resources of any kind. He was intellectually renowned but not an institutional troublemaker: he attained distinction in specialized research without seriously disturbing Oxford's established structures and interests. Moreover he was a connoisseur in fields which were cherished at Oxford, namely, hospitality, art, literature, philosophy, history, and bibliography. In term time he gave every Sunday afternoon a tea party for about twenty people, mostly students from overseas to whom he revealed a new world of culture and style based on his knowledge of books and art.[50] Though he was not interested in music and theatre, his range of interests was exceptionally wide. He visited art exhibitions and possessed a large collection of nineteenth-century landscape paintings.[51] He was widely read in English poetry, an expert on Goethe, a publishing poet himself, and capable of declaiming by heart long passages from Robert Bridges' *The Testament of Beauty* while preparing for an experiment in his laboratory. He read easily no fewer than six foreign languages.[52] Though much of his research concentrated on the mechanisms of neurophysiological phenomena, Sherrington was also a philosopher concerned with the relation between the human mind and body. This interest, first shown in his presidential address to the British Association in 1922, was made more manifest in his Rede lecture of 1933 on the brain and received its richest expression in his famous Gifford lectures of 1937–8 on man and his nature.[53] His viewpoint was consistent: he believed in Cartesian dualism, accepting that the mind is immaterial, not energy, not spatial, and not

---

[47] Carr-Saunders to Sherrington, 18 June 1937, Sherrington Papers.
[48] Eccles and Gibson, *Sherrington*, 92.
[49] S. Cooper to Sherrington, 30 July 1933, Sherrington Papers.
[50] Eccles and Gibson, *Sherrington*, 43–5.
[51] M. F. Shaffer, 'Sir Charles Sherrington: A Chance Encounter', *Oxford*, 43 (1991), 81–2.
[52] C. S. Sherrington, *The Assaying of Brabantius and Other Verse* (London, 1925); id., *Goethe on Nature and on Science* (Cambridge, 1942); Liddell, 'Sherrington', 257–8.

subject to mechanical laws. This interest in philosophy of mind was continued by Eccles who followed his great master in winning a Nobel prize and in delivering Gifford lectures in which he took his cue from *Man on his Nature*.[54] Sherrington's own concern with philosophy of mind was connected with his interest in history and in bibliography. Though his account of Jean Fernel, the sixteenth-century Paris physician, was published in 1946, Sherrington began work on it in 1926. Bibliography was a consuming passion with him. A trustee of the British Museum, Sherrington was its best benefactor from 1935; he also gave generously to the Royal Colleges of Surgeons and Physicians. When Sherrington retired from his Oxford chair in 1935 and cleared out his desk, just one of his drawers contained no fewer than fifty-eight medical incunabula.[55] His interest in history and bibliography infected a number of his pupils who themselves combined careers as biomedical scientists with historical and bibliographical work, the best known being Fulton, Hugh Sinclair, Kenneth Franklin, James Olmsted, and Liddell.[56] The most prolific was Fulton who from 1930 not only published extensively on neurophysiology but also poured out a succession of bibliographies, biographies (the most impressive being that of Harvey Cushing), and selected readings in the history of physiology. In 1951 Fulton resigned from his chair of physiology at Yale University to become professor of history of medicine there and editor of the *Journal of the History of Medicine*.

The years after 1932 were not easy for either Sherrington or Oxford physiology. His wife died after a long illness in 1933, then Sherrington himself was absent for 1933–4 with crippling arthritis, and early in 1935 he announced his retirement.[57] His chair was not easily filled. It was offered to

[53] C. S. Sherrington, *The Brain and its Mechanism* (Cambridge, 1933); id., *Man on his Nature: The Gifford Lectures Edinburgh 1937–8* (Cambridge, 1940); R. Smith, *Inhibition: History and Meaning in the Sciences of Mind and Brain* (London, 1992), 179–90.

[54] J. C. Eccles, *The Human Mystery* [Gifford lectures] (New York, 1979); id., *The Brain and the Unity of Conscious Experience* (Cambridge, 1965); id., *Facing Reality: Philosophical Adventures by a Brain Scientist* (London, 1970); id. and K. R. Popper, *The Self and its Brain* (Berlin, 1977).

[55] C. S. Sherrington, *The Endeavour of Jean Fernel: With a List of the Editions of his Writings* (Cambridge, 1946); Eccles and Gibson, *Sherrington*, 161, 163, 228–31; Cohen, *Sherrington*, 13.

[56] Some historical works by this quintet are: Fulton, *Cushing*; id., *Selected Readings in the History of Physiology* (London, 1930); J. M. D. Olmsted, *Claude Bernard, Physiologist* (New York, 1938); id., *Charles-Édouard Brown-Séguard: A Nineteenth-Century Neurologist and Endocrinologist* (Baltimore, 1946); id. and E. H. Olmsted, *Claude Bernard and the Experimental Method in Medicine* (New York, 1962); E. G. T. Liddell, *The Discovery of Reflexes* (Oxford, 1960); Sinclair and Robb-Smith, *Anatomical Teaching in Oxford*; in 1965 Sinclair realized £90,000 from the sale of some of his old medical books, *The Times*, 28 June 1990; K. J. Franklin, *A Short History of Physiology* (London, 1933); Franklin, 'Medical School'; id., 'A Short History of the International Congresses of Physiologists', *Annals of Science*, 3 (1938), 241–335; id., *Joseph Barcroft, 1872–1947* (Oxford, 1953); id., *William Harvey, Englishman, 1578–1657* (London, 1961).

Edgar Adrian; but, satisfied with his Royal Society Foulerton professorship and preferring to wait for a Cambridge chair, he refused. As a *pis aller*, John Mellanby, then aged 58, was appointed with effect from January 1936.[58] Unlike Sherrington he was not a visible public performer: he had no published lectures, no important office-holding, no reviews, and no textbook to his name. Initially he was ignored in the planning of the new science buildings, but he protested vehemently that the 1884 building, with which Sherrington had been generally content, often contained four to five times as many students as it was designed for. By 1938 the vice-chancellor accepted publicly that it was seriously out of date and that the University was paying the penalty of having been a pioneer in physiology. The University set aside £30,000 from its Higher Studies Fund in June 1938 to erect part of a new physiology building, but Mellanby was hostile to this move: he preferred a whole new building. In spring 1939 he changed his mind, but his own death in July 1939 and the outbreak of war led to the postponement of any new accommodation for physiology.[59] Mellanby's successor was not easily found. Eccles, who had left Oxford for Australia in 1937, Sir Edward Mellanby (secretary of the Medical Research Council), Charles Best, professor of physiology at Toronto, and Liddell were considered. Early in 1940 by four votes to three the electing committee invited Edward Mellanby to take the chair but he declined. Later that month Liddell was elected by four votes to none, with two members abstaining.[60] The chair which Sherrington had graced was spurned by outsiders; and curiously Eccles, a fellow of Magdalen from 1934 and Sherrington's leading research associate, was apparently not considered in 1935 and was rejected in 1939. By spring 1940 Veale, the registrar, had forgotten Sherrington's Nobel prize when he wrote a damning indictment of Sherrington's inattention to overcrowding in his main teaching laboratory: 'the plain fact is that the condition of the laboratory is a disgrace which ought to have been dealt with years ago.'[61]

---

[57] Eccles and Gibson, *Sherrington*, 72.
[58] For the 1935 manoeuvres see OUA, UD/M/41/1, 24 May, 13 June, 22 Oct. 1935; on John Mellanby (1878–1939), J. B. Leathes, 'Mellanby', *ON* 3 (1939–41), 173–95.
[59] Veale to Lodge, 12 Sept. 1936; Veale to Mellanby, 14 Dec. 1937, 1 Apr. 1938; Veale memorandum, 3 May 1940, all in OUA, Department of Physiology, 1933–46, UR/SF/PHY/1.
[60] For the manoeuvres of 1939–40 see OUA, UD/M/41/2, 10 Dec. 1939, 10 and 31 Jan. 1940; on Edward Mellanby (1884–1955), professor of pharmacology, University of Sheffield, 1920–33, secretary of Medical Research Council 1933–49, see H. H. Dale, 'Mellanby', *BM* 1 (1955), 193–222; Charles Herbert Best (1899–1978), professor of physiology, University of Toronto, 1929–65, co-discovered insulin in 1921.
[61] Veale, memorandum, 3 May 1940, OUA, UR/SF/PHY/1.

5.2 The Dunn Trustees and Pathology

Many laboratory-based scientists at Oxford between the wars wanted better physical accommodation and often met difficulties in obtaining it. They had the brains but thought they lacked appropriate bricks and mortar. In the case of pathology up to 1934 the opposite situation had developed: from the mid-1920s Georges Dreyer, the professor of pathology, presided over a department which the Dunn Trustees had endowed with a new building but his own research faltered badly. Under Howard Florey, Dreyer's successor, the department achieved world-wide fame and it is tempting to make a black-and-white contrast between them. Resisting such temptation I shall argue that until the early 1920s Dreyer's department was worthy of the external patronage it received, but that the standard denigratory view of Dreyer applies to the subsequent years of his reign.

Dreyer was the first full professor of pathology at Oxford but not the first to teach it: in the 1890s Burdon-Sanderson lectured on it, while James Ritchie taught bacteriology. As often in Oxford's history, private endowment was the spur to action by the University: in 1901 a new building, half paid for by Dr Ewan Fraser, a Balliol medical graduate, was opened and Ritchie was made reader. In 1904 Oxford medical graduates and Edward Whitley made substantial gifts to pathology so that when Ritchie resigned in 1907 his post was elevated to a permanent chair to which Dreyer was elected. He was not universally popular in Oxford: he was a foreigner, a Dane who initially spoke little English. But he researched vigorously on the formation of antibodies that agglutinate bacteria which cause infections, being ably assisted by two loyal Oxonians, Ainley-Walker and Alexander Gibson, both of whom were Christ Church graduates. Ainley-Walker was the first person to be elected to a medical tutorship at an Oxford college (University, 1903). He was Dreyer's chief collaborator until 1922 when he published his last research paper. Unambitious and modest, he did not seek to rise to greater heights in the academic world: he was a devoted college man and thoroughly content with the rewards of teaching, tutoring, and administrating. Alexander Gibson was a pathologist at the Radcliffe Infirmary and enjoyed a large consultant practice.[62]

The work of this triumvirate before the war was focused on developing accurate quantitative methods and on establishing a series of constants or

---

[62] Ewan Richards Fraser (1867–1930); James Ritchie (1864–1923), reader in pathology 1901–7, superintendent of Royal College of Physicians, Edinburgh, Laboratory 1907–20, professor of bacteriology, University of Edinburgh, 1913–23; obituaries of Dreyer in *OM* 53 (1934–5), 13–14, *Journal of Pathology and Bacteriology*, 39 (1934), 707–23; *BMJ* (1934), 2: 376, 946, and by S. R. Douglas, *ON* 1 (1932–5), 569–76. On Ernest William Ainley-Walker (1871–1955) see obituaries in *Journal of Pathology and Bacteriology*, 71 (1956), 239–46 and *Lancet* (1955) 2: 513–14, 623; on Alexander George Gibson (1875–1950), *OM* 68 (1949–50), 248–9; Edward Whitley (1879–1945).

standards in human and animal physiology to act as a basis for assessing normality and hence to allow pathological states to be detected. This interest in pathological constants was institutionally recognized in 1915 when Dreyer persuaded Walter Morley Fletcher, secretary of the Medical Research Committee (later Council), to set up at Oxford a bacteriological standards laboratory which provided for Britain and the dominions standardized reagents for serological diagnosis.[63] The inspiration for this venture was the State Serum Institute, Copenhagen, where Dreyer had been assistant director. Fletcher provided the means. He wished to develop pre-clinical medicine at Oxford and at that time saw in Dreyer *qua* medical scientist the nearest thing to Hopkins at Cambridge. The war provided an irreproachable opportunity for Dreyer and Fletcher to find common cause. During the war the standards laboratory produced in bulk and distributed standard agglutinable cultures and standard sera for those bacteria which cause enteritis and dysentery. After the war, in line with Dreyer's obsession with biological standards and their application in medicine, the laboratory was financed by the Medical Research Council as a national institution. It kept pure cultures of certain pathogenic bacteria for reference purposes: they were used to check the antisera employed elsewhere, to identify unknown organisms, and to provide known organisms to help to identify antibodies in blood. The person in charge of the standards laboratory was A. D. Gardner, an Oxford graduate who had been taught by Dreyer and had been lecturer in pathology at St Thomas's Hospital, London. In 1915 he accepted Dreyer's invitation to run the standards laboratory under Dreyer's aegis: for Gardner it was an opportunity to pursue research, to contribute to the war effort, and to return to his beloved Oxford. Gardner ran the standards laboratory until 1937. His return to Oxford in 1915 turned out to be propitious: he ended his long career there as Regius professor of medicine after the Second World War.

Until the early 1920s Dreyer's career prospered. Elected FRS in 1921 he was even more highly thought of by Fletcher. They were so friendly that they chaffed each other mercilessly. This connection with such a powerful patron stood Dreyer in good stead in 1922 when the Dunn Trustees offered an endowment of £100,000 for pathology at Oxford. Three schemes from Oxford were put to the Trustees; but it was Fletcher's advice, given through the Medical Research Council, which ensured that their philanthropy was

---

[63] C. M. P. Bradstreet, 'Fifty Years of "Oxford" Standards', *Lancet* (1965), 1: 1264–5; A. D. Gardner, 'Some Recollections', typescript, library, University College, Oxford, 103–14; on Walter Morley Fletcher (1873–1933), secretary of the Medical Research Committee 1914–20, Medical Research Council 1920–33, see T. R. Elliott, *ON* 1 (1932–5), 153–63; Webster, 'Medicine', 321; Dreyer, *Dreyer*, 118–26; on Arthur Duncan Gardner (1884–1978), reader in bacteriology 1936–48, Regius professor of medicine 1948–54, see *The Times*, 30 Jan. 1978 and *Lancet* (1978), 1: 456–7. The standards laboratory was moved from Oxford to Colindale, north London, in 1946: Thomson, *MRC* ii. 253.

directed to pathology. Oxford was not the only beneficiary of the Dunn Trustees after the end of the first war. Until 1918 the Trust gave small amounts of money to many good causes. Then this policy of amelioration was replaced by preventive philanthropy, in which biomedical research and education were prominent. Between 1919 and 1925, the Trustees made gifts totalling £464,000 for medical science, every investment involving Fletcher, a lodestar of benefactors, as adviser. The two biggest bequests were for biochemistry at Cambridge (£165,000; 1920) and pathology at Oxford (£100,000; 1922).[64]

The first proposal put from Oxford to the Dunn Trustees was a scheme for radiology instigated not by a medical professor but by the Dr Lee's professor of chemistry, Frederick Soddy. Armed with his Nobel prize conferred in 1921, Soddy was hoping to attract to Oxford a laboratory to be established in England by the new Radium Corporation of Czechoslovakia, the broad functions of the laboratory being to house radium and other radioactive materials and to prepare and sell emanation from them for medical and other purposes. The idea of a therapeutically oriented radioactivity centre at Oxford appealed strongly to Garrod, the Regius professor of medicine, and to Sherrington, who both gave it priority over an alternative scheme for a new building for pharmacology which would cost £60,000 according to J. A. Gunn, the professor of pharmacology, but £50,000 in Sherrington's view. The radioactivity plan was put to the Dunn Trustees but rejected by them in December 1921 as too remote from their concerns.[65] The University then negotiated with the Trustees about a scheme for pharmacology drawn up with Orcadian economy by Gunn. The Trustees were favourably impressed, but at that point Fletcher intervened to advise them that pathology at Oxford would be a better investment than pharmacology and that, as a consolation prize, the existing pathology building could be converted into one for pharmacology. The senior medical professors accepted this variation of the pharmacology scheme. In June 1922 the Dunn Trustees formally offered the University £100,000 for a Sir William Dunn school of pathology, £80,000 to be spent on a new building, with £20,000 kept in reserve and invested to provide a maintenance fund. They also offered £3,000 for refitting the existing pathology building as a home for pharmacology. Their benefaction was accepted by the University in November 1922 but did not come into full

[64] R. E. Kohler, *From Medical Chemistry to Biochemistry: The Making of a Biomedical Discipline* (Cambridge, 1982), 89; id., 'Walter Fletcher, F. G. Hopkins, and the Dunn Institute of Biochemistry: A Case Study in the Patronage of Science', *Isis*, 69 (1978), 331–55.

[65] Sir J. Colman to vice-chancellor, 8 Nov., 20 Dec. 1921; Garrod to vice-chancellor, 10 Nov. 1921; Sherrington to vice-chancellor, 25 Nov. 1921; Gunn memorandum, n.d. [autumn 1921]; all in OUA, Pharmacology. Dunn Trustees, UR/SF/PHA/3; Archibald Edward Garrod (1857–1936), Regius professor of medicine 1920–8; James Andrew Gunn (1882–1958), professor of pharmacology 1917–37.

effect until February 1924 owing to legal proceedings pertaining to the Trustees.[66]

As soon as it was clear that the Dunn benefaction would be for pathology, in late 1922 Dreyer began work in earnest, in part at Fletcher's instigation, on a vaccine called diaplyte that would prevent tuberculosis in humans. As Ainley-Walker and Gibson had just ceased to be research collaborators of Dreyer, he worked mainly on his own on the diaplyte project. The enthusiasm of his youthful assistant Roy Vollum, who had arrived in Oxford as a Rhodes scholar in 1921, was no substitute for the steadying influence of Ainley-Walker and Gibson. The vaccine seemed to work well *in vitro* and on animals and, in response to publicity, Dreyer rushed into print solo in 1923. Alarmed by his hastiness and optimistic publicity, the Medical Research Council immediately urged caution and quickly arranged for Dreyer's vaccine to be subjected to controlled and extensive clinical trials without waiting for further experimental work from him. These trials showed that the vaccine was ineffective for humans.[67] Having been a blue-eyed boy of the MRC up to mid-1923, Dreyer was then suddenly found wanting. He was humiliated and the MRC embarrassed and disappointed: if Dreyer had been hustled into premature publication, he showed lack of judgement; if he had not been so hustled, he was irresponsible. After this notorious flop, Dreyer's research publications were few, with nothing appearing in the five years from 1926.

As a substitute for research Dreyer occupied himself with the mechanics of erecting and furnishing the new building which was finished in 1926 and formally opened in 1927: he had little interest in what to do in it. He saw the Dunn building not as providing an opportunity for more research but as something which would provide a new and obsessive role for him as its superintendent and designer: he supervised the builders, climbing ladders and scaffolding to inspect their work, and he planned in minute detail the furnishings and fittings.[68] With the completion of the Dunn school of pathology, Dreyer lost another role so for solace he turned more and more to conviviality and home life. Though the Dunn building was a monument to him after his death, it became almost a mausoleum a few years before it.[69] But his pathological palace was not entirely deserted. The Medical Research Council continued to finance the standards laboratory which was run as a

[66] *HCP* 122 (1922), 197–8; *OUG* 53 (1922–3), 177; 54 (1923–4), 436–7.

[67] Thomson, *MRC* ii. 112; G. Dreyer, 'Some New Principles in Bacterial Immunity, their Experimental Foundation, and their Application to the Treatment of Refractory Infections', *British Journal of Experimental Pathology*, 4 (1923), 146–76; editorial, 'Vaccine Treatment of Tuberculosis by a New Method', *BMJ* (1923), 1: 1059–60 hailed Dreyer's method as a milestone; Dreyer, *Dreyer*, 182–92; Roy Lars Vollum (1899–1970), departmental demonstrator in pathology 1924–7, university demonstrator 1927–38, demonstrator in bacteriology, Radcliffe Infirmary, 1938–68.

[68] Dreyer, *Dreyer*, 207–16.

[69] Macfarlane, *Florey*, 231–3.

virtually autonomous enterprise by the astute Gardner. A few Rhodes scholars were attracted from the USA and Australia in the early 1930s by the facilities of the Dunn building, but there were no British or foreign researchers. Dreyer was not often seen in his laboratory, saw his research pupils once or twice a term or not at all, and created an atmosphere in which talk at tea-time about research was taboo.[70] His department was not bursting with eager researchers, a fact that was not lost on Buzzard, the Regius professor of medicine. He deplored the decline of research in the school of pathology after 1925, its inability to attract medical students, and the contrast between the fine building and poor, outdated, or unrepairable apparatus in it.[71] This peculiar situation ended unexpectedly in summer 1934 when Dreyer died suddenly from a heart attack while on holiday in his native Denmark.

5.3 The Rockefeller Foundation and Biochemistry

The most striking feature of biochemistry at Oxford between the wars was that it was brought into existence as a subject separate from physiology, of which it had previously been a part, by a mixture of individual benefaction and corporate philanthropy. Though biochemistry was supported by both Sherrington and Garrod, the Regius professor of medicine, and received the general backing of the medical faculty, their approval would have been nugatory had not Edward Whitley and the Rockefeller Foundation provided most of the requisite funding for a chair of biochemistry (established 1920) and a new building (opened 1927) respectively. Though Oxford was the fourth British university to have an independent chair of biochemistry, following Liverpool (1902), Cambridge (1914), and Edinburgh (1919), its fledgeling department of biochemistry played second fiddle to that at Cambridge led by F. G. Hopkins, a Nobel laureate in 1929. In some ways the Oxford department was an outpost of Cambridge: the second professor, Rudolph Peters, was a protégé of Hopkins and best known for his research on the Hopkinsian field of vitamins; and the Rockefeller benefaction for Oxford biochemistry in 1923 was an explicit counterpart to the gift made in 1920 to Cambridge biochemistry by the Trustees of the estate of Sir William Dunn. Crucially Cambridge fared better financially than Oxford: the Dunn Trustees provided £165,000, which included £60,000 as endowment for research, for biochemistry at Cambridge, while the Rockefeller Foundation

[70] M. F. Shaffer, 'The Sir William Dunn School of Pathology: Before Penicillin', *Oxford*, 45 (1993), 69–72; R. S. Smith to author, 18 Mar. 1992; Morris Frank Shaffer (1910–94) and Robert Sydney Smith (b. 1906) were Rhodes scholars in pathology 1931–4 and 1930–2, gaining a D.Phil. (1934) and a B.Sc. (1932) respectively.

[71] Buzzard to curators of Chest, n.d. [Nov. 1935], OUA, Department of Pathology, 1930–48, UR/SF/PAT/1.

gave £75,000 to biochemistry at Oxford for a new building, teaching staff, and laboratory assistance. Teaching was the primary consideration in the Oxford benefaction so no explicit provision was made as at Cambridge to support research. As a final degree subject, biochemistry prospered more at Cambridge than Oxford: in 1924 it was separated from physiology in part two of the natural sciences tripos at Cambridge. At Oxford, in contrast, biochemistry remained subordinate to physiology and to chemistry in the final degree examinations. Most of the biochemistry teaching was done for physiology finals and for the BM examination; few chemistry undergraduates availed themselves of the opportunity created in 1930–1 of being able to take biochemistry as a supplementary subject as part of their finals.[72]

Before the First World War chemical physiology was taught at Oxford in the department of physiology mainly by J. S. Haldane and then Walter Ramdsen, who migrated to Liverpool in 1914 to be the second occupant of the Johnston chair of biochemistry. After the war Sherrington prodded the medical faculty at Oxford to agree that the time was ripe for an independent chair and a separate laboratory for 'that branch of chemistry which studies the chemical conditions and activities of animal life', a branch of science which it claimed was known as biochemistry.[73] Proposing was one thing, disposing another. Fortunately for the University Edward Whitley came to its aid financially.[74] Whitley belonged to a wealthy and influential Liverpool family which made its money from the brewing firm of Greenall Whitley; and his father had been mayor of Liverpool and an MP. He attended Trinity, Oxford, graduating with a third in physiology in 1902. Already keen on biochemistry, by 1904 he was at Liverpool University doing research on photosynthesis with Benjamin Moore, the first occupant of the Johnston chair of biochemistry. In 1906 Moore and Whitley launched the *Biochemical Journal*, and they co-edited the first six volumes. Whitley met the early deficits of this important periodical which in 1911 was purchased by the new national Biochemical Society. At the same time he was a vehement left-winger and Fabian socialist, who was not embarrassed to canvass in Liverpool's Irish slums in a big chauffeur-driven car which was air-conditioned in order to nurture hot-house flowers inside it. In 1911 Whitley moved to Oxford. He never lost his Wellsian socialist enthusiasm,

---

[72] For biochemistry at Oxford and Cambridge see Kohler, 'Fletcher and Biochemistry'; Weatherall and Kamminga, *Dynamic Science*; Ord and Stocken, *Oxford Biochemistry*; and generally Kohler, *Biochemistry*; on Rudolph Albert Peters (1889–1982), professor of biochemistry 1923–54, see R. H. S. Thompson and A. G. Ogston, *BM* 29 (1983), 495–523, and R. A. Peters, 'Forty Five Years of Biochemistry', *Annual Review of Biochemistry*, 26 (1957), 1–16.

[73] Med./M, 6 Dec. 1919; *HCP* 115 (1920), 47–8 (quotation).

[74] On Whitley see Sherrington to Peters, 14 Dec. 1945, 6 Apr. 1946, Peters Papers; R. A. Peters, 'Whitley', *OM* 64 (1945–6), 322–4; Whitley's father was Edward Whitley (1825–92), a solicitor, mayor of Liverpool 1868, Conservative MP for Liverpool 1880–92.

being an original promoter of the *New Statesman* in 1913 and in 1931 chairman of the combined *New Statesman and Nation*.

After the First World War, Whitley volunteered himself to teach biochemistry in Sherrington's department. Hearing from Sherrington that the University claimed it had no extra money for biochemistry, Whitley intimated to Sherrington his desire to offer £15,000 for a chair in it, subject to two unusual conditions, the first being that the first holder would be selected by himself and the second that the chair would be attached to Trinity, his old college. The second condition was less troublesome than the first. At a Hebdomadal Council meeting in March 1920, Whitley's formal offer dated 1 March was attacked by Lewis Farnell, rector of Exeter, on the grounds that it was absurd to have an amateur like Whitley nominating the first occupant and that the benefaction would involve extra running costs which would fall on the University. Whitley's benefaction was strongly defended by Sherrington who stressed that Whitley was an authority on biochemistry and that the physiology department would absorb the running costs of biochemistry as it had done in the past. Whitley soon laid down a third condition, namely that his initial offer of an endowment of £15,000 was to be reduced to £10,000, Whitley insisting that the University find £400 a year towards the endowment of the chair. Rather unusually for a science professor anywhere, Sherrington quickly made it clear to the Council that he would relinquish £400 p.a. from his departmental grant from the University which could then spend it on the biochemistry chair.[75]

Whitley's nominee for the chair was Benjamin Moore, his intimate friend, former research colleague, and fellow socialist, whose aim of founding a national journal devoted to biochemistry had been made possible by Whitley's personal and financial aid.[76] Moore was a restless and flamboyant character whose impatience with detail and tendency to speculate had not been reined in at Liverpool where his post was basically a research chair. Free from service courses to medical students, Moore developed an idiosyncratic view of biochemistry and medicine while at Liverpool. With Whitley he studied photosynthesis in marine algae in an attempt to show that the whole living world depended on the building up of the energy of sunlight into the chemical energy of minute colloidal forms of life which he thought constituted living structures less developed than bacteria and protozoa.

[75] Jenkinson to Curzon, 10 Mar. 1920, archives, Brasenose College, Oxford; *HCP* 115 (1920), 113–14, 147–8; Lewis Richard Farnell (1856–1934), rector of Exeter 1913–28, vice-chancellor 1920–3.

[76] On Benjamin Moore (1867–1922), professor of biochemistry 1920–2, see Kohler, *Biochemistry*, 55–6, 89; L. Hill, 'Moore', *Nature*, 109 (1922), 348; F. G. Hopkins, 'Moore', *Biochemical Journal*, 22 (1928), 1–3. For his medical socialism, see B. Moore, *The Dawn of the Health Age* (London, 1911). For his biochemical work at Liverpool see B. Moore, *Biochemistry: A Study of the Origin, Reactions, and Equilibria of Living Matter* (London, 1921); id., *The Origin and Nature of Life* (London, 1913), 117, 190; and R. A. Morton, 'Biochemistry at Liverpool 1902–1971', *Medical History*, 16 (1972), 321–53 (323–8).

Though his results were criticized, Moore was one of the first in Britain to apply physical chemistry, which he had learned from Wilhelm Ostwald during a year at Leipzig, to biological problems. As a pundit on medicine and as a socialist, Moore advocated a state-supported medical service.

In 1914 Moore resigned his Liverpool chair to join the new National Institute for Medical Research at Hampstead, London, where during the war he worked on TNT poisoning and miners' phthisis. He enjoyed his Oxford chair for only eighteen months; he died in March 1922. His Hibernian combativeness and rebelliousness against orthodoxy in science and in social affairs were assuaged by ill-health. Even so he did some research on photosynthesis, undertaking a trip to Geneva, paid for by Whitley, to study the synthesis of life in the deep water of the Lake. At the same time Whitley gave to Moore the income of the £5,000 saved from his initial offer of £15,000 as the endowment of the chair.

After Moore's death Archibald Garrod, the Regius professor of medicine, was officially put in charge of biochemistry until Rudolph Peters was appointed to the Whitley chair in January 1923. Garrod was not unqualified for the task. For many years he had been interested in a group of congenital metabolic diseases in humans which he called inborn errors of metabolism. As a father of biochemical genetics, Garrod put forward the novel idea that some diseases were caused by the hereditary deficiency of certain enzymes; he did so well before the idea of vitamin deficiencies had become clearly established. Shortly after Moore's death Garrod persuaded the University that the next professor of biochemistry should be head of a separate department and no longer dependent on the professor of physiology. Garrod was also influential in the election of Peters to the Whitley chair on 16 January 1923. Along with Hopkins and Sherrington, the key figures in the eight-strong electoral board, Garrod saw to it that Peters, a protégé of Hopkins, was preferred to George Barger, a clinical biochemist who was then professor of chemistry in relation to medicine at Edinburgh.[77] Garrod's final service to Oxford biochemistry was to secure in late 1923 a grant of £75,000 from the Rockefeller Foundation.

The new post-war aims of the Rockefeller Foundation have been well analysed by Kohler.[78] Preventive philanthropy was replaced by broader and non-interventionist schemes in which the patronage of scientific medicine was prominent. As the Dunn Trustees had already endowed pathology at Oxford and biochemistry at Cambridge, the Rockefeller officials proposed to develop biomedical sciences at Oxford and Cambridge even further by

[77] On Garrod see *BMJ* (1936), 1: 731–3; R. A. Peters, 'Garrod', *American Journal of Human Genetics*, 10 (1958), 1–2; H. Krebs, 'Sir Archibald Garrod', in Dewhurst, *Oxford Medicine*, 127–35; Peters, interview for Biochemical Society, 6 Mar. 1981, transcript in Peters Papers, A42; George Barger (1878–1939), professor of chemistry in relation to medicine, University of Edinburgh, 1919–37.

[78] Kohler, *Biochemistry*, 90–1; and generally id., *Partners in Science*.

making grants for pathology at Cambridge and biochemistry at Oxford. Though the Rockefeller officials were committed to rounding out and completing the fundamental medical sciences at Oxford and Cambridge, which they believed were attended by the best type of English youth, they were realistic enough to ensure that Cambridge fared better. Rockefeller offered £133,000 for pathology to Cambridge where £33,000 was soon raised to supplement that endowment. At Oxford Peters had originally applied for £144,000 but, after advice from Garrod and Sherrington, the University asked Rockefeller for £106,000 to support biochemistry. Garrod used all his charm in his negotiations with the Foundation's division of medical education, and succeeded in securing £75,000 (£55,000 building; £20,000 endowment) from the Foundation provided that the University would contribute at least £1,250 per year, which represented a capital sum of £25,000.[79] It did so, the new building being occupied in 1926 and officially opened in 1927. In late 1923 Garrod had been sanguine that an individual would provide £20,000 to complement the Rockefeller grant and relieve the University of much of its contribution; but no person came forward.

Like Sherrington, Peters was new to Oxford when he assumed his chair. Having been an undergraduate at Gonville and Caius College, Cambridge, where he was inspired by Joseph Barcroft and thrilled by Hopkins, Peters had worked during the war under Barcroft at the Chemical Defence Experimental Establishment at Porton, Wiltshire, where he became decidedly keener on physiological biochemistry through his work on poisonous gases and the mechanisms of their action. After the war Peters returned to Gonville and Caius as tutorial fellow in medicine/physiology and was soon recruited by Hopkins as an adjutant in biochemistry. Though Peters quickly became a protégé of Hopkins, he was also indebted to Barcroft who had suggested to Peters his first research problem and recruited him to Porton. From 1919 to 1923 Peters's research was focused on three topics derived from or exacerbated by his war-work: first, the mechanisms of action of toxic gases, secondly, surfaces and interfaces, and thirdly in 1922 the vitamin B complex as a result of trying to grow a species of protozoan in media containing yeast extracts. With Hopkins's encouragement, Peters began work on the isolation of vitamin B.[80]

When Peters arrived in Oxford he found he had just one room for research. He decided to focus it on the vitamin B complex because working on the factor that cured pigeons of beri-beri was feasible in the limited

[79] For the Rockefeller negotiations, Peters to Garrod, 8, 12, and 16 Feb. 1923; Sherrington to Garrod, 12 Feb. 1923; Garrod to Peters, 24 Sept. 1923, all in Peters Papers, B1–3; Rockefeller Foundation Record Group 1.1, Series 401A, Oxford University, Biochemistry file, folder 279, box 21, Rockefeller Archive Centre; *HCP* 126 (1923), 213–14; 127 (1924), 31–3, 109.

[80] Thompson and Ogston, 'Peters', 497–501; Franklin, *Barcroft*, 79, 103–4; Peters, 'Biochemistry', 2–7, 12–13.

accommodation available.[81] Until 1927 the vitamin B work did not dominate the department's research. Though Peters had four collaborators by then, other topics were also being pursued. In the mid-1920s, for example, Cyril Carter, who was to be elected medical fellow at Queen's in 1927, began work on photo-oxidation, being funded initially by Whitley who demonstrated in the department until 1936 when ill-health induced him to move to Torquay.[82] Peters's main task in his opening years was to oversee the planning and building of his new department. The key to his approach was that he believed that a good biochemist had to be above all a good physiologist. Moreover at a time when many chemists at Oxford had little time for biochemistry, Sherrington supported it strongly. For Peters it was desirable as well as expedient to align his small and new department with physiology. Thus the new biochemistry building was connected with a revamped physiology department on all floors by passages; and the two departments shared a new large lecture-room, a conjoint library and writing-room, a joint tea-room, and had adjacent animal rooms. There was indeed the closest contact with physiology consistent with the independence of each department. Moreover the Rockefeller benefaction benefited physiology greatly: as a result of the new arrangements with biochemistry, physiology gained some refashioned accommodation which cost the University £3,000.[83]

Once the new building had been opened the research of the department was increasingly focused on vitamin B. By 1932 Peters and collaborators had obtained crystalline vitamin B1 so they then turned their attention to physiological biochemistry. By 1936 they had shown for the first time the precise mode of action of a vitamin: vitamin B1 was a catalyst needed to remove by oxidation pyruvic acid, one of the products of carbohydrate metabolism. This achievement impressed the Rockefeller Foundation who gave £2,400 to Peters in 1939 for research on brain metabolism.[84] The increasing focus on vitamin research is well shown by the contrasting careers of two demonstrators, Cyril Carter and Ernest Walker. In the late 1920s Carter turned to research on vitamin B6 and prospered, becoming the department's first reader in 1948. In contrast Walker, who had been Peters's assistant at Cambridge, moved with him to Oxford and until 1928 worked on poison gases. When Peters began to focus the department's research more strongly on vitamins, Walker lost his research drive and in compensation became the general factotum of the department.[85]

[81] Peters, interview, Peters Papers, A42.
[82] Ord and Stocken, *Oxford Biochemistry*, 7; obituary of Carter, *The Times*, 18 and 21 May 1974.
[83] Ord and Stocken, *Oxford Biochemistry*, 6–7; 'The New School of Biochemistry at Oxford', *Nature*, 120 (1927), 634–5; R. A. Peters, 'Department of Biochemistry, University of Oxford', *Methods and Problems of Medical Education*, 18 (1930), 109–18.
[84] *OUG* 70 (1939–40), 58, 246–7.
[85] Ernest Walker (1900–42), obituary by Peters, *Biochemical Journal*, 37 (1943), 449–50.

Peters had little sympathy with an exclusively chemical or physical approach to the study of living matter but he encouraged the use of chemical and physical methods when he thought them useful. His collaborator J. R. P. O'Brien was sent to Graz to learn Fritz Pregl's techniques of quantitative micro-chemical analysis.[86] More importantly a Svedberg ultra-centrifuge was installed in the department in 1937, which made Oxford one of the three centres in Europe to possess this device which made possible the study of large molecules in solution and the separation of proteins (the other centres were Uppsala and the Lister Institute, London).[87] Its installation indicated a break with the tradition inherited from physiology, which saw the value of making one's own apparatus which could be repaired without specialized technical help. The impetus for the ultra-centrifuge came from John Philpot, a Balliol chemist and physiologist who had used a Rockefeller medical fellowship to work on pepsin at Prague and Uppsala where in early 1934 he spent several months in Svedberg's laboratory. Having been made a fellow of Balliol in biological sciences in 1934 and wishing to continue the pepsin work, he pressed for an ultra-centrifuge at Oxford. Peters supported the plan and by spring 1935 had persuaded the Royal Society to offer him £2,000 to purchase on loan a Svedberg ultra-centrifuge, provided his department paid for its transport, installation, and maintenance. The machine, as Peters called it, was not mass-produced: it was constructed to order by Svedberg, who spent a year building it. As it gave a maximum centrifugal force of 500,000 times gravity, it needed not only special housing but also concrete foundations to stop it from walking about. The total cost of installing the centrifuge and the accompanying extensions to the department came to about £10,000, of which about £8,000 was provided from unused Rockefeller money, the rest coming from the University. In spring 1937 Svedberg came to Oxford to receive an honorary degree and to open the centrifuge: he liked what he called Philpot's installation. Though there were to be little more than two years before war broke out, the ultra-centrifuge facilitated research on the physical properties of macro-molecules by Philpot and by Alexander Ogston, a Balliol chemist elected a fellow of Balliol in natural science (physiology) in 1937.

Peters was an accomplished violinist devoted to chamber music. For him biochemistry was the study of nature's counterpoint by scientific artists. His approach to research was leisurely in that he did not insist that his colleagues keep their research going during the vacations.[88] Unlike Hopkins he was not

---

[86] Thompson and Ogston, 'Peters', 510; John Richard Percival O'Brien (b. 1906), fellow of Pembroke 1955–74; Fritz Pregl (1870–1931), head of Institute of Medical Chemistry, Graz.

[87] On the centrifuge, Ord and Stocken, *Oxford Biochemistry*, 8–9; *OUG* 66 (1935–6), 13; *HCP* 161 (1935), 177; R. A. Peters, 'Historical Remarks 1966', Peters Papers, B24; Svedberg to Peters, 2 Mar. 1937, Peters Papers, G27; The Svedberg (1884–1971), *DSB*.

[88] Peters, 'Biochemistry', 16; Ord and Stocken, *Oxford Biochemistry*, 66.

a hustler who attracted big grants and lots of research fellows, though he did receive continuous support from the Medical Research Council partly in exchange for his deep involvement from 1930 in the work of its Accessory Food Factors Committee. In the two years before the Second World War, he was the most successful pre-clinical professor in gaining research support from the Nuffield benefaction: his department received almost £1,000 p.a. which was spent mainly on Philpot's centrifuge and on research into vitamin B1.[89] In the 1930s, when the teaching of biochemistry to undergraduates expanded, that very success adversely affected the money available for research.[90] This happened because the cost of teaching each undergraduate per year was at least twice the annual fee received by the department. Thus the extra students taught required a greater subsidy from departmental funds, leaving less money available for research. At the same time Peters could not offer or obtain postgraduate studentships for one to two years of research. Consequently he had few B.Sc. or D.Phil. students in his department, at best four to five at any one time, most being supervised by him. In any event, Peters does not seem to have believed in empire-building. He thought it essential to work regularly at the laboratory bench and to know well what he called his folk. Moreover Peters certainly admired and perhaps copied Sherrington, whose influence was wide and reputation very high, but he never had hordes working with him.[91] Until the Second World War most of Peters's researchers and teaching staff were Oxonians, five of whom (Carter, Philpot, Sinclair, Ogston, and R. H. S. Thompson) were college fellows. Though Peters told Krebs, his successor, in 1954 that he (Peters) had failed to pull the colleges into a pro-science attitude, they nurtured biochemistry before the Second World War by electing to fellowships five men whose research was mainly biochemical, though their college posts were in physiology, biology, and medical science.[92] Some of those demonstrators who were not college fellows departed: P. C. Raiment and G. L. Peskett, survivors from Moore's reign, resigned in 1926 and 1936 to become professor of physiology at Cairo and an Isle of Wight general practitioner respectively. But R. B. Fisher stayed for twenty-four years from 1935 without a college fellowship until he resigned in 1959 to assume the Edinburgh chair of biochemistry.[93]

Peters's ambivalent attitude to discipline-building was revealed in his failure to recruit refugee biochemists dismissed by or fleeing from the Nazis, whereas Hopkins at Cambridge took in six, one of whom was Hans Krebs, who was to be a Nobel prize-winner in 1953. There was, however, at one

[89] Thomson, *MRC* ii. 76; *OUG* 69 (1938–9), 365–76; 70 (1939–40), 377–92.
[90] *HCP* 164 (1936), 62–5.
[91] Peters, 'Biochemistry', 14–15.
[92] H. Krebs, *Reminiscences and Reflections* (Oxford, 1981), 90.
[93] Reginald Brettauer Fisher (1907–86), university demonstrator in biochemistry 1935–59, professor of biochemistry, University of Edinburgh, 1959–76.

time a good chance that Krebs might have landed in Oxford.[94] By April 1933 Krebs knew that he would be dismissed from his post at Freiburg on 1 July 1933. Next month Hopkins, who was already keen to recruit Krebs, assured him that Oxford would welcome a biochemist but warned him that Peters was vacillating. Krebs then wrote to Peters but received no reply by 19 June when he left Freiburg. When Krebs arrived in England he went to Oxford but Peters and Balliol had nothing definite to offer. Then Krebs visited Cambridge where on 25 June Hopkins gave him a warm welcome and offered a Rockefeller grant of £300 p.a. which Krebs accepted on the spot. Three days later Krebs received a letter from Peters saying that Balliol would offer £100 p.a. provided the Rockefeller grant could be transferred to Oxford. Thus Peters was dilatory, vacillating, and unable to find money in time for Krebs whose knowledge of Oxford's complexities and of English was admittedly limited. He was learning English then and took some time to appreciate its idioms. During his two-year stay at Cambridge he once reported triumphantly that he just pulled Dr Marjory Stephenson's legs!

5.4 Striking Personality 1: W. E. Le Gros Clark

Clark was the first of three new brooms to be appointed to biomedical chairs at Oxford in the short space of three years. In autumn 1934 he assumed the chair of human anatomy, being followed in 1935 by Florey (pathology) and in 1937 by Burn (pharmacology). All three were discipline-builders who put a new or an increased emphasis on laboratory-based experimental research. In Clark's case he immediately set about changing the existing focus of research of his department from craniology, pursued as a key aspect of ethnology, to experimental work on neurology and on the reproduction and embryology of primates. At the teaching level neurology and embryology were introduced at the expense of physical anthropology which lost its previous prominence. Clark had to face, and faced with outraged determination, the related problems of a department packed with thousands of skulls which had been lovingly collected by Arthur Thomson, his predecessor, and of staff inherited from him. Clark pitched into attack immediately: he launched and won the so-called battle of the skulls; and he ejected from his department Beatrice Blackwood, a zealous craniologist and ethnologist. Simultaneously Clark redefined human anatomy in general and

---

[94] Krebs, *Reminiscences*, 65–8, 84–6, 92; H. Kornberg and D. H. Williamson, 'Krebs', *BM* 30 (1984), 349–86 (360–4); F. L. Holmes, *Hans Krebs*, i: *The Formation of a Scientific Life, 1900–1933* (New York, 1991), 432–4; id., *Hans Krebs*, ii: *Architect of Intermediary Metabolism, 1933–1937* (New York, 1993), 3–4; Hans Adolf Krebs (1900–81), professor of biochemistry, Oxford, 1954–67.

physical anthropology in particular. As Clark himself said, his subject at Oxford moved from the 'museum stage' of its development to an experimental one.[95]

Arthur Thomson, who resigned from the chair of human anatomy at the end of 1933 aged 75, was an Edinburgh-trained Scot who had launched the teaching of the subject in 1885 when he was appointed lecturer. In 1893 a proper building, costing £7,000, replaced the notorious small tin shed in which Thomson had begun teaching and he was promoted to professor.[96] Before the First World War Thomson construed human anatomy in two ways which were rarely combined: he was active as an artist and as a physical anthropologist. At heart he was a painter more devoted to his brush in his studio than to the knife in the laboratory. From 1900 to 1934 he doubled as professor of anatomy at the Royal Academy. A good ambidextrous drawer, his book on anatomy for arts students was a standard work which went through five editions between 1896 and 1930.[97] Prouder of his pictures hung in the Royal Academy than of his published research, he was probably more at home in teaching the form of the human body to art students than its structure to medical ones. His closest friends were artists and his favourite London resort was the Arts Club. His penchant for illustrations was best shown in the enlarged stereoscopic photographs which adorned his textbook on the human eye.[98] Thomson combined his expertise as an artist with a consuming passion for physical anthropology revealed in his obsession with craniological collecting and anthropometric research. Before the First World War he began to fill his department with ethnological collections. In 1913, for example, he secured the transfer of about 1,100 human crania from the department of zoology and comparative anatomy.[99] Though Thomson had been president of the Anatomical Society, his research looked less to medicine than to anthropology and archaeology. This propensity was best shown in the book he wrote with D. R. MacIver on the ancient races of the

---

[95] W. E. Le Gros Clark, 'The Scope and Limitations of Physical Anthropology', *Nature*, 144 (1939), 430–1; for stimulating general discussion see J. V. Pickstone, 'Museological Science? The Place of the Analytic/Comparative in Nineteenth-Century Science, Technology and Medicine', *History of Science*, 32 (1994), 111–38, and id., 'Ways of Knowing: Towards a Historical Sociology of Science, Technology and Medicine', *British Journal for the History of Science*, 26 (1993), 433–58.

[96] Strictly speaking Arthur Thomson (1858–1935) occupied three chairs at Oxford, being extraordinary professor of anatomy 1893–1907, university professor 1907–19, Dr Lee's professor 1919–33. On Thomson, president of Anatomical Society 1906–9, see *Journal of Anatomy*, 69 (1934–5), 298–302; *Lancet* (1935), 1: 405–6; *BMJ* (1935), 1: 334.

[97] A. Thomson, *Handbook of Anatomy for Art Students* (Oxford, 1896, 1899, 1906, 1915, 1930) contains photographs not only of professional models but also of 'better known athletes of this University' (pp. viii–ix).

[98] A. Thomson, *The Anatomy of the Human Eye, as Illustrated by Enlarged Stereoscopic Photographs* (Oxford, 1912).

[99] *HCP* 98 (1914), 145–7.

Thebaid, a work based on the anthropometric examination of about 1,500 ancient Egyptian crania.[100]

In the early 1920s Thomson, like Dreyer, perpetrated an embarrassing gaffe in his published research. As an opening shot in what he announced as a detailed study of the human ovum, he claimed that cells in tissues from the uterus constituted the earliest known stage in the development of the fertilized human egg. To Thomson's chagrin, this claim was widely rejected: the alleged embryo was nothing more than a blocked gland filled with cellular debris.[101] He then abandoned research in human embryology, and in compensation became even more enthusiastic about physical anthropology which for him meant recording measurements of human skulls and skeletons, both ancient and modern, in the hope that small differences in their sizes, shapes, and markings would give insight into the history of racial origins and diversification. In Thomson's case this programme was expressed in an insatiable desire to collect human skulls which were deposited in his department; and, given his illnesses and advancing age, through the work of Leonard Buxton and Beatrice Blackwood, two craniologists in his department.

A classics graduate, Buxton took the Oxford diploma in anthropology (established 1907) through which he was inspired by Thomson to become a physical anthropologist.[102] Appointed by Thomson as demonstrator in physical anthropology before the first war, he returned after war service to become the subject's leading exponent at Oxford in the 1920s and 1930s, being elected reader in 1927 and a fellow of Exeter in 1933. Until the mid-1930s he was easily the most active publisher, via books and articles, in the department of human anatomy. Aided by the Albert Kahn travelling fellowship from 1922, he became a globe-trotter who was an expert on humans in their present-day diversity. But he did not neglect ancient humans. In 1925 he went on the Kish excavation and secured access to all the resultant material, which he inspected at Oxford before half of it went to Chicago. It was Buxton who in 1925 attracted from Christ Church David Talbot Rice to study the human remains he had excavated in Mesopotamia. Though Buxton's publications were diverse, he remained devoted to anthropometry. He published with Thomson in 1923 a study of the nasal index of humans, claiming that variations in the shape of the nasal opening were produced

[100] A. Thomson and D. R. MacIver, *The Ancient Races of the Thebaid: Being an Anthropometrical Study of the Inhabitants of Upper Egypt from the Earliest Prehistoric Times to the Mohammedan Conquest* (Oxford, 1905); David Randall MacIver (1873–1945), *DNB*.

[101] A. Thomson, 'The Maturation of the Human Ovum', *Journal of Anatomy*, 53 (1919), 172–208; *OUG* 50 (1919–20), 686.

[102] On Leonard Halford Dudley Buxton (1889–1939) see *Nature*, 143 (1939), 506–7; *Stapledon Magazine*, 9 (1939), 285–7. In the 1920s Buxton published *The Eastern Road* (London, 1924); *Primitive Labour* (London, 1924); *The Peoples of Asia* (London, 1925); *China: The Land and the People* (Oxford, 1929); and *From Monkey to Man* (London, 1929). David Talbot Rice (1903–72) became professor of history of fine art, Edinburgh, in 1934.

mainly by different habitats. Ten years later he began what he intended to be a series of papers on the standardization of anthropometric technique in Britain and Ireland, still convinced that craniology was the leading sector of physical anthropology.[103]

Buxton's co-worker, Blackwood, an Oxford graduate in English and protégée of Thomson, joined him in 1920 as a research assistant. Buxton had been to Malta and had returned laden with voluminous anthropometric data. Not content with this haul, he began an anthropometric survey of rural Oxfordshire, assisted by her, the local clergy, and the Young Men's Christian Association. Next year she joined him as a demonstrator in physical anthropology in the department of human anatomy and, on his elevation to a readership, she became a university demonstrator in ethnology. Like Buxton she studied both present-day and ancient humans. Though she spent much time in the department arranging and cataloguing hundreds of skulls, she was sufficiently promising as a cultural anthropologist to receive not only a grant from the Rockefeller Foundation but also one from the National Research Council, USA, to do field-work in the Solomon Islands for a whole year.[104]

Thomson had little inclination to undertake such research. He concentrated on gaining more space for the growing collection of human skulls in his department. In 1923 he made an unsuccessful informal approach to the Rockefeller Foundation for £5,000 to pay for extra accommodation for physical anthropology: 'nothing doing' was the immediate response. Undeterred by this rebuff, Thomson turned next year to the University and persuaded it to spend £4,000 on providing extra accommodation exclusively for physical anthropology which he deemed 'a living and important branch of study'.[105] His obsession with collecting crania reached its apogee in 1931 when he acquired on permanent loan over 600 human skulls which had been assembled by George Williamson of the Army Medical Department. This collection had been temporarily moved to Oxford in 1910 at Thomson's request and measured by Francis Knowles, his pupil and assistant, before being returned to London. For Thomson having this collection in Oxford,

---

[103] L. H. D. Buxton and A. Thomson, 'Man's Nasal Index in Relation to Certain Climatic Conditions', *Journal of the Royal Anthropological Institute*, 53 (1923), 92–122; L. H. D. Buxton and G. M. Morant, 'The Essential Craniological Technique, Part 1: Definitions of Points and Planes', ibid. 63 (1933), 19–47; *OUG* 64 (1933–4), 210.

[104] Beatrice Mary Blackwood (1889–1976), second in English 1912, diploma in anthropology 1918, research assistant to Thomson 1918–21, departmental demonstrator 1921–4, Laura Spelman Rockefeller fellow 1924–7, university demonstrator 1927–59. The main fruit of her year in the Solomon Islands, 1929–30, was *Both Sides of Buka Passage: An Ethnographic Study of Social, Sexual, and Economic Questions in the North-Western Solomon Islands* (Oxford, 1935) dedicated to her 'chief' Thomson, anatomist, anthropologist, and artist, to whom she owed her first and greatest debt.

[105] R. M. Pearce note, 8 Oct. 1923, Oxford Biochemistry file, Rockefeller Archives; *HCP* 127 (1924), 171.

permanently and not temporarily, was 'the happy consummation of a project which I have long had at heart'.[106] For Clark the collection of thousands of human skulls, including no fewer than 3,000 in just one room, made an imposing array; but it was useless because there were too few skulls from any one geographical region for valid statistical analysis of contrasting racial characteristics to be made.[107] More importantly the collection of skulls represented a view of human anatomy which Clark deplored and was determined to defeat.

In temperament Clark was impatient in controversy and gave to public expositions of his views a sense of conviction and authority. He came to Oxford in 1934 having had no previous connection with it, but he had the benefits of ten years' experience as a professor at two London hospital medical schools, St Bartholomew's and St Thomas's. He brought with him developed views about the inadequacies of very detailed and descriptive topographical anatomy of the human cadaver. In his opinion the low status of anatomy as a pre-clinical science was the result of its exclusive concern with dissection of dead bodies and with descriptive morphology. A few anatomists, however, had in Clark's view escaped from the morphological trap: in their publications Frederic Wood Jones, Arthur Keith, and Grafton Elliot Smith took a large view of anatomy so that it embraced comparative anatomy, human evolution, and neurology. Not surprisingly Clark's interests became focused on these themes. By 1934 he had acquired a high reputation for his work in three fields: the comparative neurology of the lower primates, which he had begun in the early 1920s when he was the principal medical officer in Sarawak; the evolution of primates and humans, of which his *Early Forerunners of Man* (1934) was the first major fruit; and cerebral functional organization via his work on the fibre connections of the thalamus, a large oval mass of grey matter buried in the middle of the brain. The thalamus research, which led to Clark's election as FRS in 1935, showed his dissatisfaction with exclusively descriptive neuroanatomy.[108]

It also revealed Clark's commitment to experimental anatomy, a commitment he shared with his close friend Herbert Woollard, who followed Clark at St Bartholomew's as professor and ended his short life as professor of anatomy at University College, London. In 1927 Woollard, an experimental

[106] *OUG* 62 (1931–2), 680; Francis Howe Seymour Knowles (1886–1953) graduated in law 1907, took the diploma in anthropology 1909, and as physical anthropologist to the Geological Survey, Canada, published *Physical Anthropology of the Roebuck Iroquois* (Ottawa, 1937).

[107] Le Gros Clark, *Exploration*, 130–1.

[108] Ibid. 73–4, 167–8, 154; S. Zuckerman, 'Clark', *BM* 19 (1973), 217–33; obituaries of Clark in *BMJ* (1971), 3: 118–19, *Nature*, 232 (1971), 429–30; W. E. Le Gros Clark, *Early Forerunners of Man: A Morphological Study of the Evolutionary Origin of the Primates* (London, 1934) used not just bones and teeth but the nervous and uro-genital systems to classify animals into groups; Arthur Keith (1866–1955), on whom see Clark, *BM* 1 (1955), 145–62; Frederic Wood Jones (1879–1954), on whom see Clark, *BM*, 1 (1955), 119–34; Grafton Elliot Smith (1871–1937), on whom see J. Wilson, *ON* 2 (1936–8), 323–33.

anatomist who even did experiments on himself, published his *Recent Advances in Anatomy*, which revived, reoriented, and widened the hoary subject, by stressing the importance of experimentation in anatomical research and by taking a functional view of neurology, cytology, and embryology.[109] Woollard's views were unpopular with the older advocates of descriptive and observational anatomy, but they reinforced Clark's belief that too many anatomists had drifted from the study of form and structure in connection with their functional significance in the living organism, thus ceding experimentation on living entities to physiologists and creating an artificial contrast between anatomy and physiology. Clark thought that the chief villains responsible for this regrettable situation were those Victorian anatomists who, inspired by Darwinian evolution, had focused their attention on morphology to the exclusion of experiment. Such anatomists had neglected an older tradition in which anatomists employed experiment, the nonpareil being William Harvey with his famous *Exercitatio anatomica de mortu cordis et sanguinis in animalibus* of 1628. Clark saw anatomy and physiology as mutually complementary, especially as experimental anatomists disturbed function to learn about structure and physiologists disturbed structure to learn about function. He went so far as to claim that some great physiologists were anatomists, adducing Sherrington's work on the anatomy of the nervous system. This was a fair point because Sherrington always tried to turn anatomical facts into physiological language. For Clark anatomy had lost prestige as a pre-clinical science because too often it had lacked a physiological outlook in that it had ignored functional processes viewed as integral elements in the complete living and evolving human body. He admitted that topographical anatomy was essential for medical students but it was often far too detailed and did not focus enough on topographical relations and on the functional significance of structure.[110]

Clark did not relinquish these views when he assumed the chair of human anatomy at Oxford. On the contrary, the recent development of such experimental techniques as tissue culture made him an even more fervent advocate and exponent of experimental anatomy. In 1939 he proclaimed in a textbook on anatomy that it was fundamentally concerned with the structural organization of the living body, with the conditions of such organization during embryonic and postnatal development, and with the processes by which it is maintained in maturity.[111] This wide and long-held

---

[109] Herbert Woollard (1889–1939), professor of anatomy, St Bartholomew's Hospital, 1929–36, professor of anatomy, University College London, 1936–9, on whom see Clark, *Nature*, 143 (1939), 231–2; H. Woollard, *Recent Advances in Anatomy* (London, 1927).

[110] Clark, *Exploration*, 72–3; W. E. Le Gros Clark, 'The Scope of Teaching and Research in Anatomy', *BMJ* (1936), 2: 413–15; and, generally, *BMJ* (1934), 1: 192–200 for report of BMA Committee on Medical Education.

[111] W. E. Le Gros Clark, *The Tissues of the Body: An Introduction to the Study of Anatomy* (Oxford, 1939).

view went far beyond the traditional approach which confined itself to the description of the structural complexity of mature organisms. For Clark experiment was a key element in studying structural organization. At the same time his 1939 textbook dispensed with topographical minutiae and replaced them with discussions of the growth, differentiation, and structural adaptation of living tissues.

When Clark accepted the Oxford chair he knew he would inherit a small staff, including three readers (Odgers in human anatomy, Parker in applied anatomy, and Buxton in physical anthropology) in addition to two university demonstrators (Alice Carleton in histology, Blackwood in ethnology/ physical anthropology).[112] When Clark arrived, he had no problem with Parker, a surgeon at the Radcliffe Infirmary: by a fortunate coincidence he had reached retirement age. Odgers was a topographical anatomist who did not neglect the surgeon's point of view; like Carleton he had his departmental uses, especially from 1935 when he responded to Clark's persuasion by focusing his research on embryology. It was the physical anthropologists who irritated Clark because their approach was in his view *passé* and they were obstacles to his aim of building up experimental anatomy, especially in research. Clark deprecated the cavalier way in which physical anthropologists had drawn large conclusions regarding race from the scanty evidence provided by studying nothing but skeletal material mainly with measuring tape and callipers: they had failed to fulfil his chief desiderata, that proper physical anthropologists should be good general biologists, familiar with genetics and taxonomy, and effective field-workers studying current populations. Though Buxton and Blackwood had pursued field-work, they were not general biologists and were therefore found sadly wanting by Clark, who wanted to eject them. Presumably Clark realized that a group of anthropologists led by Marett and Myres wanted to give physical anthropology a recognized and statutory place in the department of human anatomy. Though Clark was determined that his department should not be vitiated in this way, it soon became clear to him that Buxton could not be sacked: he was a college fellow, widely liked in the University, and a popular curator of the University Parks. But Buxton could be squeezed and discouraged: by 1937 craniological research in the department was petering out. Subsequently Buxton's unexpected and sudden death in March 1939 totally removed the major craniological thorn from Clark's flesh. The minor thorn of Blackwood, a cultural and physical anthropologist who was not an anatomist, was more easily dealt with. Clark drew the University's attention to her anomalous position and saw to it that in 1936 she was given four terms' study leave at the University's expense to work in New Guinea. On

---

[112] Paul Norman Blake Odgers (1877–1958), reader in human anatomy 1931–45, on whom see *BMJ* (1958), 2: 1477–8; Arthur Percy Dodds Parker (1867–1940), reader in applied anatomy 1927–34, on whom see *BMJ* (1940), 2: 509.

her return she was transferred to the staff of the Pitt Rivers Museum.[113] In autumn 1934 Blackwood had been replaced by Solly Zuckerman, who arrived from Yale in 1934 as a Beit research fellow and departmental demonstrator.

Clark's recruitment of Zuckerman was a master-stroke.[114] A South African Jew, Zuckerman had arrived in 1928 in London where he worked as anatomist to the Zoological Society and as a demonstrator in anatomy at University College. From these bases he acquired a precocious and controversial reputation with his pioneering book of 1932 on the social life of monkeys and apes, followed next year by a volume on the functional affinities of man, monkeys, and apes. The latter work must have appealed greatly to Clark: here was an anatomist who had departed from the usual morphological method of assessing affinities between man and other primates by looking for similarities of structure. Instead he was concerned with functional similarities, with the bearings of physiology and behaviour on questions of taxonomy and evolutionary history, and with trying to use the results of experimental biology to illuminate taxonomic problems. No wonder that Clark, while at St Thomas's Hospital, had tried to lure Zuckerman to join him but Zuckerman had declined. At Oxford Zuckerman soon became a university demonstrator and star researcher. As a publisher he was prolific, producing sixteen papers solo in his first year at Oxford and maintaining that publication rate, either solo or in harness, up to the outbreak of war. He was especially good at attracting and organizing researchers, his main field being endocrinology (the study of ductless glands) with particular reference to sex hormones and the menstrual cycle in baboons. He caused consternation and, in some quarters, irritation by importing into the department these ferocious-looking creatures who lived in large cages and in summer disturbed neighbouring departments with their ceaseless din. The experimental research with monkeys was a far cry from collecting crania and it attracted favourable attention in Oxford and Birmingham. Oxford's vice-chancellor, Lindsay, applauded the work of Zuckerman on the reproductive and endocrine functions of monkeys not least because in these respects he thought them unique among animals in resembling humans. But no college fellowship came Zuckerman's way, though he did become a member of Christ Church common-room and enjoyed its social life. Birmingham was

---

[113] W. E. Le Gros Clark, 'The Scope and Limitations of Physical Anthropology', *Nature*, 144 (1939), 430–1; Clark memo, 5 Nov. 1934; Clark to Margoliouth, 21 Jan. 1935; Blackwood memo, 26 Feb. 1936, all in BS/R; John Linton Myres (1869–1954), Wykeham professor of ancient history 1910–39, president of Royal Anthropological Institute 1928–31.

[114] Obituaries of Zuckerman, *Independent*, 2 Apr. 1993; S. Zuckerman, *From Apes to Warlords: The Autobiography (1904–1946) of Solly Zuckerman* (London, 1978), 84–105; id., *Functional Affinities of Man, Monkeys, and Apes. A Study of the Bearings of Physiology and Behaviour on the Taxonomy and Phylogeny of Lemurs, Monkeys, Apes, and Man* (London, 1933); id., *The Social Life of Monkeys and Apes* (London, 1932); *OUG* 68 (1937–8), 26.

more practical in its approach: the University there invited him in October 1938 to apply for its chair of anatomy and, after protracted negotiations, he finally accepted it in November 1939.

Clark's espousal of experimental anatomy and his contempt for craniology soon became apparent in the ways he revamped the facilities of his department and relieved it of its craniological impedimenta.[115] His approach was to rearrange available space in order to expand experimental research. His chief problem was the large volume of skulls and skeletons, so lovingly collected by Thomson, Buxton, and Blackwood, who had made the department into 'little more than a charnel house'. The answer was to eject such material so that his own anatomists had somewhere to work. He therefore launched the battle of the skulls. His plan, which eventually prevailed, was to transfer on permanent loan the large collection of 3,000 human crania from his department to the Natural History Museum, London, where they would be added to the 10,000-strong skull collection there. In this way Clark gained the only room that could be converted into a teaching laboratory. He abolished the separate rooms where women undergraduates did human dissections under a woman anatomist, brought women into the main dissecting room, and thereby gained a research laboratory. A third adaptation concerned the studio at the top of the building which Thomson had used for painting and photography. As it was well lit and well ventilated, it was suitable for conversion to an animal house, with the dark room as a kitchen for preparing food. A fourth conversion was the result of replacing open coal fires by gas fires: the coal cellar became redundant and was made into another research laboratory. At the same time Clark introduced facilities for experimental anatomy, such as an operating theatre, a sterilizing room, apparatus for cerebral operations, and a crane for lowering cages from the animal house to the basement. Clark's scorn for anthropometry was nicely shown in 1937 when he threw out various instruments acquired in 1895 by Thomson from Francis Galton's laboratory where they had been used to test and measure the efficiency of the mental and bodily powers of individuals: they landed in the Oxford Museum of History of Science. On occasion Clark was simply iconoclastic. The entrance hall to his department was adorned with life-size plaster casts of Venus de Milo, Antinous, the fighting gladiator, and Goodsir's representation of a dead and partly dissected human body, which had been donated by Henry Acland in 1893 to celebrate the opening of the new anatomy building that year. Feeling that these casts were incompatible with his own views about anatomy, he had them surreptitiously removed to hidden recesses in the basement; no protests having been made in six months, he then had the casts broken up.

[115] Clark, memorandum, n.d. [autumn 1934], Med./R (q): *HCP* 166 (1937), 166–8; Clark, *Exploration*, 130–6; Atlay, *Acland*, 478; *OUG* 65 (1934–5), 213–14; 66 (1935–6), 234–5; 68 (1937–8), 487.

Despite some successes in what he regarded as reconditioning, reconstructing, and rearranging his department, Clark was dissatisfied with his lot.[116] He was always short of space simply because his emphasis on experimental anatomy put a premium on accommodation: he became fed up with making *ad hoc* adjustments to the existing building and his preliminary plans for a new one were shelved because of the imminence of war. The colleges continued to show their indifference to anatomy: only one member of Clark's staff was a college fellow (Buxton) and he had been elected *qua* physical anthropologist at Exeter where the rector, Marett, promoted anthropology in general. The University provided for him no more staff than it had done for Thomson at the end of his reign; and though its annual grant increased from 1934 by about £1,000 per year to £1,500 on average, most of it was spent on wages and materials pertaining to teaching. Apropos capital expenditure, the University spent £1,500 in 1935 to provide facilities for the teaching of embryology and neurology and for experimental research. But to promote research Clark had to rely on external grants, applying for which took up much time and caused perennial anxiety. The Nuffield medical benefaction was some use to him but more use to Zuckerman: before war broke out, Clark received almost £700 for departmental research expenses, but Zuckerman secured just over £1,000 for his endocrinology work. From outside Oxford during the five years before the second war, the Rockefeller Foundation with almost £1,700 was easily Clark's best supporter, the Eugenics Society being second with a grant of £400 in 1935 for research on the reproduction and embryology of primates.

There was another feature of anatomy at Oxford which Clark deeply resented.[117] He had tried to widen anatomy to give it more educational value in consonance with Oxford's traditions and prestige; and he had correlated more closely lectures and practical dissection. Yet he thought the standard of anatomy in the first MB was deplorably low. He attributed this parlous situation to the way in which college tutors saw anatomy as less important than physiology and biochemistry, on which they focused their students. Moreover these tutors encouraged their students to take during the Oxford vacations condensed anatomy courses, including dissection, in the London medical schools. To Clark's amazement, Oxford medical students were encouraged to take anatomy courses outside the University. In 1938 his proposed solutions to the problem of low status of anatomy in the medical course at Oxford were to raise the standard of the first MB and to include anatomy in physiology finals. A year later he was pressing for an

[116] Clark, *Exploration*, 140, 157; *HCP* 166 (1937), 166–8; *OUG* 66 (1935–6), 14; 69 (1938–9), 365–76; 70 (1939–40), 377–92.
[117] Clark, memorandum on position of anatomy in medical curriculum, n.d. [Jan. 1938], memorandum on an honours school of anatomy, n.d. [Jan. 1939], Med./R.

independent honours school of anatomy, devoted to structural organization and on a par with physiology.

Clark had experienced such difficulties and frustrations at Oxford that, when Woollard died suddenly in January 1939, Clark applied for the vacant chair of anatomy at University College London and was elected to it in spring 1939. After all, the department there was regarded as the best in England; and it had four features which were lacking at Oxford, namely, adequate endowment, excellent accommodation, ample staff, and propitious prospects. Moreover, the chair at University College London would provide the best platform from which Clark could take upon himself the mantle of Woollard, a long-standing and close friend and professional colleague. Clark also knew that Zuckerman, his star researcher, had been offered the chair of anatomy at Birmingham and was irreplaceable. It was only the contingency of the outbreak of war which induced Clark, on the advice of the provost of University College, to withdraw his resignation which was to have taken effect in autumn 1939.[118]

5.5 Striking Personality 2: H. W. Florey

Florey enjoyed only four and half years as professor of pathology before the Second World War broke out but in less than a quinquennium he transformed the school of pathology. As a professor who developed research rapidly and effectively, Florey was not aided by the colleges: in 1939 only one of his non-professorial staff was a college fellow.[119] Florey's procedure was to outflank the colleges. For teaching staff he turned to the University: in May 1935 he inherited just one university demonstrator (Vollum); by 1937–8 he had three university demonstrators (Vollum, Maegraith, Edward Duthie) and a departmental one (Chain). But for support for researchers he pestered external bodies, especially the Medical Research Council and the British Empire Cancer Campaign. He also attracted several Rhodes scholars, a few people of independent means, the odd Beit research fellow, a couple of academic refugees from the department of zoology (J. R. Baker and Medawar), and an affluent Polish biochemist, Ruth Schoental. The result of his entrepreneurship was that the number of non-staff researchers expanded rapidly: by the late 1930s, at a time when the total number of undergraduates per year taking courses in pathology did not exceed fifteen, there was a maximum of twenty-five non-staff researchers in addition to the staff of six.[120] By the standards of Oxford, Florey's department was extra-

[118] Clark, *Exploration*, 140.

[119] Brian Gilmore Maegraith (1907–89), Australian Rhodes scholar, fellow of Exeter 1934–40, university demonstrator 1937–44, professor of tropical medicine, University of Liverpool, 1944–72.

[120] *OUG*, 69 (1938–9), 229; 70 (1939–40), 253.

ordinary: in the late 1930s, in a university which prided itself on its teaching, the ratio of researchers to undergraduates was 2:1. This fact is the key to understanding Florey's reign and, if appreciated, it gives extra nuance to the excellent accounts of the work on penicillin by Florey's team during the war.

Florey had great drive and commitment to research.[121] He was free from cant, humbug, and pretentiousness. Always direct in his dealings, he did not waste his own time or that of other people. He was easily piqued, with an abrasive and rasping tongue. He was reserved and mordant, a combination which some people found difficult to handle. He wanted to succeed in all he undertook, especially research. As a bushranger of research, he went flat out to win. Trading on the directness and gaucherie expected from a rough Aussie, he was a great finisher of research projects which he pursued with dedicated ferocity. The best example of his power to pursue and organize research with an unswerving sense of direction is, of course, the penicillin project for which he was knighted in 1944 and for which, with Chain and Fleming, he received a Nobel prize for physiology and medicine in 1945; but that power was already evident from the mid-1930s. It was fuelled by Florey's unhappy marriage: from 1926 research was to be his sole emotional outlet for decades. As professor at Oxford, Florey was not distracted from research by office-holding or by teaching hordes of undergraduates. Though he was disappointed by his failure in the 1930s to make pathology a final honours school at Oxford, that failure meant that he was not lumbered with teaching many courses and many students. Nor was he diverted by involvement in diagnostic pathology and morbid anatomy pursued for clinical reasons often in a hospital. Indeed it had been said in the early 1930s that Florey, who had never served his time in a hospital laboratory or done a human post-mortem examination, was not a pathologist at all.

Florey was essentially a physiologist who was a protégé of Sherrington to whom he was indebted for guidance and patronage.[122] Sherrington began to notice Florey in summer 1923 when the ambitious Rhodes scholar gained a first in finals in physiology, as did his friend John Fulton. As a reward for Florey's excellent work, Sherrington appointed him as a demonstrator in mammalian physiology for 1923–4. Sherrington also had two plans for Florey as a researcher. The first, concerning research with Fulton on the nervous control of muscle, was rejected by Florey who was not keen on

---

[121] On Florey's personality see L. Bickel, *Rise up to Life: A Biography of Howard Walter Florey who Gave Penicillin to the World* (London, 1972); E. P. Abraham, 'Florey', *BM* 17 (1971), 255–302; and Macfarlane, *Florey*, esp. 2, 145–8, 199, 209, 216. For Florey's high reputation in 1934, Florey Papers, box 291, file 5 (application for Oxford chair) and file 11 (letters of congratulation).

[122] For Florey's debts to Sherrington, Abraham, 'Florey', 257; Macfarlane, *Florey*, 64, 72, 77, 80, 99–100, 234–7; Sherrington to Florey, 12 July 1923, 4 Feb. 1925, 13 Nov. 1926, 12 Aug. 1928, 6 Apr., 4 May 1929, 3, 26 Oct. 1934, Florey Papers, box 290, file 2; Matthew John Stewart (1885–1956), professor of pathology, University of Leeds, 1918–50.

neurophysiology. The second, which Florey accepted, concerned capillaries on which he researched for a year. It was Sherrington who suggested and then backed Florey's successful application in May 1924 for the John Lucas Walker studentship in pathology at Cambridge from summer 1924. It was Sherrington who suggested and then backed Florey's successful application in summer 1925 for a Rockefeller fellowship which on Sherrington's advice he spent in the USA and not Spain from September 1925 to May 1926. In autumn 1925 Sherrington warmly recommended Florey to the London Hospital, who offered him a research appointment which he accepted. In the mid- and late 1920s Sherrington introduced Florey at the Physiological Society, sent an early paper by Florey to be read at the Royal Society, took Florey to dine with him at the Royal Society anniversary dinner, and kept him informed about posts in pathology. Furthermore Sherrington encouraged Florey to be an experimental pathologist: they agreed that in the mid-1920s the time was ripe for taking an experimental approach to pathology using physiology as a major resource. His final service to Florey was to act as his adviser and promoter in autumn 1934 before Florey was elected to the chair of pathology at Oxford in January 1935. Knowing that the Oxford chair had been Florey's mecca for many years, Sherrington told him that he would do anything that Florey might suggest. Certainly he lobbied three Oxford members of the electoral board on Florey's behalf, he persuaded Lincoln (the college to which the chair of pathology was attached) to nominate as its second representative Edward Mellanby, an old friend of Florey's from his time at Sheffield and then secretary to the Medical Research Council. Sherrington also provided a glowing testimonial which stressed Florey's gift for research and his ability to guide and stimulate research collaborators. Though Sherrington was not a member of the electoral board, his success in persuading Lincoln to put Mellanby on it turned out to be decisive. Just when it seemed that Matthew Stewart, professor at Leeds and a traditional Scottish morbid anatomist, would be elected, Mellanby, whose train had broken down, arrived late and swung the decision in favour of Florey who was widely regarded as Britain's leading experimental pathologist.

Sherrington hoped that, once Florey had put the laboratory in order, he would not be hampered by stringency of means or by the worry and interruption of going outside for funds which should be available inside. This hope was to remain not completely fulfilled: Florey was forced into being what he called an 'academic highway robber' who made frequent applications for grants from all sorts of bodies.[123] In his dealings with the University, Florey stressed that under Dreyer research was focused on bacteriology, whereas he was encouraging a greater diversity of work,

[123] Sherrington to Florey, 10 Jan. 1936, Florey Papers, box 290, file 2; Macfarlane, *Florey*, 296.

embracing experimental pathology, chemical pathology, and bacteriology.[124] He tried to embarrass the University by drawing attention to the potential danger of the animal houses he had inherited. In 1936 he challenged the University to decide whether it wanted a department of pathology containing about twenty-five researchers, making full use of its facilities, or a half-empty and cheaper one in which no more than seventeen researchers would only partly exploit its resources. Subsequently he again tried to embarrass the University, this time by claiming that Beatrice Pullinger, a university demonstrator whom he had brought with him from Sheffield in 1935, was resigning because of inadequate technical assistance; and he argued that it was spoiling the ship for a ha'p'orth of tar to permit laboratory conditions which resulted in research being brought to naught because experimental animals were going septic on account of inadequate cleaning of the laboratory. The result of such sustained pressure was that the University spent £1,600 on restructuring and re-equipping the laboratory, provided more university demonstrators (three by 1937–8), and increased modestly the number of laboratory attendants.

The main external supporters of research were the Medical Research Council, the British Empire Cancer Campaign, and the Rockefeller Foundation; but, of course, not all of Florey's relentless requests were met. At the MRC Mellanby retained his high opinion of Florey, proposing him for election to the Royal Society in 1937. Their mode of communication was identical: each adopted a brusque and forthright manner to stimulate argument, without any consideration for the feelings or position of those to whom he was speaking. Even so, by 1939 Mellanby's patience with Florey was so thin that he rebuked Florey for not giving enough attention to the standards laboratory which the MRC financed.[125] In 1939 the British Empire Cancer Campaign ended its support of cancer research done in Florey's department by Chain and Berenblum, who had received annual grants and expenses from 1936. The Campaign rejected Florey's 1938 scheme of having an Oxford research centre of the BECC who, it was hoped, would dole out a big annual block grant.[126] In his dealings with the Rockefeller Foundation, he was circumspect until autumn 1939 partly because any medical project put to it from Britain was automatically referred to the MRC; but he did secure modest support from the Foundation's Natural

---

[124] For Florey's passages at arms with the University see *HCP* 162 (1935), 85–91; 164 (1936), 66–8, 95; 166 (1937), 17–20; Beatrice Pullinger, university demonstrator 1935–7, who left Oxford to work for the Imperial Cancer Research Fund at Millhill, London, felt bludgeoned by the cohort of Australian bandits in Florey's department: Macfarlane, *Florey*, 275–6.

[125] Macfarlane, *Florey*, 297–8; Dale, 'Mellanby', 193–222.

[126] Florey to secretary, British Empire Cancer Campaign, 15 Oct. 1938, and Veale memorandum, 17 Oct. 1938, OUA, Department of Pathology, 1930–48, UR/SF/PAT/1. Isaac Berenblum appeared in 1936 as a Beit research fellow and stayed until 1950 when he became head of experimental biology, Weizmann Institute of Science, Rehovoth, Israel.

Sciences Division for his research on lysozyme. In autumn 1939 he bypassed the MRC for a second time in securing £1,000 capital and £1,670 p.a. for five years from the same Division for a large biochemical project to be undertaken by Chain and himself.[127] For the first time at Oxford Florey experienced no financial restraint on his research and saw to it that what soon became the penicillin project made rapid progress. On his own doorstep, however, Florey did not benefit greatly from the huge Nuffield medical benefaction.[128] Having advised Cairns extensively and having been assured by him in 1936 that at least £100,000 was earmarked for basic biomedical sciences, he found himself in 1938 fighting to secure £7,500 from the Nuffield Research Fund for them. Florey felt cheated by Cairns and Witts about the funding of his department. In the two years before the war Florey secured from the Nuffield Research Fund just over £900 p.a. to recast the teaching of bacteriology and to support two researchers. Thus in Florey's case the Nuffield benefaction did not match the £1,200 per year then generated for his department by the Dunn endowment. In summer 1939 the Nuffield Research Committee turned down an application from Florey in what he regarded as contemptuous terms which led to what he called, with laconic understatement, a row with the main Nuffield Committee.

Florey was a grant-beggar not only because he could manage several research projects simultaneously but also because some of them involved team research in which people with different skills attacked a problem from various directions. Initially he had some difficulty in operating as he wished because he inherited staff from Dreyer, including two readers (Ainley-Walker in pathology, Gibson in morbid anatomy). He also inherited the autonomous standards laboratory, financed by the MRC and directed by Gardner. Florey's solutions to these problems were to encourage Walker to resign from the pathology staff, to promote Gibson's move to being Nuffield reader and director of pathology at the Radcliffe Infirmary, and to keep the standards laboratory even though he did not like it. Florey needed the aid of Gardner, a good bacteriologist, a college fellow, and an astute university diplomat. In 1936 Florey engineered the appointment of Gardner as reader in bacteriology and head of bacteriology in Florey's own department, leaving the standards laboratory to be run by R. F. Bridges, a colonel recently retired from the Indian Medical Service and a chum of Gardner.[129]

[127] Macfarlane, *Florey*, 276–7, 302–3.
[128] OUA, Medical School: Radcliffe Infirmary: Finance and Relations with, Nov. 1937–Aug. 1946, UR/SF/MD/13/11, esp. Florey to Veale, 24 Jan. 1938; Cairns to Florey, 8 Aug. 1936, OUA, Post-graduate Medical School, Dec. 1932–Dec. 1936, UR/SF/MD/13. File 1; Macfarlane, *Florey*, 296–7; Fraenkel, *Cairns*, 104–5, 124, 127; Florey Papers, box 141, files 5 and 7.
[129] Ainley-Walker and Gibson departed in 1937. For Florey's policy to Dreyer's left-overs and rubbish, Macfarlane, *Florey*, 239–42; Gardner, 'Recollections', 199–208. For Gardner's high standing as bacteriologist see his *Bacteriology for Medical Students and Practitioners* (Oxford, 1933, 1938, 1944, 1953).

Though Florey is famous for directing his penicillin team during the Second World War, he had previously led team research at Oxford on two major projects. The first was concerned with lymphocytes on which Florey assembled no fewer than seven researchers, including two recent recruits, A. G. Sanders and N. G. Heatley.[130] The second project focused on lyzosyme, an enzyme which occurs in mucus: it was known that it killed, dissolved, or inhibited bacteria harmful to humans. From 1929 Florey was interested in the biochemical study of lyzosyme and other naturally occurring substances with anti-bacterial properties. In 1933 he began work on lyzosyme in the context of the larger problem of natural immunity, but for two years he failed to make progress on its chemical nature because of his lack of biochemical skill. Once in Oxford he recruited from the chemistry department two organic chemists, E. A. H. Roberts and E. P. Abraham, who purified and crystallized lyzosyme respectively by 1937. At this time chemotherapy, in the form of sulphonamides, was all the rage but their advent only strengthened Florey's interest in naturally occurring anti-bacterial substances. He therefore encouraged Chain to work on the precise mode of action of lyzosyme on bacteria. By 1940 Chain and three collaborators (Epstein, R. V. Ewens, and Duthie) had shown that lyzosyme was an enzyme and had determined the nature of the substance in the bacterial cell wall that this enzyme broke down.[131]

The lysozyme work showed the existence of a naturally occurring substance which was lethal to certain bacteria yet non-toxic. As it proceeded, Chain and Florey had many discussions about a systematic study of all known anti-bacterial substances. It seems it was Chain who in 1937 searched the literature to try to discover other enzymes besides lysozyme with anti-bacterial properties, collected 200 references, and discovered early in 1938 Fleming's paper on penicillin. Initially Chain thought that penicillin was a sort of mould lysozyme so that investigating it would be a parallel exercise to the ongoing work on lysozyme. He and Epstein therefore began preliminary experiments on the purification of penicillin in 1938. Florey encouraged this

---

[130] Macfarlane, *Florey*, 274–5; Norman George Heatley (b. 1911), a biochemist who had worked for his Cambridge Ph.D. under Joseph Needham on the application of micro-chemical methods to biological problems, came to Oxford in autumn 1936 at Chain's suggestion and was initially funded by the MRC; Arthur Gordon Sanders (1908–80) came in autumn 1936 from St Thomas's Hospital, London, and enjoyed private means; obituary in *The Times*, 12 Sept. 1980. The eight-strong lymphocyte team was: Florey; Heatley; Sanders; P. B. Medawar; Jean Medawar; J. M. Barnes, a Nuffield research fellow; R. H. Ebert, a Rhodes scholar; and A. Cruikshank, a B.Sc. student.

[131] For the lysozyme work, see Abraham, 'Florey', 259–62; id., 'Chain', 47–59; Macfarlane, *Florey*, 252–3, 276–81; Edward Penley Abraham (b. 1913) and Roberts were organic chemists from the Dyson Perrins Laboratory, Oxford; L. A. Epstein (later Falk) was an American Rhodes scholar and D.Phil. student from 1937; R. V. G. Ewens, biochemical assistant to Chain; Edward Stephens Duthie (1906/7–59), demonstrator in pathology at Sheffield 1936–7, worked on mucus secretion, attracted Florey's attention, and came to Oxford in 1937 to be university demonstrator in pathology, obituary in *Lancet* (1959), 2: 356–7.

work because it fitted into his long-standing interest in the inhibition of one sort of bacterium by another, and into his recent concern about his daughter's boils, which were caused by staphylococci which he knew were attacked by penicillin. By late 1939 the penicillin project, mainly funded by the Rockefeller Foundation, was launched.[132]

Two features of the penicillin work during the Second World War were related to Florey's research practice before it. First, as an organizer of team research he had enjoyed the experience of the lymphocyte project and, more vicariously, the lysozyme work. Though Chain was undoubtedly the instigator of the penicillin work both chemically and biochemically, it was Florey who in late 1939 organized a large-scale enterprise in which four main lines of attack were pursued from the start. Florey was a master of logistics, Chain was not: when Florey was away in the USA for three months in 1942, Chain was left in charge of penicillin production which soon ground to a halt. As Macfarlane rightly said, if Florey was the founder and conductor of the orchestra, then Chain was the brilliant soloist in a concerto he chose. He was a quixotic and mercurial virtuoso, quite capable of successfully proposing marriage to Anne Beloff in the unusual environment of a stranded boat in the Baltic Sea.[133] But it was Florey who was captain, secretary, and treasurer of the penicillin team. Florey assembled it, kept it going, and refused to be distracted by the strained relations between Chain and Heatley, especially during 1939–41. Florey had, as the leading members of his group, Chain and Abraham working on the biochemistry and chemistry of penicillin, Heatley on its production, Gardner and Orr-Ewing on its bacteriology, Sanders on its large-scale extraction, Florey and M. A. Jennings on its pharmacology and biology, and Florey, C. M. Fletcher, and Mrs Florey on the clinical testing. This team effort was supported by at least thirteen technical assistants.[134]

Secondly, by early 1940 Florey had available in his department a cohort of researchers whose various skills were put to profitable use in the penicillin team. That was so because Florey took a wide view of what experimental pathology entailed, so that it included biochemistry, and because he had

---

[132] Abraham, 'Chain', 47–59; id., 'Florey', 262–9; Macfarlane, *Florey*, 282–303.

[133] Macfarlane, *Florey*, 357–9; R. W. Clark, *The Life of Ernst Chain: Penicillin and beyond* (London, 1985), 133; T. I. Williams, *Howard Florey: Penicillin and after* (Oxford, 1984), 379.

[134] H. W. Florey, E. Chain, N. G. Heatley, M. A. Jennings, A. G. Sanders, E. P. Abraham, and M. E. Florey, *Antibiotics: A Survey of Penicillin, Streptomycin and Other Antimicrobial Substances from Fungi, Actinomycetes, Bacteria, and Plants*.(London, 1949), ii. 631–71 (639); H. W. Florey and E. P. Abraham, 'The Work on Penicillin at Oxford', *Journal of the History of Medicine and Allied Sciences*, 6 (1951), 302–17; Dr Jean Orr-Ewing (1897–1944), second in physiology 1919, tutorial fellow at Lady Margaret Hall 1938–43, worked in the standards laboratory under Gardner; Margaret Augusta Jennings (1904–94), an Oxford histologist, arrived in autumn 1936 from the Royal Free Hospital, London, and in 1967 married Florey; Charles Montague Fletcher (b. 1911), son of Walter Fletcher and a Nuffield clinical research student 1940–2, joined the penicillin team in 1941; Mary Ethel Florey (1901–66).

attracted or recruited several researchers who ensured that in the department several lines of research were being pursued simultaneously. Of the leading members of the penicillin team, only Gardner and Orr-Ewing were survivors from Dreyer's reign. Florey himself, Chain, Heatley, Sanders, Jennings, and Abraham had all arrived in Oxford after Dreyer's death. Chain was recruited by Florey in 1935 to organize the biochemical side of research in his department. Chain suggested to Florey that Heatley, whom Chain had met in the biochemistry department at Cambridge, should be invited to Oxford to develop micro-biochemical methods in connection with the study of the metabolism of cancer tissue. Heatley accepted Florey's offer, arrived in September 1936, and soon showed his skill by devising a new micro-respirometer. Sanders, a pathologist, and Jennings, a histologist, were both attracted to Florey's department in 1936, the former to do a D.Phil., the latter to work with Florey on mucus secretion. Though Abraham was not a member of Florey's department before the war, he had impressed Florey by crystallizing lysozyme in 1937. When war broke out he was in Sweden studying at the Biochemical Institute, Stockholm, under Hans von Euler, the prince of enzymology. On returning to Oxford, Abraham was told by Robinson, professor of organic chemistry, that Florey had an interesting proposal to make to him. In 1940 Abraham joined the research staff of Florey's department to work with Chain.[135] Thus in early 1940 Florey had available in his department five researchers who were to play important roles in the penicillin work but had made their way there for other reasons. Of the five, two were biochemists (Chain, Heatley) and one (Abraham) an organic chemist interested in labile substances. Florey had the prescience in the mid-1930s to see the importance of biochemistry to experimental pathology so he found posts for Chain and Heatley in his own department because he wanted their particular skills in enzymology and micro-methods respectively; and he wanted them full-time, so apparently he never tried to borrow a biochemist from Peters's department at Oxford. It was indeed characteristic of Florey's wide approach to experimental pathology that within five months of assuming the Oxford chair he had recruited as his first protégé a refugee biochemist, Chain, who was to share with him a Nobel prize ten years later.

5.6 Striking Personality 3: J. H. Burn

In pharmacology, as in pathology and human anatomy, there was a change of professor in the mid-1930s but the effects were not as noticeable because J. A. Gunn was replaced by J. H. Burn in 1937 and the imminence of war

---

[135] E. P. Abraham to author, 13 July 1987, 30 Apr. 1992; Hans Karl August Simon von Euler-Chelpin (1873–1964), Nobel prize 1929, director of Biochemical Institute, Stockholm, 1929–41.

gave him little time to be a new broom who swept vigorously. Burn's department became a world centre for research and teaching only after the Second World War, though there were hints before 1939 of what was to come. The story of pharmacology between the wars mainly concerns the reign of Gunn, who resigned the chair of pharmacology in 1937 to be professor of therapeutics in the Nuffield Institute for Medical Research of which he had been director since 1935. That said, from 1937 there was a shift in the concerns of the department from therapeutics and history of medicine towards research on fundamental pharmacological and physiological problems concerning the autonomic nervous system.

The teaching of pharmacology at Oxford, like that of pathology, was the brain-child of Burdon-Sanderson, the Regius professor of medicine: in 1897 W. J. S. Jerome was appointed lecturer in pharmacology and materia medica. In 1912 the subject was upgraded. Gunn, an Orcadian who was assistant to Sir Thomas Fraser in the department of materia medica at Edinburgh, was lured to Oxford to be reader and to launch laboratory work in new accommodation in the attic floor of the University Museum. The unassuming Gunn so impressed his colleagues that in 1917 he was given the title of professor. Until 1919 he *was* the department, but by 1926 he enjoyed the services of three demonstrators, M. H. MacKeith, K. J. Franklin, and T. B. Heaton, all of whom were Oxonians who had read physiology as undergraduates and had been elected to college fellowships.[136] Next year he inherited the old pathology building as a result of the opening of the new Dunn school of pathology. This was a consolation prize for Gunn whose scheme for the development of pharmacology at Oxford appealed strongly to the Dunn Trustees until Fletcher of the Medical Research Council turned their attention and munificence towards Dreyer and pathology.

This was a set-back for Gunn who wanted a new research laboratory costing £60,000.[137] At the time of his application to the Dunn Trustees, after a decade in Oxford, his chair was still not an established one: his salary was made up of contributions from the University, the Rhodes Trustees, and various colleges. It was only in 1926 that his chair became a permanent one and he became a professorial fellow of Balliol, where he was affectionately styled the chief poisoner. The previous year he suffered an all but fatal infection which left him blind in one eye and sometimes debilitated. He had to learn how to do practical work using monocular vision. In tempera-

[136] William John Smith Jerome (1839–1929); on Gunn see *BMJ* (1958), 2: 1107–8; Thomas Richard Fraser (1841–1920), professor of materia medica, University of Edinburgh, 1877–1918; Malcolm Henry MacKeith (1895–1942), demonstrator in pharmacology 1923–33, fellow of Magdalen 1922–33; Trevor Braby Heaton (1886–1972), fellow of Christ Church 1920–54 and Dr Lee's reader in anatomy 1919–54; Franklin was fellow of Oriel 1924–47; *HCP* 114 (1919), 91–2; 129 (1924), 75, 353–4.

[137] Gunn memorandum, n.d. [autumn 1921], OUA, Pharmacology. Dunn Trustees, UR/SF/PHA/3.

ment Gunn was a modest man who did not try persistently to be a discipline-builder in the University. He had no stomach for supplicating to secure funds from external sources: thus his department was dependent for income on its annual grant from the University and on the fees paid by undergraduates.[138] The former did increase in the last ten years of Gunn's reign from around £750 to £1,250 p.a., but given the paucity of undergraduates (twenty-five was the biggest class in the early 1930s) fee-income stabilized at about £100 p.a.

Gunn did not attract many postgraduates, though he was a distinguished researcher (sometimes in collaboration with colleagues) mainly on amines and adrenaline: he was searching for amines with a longer action than adrenaline but not as toxic as amphetamine. His chief roles, however, were in editing, authoring, and serving institutions. While professor of pharmacology at Oxford, Gunn co-edited three editions of Cushny's famous textbook on pharmacology and theurapeutics and produced five editions of his own textbook.[139] He was a stalwart member of the Pharmacopoeia Commission set up in 1928. Above all he was the chief founder of the British Pharmacological Society in 1931.[140] He took the initiative in sending out a circular proposing the establishment of such a society, he chaired the first meeting held at Oxford, and his colleague MacKeith was elected secretary and treasurer. Gunn also saw to it that the fledgeling Society returned for its annual meeting to Oxford in 1933 and 1934.

In 1935 Gunn was made director of the new Nuffield Institute for Medical Research at Oxford. Two years later he resigned from the pharmacology chair in order to be first occupant of the new chair of therapeutics, while retaining his directorship of the Nuffield Institute. From 1937 he had as his assistant director Kenneth Franklin, an Oxford classicist turned physiologist who in 1924 had been elected fellow in medicine at Oriel and demonstrator in pharmacology. The most prolific researcher in Gunn's staff, Franklin studied circulatory physiology, working on veins up to 1935 and then on fetal circulation. In the 1920s he became a friend of both Sherrington, under whom he had demonstrated for a year, and John Fulton, who encouraged him to use his love of classics as a translator of the works of such pioneer physiologists as Lower and Fabricius. This work broadened into history of medicine, with a book on the history of physiology which had a telling epigraph: 'He who calls what has vanished back again into being enjoys a

---

[138] Gunn to Veale, n.d. [early Nov. 1930], OUA, Pharmacology, Department of, 1925–50, UR/SF/PHA/1.

[139] A. R. Cushny, *A Text-book of Pharmacology and Therapeutics; or, The Action of Drugs in Health and Disease*, rev. C. W. Edmunds and J. A. Gunn (London, 1928, 1934, 1936); J. A. Gunn, *An Introduction to Pharmacology and Therapeutics* (London, 1929, 1931, 1932, 1934, 1936).

[140] W. F. Bynum, *An Early History of the Pharmacological Society* (London, 1981), 1–2, 7–8, 10–11, 19–25.

bliss like that of creating.' Though disappointed by his failure to succeed Gunn in the pharmacology chair, Franklin remained loyal to Oxford's medical school, serving for eleven years as its dean between 1934 and 1946. In 1947 he moved to St Bartholomew's Hospital, London, as professor of physiology and gave full vent to his interest in history of physiology with biographies of William Harvey and Joseph Barcroft and translations of two books by William Harvey.[141]

Burn was elected to the chair of pharmacology in July 1937, the other short-listed candidates being Franklin, to whom it was to be offered if Burn refused it, and A. J. Clark, professor of pharmacology at Edinburgh.[142] Burn brought to Oxford a distinctive physiological approach to pharmacology and a great reputation as a leader and organizer of research. His physiological approach derived from his years at Cambridge where he read physiology as an undergraduate and did research under Joseph Barcroft. It was confirmed by two periods with H. H. Dale, who in 1936 shared a Nobel prize for discoveries pertaining to the chemical transmission of nerve impulses. Burn's first spell with Dale, in 1914 at the Wellcome Physiological Research Laboratories, London, lasted only six months but was sufficient to give him an inspiring introduction to pharmacology. The second spell, which lasted five years from 1920, was as a member of Dale's department of pharmacology at the National Institute for Medical Research, Hampstead, London, where he collaborated with Dale in research on biological standardization, insulin, and histamine. Having had the best apprenticeship possible in physiological pharmacology, Burn left Hampstead in 1925 to become first director of the Pharmacological Laboratories of the Pharmaceutical Society, Bloomsbury, London. In his twelve years there he showed he was an energizing director of research, attracting forty-four co-workers of whom thirty came from abroad. Though his research was primarily concerned with pharmacological standardization, he also worked on the nature and function of sympathetic nerves, being aided from 1933 by Edith Bülbring, a German refugee who with Dale's help became Burn's assistant in his standardization laboratory.[143]

Once settled in his Oxford chair, Burn found himself satisfied with his accommodation but not his staff. He therefore brought to his department two colleagues from Bloomsbury. One was Bülbring who became a departmental demonstrator and continued to be Burn's chief research collaborator. Indeed until 1948 she remained very much under Burn's wing as his

[141] K. J. Franklin, *A Short History of Physiology* (London, 1933); J. De Burgh Daly and R. G. Macbeth, 'Franklin', *BM* 14 (1968), 223–42.
[142] OUA, Meetings Miscellaneous 1937–53, UDC/M/41/2, 16 June, 12, 26 July 1937; Alfred Joseph Clark (1885–1941), professor of pharmacology, University of Edinburgh, 1926–41.
[143] E Bülbring and J. M. Walker, 'Burn', *BM* 30 (1984), 43–90; on Edith Bülbring (1903–90) see T. B. Bolton and A. F. Brading, *BM* 38 (1992), 69–95; on Henry Hallett Dale (1875–1968) see W. Feldberg, *BM* 16 (1970), 77–173.

main research assistant. Only in 1948, perhaps encouraged by her appointment as a university demonstrator in 1946, did she begin work on the physiology of smooth muscle, in which field she became renowned. The other import was Harold Ling, a laboratory technician who had worked with Burn for twelve years in Bloomsbury. At Oxford he was head technician, administrator, and accountant in the department. His ability to help in furthering research was so remarkable that his name appeared as that of co-author on several of Burn's papers. Aided by Bülbring and Ling, Burn quickly developed several lines of research, using young Oxonians and workers from abroad who attacked problems such as the mode of action of local anaesthetics, the relation of sympathetic nerves to skeletal muscle, stimulants of the central nervous system, and the effect of hormones on resistance to poisoning by carbon tetrachloride. This explosion of research was possible because Burn transformed the finances of the department between 1937 and 1939. Though its income from fees was no more than before, the University increased the annual grant from £1,250 to £1,620. Burn also secured £600 from the Rockefeller Foundation for work on the central nervous system and hormones. From the Nuffield benefaction Bülbring received £675 over two years and the department one grant of £300 for research expenses.[144] Consequently the department's income in 1938–9 was more than double what it had been just two years before.

Burn imported from Bloomsbury not only two essential helpers but also a tried social function, that of daily departmental lunches in the library to which visitors were always invited.[145] He acted as paterfamilias, keeping the conversational balls rolling and himself informed about what his researchers were doing. He justified the existence of these lunches, and having a cook on his pay-roll, by the claim that the physiological pharmacology he promoted involved long-drawn-out experiments by researchers who could not waste time and had to eat (minus white coats) on the premises. As Burn had at most only two college fellows, Franklin and Heaton, on his staff, it was not difficult for him to encourage in his department the sort of family atmosphere which hitherto had been mainly the preserve of the colleges. As an added attraction after lunch Bülbring sometimes gave a recital on the grand piano which Burn kept in the library. This musical offering was a departmental affair, there being no evidence that Chain, another accomplished pianist, ever came over from pathology to join her in playing duets.

Burn soon changed the teaching of pharmacology for medical students, making it more experimental and strongly based on physiology. In research he had a lifelong preoccupation with fundamental pharmacological and physiological problems concerning the autonomic nervous system. His

[144] *OUG* 70 (1939–40), 250–2, 377–92.
[145] For the features of Burn's reign 1937–9 see Bülbring and Walker, 'Burn', 52–9, and Bynum, *Pharmacological Society*, 12, 16, 27.

predilection for a physiological approach to pharmacology made his department yet another satellite of physiology, the dominant biomedical science at Oxford, the others being biochemistry under Peters, pathology under Florey, and human anatomy under Clark. Burn, Peters, and Florey were all trained as physiologists at Cambridge or Oxford and brought that inheritance to bear on their respective departments. Though Clark was not trained as a physiologist, his approach to human anatomy emphasized the study of structure in relation to function. In terms of focus of research, Burn's interest in the autonomic nervous system and Clark's in functional neurology were quite compatible with the neurophysiology which had long characterized Oxford physiology. Like Sherrington, Peters, Florey, and Clark, Burn was strongly research-oriented. An intense and impulsive man, he put departmental research before everything else but he did not neglect external responsibilities. Like Gunn he served the Pharmacological Society nobly, being secretary and treasurer for eleven years from 1934 and in 1938 hosting its annual meeting at Oxford not least to show the transformation he was beginning to effect in his department. Unlike Gunn, Burn engendered constant activity and even *frisson* in his department. No wonder it was said, when he was absent, that the department fiddled while Burn roamed.

6
Small Sciences

Geology, botany, zoology, mineralogy and crystallography (henceforth mineralogy), astronomy, history of science, and history of technology (under the aegis of anthropology and the Pitt Rivers Museum) were subjects which existed almost totally outside the colleges but they were recognized by the University with a chair, a readership, or a curatorship. The professors were supported by small cohorts of departmental and University demonstrators, varying in number between two and six, most of whom never became fellows of colleges or waited a long time for their election. In botany, for example, W. O. James and A. R. Clapham, two distinguished demonstrators, were not elected to college fellowships and eventually moved to chairs elsewhere. James served for almost thirty years as a demonstrator in botany at Oxford, being made reader in 1947 and FRS in 1952: in 1959 he assuaged his bitterness by migrating to Imperial College, London, as professor and head of department. Roy Clapham was demonstrator for fourteen years before his elevation to the chair at Sheffield. Another botanist, A. H. Church, was elected FRS in 1921 but no college then recognized his talents.[1] Of those who waited a long time perhaps the best example was John Baker. He was made departmental demonstrator in zoology in 1923, university demonstrator in 1927, reader in cytology in 1955, and FRS in 1958; but he did not become a fellow of New College until 1964 when he was 64 years old and near retirement.[2]

These seven subjects were not only peripheral to the colleges: they were small, attracting few undergraduates or in some cases not offering regular teaching. Though geology, botany, and zoology were subjects in which finals were taken, the number of finalists in zoology and in botany rarely reached ten and in geology sank on occasion to zero. Astronomy was available in finals but recruited no one for decades. Mineralogy and crystallography

---

[1] Arthur Roy Clapham (1904–90), departmental demonstrator 1930–1, university demonstrator 1931–44, professor of botany, University of Sheffield, 1944–69; A. J. Willis, 'Clapham', *BM* 39 (1994), 73–90; William Owen James (1900–78), departmental demonstrator 1927–8, university demonstrator 1928–59, professor of botany, Imperial College, London, 1959–67; A. R. Clapham and J. L. Harley, 'James', *BM* 25 (1979), 337–64; Arthur Harry Church (1865–1937), research fellow at Jesus 1908–12, lecturer in botany 1908–30; A. G. Tansley, 'Church', *ON* 2 (1936–8), 433–43.

[2] John Randal Baker (1900–84); E. N. Willmer and P. C. J. Brunet, 'Baker', *BM* 31 (1985), 33–63.

were not finals subjects but as options for finalists in natural sciences attracted some takers. In the museum-based subjects, history of technology was regularly taught in the anthropology diploma course; but history of science was only irregularly taught via occasional lectures. Though these seven subjects were small in staff and student numbers, their staff were usually active in research and in some cases distinguished. Without college duties and unburdened by large numbers of undergraduates, it was possible for dedicated people and indeed a department to establish a distinctive and distinguished identity in research. That happened particularly in zoology, which was peculiar in that it enjoyed two chairs, the Linacre professorship of zoology and comparative anatomy and the Hope professorship of entomology. There is little doubt that zoology was the most remarkable of these small departments: of those who in 1923 were in it, as either staff or students at various levels, nine were or were to become FRSs. It deserves and receives a separate chapter. Though the chief problem for geology and mineralogy was survival, the former produced Arkell, the world expert in Jurassic geology, and the latter provided the first research home in Oxford for the young Dorothy Hodgkin, later a Nobel prize-winner.[3]

In these small non-vocational subjects a few senior posts, mainly readerships, were created as rewards for personal research and sometimes as a means of conferring responsibility on an individual for all aspects of his subject including teaching. Zoology was the main beneficiary, with readerships in animal ecology (C. S. Elton, 1936) and genetics (E. B. Ford, 1939); the other readerships were in ancient astronomy (J. K. Fotheringham, 1925), chemical crystallography (T. V. Barker, 1927), and history of science (R. T. Gunther, 1934). Zoology was also prominent in institutional innovation. In 1932 Elton launched the Bureau of Animal Population, the most important centre of research in animal ecology in Britain between the wars; after two trial periods, in 1936 it was given limited support by the University. In 1938 the Edward Grey Institute of Field Ornithology was created as a national centre for the study of bird behaviour mainly as a result of the efforts of B. W. Tucker and E. M. Nicholson.[4] In 1935 the Museum of the History of Science was officially established in the old Ashmolean building in Broad Street through the ceaseless and pugnacious endeavours of Gunther. These three institutional innovations had several features in common. All were launched and sustained by persistent, productive, and credible activists who exploited a local pressure group to give visibility to a given project. Elton was a key figure in the Oxford Exploration Club established in 1927. Tucker

---

[3] William Joscelyn Arkell (1904–58); L. R. Cox, 'Arkell', *BM* 4 (1958), 1–14; Dorothy Crowfoot Hodgkin (1910–94), Nobel prize for chemistry 1964.

[4] Bernard William Tucker (1901–50), departmental demonstrator 1925–7, university demonstrator 1927–50, reader in ornithology 1946–50; Edward Max Nicholson (b. 1904), on whom see *Who's Who*.

and Nicholson were prominent in the Oxford Ornithological Society, founded in 1921, which spawned the Oxford Bird Census in 1927 and in 1932 the British Ornithological Trust, with Nicholson as secretary. The Friends of the Old Ashmolean were assembled by Gunther in 1928 to press for the creation of a museum of the history of science in Oxford in a restored old Ashmolean building and not just to augment the collections in it. All three new institutions were heavily dependent on external endowments, not only before but after they were recognized as university institutions: the input of the colleges, apart from a research fellowship at Corpus for Elton and small donations of cash and instruments to the Museum of the History of Science, was negligible. In its early years the Bureau of Animal Population was supported mainly by the Royal Society of London, the New York Zoological Society, the Agricultural Research Council, ICI, and the Carnegie Corporation; in one year it was financed from no fewer than thirteen external sources, after Elton had written over a hundred different begging letters. The Grey Institute was established by an endowment of £3,000 from the appeal launched to commemorate Viscount Grey, a keen ornithologist and chancellor of the University when he died in 1933.[5] The Museum of the History of Science, based on the donation by Lewis Evans of his valuable collection of scientific instruments, depended heavily on money from the Goldsmiths' Company and other city companies interested in displays of fine craftsmanship.[6] Finally the birth-pangs of all three new institutions were somewhat eased by the organizational skill of the University's registrar, Douglas Veale, who as an ally working behind the scenes gave advice and suggested workable compromises.

As well as gains there were institutional losses between the wars. The most important was the closure in 1935 of the Radcliffe Observatory owned and run by the Radcliffe Trustees who moved it to Pretoria. Though the Radcliffe Observatory was only associated with the University by custom, and not by statutory right, some Oxonians felt that, like the Radcliffe Camera and the Radcliffe Library, it was by a long connection an integral part of the University. Accordingly the University took the Radcliffe Trustees to court about their proposed removal scheme. After a long action the Trustees won their case in July 1934, to the general acclaim of most British and dominion astronomers but to the chagrin of Oxonians such as Lindemann who deplored a lost resource and the prospective loss of future endowments for Oxford science. Thus in 1935 Oxford lost its older observatory, active from 1772, leaving it with the University Observatory, opened in 1875 and directed by the Savilian professor of astronomy. A second loss concerned the Science Room, opened in 1914 in a quiet alcove of the Radcliffe Camera,

[5] Edward Grey, Viscount Grey of Falloden (1862–1933), chancellor of the University 1928–33, *DNB*.
[6] Lewis Evans (1853–1930).

which was then a reading room of the Bodleian Library. The Room was devoted to history of science and was the brain-child of Charles Singer who was supported by Sir William Osler, the Regius professor of medicine, and Falconer Madan, the librarian of the Bodleian. Singer's group prospered until 1919 when Madan retired and Osler died. By 1920 the disappointed Singer had transferred his allegiance to University College, London, and no more was heard of his Science Room. This short-lived experiment in promoting history of science failed because Singer lacked a supporting pressure group, did not exploit local pride and opportunity, and had not the stomach to take on the hostile Cowley, Madan's successor. Crucially Singer procured no external funding for his experiment: instead he paid a total of £500 to the Bodleian for the privilege of promoting history of science. Then Cowley bundled him out and appropriated for general Bodleian purposes the balance of £234.[7] A third institutional loss occurred in 1941 as a result of the persistently low recruitment of undergraduates in mineralogy and the growing importance of crystallography: when H. L. Bowman resigned from the chair of mineralogy in 1941, it was abolished and two readerships were created, one in mineralogy in the department of geology and the other in crystallography in the department of inorganic chemistry.

Between the two world wars some subjects gained institutionally, some lost, and others such as geology, botany, and history of technology neither gained nor lost. Thus the septet of peripheral, small, and non-vocational subjects was not homogeneous: though they shared some common features, their careers varied considerably, showing that each of them was also *sui generis* and heavily dependent on the discipline-building capacity (or lack of it) of its chief practitioners.

6.1 Survival: Geology

The most salient fact about geology was that on occasion it endured the embarrassment of having fewer finalists than staff. In 1928 there was just one finalist—who gained a fourth-class degree. Two years later there was not even one finalist. Such low numbers allied to low calibre were perceived at the time as embarrassments: from the late 1920s most entrepreneurial scientists at Oxford used the existence of increased or increasing numbers of finalists as a strong argument in trying to extract more money from the University. The geologists were never in that position between the wars. Their classes, too, were small. In 1933–4 courses on igneous and tectonic geology attracted only three students; those on glaciation and Pleistocene geology just two. In 1929 Arkell gave a class on British Jurassic geology to

---

[7] Charles Joseph Singer (1876–1960), *DNB*; Falconer Madan (1851–1935), librarian of the Bodleian 1912–19; Arthur Ernest Cowley (1861–1931), librarian of the Bodleian 1919–31.

just one student. Between the wars there were only three D.Phil.s taken in geology.[8] Given such small student numbers, the University saw fit to provide from the late 1920s only a trio of university demonstrators (Douglas, Sandford, and Bayzand), none of whom was a college fellow. In addition to the formal staff, there were two wealthy guests in the department, one of whom was a college fellow: Arkell, elected a senior research fellow of New College in 1933, worked in the department, sometimes taught there, and led odd field trips; the other guest was Donald Baden-Powell who worked on Pleistocene molluscs.[9] Yet another aspect of geology, its annual income, was small: though an individual such as Douglas gained intellectually and financially from an oil-company consultancy, the department did not attract external funding until 1937 when Shell-Mex gave £25,000 earmarked for geology as its contribution to the university Appeal. The geology department was totally dependent for income on student fees and a grant from the University: when the former was low, its annual income could be as little as £300.

In addition to these problems geology was burdened with the University's oldest science professor, W. J. Sollas, who died in harness in 1936 aged 87. Appointed in 1897, Sollas's best work had been done before 1914 on the borderline between anthropology and palaeontology, epitomized by his book on ancient hunters. After the First World War, a period on which most of his obituarists are apologetic or reduced to silence, Sollas became remote, contenting himself with updating or completing pre-war works. Though he was a good climber until his late seventies and a graceful diver at Parson's Pleasure on the river Cherwell until his early eighties, he was disarming, unpredictable, and dangerously eccentric as head of department, especially after 1928 when his second wife died and he became extremely deaf. Sollas was a legendary figure in Oxford but was ill suited to running a department: as Sandford rightly said, Sollas belonged to a type fast vanishing under modern conditions of life. The first indication of his strangeness was his choice of the names Hertha and Igerna for his two daughters. Subsequently when his first wife died he replied to a letter of condolence saying that his wife was not dead but had translated herself into another sphere. Bigamy was apparently not prohibited in that sphere: Sollas soon acquired a second terrestrial wife. As an administrator he once tried to pay a demonstrator with a microscope in lieu of money and was not above

---

[8] *OUG* 60 (1929–30), 652; 65 (1934–5), 225–6. The three successful D.Phil. candidates were Kenneth Stuart Sandford (1899–1971) 1923, Arkell 1928, Richard B. McConnell 1937: E. A. Vincent, *Geology and Mineralogy at Oxford 1860–1986: History and Reminiscence* (Oxford, 1994), 226.

[9] James Archibald Douglas (1884–1978), professor of geology 1937–50; Sandford, reader in geology 1948–67, supernumerary fellow of University College 1965–7; Charles John Bayzand (1878–1958); Donald F. W. Baden-Powell (1897–1973), obituary in *Yearbook of the Geological Society for 1973* (1974), 33–4.

spending all available funds for a coming financial year on a large mounted model of a bison. By the mid-1930s Sollas had become a danger to himself. One day he was found at home with his face cut and bleeding. According to Sollas he had swum the Bristol channel and had been bitten by a salmon, when clearly he had either walked through a french window or put his head through a pane of glass. Shortly afterwards the University took the then rare step of appointing a deputy professor (Douglas) to run a science department.[10]

In such a small enterprise as geology, the Burdett-Coutts scholarship for two years of postgraduate work, available to Oxford graduates only, was disproportionately important in launching research careers: of the four regular teachers of geology at Oxford from the late 1920s, three (Douglas, 1905, Sandford, 1921, and Arkell, 1925) were Oxford undergraduates who had gained firsts in finals and then immediately won the Burdett-Coutts award. The other teacher, Bayzand, had come up the hard way, starting as a laboratory technician in 1896 and taking his degree in 1921 as an ex-serviceman: not a brilliant researcher, he became the bluff general factotum of the department which he served for almost fifty years. The senior demonstrator was Douglas who had taught continuously since 1905, with interruptions for research in Bolivia and Peru 1910–12, for war-work in charge of mines and corps of miners in France, and for peacetime work in Persia. In 1924 he became a long-serving palaeontological consultant to the Anglo-Persian Oil Company and soon established a national reputation as an expert on Persian palaeontology, especially permo-carboniferous fauna: by 1925, for instance, he had determined the age of Asmari limestone which was the chief source of oil. Douglas did a lot of teaching and, as the senior demonstrator, acted as Sollas's second in command.[11] Sandford was initially an expert on Thames gravel, but in 1924 he went to Spitsbergen as geologist and glaciologist to the University's Arctic expedition of that year. Subsequently he was a founder of the Oxford University Exploration Club in 1927 and remained an active member, becoming its senior treasurer in 1935. In the mid-1920s Sandford's knowledge of Pleistocene river terraces led to his being invited by the British School of Archaeology (University of London) to work on the Pleistocene deposits of the Nile valley and their relations to early man. He soon correlated the palaeolithic planes of stratification in Egypt with the standard European succession. This coup led to his being appointed director of field research for the Oriental Institute, University of Chicago, which paid for a survey of the traces of palaeolithic man in 1,200 kilometres of the Nile valley,

---

[10] On Sollas see Vincent, *Geology at Oxford*, 26–47; K. S. Sandford, 'Sollas', *OM* 55 (1936–7), 117; A. Smith Woodward and W. W. Watts, 'Sollas', *ON* 2 (1936–8), 265–81; *OUG* 67 (1936–7), 111.

[11] On Douglas see Vincent, *Geology at Oxford*, 34–63; J. M. Edmonds, 'Douglas', *Nature*, 274 (1978), 196.

Arkell acting as a co-researcher for four winter seasons from 1926. In 1932 Sandford extended his Nile work into a major adjoining desert when he went on Major Bagnold's Libyan desert expedition, being supported from the Murchison fund of the Royal Geographical Society. Thus Sandford's long stint of work in north Africa from 1925 to 1933 was supported by three bodies outside Oxford. That happened because he had acquired an international reputation, being secretary from 1926 of the commission on Pliocene and Pleistocene terraces of the International Geographical Union.[12]

Whereas Douglas and Sandford were dependent on external financing of their research when it involved field-work abroad, Arkell was not.[13] A man of independent means, with homes near Oxford and on the Dorset coast, Arkell did some teaching but his forte was virtually full-time research. Though he remains famous as an authority on Jurassic stratigraphy and palaeontology, it should not be forgotten that between 1928 and 1934 he published extensively with Sandford on prehistory and that in the late 1930s he worked on the tectonic geology of south-east Dorset. That said, early in his career Arkell conceived the idea of revising Albert Oppel's work on the Jurassic rocks of England, France, and Germany, and extending it to Europe and the rest of the planet. His *Jurassic System in Great Britain* (1933) was the first stage of a programme finally completed in 1956 with his *Jurassic Geology of the World*. The big 1933 book, with its sophisticated adaptation of Oppel's notion of zones, its use of ammonites as guide fossils, and its long-distance correlations of strata, impressed Sollas and others. Arkell turned down an offer from Cambridge of a paid post; yet no equivalent offer was made to him from Oxford which Sollas wanted to make into a mecca for researchers in Jurassic geology. Sollas also hoped that Arkell, if recognized in his own university, would attract more undergraduates to study geology and increase the fee-income of the department. From 1933 Arkell was a research fellow of New College but nothing more than a guest researcher in the department which he adorned. Even so he expressed strong views in 1934 on the planning of the Science Area and the place of geology in it. He thought the provision for research the most deplorable part of the department: researchers worked in nothing more than stalls, divided by thin partitions in a sort of penthouse tacked on to an outer wall of the

---

[12] On Sandford see *Who Was Who*; Vincent, *Geology at Oxford*, 40–2, 55–8, 65–7, 129–36; *Proceedings of the Geologists' Association*, 83 (1972), 117–18; *HCP* 161 (1935), 153–5; Ralph Alger Bagnold (1896–1990) led several expeditions to the Libyan desert 1925–32: see M. J. Kenn, 'Bagnold', *BM* 37 (1991), 55–68.

[13] On Arkell see Cox, 'Arkell'; *DSB*; W. J. Arkell, *The Jurassic System in Great Britain* (Oxford, 1933), 681 pp., 42 plates; Sollas to Margoliouth, 25 Apr. 1933, BS/R/1/2; Sollas memo, n.d., considered Oct. 1934, BS/R/1/3; Arkell to Veale, 23 Feb. 1934, and Arkell memo, n.d. [May 1934], OUA, Sites. Committee on Building Requirements, UR/SF/S4. Albert Oppel (1831–65), *DSB*, divided Jurassic rocks into thirty-three zones, each of which was characterized by several typical species of fossils, mostly ammonites.

Museum building. Arkell wanted a new geology building separate from 'the antiquated Museum, with its ill-lit passages and steps, its dusty collections, and above all its atmosphere of makeshift, of adaptation superimposed on adaptation'.

Though not in favour of a new building, Douglas was also seething: the Oxford department was the worst equipped in any British university because the University had ignored it, and its inadequacy was a byword among geologists. Only one of the research stalls had a sink and water. Douglas felt that, though there were openings for geologists in oil-fields, in colonial survey work, and in British waterworks, the poor accommodation and equipment at Oxford deterred all but a handful of undergraduates from reading honours geology. From 1934 the geology department endured a further tribulation: in summer its inhabitants were constantly disturbed and driven frantic by the ceaseless din emanating from Solly Zuckerman's colony of baboons housed in the human anatomy department which was opposite their windows only a few yards away. Sollas did not fully apprehend the nature of this aural problem: when Douglas complained to him about it, Sollas retorted, 'Balloons! They are probably something to do with the Air Force.'[14]

After Sollas's death in autumn 1936 the University decided, on the advice of the Biological Sciences Board, that it would be disastrous to elect to the geology chair a specialist in either petrology or palaeontology who lacked a wide knowledge of geology as a whole. Only five candidates, Douglas, Arkell, J. V. Harrison, H. A. Baker, and C. T. Madigan, thought it worth applying for the chair. In March 1937 Douglas, the senior demonstrator and deputy professor with a safe pair of hands, was elected in preference to Arkell, twenty years his junior, who resigned his New College fellowship in 1940.[15] After war-work and illness, in 1947 he accepted enthusiastically the offer of a senior research fellowship at Trinity, Cambridge, moving his home to Cambridge in 1948. Thus in 1937 Oxford lost the opportunity of making its department of geology into a mecca for students of Jurassic rocks. Twenty-one years later Arkell tried again: he bequeathed his earlier fossil collections and his vast library of world Jurassic literature to Oxford in the hope that the geology department would become a centre for work on Jurassic palaeontology.

---

[14] Douglas to Veale, 20 Apr. 1934, UR/SF/S4; Douglas to McWatters, 27 Nov. 1937, OUA, Science Research Studentships, 1937, UC/FF/500/2; Douglas, 'Sollasiana', tape 1976, University Museum, Oxford.

[15] *OUG*, 68 (1937–8), 27; OUA, UDC/M/41/1, 17 Mar. 1937; Herbert Arthur Baker (1885–1954), director of the Geological Survey of Newfoundland 1926–9; Cecil Thomas Madigan (1889–1947), Rhodes scholar, first in geology 1919, lecturer at University of Adelaide 1922–40; John Vernon Harrison (1892–1972), university demonstrator in geology 1938–55, reader 1955–9.

Douglas enjoyed little more than a couple of years as professor before the Second World War broke out.[16] He had long made plans for revitalizing the department, the first fruit of which was the donation of £25,000 by Shell-Mex in 1937 for new accommodation for geology. This gift, the result of Douglas's strong connection with the oil industry, made possible the proposed adaptation of the old Clarendon Laboratory for the purposes of geology, a conversion not implemented by 1939 because the new Clarendon Laboratory was completed only in autumn of that year. Thus the early years of Douglas's period of reconstruction saw little improvement in accommodation: researchers still worked in stalls, with matchboard partitions but without back walls or doors; they were still subjected to the screeching of Zuckerman's fighting baboons all day long; and Douglas thought it worthy of record that in 1937–8 water and a sink were laid on for a new colleague, J. V. Harrison. Apropos staff, Arkell was still invited to teach as a guest but the main change was the arrival of Harrison, a non-Oxonian who had worked for almost twenty years for the Anglo-Persian Oil Company, to which Douglas was palaeontological consultant, and had collaborated with him. He extended the teaching of petrology and revived field excursions to classic geological sites such as the Isle of Arran. Douglas also secured a much bigger annual grant from the University. Yet problems remained. The conversion of the old Clarendon Laboratory was completed only in 1949, one year before he retired. The relation with mineralogy remained to be worked out in 1939. *Qua* professor Douglas never saw a geologist become a college tutorial fellow: that happened as late as 1964 when St Edmund Hall elected Oxburgh.[17] Lastly the contrast with Cambridge was not encouraging, especially with respect to undergraduate and postgraduate numbers and to accommodation.

6.2 Indexes and X-rays: Mineralogy and Crystallography

Prima facie the department of mineralogy and crystallography was in an even worse position than that of geology. It shared geology's low recruitment of undergraduates, but at best enjoyed the services of only two university demonstrators in the 1930s. Unlike geology it was not a final honours school: it was available only as a supplementary subject in chemistry and geology finals. In 1941 the department suffered the ultimate indignity when, on Bowman's resignation, his chair was abolished somewhat autocratically and the department was split, with mineralogy going to

---

[16] *Oxford*, 5 (1938), 49; *OUG* 69 (1938–9), 222–4; Vincent, *Geology at Oxford*, 46–52, 56–8.
[17] Ernest Ronald Oxburgh (b. 1934), fellow of St Edmund Hall 1964–78, rector of Imperial College, London, 1993– , on whom see *Who's Who*.

geology and crystallography to inorganic chemistry.[18] At the same time readerships were created in principle in mineralogy and crystallography. Yet the department proved to be a propitious home, in its eventide phase, for the development of X-ray crystallography as pursued by H. M. Powell (known as 'Tiny' owing to his diminutive stature) and by Powell's first research student, Dorothy Crowfoot.[19] In the 1930s these two researchers in X-ray crystallography laid the foundations of their future success: both became FRS, both stayed in Oxford until retirement, and Hodgkin, as she was then known, was awarded a Nobel prize in 1964. She is the only British woman to win a Nobel prize for science. Though Powell was a university demonstrator in the mineralogy department from 1935, she was an unpaid guest researcher there just like Arkell in geology. Their fates were, however, different: Oxford retained her but lost Arkell to Cambridge. Thus the department of mineralogy was paradoxical: while it moved from marginality to extinction institutionally, it provided a not unfavourable habitation for two innovative researchers, Powell and Hodgkin.

Bowman, who assumed the chair of mineralogy in 1909 aged 35, could have soldiered on to an advanced age like Sollas but chose to resign in 1941 aged 67, then the standard retiring age for professors appointed after 1924. He cannot therefore be regarded as a geriatric eccentric or failure. He was an Oxonian to the core, having been an undergraduate (taking finals in both chemistry and physics), a demonstrator, and a professor there, with just a short period abroad studying at Munich under Groth. Bowman was a reticent, shy, and urbane man, a keen supporter of the Oxford Orchestral Society and a string-quartet player. He was not a fund-raiser, an empire-builder, or after the first war an active researcher. His department, like geology, was funded from a university grant and student fees; in the 1930s he enjoyed the services of at best two university demonstrators, neither of whom was a college fellow, and one technician; and in the 1920s he revised the standard textbook of Miers who had inspired him as an undergraduate. Though this was a work of homage, Bowman introduced a chapter on X-ray methods, as befitted a professor whose chair encompassed crystallography from 1927. Characteristically he appreciated their importance but did not use them himself. As a teacher Bowman found himself on occasion facing just one student. As an administrator of his department, he existed on such a low budget and had such distaste for grantsmanship that he made do with apparatus for research improvised from pram wheels and second-hand motor-cycle gearboxes.[20]

---

[18] *HCP* 172 (1939), 199, 229–30, 302–3; 173 (1939), 52–5, 97–9; Vincent, *Geology at Oxford*, 19–24.

[19] Herbert Marcus Powell (1907–91), reader in chemical crystallography 1944–64, professor 1964–74, fellow of Hertford 1964–74; Powell Papers, A3, B24.

[20] On Bowman see *Nature*, 149 (1942), 662; *OM* 60 (1941–2), 307–8; H. A. Miers, *Mineralogy: An Introduction to the Scientific Study of Minerals*, rev. H. L. Bowman (London, 1929),

Until about 1930 the work of Bowman's department was chiefly focused on crystal indexes, which involved studying the external characteristics of crystals irrespective of their chemical composition and internal structure; then there was a shift, by no means total, to X-ray crystallography which focused on the internal structure and chemical composition of compounds, while neglecting their external features. That relative usurpation is the main feature of the department's research between the wars. The work on crystal indexes was led by T. V. Barker, like Bowman an Oxford undergraduate who had been inspired by Miers and studied under Groth. In addition, while a research student at Magdalen, Barker went to St Petersburg in 1908 to study under Federov, then director of the Imperial School of Mines.[21] When Bowman was elected to the chair of mineralogy in 1909, Barker replaced him as demonstrator and slowly gained recognition, with a fellowship at Brasenose in 1913 and a university readership in chemical crystallography in 1927. Barker proselytized on behalf of crystallography, lecturing to the Science Masters' Association, giving courses for teachers in schools, and putting on a display at the British Empire Exhibition held at Wembley, London, in 1924. Above all he was concerned to correct and amplify the methods of his revered master Federov, who aimed to establish a systematic method of classifying all known crystalline substances so that an unknown crystalline substance could be identified from a single measurable crystal without recourse to chemical analysis. While primarily a Federovian Barker was not hostile to the new X-ray crystallography: in 1924 he spent the summer vacation in Berlin learning its techniques; and he did not oppose the installation of X-ray apparatus in the department in 1928. However his career was becoming increasingly consumed by administration. He was secretary to the Delegates of the University Museum 1925–8 and the elected secretary to the University Chest, a post so onerous that he resigned his readership in 1929 and left the department's staff. His book of 1930 on systematic crystallography was his major attempt at revising the approach of Federov, who had not made clear how the characteristic angles of a crystalline substance were to be chosen. Barker provided a set of rules with which he tried to ensure that all investigators would measure the same angles of a given

---

281–99; F. Welch, 'Reflections', in G. Dodson, J. P. Glusker, and D. Sayre (eds.), *Structural Studies on Molecules of Biological Interest: A Volume in Honour of Professor Dorothy Hodgkin* (Oxford, 1981), 26–9. Henry Alexander Miers (1858–1942), *DNB*, professor of mineralogy, Oxford, 1896–1908, principal of University of London 1908–15, vice-chancellor of University of Manchester 1915–26, studied under Paul Heinrich von Groth (1843–1927), *DSB*, professor of mineralogy, Munich, 1883–1924.

[21] On Barker's career see H. A. Miers, 'Barker', *Journal of the Chemical Society* (1931), 3344–8; T. V. Barker, *Graphical and Tabular Methods in Crystallography as the Foundation of a New System of Practice* (London, 1922); id., *Systematic Crystallography: An Essay on Crystal Description, Classification and Identifi*cation (London, 1930); *OUG* 55 (1924–5), 732; 59 (1928–9), 681. On Evgraf Stepanovich Federov (1853–1919) see *DSB* and his *Das Krystalreich: Tabellen zur krystallochemischen Analyse* (Petrograd, 1920).

substance. His unexpected and premature death in 1931 did not bring his research project to an end. On the contrary it was pursued vigorously by R. C. Spiller and M. W. Porter, who began to evaluate Barker's method of crystalline classification immediately after his death with a view to preparing a comprehensive index of crystal angles for use in identifying substances by their crystalline form.[22] Their mammoth work on the Barker index of crystals, which involved thousands of calculations and measurements and international collaboration, took thirty-three years to complete. Spiller and Porter were research collaborators of Barker, only five years their senior, in the 1920s and took his *Systematic Crystallography* as their bible. Both were loyal Oxonians but in heterodox ways. Spiller began his career in the mineralogy department aged 16 as a laboratory assistant, learned much from Miers, fought in the First World War, took a first in geology in 1922 as a mature student (a rare breed then), won the Burdett-Coutts scholarship, and was then successively departmental demonstrator, university demonstrator, and reader in mineralogy. Porter's career was just as unorthodox. Resident in Oxford but lacking formal academic qualifications, she was spotted and encouraged by Miers and studied abroad. In 1919 she was elected to the Lady Carlisle research scholarship at Somerville, and next year became the first woman to gain an Oxford B.Sc. For decades she was a guest researcher in the department and then in geology until she was in her early nineties.

In 1930 the Barkerian tradition of morphological crystallography was found wanting by Tiny Powell, an Oxonian who had graduated in chemistry in 1928 and had been appointed as departmental demonstrator in 1929 to replace Barker. For the first six months of 1930 Powell went to Germany to learn X-ray techniques from Ernst Schiebold at the University of Leipzig. On his return he began research on alkyl thallium compounds and in autumn 1931 was joined for a year by Dorothy Crowfoot, an Oxonian chemist whose subject for her part two degree was an extension of Powell's research to dialkyl thallium halides. Under Powell X-ray plant was installed in 1932–3 in a different room at the north end of the Museum. Two years later Crowfoot secured via Robert Robinson a grant of £600 for new X-ray apparatus in what came to be called the X-ray laboratory. It was there that Powell and Crowfoot pursued their research. Powell studied the structure of complicated inorganic compounds. In the late 1930s his research prospered, in part through collaboration with Hume-Rothery on the structure of alloys

---

[22] Reginald Charles Spiller (1886–1953), reader in mineralogy 1944–52, obituary in *OM* 72 (1953–4), 270; Mary Winearls Porter (1886–1980), Lady Carlisle research fellow at Somerville 1919–29, honorary research fellow at Somerville, 1949–80, obituary in *The Times*, 2 Dec. 1980; R. C. Spiller and M. W. Porter, 'The Barker Index at Oxford', *Nature*, 144 (1939), 298–9; M. W. Porter and R. C. Spiller, *The Barker Index of Crystals: A Method for the Identification of Crystalline Substances*, 2 vols. (Cambridge, 1951, 1956); M. W. Porter and L. W. Codd, *The Barker Index* (Cambridge, 1964).

and with promising pupils such as A. F. Wells.[23] Such was Powell's reputation relative to that of Crowfoot in 1944 that he defeated her and other strong contenders in a contested election for the readership in chemical crystallography. In 1947 she turned the tables, being elected FRS six years before he was. His FRS did not lure any college into electing him as a fellow: that happened later when he was 57 years old, when Hertford responded to his elevation to the chair of chemical crystallography in 1964. In contrast, Crowfoot's career shows the vital importance for her of the college system. Her father J. W. Crowfoot was an Oxford classicist who pursued a distinguished career in North Africa and Palestine, as an educator and archaeologist, becoming principal of Gordon College, Khartoum, director of education and archaeology in Sudan, and director of the British School of Archaeology in Rome. Her mother was an expert on ancient textiles. Crowfoot went to Oxford simply because her father had been there. She landed at Somerville because her mother chose it and her father had met and was very impressed by Margery Fry, its principal. She decided to read chemistry because it was well taught at her school and she had been particularly impressed by two books, one of W. H. Bragg's books on X-ray crystallography and Parsons's textbook of biochemistry. Before arriving in Oxford she was thinking about X-ray work on the structure of biochemically important molecules. As an undergraduate she was excited by the lectures on complicated organic chemical structures given by Robinson, the professor of organic chemistry. Through Freddie Brewer, her final-year tutor, she was told that Powell had begun research in X-ray crystallography in the mineralogy department so she did her part two in chemistry for a year under him, graduating in 1932 with a first.

In 1932–4 she worked under Bernal at Cambridge, learning about the application of X-ray crystallography to the study of the structure of big organic molecules such as steroids and proteins, a new and exciting field which became her speciality. She was enabled to serve her intellectual apprenticeship under Bernal for two years and to gain her Ph.D. in 1936 because Somerville elected her for 1932–3 to a Harcourt research scholarship, available only to Somerville graduates every other year and worth £100 p.a., and then in spring 1933 made her a research fellow in natural science tenable for two years from October 1933. Funded by this fellowship she

---

[23] On Powell, Bowman memoranda, 24 Nov. 1931, 7 Mar. 1932, PS/R/1/3; *OUG* 61 (1930–1), 699; 62 (1931–2), 670–1; 64 (1933–4), 215–16; 66 (1935–6), 218–19. The short list for the readership in 1944 was Powell; Hodgkin; Cecil Arnold Beevers, co-inventor of the Lipson–Beevers 'strip'; Charles William Bunn (1905–90), researcher with ICI, on whom see U. W. Arndt, *BM* 37 (1991), 71–83; Ernest Gordon Cox (1906–96), then reader in chemical crystallography, University of Birmingham; Alexander Frank Wells (b. 1912), first in chemistry 1935, author of *Structural Inorganic Chemistry* (Oxford, 1945, 1950, 1962, 1975, 1984). Powell provided many diagrams for W. Hume-Rothery, *The Structure of Metals and Alloys* (London, 1936). On Hume-Rothery (1899–1968), see *DSB* and G. V. Raynor, *BM* 15 (1969), 109–39.

returned to Oxford in autumn 1934 and chose to begin her own independent work as an unpaid and unpaying freelance guest researcher in the department of mineralogy doing just what she wanted. She needed more apparatus and, realizing that neither Bowman nor Powell was adept at securing funding, she turned to Robinson who rapidly persuaded ICI to give her £600 for two X-ray tubes, a transformer set, and two goniometers (used to measure the angles of crystals). Through Powell's contact with Lindemann, alternating current was laid on via a cable from the Clarendon Laboratory. In spring 1935 Crowfoot used the new facilities to begin work on the structure of the protein insulin, using crystals supplied by Robinson. In return she helped Robinson by identifying sterol derivatives for him. In 1935 Somerville formally elected her as tutorial fellow in natural science, which gave her further financial and intellectual independence. At the outbreak of the second war she was one of only two tutorial fellows in science in the five women's colleges, the other being Jean Orr-Ewing at Lady Margaret Hall. Somerville also provided the means by which she met her future husband Thomas Hodgkin, an Oxford historian and son of R. H. Hodgkin, provost of Queen's. They first met at the London home of Margery Fry, retired principal of Somerville, who was providing a haven for her cousin Thomas Hodgkin, a communist who had recently been deported from Palestine. They married in 1937, the first of their three children being born in 1938; but Crowfoot managed to sustain her left-wing politics and to combine motherhood and research. It was in 1937 that she secured her first research student, D. P. Riley, in an unusual way. Even though she was a college fellow, as a woman she was denied membership of the Alembic Club, the University's chemical society. In 1936 she was asked by Riley on behalf of the junior section of the Club to talk about the research she had done at Cambridge. He was so impressed by her account that he asked her to supervise his research for his part two in chemistry for 1937–8. Then he became her first D.Phil. student, working on the crystal structure of proteins. By the outbreak of war she was beginning to withdraw from collaborative work with her mentor Bernal, and had established an independent reputation: in 1939 W. H. Bragg and M. Perutz were happy to supply her with X-ray data about haemoglobin for Riley's calculations, and she was widely regarded as Britain's leading authority on the determination of molecular weights using X-rays.[24] Her scientific life was focused on research: before the

[24] For Crowfoot's early career see D. C. Hodgkin and D. P. Riley, 'Some Ancient History of Protein X-ray Analysis', in A. Rich and N. Davidson (eds.), *Structural Chemistry and Molecular Biology* (San Francisco, 1968), 15–28; D. P. Riley, 'Oxford: The Early Years', in Dodson *et al.*, *Structural Studies*, 17–25; *OUG* 66 (1935–6), 218–19; Hodgkin Papers, A1–8, A13, A15–18. John Winter Crowfoot (1873–1959), *Who Was Who*; Sara Margery Fry (1874–1958), principal of Somerville 1926–31, on whom see *DNB* by Thomas Lionel Hodgkin (1910–82), *Who Was Who*; Robert Howard Hodgkin (1877–1951), provost of Queen's 1937–46, *Who Was Who*; Frederick Mason Brewer (1902–63), then demonstrator in inorganic chemistry; John Desmond Bernal

war she gave only one course of lectures and, unlike Powell, avoided a demonstratorship.

It is not generally appreciated that in the late 1930s there was a challenge launched in Oxford to Crowfoot's experimental approach to protein structure. It came from Dorothy Wrinch, a Cambridge mathematician who in 1922 married J. W. Nicholson, fellow in mathematics and physics at Balliol. Through him she obtained *ad hoc* teaching of mathematics in several women's colleges, especially Lady Margaret Hall. Her distinction was such that in 1929 she became the first woman to gain a D.Sc. at Oxford. Next year she separated from her husband, whose drunkenness had become intolerable. She was left as a single parent, caring for a daughter and subsisting mainly by continuing to teach mathematics in various women's colleges. By 1934 she was focusing her research on the application of mathematics to biology and her work caught the attention of the Rockefeller Foundation who awarded her £500 p.a. for five years from autumn 1935 to pursue it full-time. As she was a lecturer in mathematics at Lady Margaret Hall, the Rockefeller grant was paid to her not directly but through the University Chest, an act which implied some formal connection with the University. From 1936 to 1939 she developed models for protein structure called cyclols, which she worked out a priori on the basis of a suggestion made to her by Bernal, who had heard it from Astbury to whom it was suggested in the discussion after a lecture he gave in Oxford in 1933. She elevated these models from working hypotheses into theories; and then claimed dogmatically that, with her closed hollow polypeptide cage-structures, she had at last discovered the structure of globular proteins. As her models were beautiful and her reasoning elegant, the cyclol hypothesis attracted widespread interest and reinforced Wrinch's belief that she was the queen of protein chemistry. She alleged that the available X-ray evidence corroborated her theory. This was a threat to Bernal and Crowfoot who had obtained the first X-ray photographs of a protein crystal as recently as 1934. Not surprisingly they suspected Wrinch, a theoretician with little experience of experimental science, as an unwelcome trespasser who had the cheek to try to interpret

---

(1901–71), on whom see Hodgkin, *BM* 26 (1980), 17–84, esp. 30–52; Dennis Parker Riley (b. 1916), D.Phil. 1942; William Henry Bragg (1862–1942), Nobel prize-winner 1915; Max Ferdinand Perutz (b. 1914), *Who's Who*, in 1939 a Ph.D. student at Cambridge, Nobel prize-winner 1962; T. R. Parsons, *Fundamentals of Biochemistry in Relation to Human Physiology* (Cambridge, 1923); W. H. Bragg, *Concerning the Nature of Things: Six Lectures Delivered at the Royal Institution* (London, 1925). The importance of Somerville to Hodgkin is ignored by G. Hudson, 'Unfathering the Unthinkable: Gender, Science, and Pacifism in the 1930s', in M. Benjamin (ed.), *Science and Sensibility: Gender and Scientific Enquiry, 1780–1945* (Oxford, 1991), 264–86; by G. Kass-Simon and P. Farnes (eds.), *Women of Science: Righting the Record* (Bloomington, Ind., 1990), 371–6; but partly appreciated by H. Rose, *Love, Power and Knowledge: Towards a Feminist Transformation of the Sciences* (Bloomington, Ind., 1994), 117, 155–8. The Harcourt scholarship was established at Somerville in 1928 in memory of Rachel Vernon Harcourt, wife of the Oxford chemist A. G. V. Harcourt, by a gift from her ten children.

their data in their new field. Wrinch's approach became a particular threat to Crowfoot. Wrinch's address was often given in her papers as the Mathematical Institute, Oxford. It sounded impressive to foreigners and non-Oxonians, but was nothing more than six rooms in the Radcliffe Science Library. In 1938 Crowfoot published a paper on insulin's structure, in which she stated that her results had no direct relation to the cyclol theory which she classified as a theoretical speculation. Wrinch reacted very quickly by claiming that Hodgkin's data provided corroboration in great detail for the cyclol theory. In November 1938 the two prospective queens of protein chemistry spoke at a Royal Society meeting on proteins, Wrinch advocating her theory and Crowfoot implicitly but not explicitly rejecting it. Presumably Crowfoot was annoyed by the way in which Wrinch implied that she (Crowfoot) could not interpret her own data. Perhaps she was amused when Wrinch tried to lure Riley to desert one Dorothy for another and, when he declined, tried unsuccessfully to gain access to his recent experimental results before he himself had analysed them.

Suspicious of Wrinch's approach, which they regarded as dogmatically theoretical and insensitive to the peculiarities of X-ray data, and of her behaviour, Bernal, Riley, and Fankuchen took up the cudgels on behalf of the then silent Crowfoot against Wrinch in early 1939, claiming vigorously that Crowfoot's data for insulin destroyed Wrinch's cyclol model of it. In Oxford Robinson and his colleagues in organic chemistry at Oxford suspected her theory and found her personally disagreeable. In late 1938 Robinson reported to the Rockefeller Foundation that the cyclol theory was baseless. Though Wrinch's models were rejected by X-ray crystallographers and organic chemists, they did arouse the interest of prominent physicists such as Lindemann, Bohr, and especially Langmuir, a Nobel prize-winner in 1932 who was awarded an honorary doctorate at Oxford in December 1938.[25] In spring 1939 they helped to secure her election as

---

[25] On Dorothy Maud Wrinch (1894–1976) see P. G. Abir-Am, 'Synergy or Clash: Disciplinary and Marital Strategies in the Career of Mathematical Biologist Dorothy Wrinch', in P. G. Abir-Am and D. Outram (eds.), *Uneasy Careers and Intimate Lives: Women in Science 1789–1979* (New Brunswick, NJ, 1987), 239–80; Kohler, *Partners in Science*, 338–41, 346, 346–52; D. Hodgkin, 'Wrinch', *Nature*, 260 (1976), 564 which is less full than draft in Hodgkin Papers, A255–7; *OUG* 66 (1935–6), 104; 69 (1938–9), 588. For the cyclol controversy see especially D. Crowfoot, 'The Crystal Structure of Insulin, 1: The Investigation of Air-dried Insulin Crystals', *Proceedings of the Royal Society of London A*, 164 (1938), 580–602 (communicated by Robinson); ead., 'X-ray Studies of Protein Crystals', ibid. 170 (1939), 74–5; J. D. Bernal, 'Vector Maps and the Cyclol Hypothesis', *Nature*, 143 (1939), 74–5; D. P. Riley and I. Fankuchen, 'A Derived Patterson Analysis of the Skeleton of the Cyclol C2 Molecule', ibid. 648–9; D. M. Wrinch, 'The Cyclol Theory and the Structure of Insulin', ibid. 763–4; J. D. Bernal, I. Fankuchen, and D. P. Riley, 'X-rays and the Cyclol Hypothesis', ibid. 897; D. M. Wrinch, 'On the Pattern of Proteins', *Proceedings of the Royal Society of London A*, 160 (1937), 59–86 (communicated by Robinson); ead., 'The Geometrical Attack on the Problem of Protein Structure', ibid. 170 (1939), 63–4; ead. and I. Langmuir, 'The Structure of the Insulin Molecule', *Journal of the American Chemical Society*, 60 (1938), 2247–55; I. Langmuir and D. M. Wrinch, 'Vector Maps

Lady Carlisle fellow at Somerville (Crowfoot's college), where she was to receive £250 p.a. for five years from October 1939 to work on mathematics in relation to natural science. A delicate situation, involving intellectual and career conflicts between Crowfoot and Wrinch not just in Oxford in general but within the walls of Somerville, was in prospect; but it was avoided because shortly after the war started Wrinch departed for the USA, where she spent the rest of her career.

6.3 Vacillation: Botany

Though botany did not attract vast cohorts of undergraduates, its position was not as embarrassing as geology's. On occasion the number of finalists reached double figures and the size of classes usually did so. In terms of D.Phil. students botany also had the edge on geology: W. O. James, the plant physiologist, had as many D.Phil. workers in the late 1930s as geology as a whole had between the wars. By the early 1930s botany, like geology, enjoyed the services of three university demonstrators, none of whom was a college fellow, even though one of them, Clapham, a polymath, an excellent linguist, and a good conversationalist on literature, history, music, and art, was exceptionally well qualified. Like geology and mineralogy, botany offered facilities to a guest researcher who was a college fellow. He was Robin Snow, an Oxford graduate of independent means who was elected fellow of Magdalen in 1922. He worked in the department until 1930, when he married and built in Oxford a home with two laboratories where he pursued the life of a gentleman-scientist.[26] The department's average income was about £2,000 p.a., made up mainly and roughly equally from university grants and student fees, with small contributions from the Fielding herbarium bequest and from Magdalen. The staff did not enjoy lucrative consultancies but in the 1930s they began to attract research grants from the Department of Scientific and Industrial Research (DSIR). The department received only one major external endowment between the wars. In 1938–9 after court action it began to enjoy the benefits of the Druce bequest in the forms of his herbarium and £19,000, the income from which was to be devoted to taxonomic botany.[27]

---

and Crystal Analysis', *Nature*, 142 (1938), 581–3. John William Nicholson (1881–1955), fellow of Balliol 1921–30; William Thomas Astbury (1898–1961), then director of the textile physics laboratory, University of Leeds, and pioneer X-ray analyser of high polymers; Isidore Fankuchen (1905–64), in 1939 a researcher with Bernal; Niels Bohr (1885–1962); Irving Langmuir (1881–1957).

[26] On George Robert Sabine Snow (1897–1969), fellow of Magdalen 1922–60, FRS 1948, see A. R. Clapham, *BM* 16 (1970), 499–522.

[27] On George Claridge Druce (1850–1932), FRS 1927, from 1879 a pharmacist in Oxford where he became sheriff and mayor and from 1895 curator of the Fielding herbarium in the

Between the wars botany was served by four professors, the first two of whom (Vines and Keeble) did little for it. Keeble was followed by the renowned ecologist Tansley, who pursued in his decade of office the comprehensive modernization and expansion of his department in teaching and in research. In 1937 he was succeeded by Osborn, an ecologist with a good track record of discipline-building in Australia, whose efforts were balked by alleged shortage of funds and the war. All these professors had to cope with a long-running problem concerning the location of the department. It was physically situated opposite Magdalen in the Botanic Garden, which was the oldest in Britain, the sixth oldest in Europe, and beautiful. Given the Garden's historical, aesthetic, and indeed scientific importance, it was impossible to transform in a root-and-branch way the department's limited accommodation in what were called the Daubeny buildings because of the various types of vandalism that would have been entailed. Patching and small *ad hoc* improvements were the only possible yet unsatisfactory ways of bettering the department's facilities. In the 1870s there had been a concerted but unsuccessful attempt to solve the problem by moving the department from the Garden to the Museum area in order to bring botany into close contact with the other life sciences and thus breaking the long link between the department and the Garden. Vines and Keeble were content with the Garden site, Tansley vacillated about moving to the Museum area, but Osborn favoured it. By 1939 the department was still located in the Garden in what the vice-chancellor publicly admitted were intolerable and squalid conditions: its buildings were unsuitable and incapable of being made suitable except at 'a ruinous cost of both money and beauty'.[28] In 1938 the Gordian knot had been cut at last but only in principle: it was decided that a new building for botany and forestry would be erected after the war [*sic*] in the Science Area. The knot was cut in fact in 1950 when the forestry/botany building was opened. Like Douglas in geology, Osborn was frustrated with his accommodation: both were appointed to chairs in 1937, both planned for a future ravaged by the war, and both spent a disproportionate amount of time negotiating for new accommodation which they secured only towards the end of their Oxford careers. Their own research suffered and neither became FRS.

Vines was a telling example of a late nineteenth-century professor of science at Oxford who arrived with a great reputation and then did very little for his subject in both teaching and research.[29] At Cambridge he had

botany department, see *DNB* and D. E. Allen, *The Botanists: A History of the Botanical Society of the British Isles through a Hundred and Fifty Years* (Winchester, 1986), 104–18.

[28] *OUG*, 70 (1939–40), 53.

[29] Sydney Howard Vines (1849–1934), professor of botany 1888–1919, obituary in *ON* 1 (1932–5), 185–8; J. Howarth, '"Oxford for Arts": The Natural Sciences 1880–1914', in M. Brock (ed.), *The History of the University of Oxford*, vii (in press).

been an innovator particularly in the teaching of practical botany but at Oxford he lived off his past reputation, re-editing his textbooks, editing *Annals of Botany*, and forsaking research. Apparently he put male undergraduates off botany, and was even discouraging to women who were partial to the subject elsewhere. Though on occasion unwell, he was no geriatric: he was 65 years old when the outbreak of the First World War effectively ended his reign. He complained that big first-year classes physically prevented him from doing research in term time and that his department's accommodation was the worst in any British university; but his lecturer, A. H. Church, managed to surmount these hindrances.[30] That may account in part for the way in which Vines treated Church as a lackey paid a low salary which Vines took no steps to improve. Even so, and in spite of the deaths of his wife and a daughter in 1915, Church published on evolutionary adaptation and associated modifications of structure and function, using a comparative morphological approach, for which he was elected FRS in 1921.

In 1919 Keeble was elected to the chair of botany in succession to Vines, but not unanimously. One of the electors, A. C. Seward, professor of botany at Cambridge, favoured the other strong candidate, A. G. Tansley.[31] Seward actually told Keeble that he was his second choice and was clearly worried that Keeble would not put his back into his job at Oxford. Seward's hunch proved to be prescient. Initially Keeble and Church co-operated in trying to revitalize the department but Church's unsociably aloof and resentfully provocative behaviour soon brought the attempt to an end. Church paralysed all but the best undergraduates, would not publish in journals because he objected to being edited, was perversely speculative in his publications, and for years was bitterly disappointed that two volumes of his huge work on floral morphology remained unpublished. For his part Keeble created new facilities for botany at the Garden site in 1923 when Magdalen closed the Daubeny Chemistry Laboratory, part of which was converted into research rooms; simultaneously an existing room used as a museum was transformed into a physiological laboratory.[32] Otherwise Keeble pursued two interests, one old and the other new, instead of research. Before he came to Oxford he had edited the *Gardeners' Chronicle*, directed the Wisley

---

[30] A. E. Gunther, *Robert T. Gunther: A Pioneer in the History of Science 1869–1940* (Oxford, 1967), 111–12; A. Tansley, 'Church', *ON* 2 (1936–8), 433–43; A. H. Church, *Types of Floral Mechanism* (Oxford, 1908); D. J. Mabberley (ed.), *Revolutionary Botany, 'Thalassiophyta' and Other Essays of A. H. Church, with a Recollection by E. J. H. Corner* (Oxford, 1981); obituary in *OM* 55 (1936–7), 558.

[31] A. C. Seward to F. O. Bower, 20 Dec. 1919, Bower Papers, DC/2/E179. Albert Charles Seward (1863–1941), professor of botany, University of Cambridge, 1906–36.

[32] V. H. Blackman, 'Keeble', *ON* 8 (1952–3), 491–501; obituary in *OM* 71 (1952–3), 106–7; *OUG* 51 (1920–1), 556; 54 (1923–4), 541–2; *HCP* 116 (1920), 27–32; 127 (1924), 103–4; Lillah McCarthy (1875–1960), *DNB*; Alfred Moritz Mond (1868–1930), *DNB*, chairman of ICI 1926–30; F. W. Keeble, *Fertilizers and Food Production on Arable and Grass Land* (London, 1932); Reader, *ICI*, ii. 81, 99–104, 158.

Gardens of the Royal Horticultural Society, and as a civil servant during the first war was controller of horticulture for the Board of Agriculture. Once in Oxford he delighted in the Garden, seeing it as of great and increasing value to the department. He enjoyed the four 'p's, planting, pruning, potting, and propagating, far more than research in scientific botany at the bench or in the field. Lacking the collaborators he had relied on before the war when professor at Reading, he did little research at Oxford. The new interest of Keeble, a widower, was Lillah McCarthy, the actress, whom he married in 1920. They lived on Boar's Hill, Oxford, where he indulged in extensive hospitality, groomed himself flamboyantly like a Louis-Philippe courtier, rejoiced in his knighthood (1922), and concentrated on horticulture. It was through his garden that he became friendly with Alfred Mond who persuaded him in 1926 to leave Oxford in order to become agricultural adviser to ICI and director from 1929 of its Jealott's Hill agricultural research station at Bracknell. Keeble devoted himself to research and propaganda about artificial fertilizers, the aim being to boost the sales of fertilizers produced by the continuous and unstoppable Haber-Bosch process in which ICI had invested £20 million at Billingham in 1927.

Early in 1927 Tansley was elected from a good field of eight candidates to the chair of botany he had failed to gain in 1919.[33] Then aged 55 he was a man of independent means as a result of an inheritance from his father who died in 1902.[34] He had time to devote to editing and writing, and when he changed direction in his career it was by choice and not by necessity. Thus, after sixteen years as a lecturer in botany at Cambridge, he resigned in 1923 partly to devote himself more to Freudian psychology and partly to take stock of his pet subject of ecology which was not flourishing as much as he wished. When he arrived in Oxford he was known primarily as an editor and as an ecologist. In 1927 he edited two journals, the *New Phytologist*, which

---

[33] The candidates were Tansley; William Rungrose Gelston Atkins (1884–1959), FRS 1925, head of physiology, Marine Biological Association Laboratory, Plymouth, 1921–55; Frederick Thomas Brooks (1882–1952), FRS 1930, mycologist, University of Cambridge; Otto Vernon Darbishire (1870–1934), professor of botany, University of Bristol, 1919–34; Walter Stiles (1886–1966), FRS 1928, professor of botany, University of Reading, 1919–29, Birmingham 1929–51; Hugh Hamshaw Thomas (1885–1962), FRS 1934, palaeobotanist, University of Cambridge; Reginald Ruggles Gates (1882–1962), FRS 1931, professor of botany, King's College, London, 1921–42; Charles Edward Moss (1872–1930), professor of botany, University of Witwatersrand, 1917–30; OUA, UDC/M/41/1, 20 Jan. 1927.

[34] On Tansley see H. Godwin, 'Tansley', *BM* 3 (1957), 227–46; id., *Cambridge and Clare* (Cambridge, 1985), 146–58; id., 'Sir Arthur Tansley: The Man and the Subject', *Journal of Ecology*, 65 (1977), 1–26; A. D. Boney, 'The Tansley Manifesto Affair', *New Phytologist* 118 (1991), 3–21; A. G. Tansley, 'Some Aspects of the Present Position of Botany', in *British Association for the Advancement of Science: Report of Ninety-First Meeting* (London, 1924), 240–60; id. and T. F. Chipp (eds.), *Aims and Methods in the Study of Vegetation* (London, 1926); A. G. Tansley, *Practical Plant Ecology: A Guide for Beginners in Field Study of Plant Communities* (London, 1923); id., *The New Psychology and its Relation to Life* (London, 1920); for Tansley on Freud, *ON* 3 (1939–41), 241–75.

he had founded in 1902, and the *Journal of Ecology*, which he had edited from 1917. As a pioneer ecologist before the first war, he had been the first president of the British Ecological Society and had given the first systematic account of British vegetation as opposed to flora. Subsequently he had published extensively on plant ecology and had been a polemicist about education in botany especially via the Tansley manifesto affair at the end of the war. This was a discussion, conducted in Tanley's own journal, *New Phytologist*, about reconstructing botany teaching. The view taken by Tansley and his fellow reconstructors was that botany teaching was vitiated by an excessive concern with morphology, based on speculative attempts to trace the phylogenetic relations of plants in the framework of Darwinian natural selection. They called for physiology and ecology to be given as much attention as morphology, with the last to be studied without a phylogenetic basis. Tansley himself appreciated that by the mid-1920s younger botanists, dissatisfied with morphology, had turned to the specialist and useful research areas of physiology, mycology, and genetics; but he thought that in teaching botany it was essential to look at the science of plants as a whole in the broadest possible way.

In his inaugural lecture Tansley revealed that he objected to loathsome utilitarian compulsion in science, but that he favoured voluntary applied science as an antidote to sterile academicism.[35] He anticipated careers for botanists in universities, museums, schools, and the tropical colonies, stressing the importance of imperial development on scientific lines. He emphasized that research should be the result of perennial spontaneous curiosity and not of demand or expectation; and that it flourished as a result of appointing the best people and giving them their heads. He wanted plant physiology and pathology, ecology, and genetics as new foci of research in his department and pleaded for expansion on a new site of buildings, equipment, staff, and annual income. Tansley clearly expected essential moral support from the University which would enable him to develop a sufficiently practical orientation for his department for it to attract money on an adequate scale from private benefactors, public bodies, and the state itself. Though Tansey modernized his department in terms of curriculum, research spread and ethos, and postgraduate numbers, the University did not give him the support he anticipated and external funding did not flow freely to his department.

Tansley inherited from Keeble three departmental lecturers, Church, Harry Baker, and W. H. Wilkins.[36] His first move was to secure university

---

[35] A. G. Tansley, *The Future Development and Functions of the Oxford Department of Botany: An Inaugural Lecture Delivered before the University of Oxford on 22 November 1927* (Oxford, 1927).

[36] Henry Baker (1895–1958); William Henry Wilkins (1886–1966) ended as reader in mycology 1947–51, obituary in *The Times*, 27 Apr. 1966, assessment by Osborn, 19 Nov. 1941, strictly confidential, OUA, Teaching of mycology 1936–50, UR/SF/BG/3.

demonstratorships (just created in 1927) for Church (tenable for four years) and Wilkins (tenable for one year, renewed for four years from October 1928). Tansley's caution about Wilkins's capacity was apparent: Wilkins had graduated at Oxford in 1921 as a mature student aged 35 and had then taught in the department for six years without developing any line of research. Finding Wilkins's contribution to be 'arid', Tansley suggested that he focus on fungi, so Wilkins studied them at Imperial College, London, for two terms in 1927 and 1928 before returning to Oxford as its mycological expert. In 1927 and 1930 Tansley brought in, as departmental demonstrators who were rapidly made university ones, two botanists who had taken their Ph.D.s at Cambridge under F. F. Blackman, the leading plant physiologist in Britain, whose sister-in-law was Tansley's wife.[37] The first to arrive was W. O. James who was needed to cover plant physiology, in which Tansley regarded himself and Church as unqualified; and the second was the highly versatile Clapham who as Church's replacement in 1930 became the chief agent of the modernization of the teaching. Clapham was not only responsible for field classes in ecology: he lectured on morphology, genetics, cytology, ecology, taxonomy, fossil botany, and biological statistics, whereas James and Wilkins kept to their specialist areas.

Between 1930 and 1935 Tansley produced significant changes in his department. The regular staff became visible through what Tansley called 'public work', sitting on the councils of learned societies and editing journals.[38] Tansley continued to edit the *Journal of Ecology* until he retired, but in 1932 James, Clapham, and Harry Godwin took over from him the editorship of the general botanical journal *New Phytologist*.[39] His department began to function as a marriage bureau, with the Snows, the Yemms, and the Harleys meeting their future spouses there.[40] The number of finalists per year reached double figures, the number of courses per term doubled, and the number of postgraduates quadrupled to thirteen. As a consequence of Magdalen leasing the whole of the Daubeny buildings in 1932 to the University, the accommodation for physiology (1932) and mycology (1933) was improved. Tansley was certain that his prime responsibility was the overall development of his subject.[41] Though Tansley was the commanding figure in

[37] Frederick Frost Blackman (1866–1947), *DNB*, *DSB*, reader in botany, University of Cambridge, 1904–36, married in 1917 Elsie Chick, whose sister Edith had married Tansley in 1903.
[38] *OUG* 64 (1933–4), 217–20.
[39] Harry Godwin (1901–85), then lecturer in botany, University of Cambridge, professor 1960–8.
[40] Elizabeth Fitt, second in botany 1934, married John Laker Harley (1911–90), first 1933, professor of forest science, Oxford, 1969–79, on whom see D. C. Smith and D. H. Lewis, 'Harley', *BM* 39 (1994), 159–75; Marie Solari, first 1930, married Edmund William Yemm (1909–93), first 1931, professor of botany, University of Bristol, 1955–74, obituary in *Independent*, 6 Dec. 1993; Christine Mary Pilkington, first 1926, married Snow, first 1921.
[41] Tansley to vice-chancellor, 25 Nov. 1935, 21 July 1936, OUA, Botany, department of, 1930–41, UR/SF/BG1; *OUG* 65 (1934–5), 208–10.

British ecology all through the first half of the twentieth century, he could not contemplate leaving physiology to Cambridge and mycology to London and confining Oxford to ecology. He was prepared to pass lightly over taxonomy, but thought that physiology, mycology, and ecology were all central to modern botany and that each science department at Oxford had to be basically comprehensive. Tansley also believed that plant ecology was not a specialized branch of botany like physiology and mycology: it was a way of approaching the study of plants which used specialized data. For these reasons he did not focus his department's teaching and research on his own speciality: indeed Tansley was well aware that work in ecology was particularly hampered by the lack of proper laboratory facilities. That was another reason why he had only a couple of D.Phil. students in ecology in his ten years at Oxford. He preferred to encourage an ecological approach in the research of his colleague Clapham and of postgraduates such as J. L. Harley. In terms of size the buoyant research groups were those of James in plant respiration and Wilkins in mycology. Strongly backed by Tansley, James attracted support from the DSIR and the Royal Society, enhanced his reputation, and attracted postgraduates of high calibre, almost all from Oxford, such as Nora Penston, Edmund Yemm, Frederick Hora, and Arthur Bunting.[42]

Wilkins, who had benefited in 1933 from the new premises for mycology which cost £2,100, was extremely jealous of James and contemptuous of plant physiology.[43] For his part James thought that Wilkins, whose knowledge was confined to fungi, was an incompetent charlatan. Relations between the two men were so tense that Wilkins was the only member of staff who did not attend the daily tea meetings. In 1936 he complained directly to the registrar that he and James were at loggerheads about accommodation, apparatus, and assistance; and that James received preferential treatment because Tansley backed a fellow Cambridge man. Wilkins was deeply resentful: he thought Tansley's department to be nothing more than a branch in Oxford of the Cambridge school of botany and that Oxford had tottered along behind Cambridge because it had imported its botany professors from there. There is no doubt that Wilkins was prickly and

---

[42] On James's research see James's memo, n.d. [Nov. 1938], BS/R/1/5; Clapham and Harley, 'James', 339–44. Nora Lillian Penston (1903–74), first in botany 1927, D.Phil, 1930, principal of Bedford College, University of London, 1951–64, obituary in *The Times*, 5 Feb. 1974; Frederick Bayard Hora (1908–84), first in botany 1932, D.Phil. 1936, ended as reader in botany, University of Reading, 1964–73, obituary in *The Times*, 19 Apr. 1984; Arthur Hugh Bunting (b. 1917), Rhodes scholar, D.Phil. 1941, professor of agricultural botany 1956–73 and agricultural development overseas 1973–82, University of Reading, *Who's Who*.

[43] On Wilkins's career under Tansley, see *HCP* 155 (1933), 31–2; Tansley to University Chest, 8 Nov. 1932, BS/R/1/2; Wilkins's memo, 18 Jan. 1937, BS/R/1/4; Wilkins to Veale, 12 June 1936, confidential, 12 Jan. 1937, Veale to Wilkins, 15 Jan. 1937, OUA, Teaching of mycology 1936–50, UR/SF/BG/3; J. L. Harley, 'Autobiography', Harley Papers, A1, 33 and 'Botany at Oxford in the 1930s', 6 pp. insert in A1.

self-seeking, with ambitions for himself and mycology at Oxford not matched by his research reputation; it is also clear that Tansley, who had a low opinion of Wilkins's research, often rejected his financial solicitations.

As Tansley was a modernizer of his department, the question of its location was far more acute for him than for his two predecessors. For several years he was keen to move to the Science Area, but in the end he accepted staying on the Garden site and continuing to make *ad hoc* improvements there. In his inaugural lecture he had stressed that the future development of botany at Oxford necessitated a new site. By 1930 he was so frustrated with the limitations imposed by the Garden location that he approached the Pilgrim Trust directly for £100,000 to pay for a new building and an experimental garden in the Science Area, accompanied by more demonstrators, and more postgraduate scholarships, so that Oxford could begin to rival Cambridge.[44] This appeal was not approved by the University and came to naught. In compensation Veale, the registrar, tried to help. He appreciated that botany was in an impossible situation: the two possibilities, having a new department costing £100,000 and uneconomically patching up the present department for £10,000, were both unattractive to prospective donors. Veale therefore proposed that the number of botany undergraduates be cut. Tansley rejected that solution outright, but agreed that henceforth he would not beg unilaterally.

By late 1934 Tansley had accepted that a new building in the Science Area would cost so much more than altering the existing department that with great reluctance he dropped the idea of moving to it. He clearly felt that his hand had been forced. Given the sudden increase of undergraduates and researchers which he had encouraged, he was obliged to develop his department on the Garden site where extra accommodation of a sort was immediately available, and consequently to disregard the advantages of moving to the Science Area which would have involved much greater expense which was not immediately possible.[45] Having at last accepted the Garden site as a *pis aller*, Tansley pressed hard but unsuccessfully in 1936 for extensions to his department, mainly for physiology and mycology and costing £13,500, though he also wanted to buy a nearby nursery garden as a future experimental plot for genetics and ecology, leaving the famous Garden unchanged. Even this Garden was a problem because it was not under his control: it was run by a board of curators (of which he was not ex officio a member) as a gardeners' showpiece. Moreover W. G. Baker, the head gardener, was an awkward bully who disliked James and Wilkins.[46] Tansley was also unsuc-

---

[44] Tansley, Needs of the department of botany, Nov. 1930, Veale memo, 12 June 1931, Tansley to Veale, 20 June 1931, UR/SF/BG/1.

[45] Tansley to Lindsay, 21 May 1934, UR/SF/S4; Tansley memo, 21 July 1936, UR/SF/BG/1.

[46] Tansley memo, Nov. 1936, revised July 1937, strictly confidential, UR/SF/BG/1; William George Baker (1862–1945), head gardener 1888–1942, obituary in *OM* 63 (1944–5), 291–2; Allen, *Botanists*, 194.

cessful in appropriating the contested bequest of Druce, curator of the Fielding herbarium in the department. Druce, who disliked Tansley, had died in 1932 and left his library, herbarium, and a large cash sum for research into British flora.

In research Tansley led from the front. Nearly all of his monumental *British Isles and their Vegetation* (1939) was written at Oxford where friends, colleagues, and pupils helped him. It was not a flora and not a manual of field ecology: it focused on the gregarious life of plants. As such it was the culmination of that phase in plant ecology which employed broad surveys and examined the dynamic relations between major vegetational communities. The book was also polemical in that it attacked the notions of plant succession put forward by F. E. Clements who depicted plant communities as organisms. By the mid-1930s Tansley objected so strongly to the dogmatism of the Clements school, which claimed that there was only one kind of vegetational outcome or climax in a given climatic region, that he renounced his previous view that Clements's ideas were heuristically useful, and his nomenclature practically the best. In 1935 as an alternative to Clements's views Tansley introduced the idea of an ecosystem, in which climate, soils, plants, and animals functioned as part of a system, each with a functional relation to the other. In his 1939 book Tansley saw Clements's notion of climax not in climatic but ecological terms: a vegetational climax was a relatively stable equilibrium between climate, soils, plants, and animals, which Tansley called a mature ecosytem. It has been claimed that, in putting forward the concept of an ecosystem, Tansley drew on the philosophy of science adumbrated by Hyman Levy. There is a simpler explanation. At Oxford in the early 1930s Clapham coined the term ecosystem to describe the physical and biological components of an environment considered mutually and as a unit. Tansley appropriated the idea, put it into print unacknowledged in 1935, and is still receiving credit for it.[47]

Tansley was due to retire in summer 1937. His presence and work were so widely recognized by his fellow life scientists at Oxford that the biological sciences board requested the University to extend his tenure by three years. It refused but in compensation allowed Tansley to write the job specification for his successor.[48] The published desideratum was an expert plant ecologist

---

[47] A. G. Tansley, *The British Isles and their Vegetation* (Cambridge, 1939); Frederick Edward Clements (1874–1945), *DSB*; F. E. Clements, *Plant Succession: An Analysis of the Development of Vegetation* (Washington, 1916); J. B. Hagen, 'Clementsian Ecologists: The Internal Dynamics of a Research School', *Osiris*, 8 (1993), 178–95; J. Sheail, *Seventy Five Years in Ecology: The British Ecological Society* (Oxford, 1987), 62–5; Willis, 'Clapham', 81; F. B. Golley, *A History of the Ecosystem Concept in Ecology: More than the Sum of the Parts* (New Haven, 1993), 8–34; J. B. Hagén, *An Entangled Bank: The Origins of Ecosystem Ecology* (New Brunswick, NJ, 1992), 79–87.

[48] Tansley memo, 28 Oct. 1936, BS/R/1/4; *HCP* 165 (1936), 247–50; Wilkins memo, 18 Jan. 1937, BS/R/1/4; Veale to Wilkins, 3 Oct. 1941, UR/SF/BG/3. In his retirement Tansley,

who as such would have a comprehensive interest in plant life. Privately Tansley argued that there were special opportunities at Oxford for plant ecology because it was the most fundamental part of ecology, so he claimed, and good ecological work was being done at Oxford in forestry, pedology, and at the Bureau of Animal Population. Wilkins disagreed: if Oxford became a special centre for plant ecology then it was likely that plant pathology, which was not unrelated to fungi, would be left to Cambridge. Wilkins was apparently in a minority of one in proposing that mycology be the specialism of the next professor.

There were twelve applicants for the chair including three FRSs (Atkins, Stiles, and Thomas) who had been unsuccessful in 1927, the young Harry Godwin from Cambridge, and the unsquashable Wilkins. The electors chose T. G. B. Osborn, who had worked on ecology and fungi as an all-round botanist but was not an FRS.[49] As his career had been focused on the universities of Manchester, Adelaide, and Sydney, he was a total stranger to Oxford but arrived with twenty-five years' experience as a professor, a department-builder, and an adviser to state and federal governments in Australia. Within three months of assuming his chair Osborn formally complained about his poor facilities: mycology and physiology had been developed at the expense of general botany and no provision was made for ecology. His own working conditions, he alleged, were the worst of his career: he had one small study and no laboratory so his own work was at a standstill. He had come to Oxford to do original work, an acknowledged professorial duty, in ecology, the University's desideratum, but no facilities existed. Osborn deplored the *ad hoc* alterations sanctioned by Tansley in the buildings at the Garden: they often had a bad but unchangeable orientation and were too rambling and piecemeal to be welded into a coherent and working unit. Osborn's solution was to shift the department to the Science Area, leaving only the herbaria and the relevant part of the library on the Garden site.[50] The University was so embarrassed by Osborn's predicament

---

supported *inter alia* by Elton and Nicholson, founded the Nature Conservancy in 1949 and was its first chairman 1949–53: S. Bocking, 'Conserving Nature and Building a Science: British Ecologists and the Origins of the Nature Conservancy', in M. Shortland (ed.), *Science and Nature: Essays in the History of the Environmental Sciences* (Stanford in the Vale, 1993), 89–114.

[49] Theodore George Bentley Osborn (1887–1973), graduate of University of Manchester and lecturer in botany there 1908–12, professor at University of Adelaide 1912–27, professor at University of Sydney 1927–37, professor at Oxford 1937–53, obituaries in *The Times*, 6 June 1973, and *Nature*, 244 (1973), 377. Besides Osborn, Atkins, Stiles, and Thomas, the candidates were: Wilkins; Godwin; Robert Stephen Adamson (1885–1965), professor of botany, Cape Town University, 1923–50; Leo Brauner (1898–1974), a German refugee to whom Tansley had given accommodation summer 1933, professor of botany, University of Istanbul, 1933–55; Ronald D'Oyley Good (b. 1896), lecturer, University College, Hull; Robert Maclagan Corrie (1897–1970), lecturer, Dehra Dun College, India; John V. F. Phillips, professor at University of Witwatersrand.

[50] Osborn memo, 28 Oct. 1937, UR/SF/BG/1; *HCP* 168 (1937), 227, 247.

that in 1938 it agreed in principle that the teaching and laboratory aspects of botany should be incorporated in a new building to be shared with forestry in the Science Area. Meanwhile the University decided to spend there and then £24,600 on the Ashmolean Museum and £60,000 on a new physiology laboratory, giving small grants to botany for more *ad hoc* changes at the Garden site.[51] The registrar, Veale, was unhappy with the situation: realizing that mycology was being pushed too much by Wilkins who ignored Osborn if he could and that Osborn needed money to develop plant ecology, he approached the Rockefeller Foundation for support but without success.[52] Veale thought the accommodation for botany was 'shamefully defective' and the vice-chancellor was dismayed that Osborn could not do his own ecological work because he lacked a proper laboratory, research assistance, experimental greenhouses, and expenses for field-work.[53] Moreover, Baker, the head of the Garden, hated Osborn who had to wait for Baker's retirement in 1942 before he secured for himself an ex officio position on the Garden's board of curators.[54] Not surprisingly Osborn did little original work as professor at Oxford: his facilities were poor, he found discipline-building much harder at Oxford than it had been in Australia, on the outbreak of war the University postponed the erection of the new forestry botany building, and he was diverted by advisory work. The lack of a major bequest to his department ensured that botany, like geology and forestry, was treated as a Cinderella science.

Ironically the Druce bequest came too late for Osborn and his predecessor and was irrelevant to their priorities. Druce died in 1932 and in his will bequeathed his library, herbarium, and house in Oxford as the basis of a botanical research institute. After court action about the interpretation of the will, the University began to enjoy in 1938–9 some benefits from Druce's bequest which amounted to £90,000: it acquired Druce's herbarium and £19,000, which was unavailable as capital for a new building because the income from it had to be spent on taxonomic botany.[55] If Osborn was disappointed by the Druce bequest, he may not have known that a tempting opportunity was rejected just before the war. In 1938 Sir Daniel Hall, who was due to retire in 1939 from the directorship of the John Innes Research Institute at Merton, near London, wished to move it to just outside Oxford where it would become a research institute of the University which he hoped would produce £10,000 for a new building to house it. Though the Innes

[51] Veale memo, 16 Feb. 1938, UR/SF/BG/1; *HCP* 169 (1938), 225, 229–30, 297; 170 (1938), 119–21.
[52] Veale to Rockefeller Foundation, 11 Jan. 1939, UR/SF/BG/1; Osborn to Veale, 18 Feb. 1938, Wilkins to Veale, 14 June 1938, Osborn memo, 19 Nov. 1941, all in UR/SF/BG/3.
[53] Veale memo, 25 Mar. 1939, vice-chancellor memo, 23 May 1939, UR/SF/BG/1; *OUG* 70 (1939–40), 53–4.
[54] T. G. B. Osborn, 'Changes at the Botanic Gardens', *OM* 61 (1942–3), 241–2.
[55] Allen, *Botanists*, 115–18; *OUG* 70 (1939–40), 53.

Institute was well endowed and a leading centre for plant genetics, the vice-chancellor, Lindsay, turned the offer down flat because he was fearful of being embroiled in a repeat of the Owen affair. Next year Hall again broached his plan, which Veale liked, and suggested that Lord Nuffield be approached for £5,000, a move which Veale vetoed: 'we must get the imbroglio of the Nuffield College plans cleared up before we tackle him again.'[56]

6.4 A Cause Célèbre: Astronomy

De jure it was possible to take astronomy as a single honours subject but few undergraduates did so. In 1932 there was a finalist in astronomy, the first in the twentieth century.[57] Given this bizarre situation, Harry Plaskett, the Savilian professor of astronomy, was happy to have honours astronomy abolished in the late 1930s: he preferred prospective astronomers to read mathematics and/or physics for their first degree. The chair of astronomy was, therefore, a research chair. Because the Observatory contributed so little to the University it was not generously endowed with staff. By 1939 there were no university demonstrators or college fellows in astronomy, and the number of assistants began to exceed three only in the 1930s. As the Observatory received no major external bequest between the wars, its main source of income was the University which by the late 1930s was spending around £3,000 p.a. including salaries on it. Under H. H. Turner, Plaskett's predecessor, there was little postgraduate supervision. Oxford was not a great centre before 1930 for training astronomers: in 1928 Turner was reduced to revealing that O. J. B. Cole, a pre-war computer at the Observatory, had risen to become Leicester chief of police.[58] From 1932 under Plaskett postgraduate research was introduced and D.Phil.s acquired. Though the reigns of Turner and Plaskett shared some common features, the most obvious difference was that for at least his last twenty years Turner did not observe at the Observatory: he used it largely as a centre of calculation in astrography and seismology. Plaskett, however, was an observer who used a new solar telescope to pursue solar physics as his special field. Thus one major theme of astronomy is that, in spite of certain common elements, the era of Turner which ended in 1930 was different from that of Plaskett which began in 1932. The other major theme was the relation between the University Observatory, situated in the Science Area, and the Radcliffe Observatory half a mile away near the Radcliffe Infirmary. The University

---

[56] Veale to vice-chancellor, 9 Aug. 1939, OUA, Botany, John Innes School of Horticulture, 1938–9, UR/SF/BG/5.

[57] I owe this point to Roger Hutchins, Magdalen, whose unpublished typescripts on the history of astronomy at Oxford are valuable sources on which I have drawn.

[58] *OUG* 59 (1928–9), 343.

foolishly confused statutory right with custom and took the Radcliffe Trustees to court about their plan to move their Observatory from murky Oxford to the clearer skies of Pretoria. After protracted litigation the Trustees won their case in 1934, they were authorized by the court to spend £65,000 on the Pretoria telescope and buildings, and in 1935 the Oxford site was vacated.

The University Observatory, opened in 1875, was directed from 1893 to 1930 by Turner; but it was younger and by the early 1900s less well equipped than the Radcliffe Observatory established by the Radcliffe Trustees in 1772.[59] Indeed the Radcliffe Observer had a house near his observatory but Turner did not. Matters came to a head in 1907 when Congregation rejected the proposal from Turner for him to have a residence built near the University Observatory so that he would be on the spot to exploit every hour of clear sky. Turner thought that William Anson, warden of All Souls and a Radcliffe trustee, was responsible because Anson had proposed co-operation between the two observatories and the future fusion of the posts of Savilian professor of astronomy and Radcliffe Observer, which meant that there would be no need to build a second astronomer's residence. Until his death Turner resented the irreparable injury which Anson, with his insincere suggestion, had inflicted on him. The 1907 passage at arms confirmed that the two observatories would continue to do totally different work.[60] By this time the Radcliffe Observatory was the better equipped. In 1903 at a cost of £8,350 the Trustees had installed a 24-inch refracting telescope which enabled their Observer, Arthur Rambaut, to contribute to two international schemes launched by the Dutch astronomer J. C. Kapteyn, who had divided the sky into 215 selected regions on which attention was to be concentrated using powerful telescopes which revealed faint stars. First Rambaut measured the parallax of distant faint stars in nine allotted regions, and secondly the proper motions of stars down to as faint a limit as the fourteenth magnitude.[61] Though Turner was slightly younger than Rambaut and in 1907 more widely recognized (FRS 1897, president of the Royal Astronomical Society 1903–5), he made no attempt to compete directly with Rambaut as an observer. Turner's observatory was pretty much as it was in 1888, its equipment becoming more worn and obsolescent by the year; and there was no chance, he judged, of persuading the University to follow the example of the Radcliffe Trustees by updating its own Observatory. Deeply chagrined

---

[59] I. Guest, *Dr. John Radcliffe and his Trust* (London, 1991); A. D. Thackeray, *The Radcliffe Observatory 1772–1972* (London, 1972).

[60] H. H. Turner, *The Radcliffe and University Observatories* (n.p. [Oxford], 1907); Guest, *Radcliffe*, 299–30; Turner, letter of 10 Apr. 1930, *The Times*, 17 Apr. 1930.

[61] Arthur Alcock Rambaut (1859–1923), Radcliffe Observer 1897–1923, obituary in *Monthly Notices of the Royal Astronomical Society*, 84 (1923–4), 220–1; Guest, *Radcliffe*, 294; on Jacobus Cornelius Kapteyn (1851–1922), *DSB*, professor of astronomy, Groningen, see H. Hertzsprung-Kapteyn, *The Life and Works of J. C. Kapteyn*, trans. E. R. Paul (Dordrecht, 1993); E. R. Paul, *The Milky Way Galaxy and Statistical Cosmology, 1890–1924* (Cambridge, 1993).

by the University's failure to provide a house for him and not a dab hand with instruments, he decided to drop observational work and to make Oxford a co-ordinating and calculating centre for international co-operative ventures.

The first of these was one he inherited when he assumed his chair in 1894. This was the Carte du Ciel project launched in 1887. It involved eighteen observatories photographing and charting the sky on the same scale. Oxford's zone was soon photographed, measured, and reduced, though not completely printed until 1914. Though reducing observations was tedious (it involved correcting observations by taking account of circumstances which modified them), Turner extended a helping hand to the Vatican Observatory apropos its contribution to the astrographic star catalogue. In Rome photographic plates were taken and measured; the measurements were then sent to Oxford where they were subjected to laborious computation and eventually published. Turner supervised this computational work which was mainly done by two loyal assistants, Frank Bellamy and his niece Ethel Bellamy. In the 1920s Oxford helped Potsdam Observatory in a similar way. Turner thought that using Oxford as a centre of calculation and publication of the data obtained by other observatories had given Oxford international recognition and that it was useful for him in the 1920s to co-ordinate the astrographic chart and catalogue under the auspices of the International Astronomical Union.[62]

In 1913 he took over an international undertaking concerned with seismology which had previously been run by John Milne. Milne first became interested in earth tremors while living in Japan. On his return to England in 1895 he established a seismological centre at Shide, Isle of Wight, and organized observing stations over much of the world under the auspices of the British Association for the Advancement of Science to whose seismological committee he was secretary. Turner joined this committee in 1898 and, when Milne died in 1913, had been its chairman for seven years. Milne's venture was paid for mainly out of his own pocket and in his lifetime it was recognized as the pioneer of modern observational seismology. When he died Turner took over Milne's seismological work partly as a tribute to his friend and partly because he saw seismology as an aspect of astronomy concerned with the study of just one planet in the solar system. Milne's work was continued at Shide until 1919 when his widow returned to Japan and his premises were sold. At that point Turner saw a chance to make Oxford a seismological centre which would not only make observations but also co-ordinate data recorded elsewhere. With the financial help of J. E.

---

[62] J. K. Fotheringham, 'The Astrographic Star Catalogue', *OM* 52 (1933–4), 217; *OUG* 55 (1924–5), 662; Frank Arthur Bellamy (1864–1936), assistant in University Observatory 1892–1936; Ethel Frances Butwell Bellamy (1883–1961), assistant 1899–1947, obituary in *Nature*, 191 (1961), 391–2.

Crombie of Aberdeen, Turner had Milne's books, records, and instruments moved to Oxford where the apparatus was installed in the basement of the Clarendon Laboratory. In 1919 and 1922 Turner tried to make Oxford the central bureau of the International Geodetic and Geophysical Union which chose Strasbourg; but in 1922 as president of its seismological section he was asked to continue collating records of the depths and periodicities of earthquakes and the Union elevated his annual report on seismology, previously written for the British Association, to the *International Seismological Summary*, for which in part it paid. Successfully locked into a second international co-operative venture, Turner supervised the observations and computations made at Oxford mainly by Edith Bellamy and from 1922 by J. S. Hughes, the latter receiving half his salary from Crombie.[63] In 1928 the Observatory received its first major addition since it was built, when the new seismological basement was opened through the generosity of Crombie who gave £500. In 1928-9 the University began to pay part of Hughes's salary, thus conferring some recognition on Turner's seismological work though no doubt it was viewed as 'not exactly astronomy in the old sense'. For his part Turner was keen to extend astronomy in the direction of geophysics and happy to have provided from Oxford clear evidence for the notion that the earth has a liquid core of radius half that of the earth. Given Turner's commitment to seismology it was entirely appropriate that he died suddenly in Stockholm on 20 August 1930 while presiding over the seismology section of the International Geodetic and Geophysical Union.

While Turner's main activities in the 1920s were focused on astrography and seismology and office-holding in the International Astronomical Union, the International Geodetic Union, and the British Association, he also pursued and encouraged history of astronomy. With J. L. Dreyer, who had retired to Oxford, he wrote the centennial history of the Astronomical Society and tried to nurture the research of J. K. Fotheringham on ancient astronomy and chronology. An Oxford graduate in Greats, Fotheringham was primarily an ancient historian who turned to ancient astronomy. In his sixteen years at the University of London he lived mainly in Oxford and in 1918 Turner made him an assistant in the University Observatory though he was not even an occasional observer. Fotheringham soon became the apple

---

[63] F. A. and E. F. B. Bellamy, *Herbert Hall Turner: A Notice of his Seismological Work* (Newport, 1931); on John Milne (1850-1913), *DNB*, see A. Geikie, 'Anniversary Address', *Proceedings of the Royal Society B*, 83 (1911), 259-72 (263-4); James Edward Crombie (1862-1932), Aberdeen wool manufacturer and seismologist, obituary in *Aberdeen Press and Journal*, 8 Aug. 1932, left £1,000 on the death of his wife to the British Association for seismological research; *HCP* 134 (1926), 149; 137 (1927), 29-30; *OUG* 55 (1924-5), 662 (quotation); 59 (1928-9), 343-4; Turner memo, 25 Jan. 1928, PS/R/1/1; Joseph Steel Hughes (1898-1965), first in mathematics 1923. For further details on Turner's seismology see unpublished typescript by R. Hutchins, 'Adjacent Radcliffe and University Observatories' (1996), Museum of History of Science, Oxford.

of Turner's eye. Though Turner realized that Fotheringham's research could be regarded as 'rather apart from the regular Observatory work' he praised it as novel and even sensational. In 1924 Turner put friendship before judgement when he strongly backed Fotheringham in his unsuccessful application for the vacant Radcliffe observership, arguing that he was the most outstanding astronomer produced by Oxford since James Bradley, his work was specially suitable for Oxford, and his appointment would bring the two observatories closer (i.e. under Turner's control).[64] Next year Turner induced the University to make Fotheringham university reader in ancient astronomy and chronology. In that year Turner pressed for an extension to the Observatory library to facilitate the historical research of Fotheringham and Dreyer in addition to arguing for underground facilities for seismology at the Observatory to replace those granted in the Clarendon Laboratory by Lindemann.

When Rambaut died in October 1923, Lindemann saw an opportunity. He and Turner were the leading figures in a deputation which met the Radcliffe Trustees in December. Lindemann urged the Trustees to develop astrophysics at their Observatory and to appoint a physicist as their Observer, and he promised that if they did so his father would donate the necessary equipment. This was the first occasion on which Lindemann revealed that he was trying to capture the Radcliffe Observatory for physics. Turner agreed with Lindemann's suggestions and opposed any merger of his chair and the observership. Jealous of their independence and autonomy, in 1924 the Trustees appointed as their Observer not a physicist but Harold Knox-Shaw, an astronomer who had enjoyed sixteen years of clear Egyptian skies as an assistant and then superintendent of the Helwan Observatory.[65]

In his opening decade at Oxford Knox-Shaw focused on two projects. One was the publication in 1932 of the 54,000 meridian observations made between 1774 and 1798 by Thomas Hornsby, the first Radcliffe Observer. The other was winding up the Observatory's contribution to Kapteyn's international project on the proper motions of faint stars. The Radcliffe catalogue of over 32,000 proper motions was eventually published in 1934 but the observing work, the most sensitive to date in Britain, quickly

[64] J. L. E. Dreyer and H. H. Turner, *History of the Royal Astronomical Society 1820–1920* (London, 1923); Johann Louis Emil Dreyer (1852–1926), *DNB*, *DSB*, director of the Armagh Observatory 1882–1916, retired in 1916 to Oxford where he completed his 15-volume edition of the works of Tycho Brahe. J. K. Fotheringham, 'Ancient Astronomy and Chronology', *OM* 49 (1930–1), 48–50; J. L. Myres, 'Fotheringham', *Proceedings of the British Academy*, 23 (1937), 551–64; *OUG* 54 (1923–4), 621; 55 (1924–5), 662 (quotation); *HCP* 131 (1925), 227–9; James Bradley (1693–1762), Balliol undergraduate, professor of astronomy, Oxford, 1721–42, Astronomer Royal 1742–62.

[65] 11 Dec. 1923, 5 Feb. 1924, Minute Book of Radcliffe Trustees, 1881–1930, Bodleian Library, Mss.dd. Radcliffe, d.39; Guest, *Radcliffe*, 301–3; *HCP* 126 (1923), 115; Harold Knox-Shaw (1885–1970), Radcliffe Observer 1924–50, obituary by A. D. Thackeray, *Quarterly Journal of the Royal Astronomical Society*, 12 (1971), 197–201.

revealed the limitations of Oxford's climate for observing faint stars.[66] As early as November 1924 Knox-Shaw proposed in vain to the Trustees that they sell part of the Observatory site to the Infirmary and build a new observatory on Headington Hill outside Oxford. Subsequently he continued to express dissatisfaction with Oxford's weather which in 1927 was the worst for faint-star work since it began in 1905. In desperation he turned to Frank Schlesinger, its co-ordinator, who advised him to move the Radcliffe Observatory to southern France or north Africa. Then events moved quickly because Sir Frank Dyson, the Astronomer Royal and a close friend of Knox-Shaw, was appointed a Radcliffe trustee in May 1929 and he was convinced that a site in South Africa would be the best. Having visited South Africa in summer 1929, Dyson and Knox-Shaw were highly impressed by a location near Pretoria where the municipality was keen to donate a site, water, and electricity. By November 1929 the Trustees were in principle committed to the Pretoria scheme, which they thought was financially viable because they were sure that Lord Nuffield would pay them £100,000 for the Observatory site. The Trustees were so determined to go ahead with the Pretoria scheme that in July 1930 they sold the Observatory site to Nuffield for the expected £100,000 and then immediately rented it from the Radcliffe Infirmary at £1,000 p.a. for five years in order to complete current work at the Observatory and to plan in detail and execute the move to South Africa. They were aware of the legal difficulty of spending so much capital outside Britain, where the Court of Chancery had no jurisdiction, but thought it was surmountable. They were also aware that there might be some resentment in the University and they made it clear in spring 1930 that they wanted a close and friendly alliance with it.[67]

Between November 1929 and July 1930 the Pretoria scheme was widely welcomed by astronomers.[68] The Visitors of the University Observatory approved it. So did the prestigious National Committee for Astronomy.

[66] *The Observations of Thomas Hornsby Made with the Transit Instrument and Quadrant at the Radcliffe Observatory in the Years 1774 to 1798: Reduced by H. Knox-Shaw, J. Jackson, and W. H. Robinson* (London, 1932); H. Knox-Shaw and H. G. S. Barrett, *The Radcliffe Catalogue of Proper Motions in the Selected Areas 1 to 115* (London, 1934); Thomas Hornsby (1733–1810), Radcliffe Observer 1772–1810.

[67] For the genesis of the Pretoria scheme: Radcliffe minutes, 23 Jan. 1925; Reports of the Radcliffe Observer presented to the Radcliffe Trustees 1925–30, Bodleian Library, Mss. Radcliffe, e.4; Guest, *Radcliffe*, 114–15, 307–10; *HCP* 146 (1930), 22–4; Frank Schlesinger (1871–1943), *DSB*, director of Yale University Observatory 1920–41 and from 1925 director of the Yale–Columbia Observatory, Johannesburg, on which see D. Hoffleit, *Astronomy at Yale 1701–1968* (New Haven, 1992), 118–23; Frank Watson Dyson (1868–1939), Astronomer Royal 1910–33, was a special friend of Knox-Shaw: M. Wilson, *Ninth Astronomer Royal: The Life of Frank Watson Dyson* (Cambridge, 1951), 231, 237, 242.

[68] *HCP* 145 (1930), 73; 146 (1930), 179–82; Guest, *Radcliffe*, 319; *Nature*, 125 (1930), 769–71; Radcliffe Trustees Minutes, 26 Feb. 1930; Milne to Pember, 26 Feb., 25 Apr. 1930, OUA, Radcliffe Observatory. Future of, Jan. 1930–July 1932, UR/SF/AST/4; Milne to Lindemann, 2, 14 May 1930, LP, B91/9–12.

The general scientific periodical *Nature* thought it an imperative necessity for British scientific prestige and scored a palpable hit in urging that Oxford's debt to South Africa, via Rhodes scholars, needed to be repaid. In the University the Pretoria scheme was supported by Turner and by E. A. Milne, the recently elected Rouse Ball professor of mathematics, whose field was theoretical astronomy. Turner was so impressed by the advantages of the scheme and so conscious of Oxford's heart-breaking climate that he buried the hatchet with the Trustees and looked forward to co-operation with them through a research studentship for an Oxonian to work at Pretoria. Milne's support for the Pretoria scheme for positional astronomy was allied to a vision of his own pertaining to what he called 'cosmical science'. In early summer 1930 he set out his view that Oxford should establish an institute devoted to observational and theoretical astrophysics, meteorology, and geophysics, which would permit Oxford's physicists, mathematicians, and astronomers to do for 'cosmical physics' what Cambridge had done for atomic physics. Milne envisaged such extensive co-operation between the Radcliffe Observatory in Pretoria and his proposed astrophysics institute in Oxford that he wanted Knox-Shaw to be given a university post. Generally Milne advocated that the University offer specific privileges to the Trustees which would permit peaceful penetration of their scheme: if it did not cooperate with the Trustees, it would lose the moral right to be consulted by them.

Opposition to the Pretoria scheme was fomented inside and outside the University from February 1930 by Lindemann and his close friends the Earl of Birkenhead, high steward of the University, and the young Roy Harrod.[69] Lindemann believed that the proposed move from Oxford of the Radcliffe Observatory would mean a reduction in Oxford's facilities for physical science and would hamper future attempts to gain endowments. He was well aware then that the decision lay with the Trustees, that the University had no *locus standi* in the matter, and that the Radcliffe Observatory was not legally part of the University though associated with it. In spring 1930 that was the official view of the Hebdomadal Council when it ventured to ask the Trustees to consider the possibility of applying some of their funds to scientific research which could be pursued at Oxford. In Lindemann's mind, however, the Radcliffe Observatory affair was yet another example of the difficulties of developing physical sciences at Oxford and of the University's indifference, nay even hostility, to them. He was also keen to lay his hands on the proceeds of the sale of the Radcliffe Observatory, proceeds that could be used for academic purposes in connection with the

[69] 4 Feb. 1930, PS/M/1/1; Guest, *Radcliffe*, 311–14; *HCP* 145 (1930), 85, 99–110, 220, 230; 146 (1930), 178; secretary of Charity Commission to S. Clay, 12 Mar. 1930, LP, B91/5–6; Frederick Edwin Smith, 1st Earl of Birkenhead (1872–1930), high steward of the University 1922–30; Roy Harrod (1900–78), fellow in economics, Christ Church, 1924–67.

University. He wanted the Trustees' recently increased endowments to be devoted to astrophysics, geophysics, and meteorology pursued in the University, thus providing facilities from which his department would benefit.

While Harrod acted as Lindemann's bulldog in the Hebdomadal Council, Birkenhead launched Lindemann's case in *The Times*, being careful to state that the University had a moral if not legal claim to have a strong voice about the future of the Radcliffe Observatory. Lindemann busied himself not only in mobilizing support from Oxford's physical scientists but also in lobbying Earl Grey, the University's chancellor, and in attacking Turner in the columns of *The Times*.[70] Lindemann regarded Turner as an old-fashioned astronomer devoid of all consideration for the University and publicly dismissed him as *passé*. Lindemann denigrated Turner as an astronomer for pursuing interests which were subterranean not celestial and, with biting sarcasm, he alleged that 'to one accustomed to sidereal quantities...a small entity like the University of Oxford seems too petty for serious consideration'. Turner retaliated by objecting to Lindemann's depressing parochialism: Lindemann had no vision of Oxford fathering a valuable enterprise abroad; his view was limited to the City and University *tout court*. Turner stressed that until recently Lindemann and his allies had been uninterested in the Radcliffe Observatory; they had changed their minds because they were looking enviously at the cash its sale would generate. By July 1930 a bizarre situation existed. Spurred by Lindemann, who had cleverly co-opted Milne's ideas about astrophysics, the University's Board of Physical Sciences had formally opposed the Pretoria scheme which was strongly supported by Turner, the University's professor of astronomy. Moreover the dispute between Lindemann and Turner was not confined to Oxford: it had been waged in the columns of *The Times* and *Nature*. While Knox-Shaw was livid about Lindemann's views and methods, Turner presumably resented the physicist's interference in an astronomical matter and felt that his own authority as professor of astronomy had been impugned.

In response to Turner's death the University did not fill his chair immediately but suspended it for fourteen months in order to consider the position and prospects of astronomy. Simultaneously the Radcliffe Trustees were forced to delay their Pretoria scheme owing to legal and financial difficulties. This combination of events enabled Lindemann to promote his case even more vigorously.[71] He was at the back of the Attorney-General's view that there had been a long association between the Radcliffe Observatory and the University which gave it a moral though not legal claim on the

---

[70] For the polemics: Harrod, *Cherwell*, 136–40; Birkenhead letters, *The Times*, 28 Mar., 15, 21 May 1930; Lindemann letters, 23 Apr., 2 May 1930; Turner letters, 17, 26 Apr. 1930.

[71] On Lindemann's scheming summer 1930–summer 1931, Lindemann to Halifax, 1 July 1930 and reply 9 July, LP, B95/4, 24; *HCP* 147 (1930), 161–4; 148 (1931), 147–50, 169; 149 (1931), 209–13, 270; Lindemann to Veale, 23 Oct. 1931, LP, B92/15–16.

Trustees. Within the University Lindemann soon proposed a scheme in which he developed Milne's idea of an institute devoted to cosmical science. Whereas Milne had said nothing in spring 1930 about the cost and location of his proposed institute, Lindemann gave prominence to these questions by arguing that the University should erect a new university observatory, devoted to cosmical science, on a hill outside Oxford at a cost of about £60,000—using money from the Radcliffe Trust. Lindemann's meagre bait to the Trustees was that Knox-Shaw could be made a professor. While Lindemann was pursuing the plan of a hill-top site, particularly through the Oxford Preservation Trust, in February 1931 the Radcliffe Trustees decisively rejected it, perhaps being strengthened in their resolve by Knox-Shaw's elevation that month to the presidency of the Astronomical Society. Irritated by what he regarded as their foolish independence, Lindemann saw to it that the University employed lawyers to draw up a document entitled 'outline scheme for application of funds arising from the sale of the Radcliffe Observatory' to be used in court as an alternative to the Trustees' Pretoria scheme. The University's scheme of June 1931, basically Lindemann's, envisaged appropriating £60,000 of the Trustees' sale money and their 24-inch equatorial telescope. In October 1931 Lindemann confirmed to Veale, the university registrar, that the Oxford Preservation Trust would provide a site for a new university observatory on a hill outside Oxford.

Not everyone was happy with Lindemann's scheme.[72] Frank Bellamy, who had been entrusted formally with supervising the University Observatory after Turner's death, objected publicly to the biased nature of Lindemann's plan which he claimed put astrophysics first and threw the rest of astronomy to the dogs. More importantly Milne thought Lindemann's proposal inadequate because it would alienate astronomers in general and it provided no way in which the Trustees could climb down gracefully. By summer 1931 Milne still supported his idea of developing cosmical science at Oxford but had changed his mind about the Radcliffe Trustees' Pretoria scheme: clearly swayed by Lindemann, Milne had concluded that the 'whole needs of the University should come first' in front of the Pretoria scheme which put observational research first. He was now a useful if reluctant ally of Lindemann's.

The chair of astronomy was advertised in summer 1931 and attracted a small and poor field. It was widely appreciated that the Observatory, with its ancient apparatus and staff, was hopelessly handicapped; and that the restoration of astronomy at the University was a tall order. There were only five candidates, two of whom were Fotheringham and H. C. Plummer. The former, aged 57, was not a practising observer; while the latter, only

[72] F. A. Bellamy, 'A Plea for Astronomy in Oxford', *OM* 50 (1931–2), 126–7; Milne memos, n.d. [late Nov. 1930], 7 July 1931, UR/SF/AST/4; Milne to Lindemann, 25 Nov. 1930, LP, B91/21–3.

slightly younger and a non-observer since 1921, clearly wished to return and retire to Oxford which he regarded as his true home. The only feasible candidate was Harry Plaskett, professor of astrophysics at Harvard University where his chief responsibility had been to establish a graduate programme in astronomy.[73] At Harvard Plaskett had acquired a high reputation as an astrophysicist who believed that solar physics, in the form of spectroscopic research on the sun, would yield more information about stellar physics than spectroscopic studies of stars themselves. He was keen to come to Oxford because he wanted to run his own show and he believed that he would be able to pursue solar research profitably. At the personal level he and Milne had deeply admired each other's work from 1929 when they had first met. Plaskett was lured to Oxford not least because he and Milne looked forward to developing astrophysics there. He was elected in November 1931, effective on 1 January 1932, but did not assume his duties until June 1932.

As Savilian professor Plaskett faced four major problems. These were what to do with astrography and seismology, both of which were embedded in international collaborative schemes and not easily run down, how to react to the University's legal entanglement with the Radcliffe Trustees, and how to promote astrophysics *de novo*. Plaskett thought that cosmography, as he called it, was *passé* in comparison with solar physics but he inherited from Turner the astrographic work on the zones taken over from the Potsdam Observatory. It was in the loving care of Frank Bellamy, the acting director of the Observatory during the professorial interregnum. Bellamy was so determined to complete the Potsdam section of the astrographic catalogue and so resentful of Plaskett's desire to move the Observatory to astrophysics that in 1933 he began to obstruct Plaskett. Early in 1936 Bellamy formally produced a list of complaints against Plaskett and resigned as from 15 March 1936. Though the Observatory Visitors rejected Bellamy's charges, the vice-chancellor privately agreed with him that the position was very unsatisfactory. It was resolved on 15 February 1936 when Bellamy died and much of his astrographic work was soon transferred elsewhere. Seismology was less easily shed partly because it was supported financially by the British Association and the International Geodetic and Geophysical Union. Though Plaskett was not interested in the subject and regarded it as an

---

[73] Vice-chancellor's oration, 7 Oct. 1931, *OUG* 62 (1931–2), 22; H. A. L. Fisher to Lindemann, 21 Oct. 1931, LP, B92/17; 19 June 1931, UDC/M/41/1; Henry Crozier Plummer (1875–1946), obituary in *OM* 65 (1946–7), 12–13, Astronomer Royal for Ireland 1912–21, professor of mathematics, Military College of Science, Woolwich, 1921–40, was recognized late in life by being president, Royal Astronomical Society, 1939–41; Harry Hemley Plaskett (1893–1980), professor of astronomy 1932–60, on whom see W. H. McCrea, *BM* 27 (1981), 445–78; Milne to Lindemann, 5 Oct. 1931, Plaskett to Veale, 6 Dec. 1931, Plaskett to Lindemann, 12 Dec. 1931, in LP, B92/12–13, 26, 30–1. For a valuable insider's view of Plaskett's regime see M. G. Adam, 'The Changing Face of Astronomy in Oxford (1920–60)', *Quarterly Journal of Royal Astronomical Society*, 37 (1996), 153–79.

irritating diversion, which his small department could ill afford, he administered the work on the International Seismological Summary done by Ethel Bellamy and Hughes until 1946 when it was transferred to Kew.[74]

On the question of the Pretoria scheme of the Radcliffe Trustees, Plaskett sympathized with Milne's view that frontal attack ought to be avoided because it would make the University look a money-grabber; and he did not share Lindemann's virulent hostility to the Trustees or his utter confidence in the case against them.[75] Once ensconced in his chair, Plaskett attacked the University's scheme of June 1931 which Lindemann had inspired. From an astronomical point of view Plaskett accepted that the Pretoria scheme was sound and criticized the plan for a new university observatory as too expensive, as involving the unwelcome abolition of the Radcliffe observership (one of the world's few major posts), and as separating astronomy from the Science Area. Plaskett's top priority was to renovate the present University Observatory by spending *circa* £10,000 on a solar telescope, preferably using Radcliffe money. He also urged that a new Radcliffe Observatory could be built outside Oxford for £50,000 and be run by Knox-Shaw who could be made a university professor and work closely with Plaskett so that Oxford would have the benefit of both solar and stellar observational facilities. Plaskett also downplayed geophysics and meteorology, which were prominent in the 1931 scheme. All Plaskett's points, made in late 1932, were accepted by the University which instructed its lawyers accordingly.

In 1933 affidavits were filed on behalf of the University by Lindemann, who tended to ignore Plaskett's amendments, and more cautiously by Plaskett, Milne, and Veale. Lindemann had also managed to recruit three star witnesses from abroad in the form of Pannekoek from Amsterdam, Freundlich from Potsdam, and Einstein. The Trustees rested their case on purely astronomical grounds as well as their legal independence from the University, their leading spokesmen being Sir Arthur Eddington, Sir Frank Dyson, and ironically Plaskett's own father J. S. Plaskett, director of the Dominion Astrophysical Observatory at Victoria in Canada. After a long action the Trustees won their case against the University in July 1934 to the general acclaim of most British and foreign astronomers. Lindemann was furious that the court authorized £65,000 for the Pretoria scheme. His case had never been strong, as Veale well knew, and Mr Justice Bennett had spotted that Lindemann had pushed the University into wooing the Radcliffe Trust

---

[74] McCrea, 'Plaskett', 451–6; Minutes of the Visitors of the University Observatory, 1875–1975, OUA, UDC/M/3b/1, 17 Oct. 1930, 30 Jan. 1936, 28 Jan. 1937; *OUG* 62 (1931–2), 308–10; 67 (1936–7), 525–7.

[75] Plaskett to Lindemann, 27 Jan. 1932, Milne to Lindemann, 2 Feb. 1932, Lindemann to Plaskett, 4 Feb. 1932, LP, B93/1–10; *HCP* 153 (1932), 161–4. The admirable Guest, *Radcliffe*, 327 does not note the difference between Lindemann's scheme of June 1931 and Plaskett's of Nov. 1932.

more as heiress than for love.[76] In 1935 the Radcliffe Observatory was dismantled, some of its possessions but not all remaining in Oxford in the Bodleian Library and the Museum of the History of Science. The delay of four years, in part caused by the University, had serious repercussions for the new Radcliffe Observatory, whose 74-inch reflecting telescope, the largest in the southern hemisphere, became operational under Knox-Shaw only in 1948.

Justice Bennett had made it plain that the Trustees should recognize their long association with Oxford by founding and financing at £700 p.a. the Radcliffe travelling fellowship in astronomy tenable for three years and to be spent half in Oxford and half in Pretoria.[77] In response to this gesture of goodwill by the Trustees, the University cheekily and unsuccessfully applied to them for £350 p.a. (on top of the £700) to be spent on solar physics research and on meteorology in the form of salary for Gordon Dobson, reader in meteorology. The first election in 1937 was also contentious. Milne argued that Oxford candidates should not receive preference but Lindemann argued that they should, because the fellowship was all that was left to the University from the Observatory affair. Milne won. From a field of ten, including eight from abroad, Herman Zanstra from Amsterdam was elected the first fellow.

Plaskett was well aware that the University might fail in its action against the Radcliffe Trustees so he had a reserve plan which he soon implemented. His general aim was to make Oxford into a leading and possibly unique place for solar physics. He realized it was futile to try to rival Cambridge and that he had to concentrate resources on some phase of astronomy. He arrived at Oxford convinced that astronomy should concern itself with the physical nature of heavenly bodies, not their distribution; and that studying the sun, a typical star which could be observed in unequalled detail, was the best way of learning about the physical conditions on the surfaces and in the near interiors of stars in general. Plaskett also knew that solar telescopes and spectroscopes cost only about one-tenth of their stellar equivalents of comparable power. With financial expediency reinforcing intellectual conviction,

---

[76] Guest, *Radcliffe*, 325–36; *The Times*, 3 July 1934; *Observatory*, 57 (1934), 250–2; Veale to Gamlen, 21 June 1934, Lindemann to vice-chancellor, 29 June 1934, UR/SF/AST/4 (1932–4); Arthur Stanley Eddington (1882–1944), professor of astronomy, University of Cambridge, 1913–44; John Stanley Plaskett (1865–1941), director of the Dominion Astrophysical Observatory, Victoria, Canada, 1917–35, on which see R. A. Jarrell, *The Cold Light of Dawn: A History of Canadian Astronomy* (Toronto, 1988), 111–25; Antonie Pannekoek (1873–1960), professor of astronomy, Amsterdam University, 1925–41; Erwin Finlay Freundlich (1885–1964), director of Astrophysics Observatory, Potsdam, in 1930s.

[77] Radcliffe Minutes, 18 June 1935, Minute Book, 1930–46, Radcliffe Trust, London; Visitors' Minutes, 28 Jan. 1937; Herman Zanstra (1894–1972), obituary by Plaskett, *Quarterly Journal of the Royal Astronomical Society*, 15 (1974), 57–64, Radcliffe fellow 1937–42, in 1937 theoretical physicist, University of Amsterdam, to which he returned as professor of astronomy 1946–59.

he chose to pursue solar physics at Oxford using new apparatus in the existing Observatory. Though Cambridge had enjoyed a solar physics observatory from 1913, under Newall and then Stratton, Plaskett had the advantage of starting from scratch in the 1930s and was able to concentrate on certain phenomena, such as solar granulation, which he made his own.[78]

Plaskett, a Canadian who was a total stranger to Oxford, was an astute negotiator. He induced the University to pay out £3,680 on solar research by 1935, but in six separate but related amounts.[79] By making incremental requests he made it difficult for the University to refuse him. Even before he arrived in Oxford he secured £400 for a spectroscope and lamps as the first instalment of an application for a larger sum. In March 1933 he extracted £2,400 for the optical parts of a solar telescope and of a spectroscope: this was regarded by the University as extraordinary expenditure to prevent Plaskett's activities being sterilized. Two years later the new ensemble of apparatus for solar physics was completed, with grants of £300 (micrometer, 1933) and of £300 (1934) and £130 (1935) for building alterations which cost much less than had been expected. Instead of a new steel tower costing £3,500 for the telescope and a special shelter for the spectroscope, Plaskett mounted the solar telescope in an existing tower and dome which had previously housed a 24-inch reflecting telescope, and he put the spectroscope in an existing basement at a total cost of £430. Plaskett was a vice-chancellor's dream: he was able to reduce costs while at the same time giving the University unique equipment for research in his field. Plaskett's solar telescope, officially opened in June 1935, was the only one in the world to have its optical parts made of silica glass which expanded much less than ordinary glass when heated by the sun.

While Plaskett was assembling the components of his solar telescope, he launched postgraduate research in astrophysics and especially solar physics. His aim was to create a strong but small ensemble of researchers. Always a realist, Plaskett appreciated that money for research and jobs in astronomy were both scarce so his emphasis was on being small but good in postgraduate research. In any case he regarded group research *per se* with horror. His Aclandian line had the merits of putting Oxford astronomy on the international research map without greatly disturbing entrenched interests at

---

[78] H. Plaskett, *The Place of Observation in Astronomy: An Inaugural Lecture Delivered before the University of Oxford on 28 April 1933* (Oxford, 1933), 29–32; id., 'A Solar Telescope for the University Observatory', *OM* 52 (1933–4), 217–20; Hugh Frank Newall (1857–1944), first director, solar physics laboratory, University of Cambridge, 1913–28, obituary by E. A. Milne, *ON* 4 (1942–4), 717–32; Frederick John Morrison Stratton (1881–1960), second director 1928–47.

[79] *OUG* 64 (1933–4), 17–18; 65 (1934–5), 17; *HCP* 151 (1932), 93–4, 109–11; 154 (1933), 117–18; 'The New Solar Telescope', *OM* 53 (1934–5), 749–51; Observatory Visitors' minutes, 27 Jan. 1938; Plaskett to Veale, 14 Nov. 1932, Plaskett's accompanying memo, OUA, University Observatory needs, 1932–49, UR/SF/AST/1B.

Oxford. Plaskett adopted various ways of developing postgraduate research. He made his lectures on solar physics, stellar physics, and astronomical spectroscopy attractive to undergraduate physicists; and he publicized his research and invited helpers at meetings of the University's Junior Scientific Society. Through these modes he recruited Madge Adam, later his assistant director at the Observatory.[80] Again in contrast with Turner he attracted Rhodes scholars, the first being T. L. Page, his first successful D.Phil. student who became his chief assistant for a year and then took up a post at the Yerkes Observatory, University of Chicago.[81] Plaskett also benefited from his close friendship with Milne, some of whose pupils and associates spent time with Plaskett in the Observatory. In this way Plaskett recruited Oxford mathematicians such as David Kendall and visiting researchers such as S. Chandrasekhar who was developing his theory of white dwarf stars.[82] But at no time did Plaskett and Milne themselves collaborate in research and Milne's scheme of 1930 for an institute of cosmical physics remained a dream. Plaskett was determined to pursue his own line of solar physics; and, though he was happy to foster links with physics, he did not wish to become a mere satellite of Lindemann's Clarendon Laboratory. Milne was a modest and charming man who was no academic entrepreneur as far as bricks and mortar were concerned; and in any case in 1934 he moved his research away from the internal constitution of stars to relativity theory, thus shifting from astrophysics. The only way in which Plaskett and Milne collaborated was in their joint research colloquium launched in 1934. Often a dialogue between Plaskett, the observer, and Milne, the theoretician, the colloquium was yet another attractive feature of Plaskett's regime which had no analogue in his predecessor's.

By 1937 the Great Reaper had benefited Plaskett twice. The death of Frank Bellamy in February 1936 allowed Plaskett to shed astrography that very year. The decease of Fotheringham in December 1936 removed the reader in ancient astronomy whose classes in 1935 had attracted just one student each. Plaskett was still lumbered with Ethel Bellamy and Joseph Hughes who continued their work on seismology; but otherwise Plaskett, elected FRS in 1936, had a totally new establishment of staff in place in 1937 with Page as his chief assistant, Adam as his research assistant, and Theodore Durham (from the Mount Wilson Observatory) as his research associate. This new establishment was associated with the Observatory's switch of interest in 1932 to astrophysics and the setting up in the University

---

[80] M. Adam to author, 11 Apr. 1988; Madge Gertrude Adam (b. 1912), first in physics 1934, research fellow of St Hugh's 1957–79, worked at the Observatory until 1979.

[81] Thornton Leigh Page (b. 1913), Rhodes scholar 1934–7, chief assistant to Plaskett 1937–8, D.Phil. 1938, pursued a distinguished career in astrophysics in USA: *Who's Who in America*.

[82] David George Kendall (b. 1918), first in maths 1939, professor of mathematical statistics, University of Cambridge, 1962–85; Subrahmanyan Chandrasekhar (1910–95), fellow of Trinity, Cambridge, 1933–7, professor of theoretical astrophysics, University of Chicago, 1937–85.

Observatory by 1935 of effective facilities for the pursuit of solar physics. The entire capital cost of these facilities (almost £4,000) was met by the University and not by external bodies. The University was prepared to expend that sum because it had become deeply embarrassed by the obsolescence of its Observatory's equipment by 1930. Having appointed Plaskett as Savilian professor of astronomy, it felt it had to give him some facilities to prevent him from being forced into sinecurism. Plaskett then skilfully capitalized on this situation by applying in instalments for money for a new solar telescope and by ingeniously reducing costs. Lured to Oxford by the presence of Milne, Plaskett soon became the living rebuttal of the modest proposition that he had advanced in his inaugural lecture, namely, that an effective University Observatory would require expenditure out of proportion to the return it could make to the University, 'certainly from the point of view of training men [sic], and even possibly from the point of view of advancing knowledge'.[83]

6.5 The Archaeology of Science and Technology

History of science and of pre-industrial technology at Oxford existed on the basis of collections in museums, respectively the Lewis Evans collection of scientific instruments established in 1924 and given the titular dignity of the History of Science Museum in 1935, and the Pitt Rivers Museum opened in 1885. In both cases their origin and development relied heavily on gifts and not purchases. The University's role was restricted to providing a building, paying a curator, and giving annual grants which by the late 1930s amounted to £900 (Pitt Rivers) and £200 (History of Science). Though history of science and of technology were museological subjects concerned with collecting and displaying objects and with teaching in the respective museums, often using the objects as illustrative material, they were differently located in the colleges and in the University's formal examinations. Though various colleges donated objects to the Museum of the History of Science, its curator, Gunther, had no college connection via a fellowship and his lectures were not part of any formal course. In contrast Henry Balfour, curator of the Pitt Rivers Museum for forty-eight years until his death early in 1939 and a specialist in the history of technology, was a fellow of Exeter where R. R. Marett, its rector, sedulously promoted anthropology. With Marett, Buxton, and Balfour as Exeter's anthropological trinity, the college nurtured a broad view of anthropology which included technology (Balfour) as well as its physical and social aspects (Buxton and Marett respectively).[84]

---

[83] Observatory Visitors' minutes, 28 Jan. 1937; Plaskett, *Inaugural*, 32 (quotation).

[84] Henry Balfour (1863–1939), research fellow at Exeter 1904–11, 1919–39, on whom see *Nature*, 143 (1939), 291 and A. C. Haddon, *ON* 3 (1939–41), 109–15; Robert Ranulf Marett

Balfour's teaching was part of the diploma in anthropology, instituted in 1907 partly in response to changes in the Indian Civil Service examination and to the arrival of the first Rhodes scholars in 1903.[85] Until 1939 anthropology was taught only through the diploma, which was a fruitful source of field-workers: only then did the subject penetrate an undergraduate course when it became available as an optional subject in geography finals. In the late 1930s the University recognized anthropology in two other ways: in 1937 it established a regular chair in social anthropology, mainly funded by All Souls, and next year instituted the faculty of anthropology and geography.[86] At best anthropology was highly esteemed as much more than barbarology but remained institutionally marginal. At worst it was denounced as an unsuitable vehicle for undergraduate education: the ancient historian Hugh Last claimed that an acquaintance with the habits of savages was not an education.[87]

The Pitt Rivers Museum derived its name and mode of arrangement from its founder who had an abiding interest in the history of technology.[88] As an officer in the Grenadier Guards he examined ways of improving muskets which led to his forming a collection of firearms to illustrate their history and development. Noticing that the shift from simple to complicated weapons occurred via slight and not large changes, he enlarged his collection to include arts and industries from early times to the age of mass production in order to illustrate more widely this sort of evolutionary progress. Eventually the collection became so large that it was accommodated from 1874 in the South Kensington Museum. In 1883 Pitt Rivers offered his collection to the University where he hoped it would find a permanent home in which it could be properly displayed and enlarged. The University agreed to provide a special building, built as an annexe to the east end of the University Museum at a cost of just over £10,000, and to appoint a lecturer on the objects in the

---

(1866–1943), reader in social anthropology 1910–36, acting professor 1936–7, philosophy fellow of Exeter 1891–28, rector 1928–43, on whom see H. J. Rose, *Proceedings of the British Academy*, 29 (1943), 357–70; L. H. D. Buxton, reader in physical anthropology 1927–39, fellow of Exeter 1933–9.

[85] R. R. Marett, *A Jerseyman at Oxford* (London, 1941), 167–71.

[86] The first professor 1937–46 was Alfred Reginald Radcliffe-Brown (1881–1955), a Cambridge graduate who was imported from the University of Chicago.

[87] Sutcliffe, *The Oxford University Press*, 233; Hugh Macilwain Last (1894–1957), fellow of St John's 1919–36, professor of ancient history 1936–48, principal of Brasenose 1948–56.

[88] B. Blackwood, *The Origin and Development of the Pitt Rivers Museum* (Oxford, 1991), revised and updated by Schuyler Jones; D. K. Van Keuren, 'Augustus Pitt Rivers, Anthropological Museums and Social Change in Late Victorian Britain', *Victorian Studies*, 28 (1984), 171–89; M. W. Thompson, *General Pitt Rivers: Evolution and Archaeology in the Nineteenth Century* (Bradford on Avon, 1977); W. R. Chapman, 'Arranging Ethnology: A. H. L. F. Pitt Rivers and the Typological Tradition', in G. W. Stocking (ed.), *Objects and Others: Essays on Museums and Material Culture* (Madison, 1985), 15–48; M. Bowden, *Pitt Rivers: The Life and Archaeological Work of Lieutenant-General Augustus Henry Lane Fox Pitt Rivers* (Cambridge, 1991); Pitt Rivers (1827–1900).

collection. Thus E. B. Tylor, who had advised his friend to bestow his collection on Oxford, became the first British university teacher of anthropology. As man was a mammal the collection was put under the charge of H. N. Moseley, the Linacre professor of zoology, who set two undergraduates, Walter Baldwin Spencer and Henry Balfour, to work on the transfer and arrangement of the collection to Oxford in 1884.[89] Spencer soon left but Balfour stayed as Moseley's helper and Tylor's assistant, becoming curator in 1891 on Moseley's death. Balfour was therefore intimately connected with the Pitt Rivers Museum for the first fifty-five years of its existence.

He was entirely familiar with the way in which Pitt Rivers had arranged the collection and as curator he did not deviate from it. While other anthropological museums arranged objects by geographical area or by cultures, Pitt Rivers's collection showed evolutionary series of different objects. He had selected and arranged in sequence, not unique, curious, beautiful, or rare objects, but ordinary and typical specimens so as to reveal the succession of ideas by which the minds of humans in a so-called primitive state had progressed from the simple to the complicated and from the homogeneous to the heterogeneous. He thought that artefacts were susceptible to being classified into genera, species, and varieties, just like plants and animals; consequently he was keen to identify phylogenetic groups of artefacts on the basis of morphological affinity. To a zoologist such as Balfour, trained by Moseley, a protégé of Darwin and an expert on classification and phylogeny, the way in which Pitt Rivers brought together evolutionary biology and technology was appealing. As a disciple of Pitt Rivers, Balfour maintained the practice of exhibiting many specimens in each showcase, which made the Museum seem overcrowded or even overflowing with objects. He maintained the typological arrangement, which meant that whole cultures and groups of artefacts remained poorly represented or simply unrepresented. During his despotic reign Balfour respected Pitt Rivers's insistence on the educational importance of the Museum—for Oxonians and serious researchers: until 1970 it was open to the public for no more than twelve hours a week.

In the interpretation of the objects under his care, Balfour diverged from the tendency of Pitt Rivers and Tylor to emphasize evolution as unilinear. As a specialist on the distribution of artefacts and a scientific ornithologist (president of the Oxford Ornithological Society 1924–35), Balfour was aware of the vagaries attending change, took cognizance of migration, of social contact, of hybridization of ideas, and indeed believed in the separate

---

[89] Edward Burnett Tylor (1832–1917), keeper of the University Museum 1883–1902, reader in anthropology 1884–96, professor (personal chair) 1896–1909; Henry Nottidge Moseley (1844–91), graduate of Exeter, fellow 1876–81, Linacre professor 1881–91; Walter Baldwin Spencer (1860–1929), *DNB*, first in zoology 1884, professor of biology, University of Melbourne, 1887–1919; Balfour graduated in zoology in 1885.

and independent creation of the same invention. He never composed a systematic treatise on the history of technology, though his talents were recognized with an FRS in 1924 and a personal chair in ethnology (comparative technology) in 1935. He was simply too busy as a curator who was fully aware of the old adage that a museum which ceases to collect dies. Under his reign the number of objects increased from about 15,000, the gift of Pitt Rivers, to well over a million. Most of these accessions were gifts, with relatively few purchases or loans. Balfour himself was a prominent donor in his lifetime and after his death his son gave to the Museum not only the remainder of his extensive collection acquired on his many travels but also his library of 4,000 books and 6,000 pamphlets.[90] Several major donations came from Oxonians who had made careers as colonial administrators. The big increase in the range and size of the Museum's holdings was not matched by corresponding increases in space or in staff, even after Balfour's elevation to a personal chair. The exhibition area was no bigger than in 1885 when the Museum opened, while storage space was makeshift and temporary. The staffing situation became dire by 1937: Balfour's assistant died in summer 1936 and Balfour himself then became seriously ill for several months. As the Museum had no permanent assistant curator, even its routine work was brought to a standstill.[91] For years Balfour felt that the University had so under-funded and neglected the Pitt Rivers Museum that it could not fulfil properly its teaching and research functions and that the unique nature of its collections, brought together with definite purposes in view, was not being fully exploited. He had a point: only in 1970 did the University appoint a permanent assistant curator.

History of science was launched at Oxford in 1914 by Charles Singer, an Oxford zoologist who procured a Science Room in the Bodleian Library. There was a precedent in the music room of the Library but more importantly Singer enjoyed the support of two patrons. Falconer Madan, the Bodleian librarian, was sufficiently interested in history of science to be involved in the celebrations of the 700th anniversary of the birth of Roger Bacon, and no doubt was happy to accept Singer's offer of £100 p.a. for five years to pay for fittings and new books for the Science Room. The other patron was Sir William Osler, Regius professor of medicine, who wanted at Oxford a separate little department of history of science and medicine where real scholars would augment the efforts of dilettante students such as himself. He knew Singer well and in 1914 had persuaded him to move from London, where he practised medicine, to Oxford to assume the

---

[90] *OUG* 70 (1939–40), 260–2. Major Oxonian donors were: John Henry Hutton (1885–1968), Indian Civil Service 1909–36, professor of social anthropology, University of Cambridge, 1937–50; James Philip Mills (1890–1960), Indian Civil Service 1913–47; Percy Amuary Talbot (1877–1945), Nigerian Civil Service 1911–31.

[91] *OUG* 68 (1937–8), 236.

Philip Walker studentship in pathology, the duties of which were to be mainly historical. Presumably he knew that Singer, who had been publishing on medical history from 1911, had a model in mind, namely the Institute for History of Medicine and Science established at Leipzig in 1905 by Sudhoff.[92]

Though Singer was away from Oxford on military service for much of the war, his wife Dorothea presided over the science history room in which collaborators, guests, and locals worked on history of science and medicine, not infrequently using Bodleian books and manuscripts. By the winter of 1916–17 Madan and Osler were so impressed by Singer's publications and the way in which he could attract and galvanize fellow workers that they pressed successfully for the Bodleian to recognize that the experimental stage of what the curators called the science research room had passed and that it should be extended to accommodate eleven researchers under Singer as director of studies in history of science and medicine. The authorities were clearly impressed by Singer's edited volume *Studies in the History and Method of Science*.[93] This volume of essays was clearly an earnest of future co-operative developments, but they came to naught because in 1917 the Bodleian curators made it difficult for non-Oxonians to use the room and because in 1919 Singer lost his two main supporters: Madan retired in April and Osler died in December. Osler's successor Garrod was nothing like as keen on history of medicine, while Madan's successor Cowley quickly made it clear that the Science Room had no place in his plans. He bundled Singer out and after 1919 appropriated for the Bodleian half of the £500 that Singer had given specifically for the Science Room.[94] By early 1920 it had disappeared and Singer had prudently begun lecturing in history of biology at University College London, where he spent the rest of his distinguished career. One move was made to retain Singer in Oxford. In March 1920

---

[92] Charles Joseph Singer (1876–1960), second in zoology, 1900, on whom see E. A. Underwood, *Medical History*, 4 (1960), 353–8 and J. Sheppard, 'Illustrations from the Wellcome Institute Library: Charles Joseph Singer (1876–1960): Papers in the Contemporary Medical Archives Centre', *Medical History*, 31 (1987), 466–71; NS/R/1/2, p. 66, details of 1914 scheme; R. T. Gunther, *Early Science in Oxford*, xi (Oxford, 1937), 12–13; id., *The Daubeny Laboratory Register 1916–1923* (Oxford, 1924), 389–99, 500–2; Roger Bacon (1214–94); for Osler's support of Singer, H. Cushing, *The Life of Sir William Osler* (London, 1940), 1076, 1131, 1198, 1244–5, 1292, 1326–7; Karl Friedrich Jacob Sudhoff (1853–1938), *DSB*.

[93] *OUG* 48 (1917–18), 305; Madan to Osler, 14 Nov. 1916, NS/R/1/2; 13 Mar. 1917, NS/M/1/3; C. J. Singer (ed.), *Studies in the History and Method of Science* (Oxford, 1917). Among the leading researchers in Singer's room were R. Ramsay Wright (1852–1933), professor of biology, University of Toronto, 1887–1912; Walter Libby, professor of psychology, Pittsburgh, author of *An Introduction to the History of Science* (London, 1918); Reuben Levy (1891–1966), later professor of Persian, University of Cambridge; Joan Evans (d. 1977), a medievalist, half-sister of Lewis and Arthur Evans; Dreyer, expert on Tycho Brahe; Dorothea Waley Singer (1882–1964), an author of distinction, on whom see *British Journal for the History of Science*, 2 (1964–5), 260–2.

[94] Bodleian Library Records, d. 211, The Science Room 1914– ; Gunther, *Gunther*, 193–4; at the end of 1920 £292 of Singer's donation was unspent.

Goodrich, who had known Singer well as an undergraduate, procured for him a university lecturership in the history of the biological sciences tenable in the zoology department for 1920–3 at £50 p.a.[95] Though Singer did lecture at Oxford, he resigned his post in autumn 1922 in order to cultivate full-time the more propitious ground offered by the Godless House in Gower Street.

By 1919 an alternative and totally independent approach to history of science in Oxford had been launched, again with Osler's encouragement. Its promoter was Robert Gunther, then science tutor at Magdalen, an impatient, prickly, and aloof Oxford-trained zoologist who was above all a fighter. His interest in history of local science had become clear by 1904 when he published the first of three volumes on the Daubeny Laboratory at Magdalen, a concern confirmed in 1916 with his second volume on that subject and amplified in books on Oxford gardens and the Oxford country. In 1915, with the encouragement of Osler, he had begun a survey of specimens and instruments throughout the University, the rich yield soon convincing him of the importance of early science in Oxford. By 1917 Gunther had become convinced of the utility of a museum in Oxford for local scientific apparatus to ensure its preservation and to be a resource for teaching history of science. He was thus already dedicated to what he called the archaeology of science in which field and museum work would be fruitfully combined via inventories, recovery, and preservation of scientific instruments. His favoured location for a museum of Oxford science was the old Ashmolean building on Broad Street, a preference stimulated by the tercentenary of Ashmole's birth in 1917.[96]

In summer 1919 Gunther organized an exhibition of 123 old scientific instruments from Oxford in the Picture Gallery of the Bodleian Library, the location being acquired by Osler who also paid for the catalogue. One visitor was Gunther's friend Lewis Evans, brother of Arthur Evans, discoverer of Knossos and chiefly responsible in 1894 for the foundation of the Ashmolean Museum of Art and Archaeology on Beaumont Street.[97] Lewis Evans had recently retired from the lucrative chairmanship of John Dickinson,

---

[95] 9 Mar. 1920, 31 Oct. 1922, NS/M/1/3; memo on Singer's qualifications, n.d. [spring 1920], NS/R/1/2, p. 100.

[96] Gunther, *Gunther*, 153–9; A. V. Simcock (ed.), *Robert T. Gunther and the Old Ashmolean* (Oxford, 1985), 47–8, 54–61, 68, 74–5; R. T. Gunther, *A History of the Daubeny Laboratory, Magdalen College, Oxford* (London, 1904); id., *The Daubeny Laboratory Register 1904–15* (Oxford, 1916); id., *The Oxford Country, its Attractions and Associations Described by Several Authors* (London, 1912); id., *Oxford Gardens, Based upon Daubeny's Guide to the Physic Garden of Oxford; with Notes on the Gardens of the Colleges and on the University Park* (Oxford, 1912); Elias Ashmole (1617–92).

[97] Gunther, *Gunther*, 159–64, 197–9; *OUG* 53 (1922–3), 349; Lewis Evans (1853–1930); Arthur John Evans (1851–1941), keeper of the Ashmolean Museum 1884–1908, lived at Youlbury near Oxford from 1893 until his death; R. F. Ovenell, *The Ashmolean Museum, 1683–1894* (Oxford, 1986), 250–64.

paper-makers, and had assembled a splendid collection of scientific instruments. Enchanted by the exhibition, in 1920 Evans made an unsuccessful offer of his collection, worth about £12,000, to the University. Two years later he persuaded the University to agree to a temporary exhibition of his collection to be displayed in the Bodleian's Picture Gallery from October 1922 to June 1924 while it considered the terms of his offer. It was the external interference of Lewis Evans which accounted for Gunther's survival at Oxford and saved him from suffering the same fate as Singer. After all, both Gunther and Singer were victims by late 1919 of the retirement and death of Madan and Osler, their patrons. Moreover in Gunther's case his career at Magdalen faltered: in 1920 he was sacked as fellow in natural science and at the age of 51 was relegated to a research fellowship tenable for no more than seven years. In 1923 the Daubeny Laboratory was closed by Magdalen and a considerable part of it was transferred to the University's department of botany so that Gunther lost his working room there. In compensation the resentful Gunther turned more and more to history of science, publishing his first two volumes on early science in Oxford, his *Early British Botanists*, and in 1924 his requiem for the Daubeny Laboratory. Simultaneously he became more and more dedicated to founding in Oxford a museum for the history of science based on Evans's collection; and, though Evans had not specified a local habitation for his collection, he and Gunther agreed in 1922 that the old Ashmolean building was the only available site and the most appropriate.[98]

In 1924 after acrimonious controversy the University accepted the Lewis Evans collection, arranged for it to be transferred to two rooms on the upper floor of the old Ashmolean building, and appointed Gunther as curator at £75 p.a., thus for once stealing a march on Cambridge whose Whipple History of Science Museum was opened twenty years later. The opposition was considerable. Hogarth, the curator of the Ashmolean Museum, and Arthur Evans were opposed to the metamorphosis of the old Ashmolean building into the Old Ashmolean Museum because in their view their Ashmolean Museum was *the* Ashmolean Museum and a possible home for the Lewis Evans collection. Farnell, vice-chancellor 1920–3, was convinced that Oxford should not entertain a third museum because it could not adequately support the Ashmolean and Pitt Rivers Museums. The Bodleian Library used the basement of the old Ashmolean as a book-store and Cowley, the librarian, coveted not just the two rooms vacant since 1922 on its top floor but all of it for the future extension of the Library. Gunther's proposal was also opposed by Oxford University Press, which had squatters' rights on the

---

[98] Gunther, *Gunther*, 118–41, 165; R. T. Gunther, *Early Science in Oxford*, i and ii (Oxford, 1923); id., *Early British Botanists and their Gardens, Based on Unpublished Writings of Goodyear, Tradescant and Others* (Oxford, 1922).

middle floor of the old Ashmolean where it was completing the *Oxford English Dictionary*.[99]

Gunther had some support within the University. The Natural Sciences Board backed his proposal. Wells, the new vice-chancellor in 1923, did not share his predecessor's hostility to yet another museum.[100] Crucially Gunther lobbied successfully for external support and indeed stipulative interference. Curzon, the chancellor, made it clear to Wells that it would be a disaster if the collection did not come to Oxford. In response to Gunther's promptings of J. B. Carrington, a close friend who was a leading figure in the Goldsmiths' Company, the latter promised £1,000 early in 1924, if the University would provide suitable accommodation. Carrington appreciated the uniqueness, the value, and the beautiful workmanship of the collection; and the University appreciated his Company's offer of hard cash. Sir John Findlay, a Balliol man, collector of scientific instruments, and owner and editor of the *Scotsman*, was primed by Harold Hartley, chemistry tutor at Balliol, to contribute £250.[101] Above all, Lewis Evans made it clear that in exchange for giving this valuable collection to the University he wanted it to be housed under Gunther's care in the upper rooms of the old Ashmolean where it would be the nucleus of a larger collection. Gunther managed to vanquish his opponents in the Ashmolean Museum, the Bodleian Library, and the University Press in 1924 because he exploited friendships to persuade Evans to donate a collection worth £12,000 to the University and to induce City companies to give £1,300 for the housing of the collection. From private individuals, including Lady Osler, Gunther extracted £440. In contrast the formal Oxford contribution was small beer: the colleges gave a total of £209 by 1925 in response to his appeal for funds; and the University paid £700 towards the installation costs of the collection in the old Ashmolean and subsequently £200 p.a. as a maintenance grant.

Having secured part of the old Ashmolean for the Lewis Evans collection of historic scientific instruments as a permanent university institution, Gunther then opened a campaign for the establishment of a university Museum of the History of Science in the old Ashmolean.[102] By 1935 he had succeeded, and he himself had been made stipendless reader in history of science the previous year. Like all his campaigns it generated contention and ill-will, not least because Gunther gave it his full-time attention from 1927

---

[99] Gunther, *Gunther*, 163–9, 174–81; Simcock, *Gunther*, 29–32; David George Hogarth (1862–1927), *DNB*, keeper of the Ashmolean Museum 1908–27; *HCP* 127 (1924), 106–7. The *Oxford English Dictionary* (completed in 1928), 12 vols., the *Supplement*, and the *Shorter Oxford English Dictionary*, 2 vols., were all published in 1933.

[100] Gunther, *Gunther*, 169–72, 181–2; Joseph Wells, vice-chancellor 1923–6.

[101] Gunther, *Gunther*, 180, 183–93; John B. Carrington (1843–1926); John Ritchie Findlay (1866–1930) took degrees in science and in Greats.

[102] Gunther, *Gunther*, 194–5, 200–23.

when Magdalen pensioned him off at £400 p.a. He mobilized favourable external opinion by encouraging distinguished visitors and by keeping his ambitions visible via incessant publicity in *The Times* and *Nature*. He cultivated local pride by publishing no fewer than seven volumes on early science in Oxford between 1925 and 1931. In 1928 he created a pressure group, the Friends of the Old Ashmolean, perhaps maliciously modelled on the Friends of the Bodleian established in 1925 under the auspices of Cowley, one of Gunther's enemies. The Friends were devoted to establishing museums of history of science and medicine at Oxford and Cambridge, to restoring all the old Ashmolean as a historical scientific institution, and to adding to the Evans collection. Gunther used formally unveiled armorial windows, dedicated to benefactors and to seventeenth-century virtuosi such as Ashmole, to anchor the collection to the old Ashmolean. He continued to secure external endowment, gaining £411 from City companies in 1927.

There was continuing opposition to his ambitions from E. T. Leeds, keeper of the Ashmolean Museum, who resented the possible resurgence of the old Ashmolean as a museum. While the *Oxford English Dictionary* was nearing completion, the University set up a commission in 1930 to investigate Oxford's libraries. Its report of spring 1931 recommended that the whole of the old Ashmolean be used for large-scale co-operative literary research by the Press. This evoked a powerful protest from Gunther who organized a petition of 123 members of Congregation who favoured the use of the top *and* middle floors of the old Ashmolean as a history of science museum. This impasse was so unsatisfactory that in January 1933 the Hebdomadal Council set up a committee on the Lewis Evans collection. Its report of June 1933 rejected the breaking up of the collection, recommended that it form the nucleus of a University Museum of the History of Science, and thought that it could expand from the upper floor of the old Ashmolean into the middle one. One senses that Gunther secured his museum because Veale, the registrar, realized that Gunther would not tie his tongue or pen, and that he was impervious to allegations that he was aggressive, devious, deceitful, unscrupulous, and insidious in controversy.[103] Certainly that became apparent yet again in May 1933 when there was a storm about the celebration of the 250th anniversary of the opening of Ashmole's museum in 1683.[104] On Leeds's advice, the University rejected Gunther's proposal for a formal celebration so he went ahead with informal celebrations in which he referred, as he had been doing for several years, to the old Ashmolean building as the Old Ashmolean Museum. His action led

---

[103] *HCP* 149 (1931), 65–7, 173–5; 155 (1933), 213–16; Edward Thurlow Leeds (1877–1955), keeper of the Ashmolean Museum 1928–45.

[104] Gunther, *Gunther*, 237–48; Evans to Leeds, 15 May 1933, Leeds to vice-chancellor, 16 May 1933, Veale memorandum, 20 May 1933, in OUA, Lewis Evans Collection and Old Ashmolean Museum, 1924–35, UR/SF/LE/1.

to brisk public exchanges between Gunther, Leeds, and Arthur Evans. The University and the vice-chancellor boycotted what Gunther claimed was only a private party. The episode made it clear that Gunther was impervious to embarrassment and remonstrance, and that he would fight, fight, and fight again in public even against such a distinguished figure as Sir Arthur Evans who was very angry about Gunther's 'bogus celebration'.

Gunther's last five years were not easy for him as scholar or as curator. It was characteristic of him that he worked alone without consulting others and on occasion he committed errors which hostile critics deplored. Though Professor Andrade donated material to the Museum, he reviewed volume x of *Early Science in Oxford* fiercely: 'from Dr. Gunther's scholarship we crave a respite. Surely his shining temple to the Goddess of Inaccuracy should by now be complete.' As curator Gunther laboured under several handicaps. At the outbreak of war the Press still occupied the middle floor of the old Ashmolean and the Bodleian Library the basement. Gunther lacked space for the avalanche of accessions so he could not take all the instruments from the Radcliffe Observatory when it closed and all those from the University Observatory when Plaskett turned it to astrophysics. His inability to do so hurt him: after all he had long believed that local exhibits constituted the basis of the good reputation of a historical museum. He failed to keep intact the Rigaud library from the Radcliffe Observatory. Finance remained an acute problem, with an annual grant of only £200 from the University. The university Appeal of 1937 ignored Gunther's Museum; and on the whole he failed in personal approaches to Lord Nuffield, City companies, the Rockefeller Foundation, C. S. Gulbenkian, and British Petroleum. Gunther was so dissatisfied with his own financial position that in 1939 he submitted his resignation from his stipendless readership in protest and then withdrew it. He died in 1940 still occupying his two offices, being succeeded as curator by Sherwood Taylor.[105]

---

[105] OUA, Museum of History of Science, Mar. 1935–Dec. 1942, UR/SF/LE/2; Gunther, *Gunther*, 249–96; Andrade in *Nature*, 136 (1936), 378–81; Edward Neville da Costa Andrade (1887–1971), professor of physics, University College London, 1928–50; Calouste Sarkis Gulbenkian (1869–1955), oil magnate, obituary in *The Times*, 21 July 1955; Stephen Peter Rigaud (1774–1839), Radcliffe Observer 1827–39, whose large library passed to the Observatory on his death; on Frank Sherwood Taylor (1897–1956), temporary curator 1940–5, curator 1945–53, see A. V. Simcock, 'Alchemy and the World of Science: An Intellectual Biography of Frank Sherwood Taylor', *Ambix*, 34 (1987), 121–39.

7
Zoology

Though zoology was a small subject it enjoyed the benefit of two chairs: the Linacre chair of zoology and comparative anatomy, occupied to 1921 by Gilbert Bourne and then for twenty-four years by Edwin Goodrich; and the Hope chair of entomology inhabited by Edward Poulton until 1933 and then by G. D. H. Carpenter, a pupil and collaborator. They received no new buildings but in 1931 there was a considerable extension of Goodrich's department at a cost of £3,750. The Hope department was so devoted to research and to the Hope entomological collections that it taught undergraduates only intermittently. Consequently it had no demonstrators and no college tutorial fellows. The Linacre department, equally devoted to research, saw the number of its finalists steadily rise into the teens. Like botany it was small, attracted women undergraduates, and acted as a marriage bureau: Alister Hardy, Vero Wynne-Edwards, and Peter Medawar all met their future wives there.[1] As the department taught large numbers of undergraduates, by no means all honours zoologists, it enjoyed at best six university demonstrators. On the outbreak of the Second World War it was blessed with five university demonstrators (J. R. Baker, C. S. Elton, E. B. Ford, J. Z. Young, and B. W. Tucker, the first four of whom were to be elected FRS) and three departmental demonstrators (P. B. Medawar, J. A. Moy-Thomas, and H. K. Pusey, the first of whom was to win a Nobel prize).[2] All eight were Oxford graduates, Baker being the only D.Phil. among them. The colleges gave some recognition to the Linacre department: Julian Huxley was a tutorial fellow at New College 1919–25, Gavin de Beer a prize fellow at Merton 1923–38, Young a fellow of Magdalen 1931–45, Medawar a fellow by examination in animal biology at

[1] Alister Hardy, Linacre professor 1946–61, 'war' degree 1920, married Sylvia Lucy Garstang, second in zoology 1921, daughter of Walter Garstang (1868–1949), an Oxford graduate, professor of zoology, University of Leeds, 1907–33; Vero Copner Wynne-Edwards (b. 1906), first in zoology 1927, professor of natural history, University of Aberdeen, 1946–74, married Jeannie Morris, second in zoology 1927; Medawar, first in zoology 1935, told Jean Shinglewood Taylor, third in zoology 1936, married 1937, that she had the first claim on his love but not his time.
[2] John Zachary Young, university demonstrator 1933–45, professor of anatomy, University College London, 1945–74; Bernard William Tucker (1901–50), university demonstrator 1927–46, reader in ornithology 1946–50; James Alan Moy-Thomas (1909–44), departmental demonstrator 1933–9, obituaries in *Nature*, 153 (1944), 427, and *OM* 62 (1943–4), 240; Harold K. Pusey (1910–87).

Magdalen 1938–44, and Elton a senior research fellow at Corpus Christi from 1936.

Two demonstrators in the Linacre department were important innovators in the study of living creatures in their natural environments. In 1932 Elton launched the Bureau of Animal Population and in 1938 the Edward Grey Institute of Field Ornithology was established, Tucker being one of the two prime activists. Both ventures, which were heavily dependent on external funds, were taken on by the University which could receive some credit for sanctioning field studies without dipping deeply into its own pocket. In contrast the Linacre and Hope departments did not attract new external endowments, except for the odd but very useful prize. The Christopher Welch postgraduate scholarship in biology, established in 1916 and tenable for four years at £100 p.a. by a male Oxonian, enabled several zoologists to launch their research careers. Beneficiaries in the 1920s included A. C. Hardy (1920), de Beer (1921), Baker (1922), Young (1928), and O. W. Richards (1924) and in the 1930s Medawar (1935). In entomology J. M. Baldwin endowed the Poulton Fund in 1920 to promote the study of natural and social selection: in the 1920s it facilitated the research of the young Ford.[3] The annual incomes of the Linacre and Hope departments differed widely. In the late 1930s, the former received about £2,000 of which about 40 per cent came from the University and the rest from fees; but the latter subsisted on about £700 of which £500 came from the University. Though the two departments played different roles in the University, they were united intellectually. In the 1910s and 1920s, when Darwinian natural selection was under attack as the sole or chief mechanism of evolution, it was staunchly defended by Goodrich, Huxley, and Poulton. In the 1930s, when the elements of the so-called evolutionary synthesis were being assembled, these men and their pupils made Oxford into a leading centre for evolutionary studies pursued in a Darwinian framework.

7.1 The Linacre Department

The Linacre department became well endowed with staff partly as a result of external pressure exerted on the University. In 1930 the Prime Minister set up a committee, chaired by Lord Chelmsford, to report on the education and supply of biologists.[4] It concluded that the alleged shortage of biologists

---

[3] Owain Westmacott Richards (1901–84), professor of zoology and applied entomology, Imperial College, London, 1953–67; James Mark Baldwin (1861–1934), professor of psychology, Johns Hopkins University, who was disgraced in a sex scandal in 1909, was a close friend of Poulton.

[4] OUA, Education and Supply of Biologists, UR/SF/APC/3; *Economic Advisory Council: Report of Committee on Education and Supply of Biologists* (Non-parliamentary Treasury Paper) (London, 1932); G. R. de Beer, 'The Training of Biologists', *OM* 50 (1931–2), 490–1; 1 Mar.

was caused by the uncertainty and unattractiveness of careers in public services at home and in the colonies. Among other remedies it called for universities to revise their degrees in order to stimulate biology, a view which the Board of Education communicated to the University. Internally the University's Appointments Committee had found it difficult to supply male candidates for posts, especially in teaching, where ability in both zoology and botany was required. De Beer and Veale were responsible for the University's answer, which was to preserve the intense specialization of its final honours schools in zoology and botany while introducing a new and optional alternative to the existing preliminary examinations in zoology and botany which were revised by amalgamating them into one subject, biology. The revised preliminary examination could still be taken and passed directly from school; but the new alternative, honours moderations in natural science, permitted an undergraduate in his or her first year to study and be awarded a class of honours in three subjects (chosen from mathematics, physics, chemistry, geology, botany, zoology, and physiology). Through natural science mods, introduced in 1936, it became possible to take honours in up to three life sciences, i.e. physiology, botany, and zoology studied for a year, followed by two years devoted to the final degree subject. The advantages of mods in natural science were flexibility of choice, broader education, better job prospects, the discouragement of excessive specialization at school, and the boosting of biology *vis-à-vis* the physical sciences.

From 1906 to 1921 the Linacre chair was filled by Bourne, a comparative anatomist whose best work was done before the first war. His internal promotion had led to external critical comment that he was a safe humdrum choice. Certainly Bourne did not rock the boat. On the contrary he was better known in Oxford as a keen oarsman and coach to the university crew for many years than as Linacre professor. Before the war he enjoyed the services of two star embryologists, Wilfred Jenkinson and Geoffrey Smith, both of whom were killed in it. Jenkinson was the first major British experimentalist in embryology. Both he and Smith, who worked on the internal generation of form, rejected on empirical grounds Ernst Haeckel's so-called biogenetic law or recapitulation theory which claimed that individual embryological development (ontogeny) recapitulated the evolutionary history of that particular species (phylogeny).[5] After the war Jenkinson and Smith were replaced officially, and at the highest level, by Goodrich, the

1932, BS/M/1/1; 3 May 1932, BS/R/1/2; *OUG* 66 (1935–6), 248–9; *Oxford*, 2 (1936), 10–12; Frederic John Napier Thesiger, 1st Viscount Chelmsford (1868–1933), *DNB*.

[5] Gilbert Charles Bourne (1861–1933), obituaries in *ON* 1 (1932–5), 126–30, *OM* 64 (1933–4), 16; Gunther, *Gunther*, 75; G. C. Bourne, *A Text-book of Oarsmanship: With an Essay on Muscular Action in Rowing* (London, 1925); on Oxford embryologists, M. Ridley, 'Embryology and Classical Zoology in Great Britain', in T. J. Horder, J. A. Witkowski, and C. C. Wylie (eds.), *A History of Embryology* (Cambridge, 1986), 35–67; Ernst Heinrich Philipp August Haeckel (1834–1919).

long-serving Aldrich demonstrator in comparative anatomy: in February 1920 he was made professor of comparative embryology. Like Jenkinson and Smith he rejected the recapitulation theory.

Goodrich's short-lived pluralism was one way in which the Linacre department coped with staff shortages after the war. Only one college, New College, came to its help when in early 1919 it elected Huxley as a fellow. In 1919 the regular staff of Bourne, Goodrich, and Huxley was supplemented by a quintet of voluntary helpers, namely, Edward Speyer who had demonstrated before the war, the histologist H. M. Carleton, the maverick physiologist J. B. S. Haldane, Robert Gunther, then science tutor at Magdalen, and Alexander Carr-Saunders. When Bourne resigned in July 1921, his staff was Goodrich, Huxley, and Carr-Saunders, assisted by Speyer.

At the end of Bourne's reign there appeared a rebel against his morphological emphasis. Alister Hardy, who graduated in 1920, began research under Bourne on a worm.[6] In less than a year he told Huxley, his tutor, that he was 'fed up with morphological research à la Bourne' and could not contemplate teaching morphology. He therefore rejected Huxley's plan for him to be a demonstrator on the comparative anatomy of invertebrates. Hardy was so keen to forget morphology that he left Oxford in 1921 for Lowestoft where the Ministry of Fisheries had established a group of researchers to study North Sea fish. Revolting against Bourne's exclusively morphological and laboratory-based approach, Hardy was happy to study the plankton of the open sea. He became a firm advocate of zoology outside the laboratory, decrying the laboratory-based and reductionist approach which merely studied the interiors of animals. Hardy believed that zoology embraced the quantitative and experimental analysis of interactions in the natural world, that ecology and ethology were key subjects because they examined animals from the outside and in relation to their natural surroundings, and that key questions in evolutionary theory could be solved only outside laboratories. He was revered by several Oxford zoologists whose careers began in the 1920s under Goodrich, who succeeded Bourne in 1921. They were glad to welcome Hardy back in 1945 when he replaced Goodrich in the Linacre chair and ended its long association with morphology.

In 1921 Goodrich had been elected to the Linacre chair in preference to J. T. Cunningham of London, James Gray of Cambridge who wished not to be considered if the intention was to appoint a comparative morphologist uninterested in function, and Walter Garstang of Leeds who wished to develop bionomics.[7] Goodrich, an Oxonian, had several advantages. He

---

[6] Hardy to Huxley, 16 May 1921, and reply 19 May, Hardy Papers, A7; A. C. Hardy, 'Zoology outside the Laboratory', *Advancement of Science*, 6 (1949–50), 213–23; N. B. Marshall, 'Hardy', *BM* 32 (1986), 223–73.

[7] OUA, 1921 Linacre chair, MR/7/1/5; G. R. de Beer, 'Goodrich', *ON* 5 (1945–8), 477–90; A. C. Hardy, 'Goodrich', *Quarterly Journal of Microscopical Science*, 87 (1947), 317–55; James

had worked in the department since 1898 as Aldrich demonstrator in comparative anatomy and more recently as professor of comparative embryology. For years he had administrated the department and twice had acted as deputy Linacre professor. His trump card was a collective letter signed by colleagues and former pupils stressing that he was a fine researcher, the builder of the department, and a clear precise lecturer with 'philosophic outlook'. The signatories included several London professors, leading people at the British Museum (Natural History), five Oxford professors including Sherrington, and all Bourne's staff in his last year.

Goodrich was first and foremost a comparative anatomist.[8] His 1930 book on vertebrates confirmed his reputation as the leading British comparative anatomist of his day, not least because he was interested in their development as well as their structure. Untainted by the current phobia against morphology, he used it to try to discover evolutionary pathways and to try to confirm the mechanism of natural selection as the main agent of evolution. In these ways he prepared favourable ground for the evolutionary synthesis which began to gather force in the mid-1930s. This concern with Darwinian evolution was an enduring feature of Goodrich's work, even in the opening years of the century when Darwinism was under eclipse and Lamarckian views were flourishing. Just before the first war Goodrich advocated natural selection as unassailable because it was natural, observable, testable, and measurable; and he was certain that Mendelian genetics had definitely disposed of Lamarck's rival doctrine of the inheritance of acquired characteristics during the organism's lifetime. These beliefs permeated his book of 1924 on living organisms. He denied Darwinian selection had lost ground, supported Mendelism, and attacked what he regarded as the illogicality of the Lamarckians. It may be that Goodrich promoted Darwinian evolution and the evolutionary synthesis in two ways, first by his morphological research and secondly by supporting the then unfashionable view that Mendelian genetics was not incompatible with Darwinian selection.

Though Goodrich had a predilection for morphology, he was also interested in promoting evolutionary biology and in reconstructing his

Gray (1891–1975), then fellow of King's, professor of zoology, University of Cambridge, 1937–59; Joseph Thomas Cunningham (1859–1935), lecturer in zoology, east London college, 1917–26.

[8] E. S. Goodrich, *Studies on the Structure and Development of Vertebrates* (London, 1930); id., *The Evolution of Living Organisms* (London, 1912), 55–66; id., *Living Organisms: An Account of their Origin and Evolution* (Oxford, 1924), 50–1, 61, 108; S. J. Waistren, 'The Importance of Morphology in the Evolutionary Synthesis as Demonstrated by the Contributions of the Oxford Group, Goodrich, Huxley, and de Beer', *Journal of the History of Biology*, 21 (1988), 291–330, attacks the standard view that morphology made no contribution to the evolutionary synthesis, a view adumbrated by W. Coleman in E. Mayr and W. B. Provine (eds.), *The Evolutionary Synthesis: Perspectives on the Unification of Biology* (Cambridge, Mass., 1980), 174–80.

department by appointing good people to teach and research in a variety of specialisms.[9] In recruiting the men who became departmental and then university demonstrators Goodrich simply appointed star graduates from his own department (de Beer graduated 1921, Baker 1922, Elton 1922, Tucker 1923, Ford 1924, Young 1928). Between 1921 and 1928 inclusive there were ten firsts in zoology finals, of whom five became core staff in Goodrich's department, the odd man out being Ford. Of this staff only Tucker, who worked all told for two years at the Naples Zoological Research Station and as a demonstrator at Cambridge, spent any appreciable time elsewhere: the rest were based in Oxford between graduation and joining Goodrich's staff. This conspicuous inbreeding was saved from inertia or torpor by the men themselves and by the inspiration that all of them, except Young, received from Julian Huxley who was their tutor and had a wide view of biology derived partly from his very varied experiences in Germany and in the USA before 1914.[10]

7.2 Julian Huxley

A Balliol graduate, Huxley was a convinced evolutionist and a broad Darwinian selectionist who was suspicious of Lamarckism. He was not wildly enthusiastic about that sort of comparative anatomy which was obsessed with descriptive morphology and classification. For him zoology could escape from the morphological trap by becoming an experimental subject pursued in the laboratory *and* by developing field studies which examined living things in their natural environment. In his formal teaching at Oxford Huxley accordingly introduced experimental zoology, genetics, and animal behaviour to complement Goodrich's comparative anatomy. Ideally he wanted as equals to comparative morphology no fewer than nine zoological specialities, namely, vertebrate embryology, invertebrates, cytology and histology, genetics, experimental embryology, ecology, comparative physiology, animal behaviour, and evolution and systematics.[11] His research was also wide-ranging, embracing bird behaviour, embryology, broadly conceived genetics, and Arctic ecology, some of which was pursued in collaboration with colleagues, previously his pupils.

---

[9] Goodrich to Huxley, 25 Aug. (1917?), Huxley Papers; G. R. de Beer, *Evolution: Essays on Aspects of Evolutionary Biology Presented to Professor E. S. Goodrich on his Seventieth Birthday* (Oxford, 1938).

[10] Useful surveys are: M. Keynes and G. Ainsworth Harrison (eds.), *Evolutionary Studies: A Centenary Celebration of the Life of Julian Huxley* (London, 1989); C. K. Waters and A. Van Helden (eds.), *Julian Huxley: Biologist and Statesman of Science* (Houston, 1992); J. R. Baker, 'Huxley', *BM* 22 (1976), 207–38.

[11] J. Huxley, preface to C. S. Elton, *Animal Ecology* (London, 1927), pp. ix–xiii.

In his six post-war years at Oxford Huxley continued the work on bird courtship on which he had embarked before the war, culminating in his famous paper on the great crested grebe. His aim was to clarify the relative roles of natural and sexual selection in the evolution of courtship behaviour. He interpreted courtship displays as symbolic ceremonies and rituals; and he invoked the idea that emotions and attitudes are important characteristics of animals, just like their anatomical structure, for purposes of evolutionary comparison. Put crudely, behaviour in animals was as important for Huxley as comparative anatomy in evolutionary research. His work on the ordinary lives of birds in their natural environment launched at Oxford the serious study of animal behaviour, pursued in a broad but not dogmatic Darwinian framework, decades before Nikolaas Tinbergen, the ethologist awarded a Nobel prize in 1973, was lured to Oxford in 1949.[12]

As an undergraduate Huxley had been an admiring pupil of both Jenkinson and Smith, whose approach to embryology rejected recapitulation and the attempt to reconstruct phylogeny. This tradition, which focused on nothing but experimentally determined features and causes, was carried into the 1920s by Huxley who was then even more rigidly experimental than Jenkinson and simply ignored recapitulation and phylogeny. His research on differential growth rates, where his aim was to produce their empirical laws, reached its culmination in his book on relative growth in which he showed that it was heterogonic and not orthogenetic.[13] At the same time Huxley worked on genetics, mainly in collaboration with Ford, using the amphipod *gammarus chevreuxi*. Their conclusions about rate-genes showed that these modifying genes determined the expression of so-called dominant genes and that dominant genes, which had been the chief interest of geneticists previously, seldom played an important part in evolution without being modified. This laboratory work on rate-genes showed that genes could control the time of onset and the rate of development of processes in living creatures. It cast even more doubt on Haeckel's theory of recapitulation and it also assailed H. F. Osborn's principle of orthogenesis, which supposed that groups of related species were endowed with inherent capacities that would control their future evolution irrespective of the species' response to environmental challenges. Thus Huxley's work on growth rates and rate-genes attacked recapitulation and orthogenesis; and it revealed the importance of the interplay between genetic and environmental factors in evolution. In so doing it helped to revive Darwinism in the 1920s and prepared the way for the evolutionary synthesis of the next decade. It is no surprise that in the 1940s and 1950s Huxley and a clutch of his former

---

[12] For a different view see R. W. Burkhardt, 'Huxley and the Rise of Ethology', in Waters and Van Helden, *Huxley*, 127–49; Nikolaas Tinbergen (1907–88), Nobel prize-winner 1973, professor of animal behaviour 1966–74.

[13] J. Huxley, *Problems of Relative Growth* (London, 1932).

pupils were prominent in consolidating the new Darwinism. He edited *The New Systematics* (1940) to which de Beer, Ford, and Arkell also contributed, he wrote *Evolution, the Modern Synthesis* (1942) which was greatly indebted to Ford, and with Hardy and Ford edited *Evolution as a Process* (1954) which contained contributions from de Beer and Willmer.[14]

Though Huxley's research on animal behaviour, embryology, genetics is well known, his ecological interests in the early 1920s tend to be underrated because they were expressed in an expedition, and expeditions have been seen too often as agents of geographical exploration. The first major Oxford expedition after the First World War was that of 1921 to Spitsbergen which was followed by others. They were not only good fun for tough young men; they were often devoted to zoology and as such were officially recognized as Oxford University expeditions. The pattern was set by that to Spitsbergen which was strongly supported by Huxley and Carr-Saunders because it aimed not to discover new species but to study the fauna, flora, and especially ornithology (particularly distribution of birds and their habits) of the archipelago from an ecological point of view. Spitsbergen had the advantage that its species were so few that it was hoped 'to master the sum of their interactions with each other and with their environment, and so to see the ecological web as a whole'. Among its eighteen members were five Oxonian ornithologists (the leader Jourdain, T. G. Longstaff, J. D. Brown, S. Gordon, A. H. Paget-Wilkes).[15] The two zoologists were Huxley himself and Carr-Saunders. The animal ecologist chosen by Huxley was Charles Elton who was still an undergraduate (see Fig. 3).

The Spitsbergen expedition of 1921, organized by George Binney, was followed by two more to the same area also organized by him: the Merton College Arctic expedition (1923) and the Oxford University Arctic expedition (1924), to both of which Elton was chief scientist and ecologist.[16] On the

---

[14] J. Huxley (ed.), *The New Systematics* (Oxford, 1940); id., *Evolution, the Modern Synthesis* (London, 1942); id., A. Hardy, and E. B. Ford (eds.), *Evolution as a Process* (London, 1954); Henry Fairfield Osborn (1857–1935), American vertebrate palaeontologist; Edward Nevill Willmer (b. 1902), first in zoology 1924, reader in histology, University of Cambridge, 1942–65, professor 1966–9.

[15] *HCP* 117 (1920), 167–70; J. Huxley, *Memories* (London, 1970), 128–34; *Spitsbergen Papers*, i: *Scientific Results of the First University Expedition to Spitsbergen* (London, 1925), esp. pp. v, vi; Francis Charles Robert Jourdain (1865–1940), *DNB*, rector of Appleton, Berkshire, 1914–25; Thomas George Longstaff (1875–1964), *DNB*, mountaineer and explorer; James Douglas Brown (1899–1971) was reading history; Seton Paul Gordon (1886–1977), *Who Was Who*; Arthur Hamilton Paget-Wilkes, who graduated in English in 1920, suggested an Arctic expedition to Huxley.

[16] *Spitsbergen Papers*, ii: *Scientific Results of the Second and Third Oxford University Expeditions to Spitsbergen in 1923 and 1924* (London, 1929); G. Binney, *With Seaplane and Sledge in the Arctic* (London, 1925), esp. 48, 176, 225, 258; Frederick George Binney (1900–72), *DNB*, third in English 1923, businessman with Hudson's Bay Company, 1926–31 and then United Steels Company, organized the 1921 expedition as a result of a chance meeting with Huxley in Blackwell's bookshop.

FIG. 3 Eleven members of the Spitsbergen expedition (1921) on board ship at Tromsö, Norway. Left to right are: F. G. Binney, secretary and organizer; C. S. Elton; the Reverend F.C. R. Jourdain, leader; R. Pocock, cook and artist, smoking; A. H. Paget-Wilkes; J. S. Huxley; S. P. Gordon in kilt; T. G. Longstaff; V. S. Summerhayes; J. D. Brown; and H. L. Powell. Pocock, Summerhayes, and Powell were non-Oxonians. Huxley was a don; Elton (zoology), Brown (history), and Binney (English) were undergraduates. Jourdain, Longstaff, Gordon, and Paget-Wilkes were Oxford graduates. Reproduced from a volume of photographs entitled 'Bear Island and Spitsbergen. Oxford University Expedition, 1921' by permission of the Edward Grey Institute of Field Ornithology, Department of Zoology, Oxford.

1924 expedition Elton enjoyed the company of Kenneth Sandford, who was its geologist and glaciologist, and of Howard Florey, its medical officer. Just as Huxley had chosen the scientific staff for the 1921 expedition, in 1924 Elton chose his scientists for their wide and common interests in the broad problems of Arctic ecology. These three big expeditions, all of which were almost totally funded externally or paid for by the participants themselves, gave Elton useful contacts, were crucial for defining his own approach to animal ecology, and gave him valuable experience as a *maître d'hôtel* capable of making a soup tureen out of a petrol can and of cooking Eider duck in a snowstorm. There was also a small ornithological expedition in 1922 to Spitsbergen led by Jourdain and Tucker, who was still an undergraduate.

7.3 Goodrichians and Huxleyans

A good deal of the research done in Goodrich's department can be understood in terms of positive debts to him and to Huxley and of reactions against them. In the case of Goodrich he had two colleagues, de Beer and Young, who were comparative anatomists. In de Beer's works on vertebrates and on the vertebrate skull, he concentrated on descriptive anatomy, avoiding discussion of function, presenting no new principles of morphology, and eschewing an experimental approach.[17] His big book on the vertebrate skull was explicitly an application of the teaching and methods of Goodrich who had introduced de Beer to the interests of morphology. His major textbook on vertebrate zoology, which went through seven reprints and one revision, was also Goodrichian in its concern with morphology and evolutionary history. No wonder that in 1936 Hardy saw de Beer as certain to succeed Goodrich in the Linacre chair and continue the tradition of necrological morphology. Young was different from de Beer in that in his research on the nervous system he concentrated as much on function and process as on structure. For Young, Goodrich's lectures were dull but an excellent foundation for future research: Young's well-known *The Life of Vertebrates* attempted to infuse embryology, physiology, biochemistry, and palaeontology into the exclusively structural approach taken by Goodrich and de Beer. Again unlike de Beer Young was happy to extend his intellectual horizons by going outside the Linacre department. He learned physiology from Denny-Brown and Eccles, his contemporaries at Magdalen; he attended an elementary course in the biochemistry department; while at the Naples Zoological Station he was stimulated to study octopuses by Enrico Sereni, a physiologist; and in the mid-1930s he used a Rockefeller fellowship to gain

---

[17] G. R. de Beer, *Vertebrate Zoology* (London, 1928); id., *The Development of the Vertebrate Skull* (Oxford, 1937); E. J. W. Barrington, 'De Beer', *BM* 19 (1973), 65–93; Hardy to Huxley, 24 July 1936, Hardy Papers A19.

access to squid at Woods Hole in the USA. By 1939 Young was collaborating with Clark and Cairns in research on the regeneration of nerves.[18]

Two of Goodrich's staff, Elton and Baker, reacted against Goodrich's morphology.[19] Elton found it gruelling: he gained some idea of the panorama of evolution but at the cost of great boredom. He contemptuously dismissed Goodrich's lectures which 'made the whole subject seem equally unimportant'. Initially Baker was more tolerant: he thought that comparative anatomy was much too prominent in teaching but ought not to be neglected. Ironically he discovered that Goodrich was intolerant of the research on contraception that he began in 1928. Goodrich outlawed it as disgusting and commercial so Baker researched in the physiology and pathology departments, eventually devising in the late 1930s Volpar, a spermicide which was strongly recommended by the National Birth Control Association. This research, done in conjunction with Carleton of the physiology department, was financed by the Association and in its later phases was pursued in association with British Drug Houses Ltd.

By the mid-1930s Goodrich was unsympathetic to those new branches of zoology which, unlike comparative anatomy, could not easily and obviously serve the cause of evolutionary biology. That was well shown by the cases of Medawar and Michael Abercrombie, both of whom were Oxford zoologists with first-class honours degrees who attained subsequent distinction.[20] Goodrich was uninterested in Medawar's early research on cell growth using tissue-culture techniques, seeing it as irrelevant to the evolutionary gospel. Medawar had therefore to launch his research career outside the Linacre department, finding a supervisor (Heaton) in pharmacology and laboratory facilities in pathology. There he was joined by Abercrombie who felt that Goodrich's unsympathetic attitude to his brand of experimental embryology left him no option but to start his research career at Cambridge and then to return to Oxford to join Medawar in Florey's department. Naturally they did not contribute to Goodrich's Festschrift which was devoted to aspects of evolutionary biology on which no fewer than thirteen post-war colleagues, associates, pupils, and his wife found it possible to write.[21]

---

[18] J. Z. Young, *The Life of Vertebrates* (Oxford, 1950); Young, private communications to author.

[19] A. Hardy, 'Autobiography', Hardy papers, A55, p. 73; Baker to Hardy, 24 Mar. 1922, Hardy Papers, B108; *HCP* 173 (1939), 88–9, 128–30; J. R. Baker, *The Chemical Control of Conception* (London, 1935).

[20] Medawar, *Autobiography*, 55–8; Michael Abercrombie (1912–79), professor of zoology, University College London 1962–70, director of Strangeways Laboratory, Cambridge, 1970–9, on whom see Medawar, *BM* 26 (1980), 1–15 (5).

[21] The thirteen were Poulton, Huxley, Ford, de Beer, J. B. S. Haldane, Richards, Carr-Saunders, Elton, Hardy, Baker, Young, Moy-Thomas, and Tucker; Helen Pixell Goodrich, a specialist on protozoa who gained a London D.Sc. and a husband in 1913, was a dominating figure in the laboratory compared with her husband.

Most of Goodrich's core staff were indebted in various ways to Huxley as lecturer or tutor. De Beer had Huxley as his final-year tutor and subsequently lodged with the Huxleys. He resembled Huxley in being a prolific author and editor of synthetic works. As a college tutor de Beer had to range widely so his book on the pituitary gland examined it from anatomical, histological, embryological, and physiological viewpoints. His interest in experimental biology, encouraged by Huxley, was shown in his research on growth and experimental embryology, some of which was done in collaboration with Huxley.[22] De Beer's positivist approach to embryology, derived from Huxley and Jenkinson, reached its culmination in their textbook of 1934. But de Beer was also an innovator. In his book on embryology and evolution he gave the *coup de grâce* to Haeckel's theory of recapitulation and the biogenetic law. Developing Garstang's notion of paedomorphosis, de Beer claimed that in many cases adult descendants retained features of the youthful stages of their ancestors—the reverse of recapitulation. As an experimental biologist interested in tissue growth in part in its connection with Mendelian genetics, de Beer cleverly connected his embryological arguments with the research of Huxley on rate-genes. De Beer's approach drew an acrimonious response from E. W. MacBride, a prominent embryologist who was an active Lamarckian, a proponent of the recapitulation theory, and an opponent of genetics. MacBride was perceptive in asserting that de Beer's key purpose was to support the genetical theory of evolution, which was totally at odds with the recapitulation theory. Thus de Beer, Huxley, and Goodrich differed from most embryologists in the 1920s and 1930s: many were evolutionists, but few endorsed natural selection as the primary mechanism of evolution.

Baker was indebted to his tutor, Huxley, for the range of his interests and for the model of a prolific and versatile scientist who advocated eugenics and commented on the public problems of science. Baker did field-work with Elton and Ford on animal populations. He studied inter-sexuality. He developed cytological techniques, on which he wrote a standard textbook. He collaborated with Carleton, the histologist in the physiology department, in research on the Huxleyan topic of birth control by studying and producing chemical contraceptives. He made three trips to the New Hebrides where he studied breeding seasons in a climate of great uniformity and

---

[22] Juliette Huxley, *Leaves of the Tulip Tree* (London, 1986), 107, 184; G. R. de Beer, *The Comparative Anatomy, Histology, and Development of the Pituitary Body* (London, 1926); id., *Growth* (London, 1924); id., *An Introduction to Experimental Embryology* (Oxford, 1926); id., *Embryology and Evolution* (Oxford, 1930); J. Huxley and G. R. de Beer, *The Elements of Experimental Embryology* (Cambridge, 1934); E. W. MacBride, *Nature*, 123 (1930), 882–4; Barrington, 'De Beer', 68–72; Ridley, 'Embryology', 60–3; F. B. Churchill, '*The Elements of Experimental Embryology: A Synthesis for Animal Development*', in Waters and Van Helden, *Huxley*, 107–26; R. A. Baker and R. A. Bayliss, 'Walter Garstang (1868–1949): Zoological Pioneer and Poet', *Naturalist*, 109 (1984), 41–53.

concluded that many factors determine them. As a polemicist about public science, he gave radio talks for the BBC on biology in everyday life, following Huxley in arguing as eugenicist for the sterilization of congenitally feeble-minded people. He popularized the physiology of sex in a book addressed to the laity. As a firm individualist who was reluctant to supervise post-graduates, Baker believed in the liberty, the fraternity, and the inequality of humankind. Accordingly he issued in summer 1939 a counterblast to Bernal's ideas about the socialist planning of science. Vehemently opposed to creeping Bernalism, in 1940 he and Tansley helped to found the Society for Freedom in Science of which he was a key member for over twenty years.[23]

Like Baker, Elton was an undergraduate at New College where Huxley was his tutor. Elton was greatly indebted to Huxley for appointing him as ecologist to the first Spitsbergen expedition. His Arctic experiences eventually led him to found the Bureau of Animal Population at Oxford and in some ways set its agenda. If the Bureau may be regarded as godfathered by Huxley, it is also the case that, after Huxley had left Oxford, his interest in ecological expeditions was continued by Elton and Sandford who were founder members of the Oxford University Exploration Club established in 1927 by two undergraduates, Max Nicholson, a historian and ornithologist who later became director-general of the Nature Conservancy, and Colin Trapnell, a classicist and botanist. The first president was John Buchan but the donkey work of being first chairman was done by Elton. In 1932 he relinquished the chairmanship only to be made treasurer, a post he passed on to Sandford in 1934. In that year Elton assumed the editorship of the Club's publications, while continuing to act as home agent and adviser to some expeditions. Elton himself went on the Lapland expedition (1930) where he studied rodents, and that to the Faeroes (1937) was half devoted to his favourite subject of small mammal ecology, which was pursued by Francis Evans, his first D.Phil. student. As editor of the *Journal of Animal Ecology* Elton gave useful publicity to the research carried out on the expeditions.[24]

---

[23] Willmer and Brunet, 'Baker'; J. R. Baker, *Sex in Man and Animals* (London, 1926); id., *Man and Animals in the New Hebrides* (London, 1929); id., *Cytological Techniques* (London, 1933); id., *Conception*; id., 'Counterblast to Bernalism', *New Statesman and Nation*, 18 (1939), 174–5; id., *The Scientific Life* (London, 1942); id., *Science and the Planned State* (London, 1945); id. and J. B. S. Haldane, *Biology in Everyday Life* (London, 1933), 79–83; G. E. Allen, 'Julian Huxley and the Eugenical View of Human Evolution', in Waters and Van Helden, *Huxley*, 193–222; W. McGucken, 'On Freedom and Planning in Science: The Society for Freedom in Science, 1940–46', *Minerva*, 16 (1978), 42–72.

[24] *Oxford University Exploration Club: Annual Reports* (1929–39); C. Swithinbank, 'The Oxford University Exploration Club', *OM* 66 (1947–8), 516–18; Colin Trapnell (b. 1907), Greats 1929, was so inspired and helped by Tansley that he pursued a career as an ecologist in Africa; John Buchan (1875–1940), an Oxford graduate, always keen on exploration, was living at Elsfield Manor near Oxford; Francis Cope Evans (b. 1914), Rhodes scholar 1936–9, D.Phil. 1940, ended his career as professor of zoology, University of Michigan, 1959–82; F. C. Evans

Though the Club had no endowment, reserves, or headquarters, it was such an effective lobbier for funding and so skilled at using Oxonian contacts in high places that it organized no fewer than thirteen expeditions before the Second World War, its first being to Greenland (1928) under Longstaff who extended to it the ecological and ornithological work done in Spitsbergen. The second was to British Guiana (1929), organized by Nicholson who acted as chief ornithologist; its purpose was to study the ecology of the tropical rain forest. The University gave its imprimatur to the Club's expeditions, to each of which it contributed no more than £50. But the official university sanction enabled the Club to secure on occasion external funding amounting to £600 from the Percy Sladen Fund and £100 or so from each of the Royal Geographical Society, the Royal Society of London, the Royal Botanical Garden (Kew), and the British Museum (Natural History). Such funding made expeditions objects of ambition and vehicles of training for undergraduates such as Max Nicholson, Nicholas Polunin, Colin Trapnell, James Fisher, and Henry Vevers, all of whom became prominent in the worlds of natural history and conservation. The Club also facilitated the research on breeding seasons carried out by Baker, a member of the zoology staff. Having been twice to the New Hebrides privately, in 1933 Baker led a much larger and better-equipped expedition to these islands to study the breeding seasons of birds and animals of the rain forest in a uniform climate.[25]

The debt of E. B. Ford to Huxley was considerable. While an undergraduate Ford began research in 1923 on genetics with Huxley who had advised him on that subject's literature. Their well-known work on rate-genes showed that specific genes could control the time of onset and the rate of development of processes in the body. Alive to the evolutionary importance of genetics, in the 1930s Ford promoted the Mendelian aspects of evolution particularly in two books, the first of which was dedicated to Huxley. Ford's second debt to Huxley, also incurred in 1923, was that through Huxley he met R. A. Fisher, the statistician, and began to collaborate with him. In 1927 in a paper on insect mimicry, Fisher envisaged the selective modification of the effects of genes. Through his collaborative

and H. G. Vevers, 'Notes on the Biology of the Faeroe Mouse', *Journal of Animal Ecology*, 7 (1938), 290–7.

[25] *British Guiana Papers: Scientific Results of the Oxford University Expedition to British Guiana in 1929* (London, 1938); Nicholas Polunin (b. 1909), *Who's Who*, first in botany 1932, botanist to the Lapland expedition 1930 and to Hudson Strait 1931, pursued Arctic zoology for the rest of his career; James Fisher (1912–70), *DNB*, second in zoology 1935, ornithologist to the Spitsbergen expedition 1933, later edited the Collins New Naturalist series; Henry Gwynne Vevers (1916–88), second in zoology 1938, leader of the Faeroes expedition 1937, later a leading official of the Zoological Society of London, obituary in *The Times*, 27 July 1988; Trapnell was botanist to the Greenland expedition 1928; for Baker's view of the academic importance of the 1933 expedition, Baker to Margoliouth, 2 Nov. 1932, BS/R/1/2.

research begun with his father in 1917 on the marsh fritillary, Ford had become convinced that the forces of selection in nature were powerful and was to show in 1930 that fluctuation in the numbers of that butterfly affected its variations and evolution. Inspired by Fisher, who gave a basis for analysing evolution in wild populations and significance to the fritillary research, Ford conceived in the late 1920s the notion of ecological genetics. This involved the experimental study of evolution and adaptation, carried out by combining field-work with laboratory genetics, and it gave a direct means of studying evolution in action. Ford's synthesis of genetics and ecology depended greatly on Fisher's statistical methods and his genetical theory of natural selection. To Fisher's new concepts and techniques Ford brought skills in field-work, using observation on successive generations of lepidoptera, and in experiments, the best known being the marking technique he devised in the late 1930s for butterflies using dots of paint on the wings. In the laboratory Ford bred species studied in the field in order to elucidate their genetics. He was different from laboratory-based geneticists in that they rarely did field-work and were ignorant of the power of selection in nature. He was on occasion dismissive of T. H. Morgan's laboratory genetics which used the Lord of the Flies, *Drosophila melanogaster*, as experimental material: he called it drosophilosophy. Ford also deprecated those ecologists who were ignorant of the variations in nature they encountered and lacked the ability to make proper quantitative analysis of their results about populations in the wild. Working mainly solo in the 1930s, Ford's most spectacular discovery was that of industrial melanism. Characteristically Ford studied the spreading of black moths in industrial areas by doing breeding experiments. He also did breeding and field-work on the scarlet tiger moth in order to exclude 'genetic drift' as an evolutionary mechanism. He produced the first demonstration using wild material (the magpie moth which has extreme variations in its markings) of Fisher's theory of genetic dominance. His main contribution by 1939 to zoology was to detect evolutionary changes in a small number of generations of lepidoptera which had many advantages for research: they had recognizable and measurable characteristics on their wing patterns, they occurred in high densities, and they were easily collected, surprisingly robust, and capable of being bred, marked, released, and captured. Using lepidoptera Ford was thus able to detect selection in operation and to observe and analyse its immediate effects.[26]

---

[26] E. B. Ford, 'Some Recollections Pertaining to the Evolutionary Synthesis', in Mayr and Provine, *Synthesis*, 334–42; id., *Ecological Genetics* (London, 1964), pp. xi–xiii, 1–9; id., 'Scientific Work by Sir Julian Huxley', in Keynes and Ainsworth Harrison, *Huxley*, 41–5; bibliography in R. Creed (ed.), *Ecological Genetics and Evolution: Essays in Honour of E. B. Ford* (Oxford, 1971), pp. xv–xxi; obituaries, *The Times*, 23 Jan. 1988, *Independent*, 25 Jan. 1988; R. A. Fisher and E. B. Ford, 'The Variability of Species in the Lepidoptera with Reference to Abundance and

Ford was the only member of staff in the Linacre department who took advantage of the Hope department of entomology. Given his persistent interest in butterflies and moths, he naturally turned to it. Presumably he felt at home there because in 1948 he applied unsuccessfully for the Hope chair. Ford also made common cause with Poulton and Carpenter, the Hope professors. In the 1920s Poulton gave Ford unlimited research facilities, suggested some early research, and steered research grants Ford's way from the Poulton Fund established in 1920 by James Mark Baldwin for the study of evolution. Moreover Poulton's public views must have appealed to the young Ford. In 1921, for example, in an exposition of protective resemblance in insects Poulton called for experimental work to be done on melanic moths, argued that Mendelism was a very valuable reinforcement of Darwin's theory of evolution by natural selection, and urged that more experiments be done on the hereditary transmission of small variations. Indeed before the first war Poulton had alluded to the importance of industrial melanism in moths: for him melanism was incompatible with mutationism, which involved large and sudden variations, as a mechanism of evolution because a long series of intermediate varieties had been found among species which had darkened and visible transformations were occurring very gradually. Moreover, for Poulton melanism was a phenomenon which ought to be used to test natural selection as the mechanism of evolution. Though not a geneticist, Poulton had realized by 1908 the importance of Mendel's work, claiming it was compatible with natural selection. Poulton was well aware of the importance of butterflies: because of their wing colours he believed they stood at the head, not only of all insects, but of all living things as registers of subtle evolution. Generally Poulton was a dedicated and vociferous believer in Darwinian selection, even when it was most assailed, and he went out of his way to denounce Lamarckism. Given all these beliefs it is not surprising that Poulton used the Poulton Fund to give a publishing subvention to Ford and Fisher for their joint paper of 1928 on variability of lepidoptera. Poulton was also interested in the high degree of polymorphism exhibited by mimetic species of lepidoptera. As a Darwinian, Poulton meant by a species a group of individuals which in a given location and time was isolated reproductively. Ford's main theoretical contributions

Sex', *Transactions of the Entomological Society of London*, 76 (1928), 367–84; E. B. Ford, *Mendelism and Evolution* (London, 1931); id., *The Study of Heredity* (London, 1938); J. F. Box, *R. A. Fisher: The Life of a Scientist* (New York, 1978), 180–3, 303–10; Ronald Aylmer Fisher (1890–1962); R. A. Fisher, *Statistical Methods for Research Workers* (Edinburgh, 1925); id., *The Genetical Theory of Natural Selection* (Oxford, 1930); J. R. G. Turner, 'Random Genetic Drift, R. A. Fisher, and the Oxford School of Ecological Genetics', in L. Kruger *et al.* (eds.), *The Probabilistic Revolution*, ii: *Ideas in the Sciences* (Cambridge, Mass., 1987), 313–54; on Thomas Hunt Morgan (1866–1945), R. E. Kohler, *Lords of the Fly: Drosophila Genetics and the Experimental Life* (Chicago, 1994); Ford followed his father Harold Dodsworth Ford (third in history 1888), a Cumberland clergyman, to Wadham.

in the 1930s were his definitions of polymorphism and of species which were Poultonian in their rejection of a morphological approach. Of course Poulton was not an active geneticist, had no mathematical or statistical ability, and did not indulge in practical ecology; but Ford's debts to him were far greater than Ford recorded in his recollections about the evolutionary synthesis. Ford also collaborated in the early 1930s with Geoffrey Carpenter who succeeded Poulton in the Hope chair in 1933. Their joint book brought together the subjects of mimicry and genetics, which for Poulton were always unconnected. It is significant that its chief focus was one dear not only to Ford but to two Hope professors of entomology and that he never again indulged in joint authorship of a book.[27]

Tucker was like Ford in that he arrived at Oxford already devoted to his life's work, in his case ornithology on which he had burst into print as a schoolboy. At Oxford he met and collaborated with Jourdain, with whom he founded in 1921 the Oxford Ornithological Society while still an undergraduate. As its secretary Tucker soon became its driving force and co-opted as a senior member his tutor Huxley. Tucker and other members of the Society helped Huxley in his research on the courtship and breeding habits of birds.[28] Though Tucker was not as close to Huxley as de Beer, Baker, Elton, Ford, and Hardy, it is tempting to ascribe to Huxley Tucker's transformation from an egg collector in his early days at Oxford to a wide-ranging ornithologist who studied birds in the field, the laboratory, the museum, and the library.

Huxley was an important figure for two other Oxford zoologists, Alister Hardy and Vero Wynne-Edwards. Hardy found Huxley more than just an inspiring tutor. In vacations he stayed with Huxley. It was Huxley who introduced Hardy to Stanley Gardiner who was recruiting staff for the new Ministry of Fisheries laboratory at Lowestoft to study North Sea fish.[29] As a winner in 1921 of the Oxford biological scholarship to study at the Naples Zoological Research Station, Hardy had become interested in marine plank-

---

[27] Ford to author, 24 Nov. 1986; Ford Papers, A12; *OUG* 59 (1928–9), 684; E. B. Poulton, 'The Inspiration of the Unknown', *Transactions of the South Eastern Union of Scientific Societies* (1921), 1–34; id., *Essays on Evolution 1889–1907* (Oxford, 1908), pp. vii–xxxvi, 50–4, 62–5, 308–10, 363; G. D. H. Carpenter, *Mimicry with a Section on its Genetic Aspect by E. B. Ford* (London, 1933) was dedicated to Poulton; for mimicry and genetics, J. R. G. Turner, 'Fisher's Evolutionary Faith and the Challenge of Mimicry', *Oxford Surveys in Evolutionary Biology*, 2 (1985), 159–96; for mimicry and evolution, W. C. Kimler, 'Mimicry: Views of Naturalists and Ecologists before the Modern Synthesis', in M. Grene (ed.), *Dimensions of Darwinism* (Cambridge, 1983), 97–128; J. R. G. Turner, '"The Hypothesis that Explains Mimetic Resemblance Explains Evolution": The Gradualist–Saltationist Schism', in Grene, *Darwinism*, 129–69.

[28] Obituaries of Tucker in *British Birds*, 44 (1951), 41–6; *Ibis*, 93 (1951), 300–5; J. Huxley, 'Some Further Notes on the Courtship Behaviour of the Great Crested Grebe', 'Some Points in the Breeding Behaviour of the Common Heron', *British Birds*, 18 (1924–5), 129–34, 155–63.

[29] Hardy, 'Autobiography', Hardy Papers, A55, pp. 81–5, Memories of Huxley, B125; John Stanley Gardiner (1872–1946), professor of zoology, University of Cambridge, 1909–37, a key adviser to the Ministry of Fisheries.

ton in the Mediterranean so he was keen to study North Sea plankton in relation to fisheries. Oxford was vital for Hardy: he met his future wife there; the biological scholarship gave him a new interest, marine ecology; and his tutor, Huxley, provided for him the job which enabled him to develop his work on the open sea. No wonder that Hardy was happy to return to Oxford in 1945 as Goodrich's successor in the Linacre chair.

Wynne-Edwards ended his career as professor of zoology in the University of Aberdeen where he wrote his controversial book on animal dispersion which attacked conventional neo-Darwinism.[30] While at school he was so impressed by a lecture given by Huxley on the Spitsbergen expedition (1921) that he went to New College in order to have Huxley as his tutor. With his research emphasis, his capacity for argument, and his voracious reading in three or four languages, Huxley was indeed inspiring and became a benevolent friend and supporter. When he left Oxford in 1925, he was replaced as Wynne-Edwards's tutor by Elton whose influence was specific and enduring. From Elton Wynne-Edwards derived an interest in population ecology that was to dominate his career. He was also indebted to Elton for recommending Carr-Saunders's seminal book of 1922 on the population problem in which it was argued that some human societies had lived at or near their optimal density, attuned to the resources of their territories, and had managed them for the common good. Wynne-Edwards's work on sociality in animals applied to them the same ideas that Carr-Saunders had advanced about humans. Their writings are important because they show that within an evolutionary framework it was possible for Oxonians to develop ideas about mechanisms of evolution which differed from those enshrined by Huxley and his allies in the so-called evolutionary synthesis of the late 1930s.

Another Oxford zoologist, Owain Richards, reacted against the synthetic tendencies of Huxley, his tutor, and the holistic approach of Elton to animal ecology.[31] In 1936 Richards and a colleague at Imperial College, London, produced their book-length study of the variation of animals in nature. Richards and Robson argued that there was no evidence for Lamarckian processes but were cautious about the universality of natural selection: it seemed to them that self-regulation and self-organization had also contributed to evolution. As professor of zoology at Imperial, Richards persisted in stressing the difficulties of selectionist theories. In the field of animal ecology he began as a disciple of Elton, studying communities of animals while an undergraduate and as a postgraduate working in both the Linacre and Hope

---

[30] V. C. Wynne-Edwards, 'Backstage and Upstage with Animal Dispersion', in D. A. Dewsbury (ed.), *Leaders in the Study of Animal Behaviour: Autobiographical Perspectives* (Lewisburg, Pa., 1985), 486–512; V. C. Wynne-Edwards, *Animal Dispersion in Relation to Social Behaviour* (Edinburgh, 1962); Wynne-Edwards to author, 3 Aug. 1987.

[31] R. Southwood, 'Richards', *BM* 33 (1987), 543, 553–6; G. C. Robson and O. W. Richards, *The Variation of Animals in Nature* (London, 1936).

departments. By the early 1930s he rejected Elton's holistic approach in favour of studying the population dynamics of individual insect species, which became the main work of his career. Richards made an important input into the ideas on evolution developed by Elton.[32] In his *Animal Ecology* Elton drew on the early work of Richards and Robson to corroborate his own notions that natural selection was just one of several agents of evolution, that sometimes selection disappeared entirely, and that its nature and intensity were periodic and constantly varying. By 1930, when Elton had accumulated more data about fluctuations of animal numbers and about animal migration, he became convinced that selection of their environment by animals was as important as the natural selection of animals by their environment.

7.4 The Hope Department

If Carr-Saunders, Richards, Elton, and Wynne-Edwards deviated from the natural selectionist line by advocating behavioural selection, that was not the case in the Hope department where it was defended and promulgated with missionary zeal even when it was unpopular to do so. For much of the inter-war period the Hope chair was occupied by Edward Poulton who resigned at the end of 1932 aged 77 after holding it for thirty-nine years. He was succeeded in 1933 by his friend, protégé, and collaborator Geoffrey Carpenter, an Oxonian who had retired from the Colonial Medical Service in 1930 and had come to live in Oxford. The chair they occupied was peculiar in that it was not a full-time teaching post: the professor was required to reside in Oxford only a third of the year, to teach only twenty-eight hours a year, to do research, and to look after the Hope collection of insects and the Hope library. He was paid a maximum of £500 p.a. and consequently found a private income useful. In Poulton's case he had made a wealthy marriage to Emily Palmer, daughter of George Palmer, MP for Reading and head of Huntley & Palmer's biscuit company. As a wag remarked, Poulton got the biscuit and the tin when he married, dozens of the latter being used by him to store lepidoptera. He was sufficiently wealthy to pay an assistant curator of the Hope entomological collections half his own salary after the first war and on retirement he desisted from asking the University for a pension after almost forty years' service. In Carpenter's case he enjoyed such a good pension that he built himself a house on Cumnor Hill, Oxford, when he retired from the Colonial Medical Service.[33]

---

[32] Elton, *Animal Ecology*, pp. viii, 184–5; id., *Animal Ecology and Evolution* (Oxford, 1930).
[33] *OUG* 63 (1932–3), 111–12; Poulton memorandum, 21 Jan. 1934, BS/R/1/2; *HCP* 134 (1926), 69–70; 159 (1934), 31; A. Z. Smith, *A History of the Hope Entomological Collections in the University Museum Oxford with Lists of Archives and Collections* (Oxford, 1986); G. Car-

Before he assumed the Hope chair, Poulton was an ardent Darwinian and student of protective resemblance, warning coloration, and mimicry in insects. These phenomena came under heavy fire during the eclipse of Darwinian selectionism in the early years of the twentieth century because they were central to the notion of natural selection. As Hope professor Poulton was a leader in Britain in maintaining that they were facts and in supporting natural selection as their explanation. Initially he was an advocate of Batesian mimicry, i.e. the way in which an unpalatable species of insect which possesses some form of protection such as a sting is mimicked by other palatable species which resemble it while lacking protective devices so that their disguise deceives their enemies. By 1887 he had become a devotee of Müllerian protective resemblance which meant that if several species of insects, all of which are protected, resemble each other, then the resemblance reduces the inroads made on them by predators who learn about their harmful qualities.[34] One form of protection of particular interest to Poulton was warning coloration and behaviour in unpalatable or dangerous species of insects. For Poulton all these phenomena were inexplicable on Lamarckian principles because most insects cannot control their own colour. Poulton also rejected special creation as a theological dogma and argued that large single mutations could not produce the thousands of cases of mimicry and protective resemblance. In his view such facts proved the effectiveness of natural selection which was responsible for protective adaptations.

A charitable man, Poulton could forgive anything save disbelief in Darwinian evolution. Though past his prime in the 1920s, he lost no opportunity to promote Darwinism and to trounce its opponents. So from 1920 he was happy to accept £100 p.a. from his friend James Mark Baldwin, who established the Poulton Fund which would promote work on organic and social evolution as Poulton thought fit. In 1931 Baldwin gave a final gift of £1,000. In practice Poulton used the fund to promote research on mimicry, resemblance, and coloration in insects, and not on evolution in general. Even in his early seventies Poulton relished opportunities to attack opponents of Darwinian selectionism, being particularly severe on MacBride for his Lamarckism and on McAtee for denying the reality of protective adaptation in insects.[35]

penter, 'Poulton', *ON* 4 (1942–4), 655–80; Geoffrey Douglas Hale Carpenter (1882–1953), Hope professor of entomology 1933–48, obituaries in *Nature*, 171 (1953), 592–3 and *OM* 71 (1952–3), 207–8.

[34] Poulton, *Evolution*, pp. xviii–xxii, 110–19, 213; Henry Walter Bates (1825–92), *DSB*; Fritz Müller (1822–97), *DSB*.

[35] For Poulton and Carpenter versus MacBride, *Nature*, 123 (1929), 661–3, 712–13, 874; 124 (1929), 183, 225, 577–8; for Poulton and Huxley versus McAtee, *Nature*, 130 (1932), 66–7, 202–3, 848, 961–2. Waldo Lee McAtee, of the Bureau of Biological Survey, Washington, argued that mimic and non-mimic insects were eaten by birds in proportion to availability so that resemblance

Poulton's second main role as Hope professor was to nurture the Hope entomological collections and library. Poulton was a collector *par excellence*: his cronies corrupted his middle name of Bagnall to Bag-all. The results were that on his retirement the huge collections were mainly unsorted and unidentified and that the library's floor was covered with publications piled chest high with winding narrow passages between them. Though he was an untidy man, Poulton had good reasons for expanding the Hope collections.[36] He was not himself a taxonomist but he appreciated the importance of making the Hope collections into the only rival to the British Museum (Natural History) for type specimens, i.e. the original specimens upon which the descriptions of species were based. He also saw the collections as a resource for studying and promoting his own favoured research topics. The collections helped Poulton to fulfil his dream of making Oxford a great imperial university dedicated to research because they attracted donations and visitors from many parts of the British empire and enabled him to make the Hope department the centre of gravity of entomological research in the empire. For Poulton the evolutionist, the Hope collections were essential because they would allow posterity to detect and measure the rate of evolution. The emphasis on collecting also enabled Poulton to dissociate himself from two aspects of entomology which were cultivated elsewhere. First he himself had little interest in human and animal diseases which were carried by insects in the British empire: he was content to leave economic entomology to be pursued by the odd pupil such as Carpenter who studied the tsetse fly in Uganda. Secondly Poulton was no experimentalist so that when he retired his department lacked the simplest apparatus. Perhaps above all the Hope collections enabled Poulton to give the impression that his department was peopled by a large number of researchers even though it was really a one-man band. He pulled in motley locals and visitors to work unpaid on the huge numbers of specimens, especially from the empire, and then through the voluminous and widely diffused *Hope Reports* he implicitly claimed that all publications based on such specimens had been produced by his department.

Poulton usually employed two assistants, the best known being Albert Hamm, a sort of entomological Huxley in his research on the courtship and mating habits of predatory empid flies. This small core was supplemented by two sorts of local amateurs, academics who cultivated entomology as a pastime and retired people. Prominent among the former were: Edwin Waters, lecturer and then professor of romance languages and a specialist on

---

was of no survival value: *Effectiveness in Nature of the So-Called Protective Adaptations in the Animal Kingdom, Chiefly as Illustrated by the Food Habits of Nearctic Birds* (Washington, 1932).

[36] Smith, *Hope Collections*, 27–33; E. Poulton, 'The Empire and University Life', *Nature*, 72 (1905), 217–18; id., 'The Reform of Oxford University', *Nature*, 80 (1909), 311–12; id., *Evolution*, 50–4; *OUG* 65 (1934–5), 210.

micro-lepidoptera; Laurence Grensted, Nolloth professor of the philosophy of the Christian religion; Arthur Pickard-Cambridge, tutor in Greek at Balliol and an expert on spiders and beetles; and Frederick Dixey (FRS 1910), bursar of Wadham, curator of white butterflies in the Hope collections, and a convinced Darwinian who developed Müller's theory of protective resemblance with his notion of reciprocal mimicry. Even more numerous were professional people who had retired to Oxford. The best known was Harry Eltringham (FRS 1930), a boiler manufacturer who was inspired by Poulton to examine mimicry in African butterflies and later developed methods for investigating the compound eyes of insects. Among former military men the most prominent were Colonel John Yerbury, a specialist on flies, and Commander James Walker, a restorer of specimens, an expert on beetles, and a keen student of the geographical distribution of insects. Francis Woodforde was a retired headmaster who arranged almost 50,000 British lepidoptera. The ranking expert on cockroaches and a heavy publisher was Richard Hanitsch who came to Oxford in 1919 having been curator and librarian at the Raffles Museum and Library, Singapore, for twenty-four years. One of the youngest of Poulton's superannuated helpers and researchers was Carpenter who, after twenty years in the Colonial Medical Service in East Africa, retired to Oxford in 1930. By 1931 he was working full-time on the vast array of specimens he had faithfully sent to Poulton, his friend and mentor, over the years. In 1933 Carpenter was also the voluntary Hope librarian. No wonder that it was said that Poulton resigned in his favour. Two of the retired researchers not only helped Poulton in his department but also enabled him to make the Entomological Society of London an outpost of it. Of course Poulton was revered by the Society which elected him president on three occasions and in its centenary year (1933) made him honorary life president. But Walker and Eltringham seconded his efforts: each was a president, a vice-president, and a secretary.[37]

---

[37] Albert Harry Hamm (1861–1951), assistant curator 1897–1931; Edwin George Ross Waters (1890–1930), Taylorian lecturer in French 1913–26, reader in French philology 1926–7, professor of romance languages 1927–30; Revd Laurence William Grensted (1884–1964), chaplain of University College 1924–30, Nolloth professor 1930–50; Arthur Wallace Pickard-Cambridge (1873–1952), fellow in classics, Balliol, 1897–1929, vice-chancellor, University of Sheffield, 1930–8; Frederick Augustus Dixey (1855–1935), junior bursar, Wadham, 1891–1906, bursar 1906–27, on whom see Poulton, *ON* 1 (1932–5), 465–74; Harry Eltringham (1873–1941), on whom see Carpenter, *ON* 4 (1942–4), 113–28; H. Eltringham, *African Mimetic Butterflies* (Oxford, 1910) and *Butterfly Lore* (Oxford, 1923) were loyally Poultonian; John William Yerbury worked for twenty-nine years in the Hope department until killed in a car accident in 1927; James John Walker (1851–1939), obituary in *OM* 57 (1938–9), 399–400; Francis Cardew Woodforde (1846–1928); Karl Richard Hanitsch (1860–1940); Smith, *Hope Collections*, 55–61; Neave and Griffin, *The History of the Entomological Society of London*. Poulton was president 1903–4, 1925–6, 1933–4; Walker, president 1919–20, vice-president 1916–17, 1921–2, secretary 1899–1900, 1905–18, Eltringham, president 1931–2, vice-president 1914–15, 1918–19, 1926–7, 1933–4, secretary 1922–5. Dixey was president 1909–10.

Carpenter's reign was an anticlimax after Poulton's. It began badly. On Poulton's resignation the filling of his chair was suspended in the hope of informally persuading a college to give a stipendiary fellowship to the next occupant, but none came forward. Carpenter's attempts to gain a demonstratorship for Bertram Hobby, his assistant and one of Poulton's few D.Phil. students, were unsuccessful.[38] Though Carpenter had introduced experimental entomology in 1934, the University refused to add staff to a department which gave formal classes only biennially. Moreover Carpenter lacked Poulton's wide biological vision and stature. On assuming his chair Carpenter saw his task in Poultonian terms, that is, studying the coloration, habits, and enemies of insects as evidence for the Darwinian theory of natural selection. In practice this broad programme soon degenerated into a narrow one just when honours were coming thick and fast to Poulton (knighthood 1935; president of the British Association 1937). Taking his cue from Poulton's pre-war view that it was necessary to study attacks by birds on butterflies, which Poulton suspected were not as immune as was widely thought, Carpenter had become convinced before 1914 that birds *do* attack butterflies. Once ensconced in his chair he became obsessed with the imprints of birds' beaks allegedly found upon the wings of butterflies which had escaped from their predators. This preoccupation became a mania that tended to dominate the Hope department. Indeed by 1939 Carpenter had enlarged the study of beak marks on the wings of butterflies to include marks made on them by lizards and bats. This research came under heavy fire from MacBride and especially from Aaron Shull who denied that birds avoided butterflies which mimic distasteful species.[39] Shull also thought Carpenter's concept of mimicry was vitiated because it was anthropomorphic in that it saw similarities between butterflies from a human and not a bird viewpoint.

While boring the Entomological Society with beak marks, Carpenter not only tried to sort out the chaotic Hope collections and library but also to add to them. For him, as for Poulton, the acquisition of specimens for a collection was true science.[40] So he was happy in one session to accept no fewer

---

[38] For Carpenter as disciple of Poulton see G. Carpenter, *A Naturalist on Lake Victoria, with an Account of Sleeping Sickness and the Tse-tse Fly* (London, 1920), pp. xii–xvii, 195–241; id., *A Naturalist in East Africa, being Notes Made in Uganda, ex-German and Portuguese East Africa* (Oxford, 1925) was dedicated to Poulton; Carpenter memorandum, 4 Dec. 1933, BS/R/1/2. For the negotiations about the chair, UDC/M/41/1, 23 Nov. 1932, 25 Jan. 1933, *OUG* 63 (1932–3), 111–12, 177, 181, 272; for staffing difficulties, Carpenter memorandum, 4 Dec. 1933, BS/R/1/2, and Carpenter to Veale, 11 Feb. 1938, BS/R/1/4; Bertram Maurice Hobby (1905–83) became a demonstrator in 1950, obituary *Entomologist's Monthly Magazine*, 99 (1983), 179–91.

[39] Smith, *Hope Collections*, 50–1; *OUG* 70 (1939–40), 259; G. Carpenter, 'Birds Do Attack Butterflies', *Science Progress*, 30 (1935–6), 628–34 attacked the MacBride school; A. F. Shull, *Evolution* (New York, 1936), reviewed by Carpenter, *Science*, 85 (1937), 356–9; A. F. Shull, 'The Needs of the Mimicry Theory', ibid. 496–8; G. Carpenter, 'The Needs of the Mimicry Theory', ibid. 86 (1937), 157; Aaron Franklin Shull (1881–1961), *DSB*.

[40] Carpenter to Plant, 21 May 1935, UR/SF/MU/2; *OUG* 70 (1939–40), 257.

than 14,896 specimens donated mainly by non-Oxonians. In the year before the second war broke out, the Hope department inherited 782 volumes and 207 boxes of specimens from Commander Walker who died early in 1939. In order to cope with acquisitions of this magnitude, Carpenter continued the Poultonian tradition of encouraging voluntary helpers his chief coadjutors being Hanitsch, Walker, Grensted, and of course, the veteran Poulton. On one matter, however, Carpenter acted in a non-Poultonian and bizarre way. In 1938 he wanted the University to launch a bureau for the study of the geographical distribution of fauna on Pacific islands to be run by E. P. Mumford of the Pacific Entomological Survey.[41] Lured by the promise of some private finance, Carpenter ignored his own ignorance about geographical distribution of animals and of island fauna and conveniently forgot that his own department studiously avoided working on the distribution problems posed by its own collections. His proposal was reworded and scaled down by the registrar, Veale, who took the advice of Elton. In 1938 the University gave Carpenter £150 p.a. from its Higher Studies Fund to work on the distribution of fauna, especially insects, on islands, without a bureau being mentioned. Of course by that time the University already had a Bureau of Animal Population directed by Elton who had made it into the most important centre for research into terrestrial animal ecology in Britain. To Elton's ecologists we now turn.

7.5 Animal Ecology

The Bureau of Animal Population (henceforth BAP) was established provisionally in 1932 by Charles Elton who succeeded in gaining limited and less temporary support for it from the University in 1936, but its history begins with his boyhood at Liverpool where he was a keen field naturalist in conjunction with his elder brother Geoffrey, who encouraged his interest in ecology before he went in 1919 to New College where his tutor was Huxley.[42] Through Huxley, who was in general an inspiring, kind, and generous friend, Elton was appointed naturalist to the Oxford University expedition to Spitsbergen in 1921. He took with him Victor Shelford's *Animal Communities in Temperate America* given to him by Huxley. A key notion of Shelford's approach was that the distribution of animals was related to their physiological response to aspects of their physical environment. Elton's own field-work in Spitsbergen, where the number of species was so small that

---

[41] *OUA*, Zoology: Bureau of Geographical Distribution, Mar.–June 1938, UR/SF/Z/1A.

[42] On Elton generally see A. Hardy, 'Charles Elton's Influence in Ecology', *Journal of Animal Ecology*, 37 (1968), 3–8; P. Crowcroft, *Elton's Ecologists: A History of the Bureau of Animal Population* (Chicago, 1991); D. L. Cox, 'Charles Elton and the Emergence of Modern Ecology' (Washington University, Ph.D. thesis, 1979); Hagen, *Entangled Bank*, 51–65; Geoffrey York Elton (1893–1927).

their ecological relations could be seen as a whole, led him to an alternative view that it was better to identify animal communities in terms of food chains and cycles. These notions were corroborated in further Arctic expeditions of 1923 and 1924. While returning from the 1923 expedition he bought a copy of Robert Collett's book on Norwegian mammals, *Norges Pattedyr*, from which he learned about lemming migrations which occurred every three or four years. Also in 1923 Huxley drew Elton's attention to Gordon Hewitt's *The Conservation of the Wild Life of Canada* which contained graphs of the annual fur returns of the Hudson's Bay Company from which Elton learned that the numbers of snowshoe rabbits and Canadian Arctic foxes fluctuated regularly with a ten-year cycle. In 1922 he had read Carr-Saunders's study of *The Population Problem* which thrilled him probably because it argued that some human societies limited their numbers to a size that enabled them to make the fullest use of their available resources.[43]

Convinced that cyclical lemming migrations represented overflows of periodically increasing populations, Elton began empirical work in 1925 on fluctuations in animal populations which he studied in two ways. First through his friend George Binney, an official of the Hudson's Bay Company whom he had met on the Arctic expeditions, he became biological consultant to it in 1926 for five years. As well as bringing in at best a salary of £400 p.a., this post enabled him courtesy of Charles Sale, the managing director of the Company, to set up a system of recording annual changes in the numbers of animals over much of Canada and simultaneously to build up the history of cycles of population from the Company's records which went back to 1736. Through Binney Elton was also made aware of the materials compiled by the Moravian missions in Canada and gained access to them. Their fur returns began in 1834 and ran almost complete to 1925 when the Company took over the fur trade at the Moravian settlements. To complement this work on the fluctuations in the populations of various fur-bearing animals, Elton set up his own *Maus-gesellschaft* in 1925 to study fluctuations in the numbers of British mice and voles in Bagley Wood near Oxford. The work involved taking censuses, determining reproduction rates, and measuring mortality rates, for three years. For such team work Elton recruited two colleagues from the Linacre department, Baker to study breeding and Ford the parasites, Gardner from the pathology department to monitor the bacteria in the

---

[43] For Elton's early experiences and reading, C. Elton, 'How are the Mice?', Elton Papers, A32, 'Life', A33; Crowcroft, *Ecologists*, 1–4; C. Elton, *The Pattern of Animal Communities* (London, 1966), 32–7; V. S. Summerhayes and C. Elton, 'Contributions to the Ecology of Spitsbergen and Bear Island', *Journal of Ecology*, 11 (1923), 214–86 for food chain diagram; Sheail, *British Ecological Society*, 85–94; V. E. Shelford, *Animal Communities in Temperate America as Illustrated in the Chicago Region: A Study in Animal Ecology* (Chicago, 1913); R. Collett, *Norges Pattedyr* (Christiania, 1911–12); C. G. Hewitt, *The Conservation of the Wild Life of Canada* (New York, 1921); Victor Ernest Shelford (1877–1968) of zoology department, University of Chicago; Robert Collett (1842–1913), Norwegian naturalist; Charles Gordon Hewitt (1885–1920), dominion entomologist.

animals, and A. D. Middleton, a school laboratory technician, as laboratory assistant funded by the Medical Research Council (MRC).[44] When this collaborative work was in its later stages Elton's brother Geoffrey died suddenly in 1927. As a memorial to his brother, a good field naturalist and teacher, Elton resolved to found a research institute to study population ecology.

In 1927 Elton also codified his ideas about animal ecology in a rapidly written book which Huxley suggested he should write.[45] Before Elton's book appeared, animal ecology lacked its own distinctive general concepts and tended to borrow ideas in an *ad hoc* way from plant ecology: thus Shelford used Clements's botanical notion of succession. Broadly speaking animal ecology was confined to marine work, to insect pests, and to animal diseases. Elton's 1927 book speedily became a classic for terrestrial animal ecology with its notions of food chains and cycles (the ways in which species are arranged in food chains which combine to form whole food cycles), the size of food, habitat niches (an animal's place in its community and its relation to its food and enemies), and the pyramid of numbers (the greater abundance of animals at the base of food chains and the relative scarcity of animals at the end of such chains). Throughout the book Elton's key approach was to concentrate on the problem of animal numbers and their regulation. His work was warmly received by Tansley, who welcomed it as the first major English book on animal ecology and anticipated fruitful collaboration between zoologists and botanists in ecological work.

Armed with a set of viable concepts and the aim of establishing an institute for studying animal population, Elton continued his north American work for Hudson's Bay Company and in 1928 secured two new financial patrons for research on the health and diseases of rodents. One was Charles Sale to whom Elton was so indebted that *Voles, Mice and Lemmings* (1942) was dedicated to him. The other was the Empire Marketing Board which aimed to improve the quality of the Empire's agricultural products and to diminish preventible losses of produce caused by animal pests. At best it spent £250,000 p.a. on research which its secretary, Stephen Tallents, conceived widely. The research on rodents led to a new technique and a surprising result. In 1930 Richard Ranson, another school technician recruited by Elton, succeeding in breeding voles in captivity for experiments by Baker on their breeding season, which turned out to be very short.[46]

---

[44] Elton Papers, A2, Hudson's Bay Company consultancy 1926–31; C. Elton, *Voles, Mice and Lemmings: Problems in Population Dynamics* (Oxford, 1942), 159–74; A. Douglas Middleton (1904–87); Charles V. Sale, Governor, Hudson's Bay Company, 1925–31.

[45] Elton, *Animal Ecology*, reviewed by Tansley, *Journal of Ecology*, 16 (1928), 163–9; id., 'Life', Elton Papers, A33.

[46] Elton, *Voles*, 174–82; Stephen George Tallents (1884–1958), *DNB*, secretary to the Empire Marketing Board 1926–33; Richard M. Ranson (d. 1944), like Middleton, came from Stowe School, Buckinghamshire.

The year 1931 saw the trough of the depression in the human trade cycle: it seemed likely that Elton's rodent work would be torpedoed because his post with the Hudson's Bay Company and his two grants all ended in 1931. Simultaneously there was zoological depression in eastern Canada where there were no fur animals, no mice, no game birds, no lobsters, no cod, and no mackerel. This widespread crash in numbers deeply concerned Copley Amory, a conservationist. He owned an estate at Matamek River on the north shore of the Gulf of St Lawrence where he owned not only the land on which Matamek Factory, an active fishing village, was built but also a fleet of fishing vessels based there. Convinced that both industry and science would benefit from more research into biological cycles, in July 1931 Amory arranged a small international conference at his expense at Matamek and appointed Elton, whom he had met in Oxford in 1930, as its secretary and editor of its proceedings. The conference members included Reid Blair, representing the New York Zoological Society, who was so impressed by Elton's contribution, research, and plans that he arranged with Madison Grant, the Society's president, a temporary two-and-three-quarter-year grant for Elton to run a small research unit at Oxford.[47] Thus the immediate cause of the founding of the BAP at Oxford in 1932, albeit provisionally, was the Matamek conference and the interest shown by the New York Zoological Society. Armed with the imprimatur of £564 p.a. from this august body, Elton persuaded the University to establish in January 1932 on a provisional basis for three years a BAP for research into the ecology of wild animals, especially their diseases, and the co-ordination of data from published sources and from field observers. The University gave the BAP free accommodation in Goodrich's department and £100 p.a. in 1932 but its main aim in giving the BAP an authorized status was to enable it to obtain more money from outside bodies.[48] That process had begun in autumn 1931 when the Royal Society awarded Elton £500 p.a. for four years for vole research. It grew quickly in 1932, 1933, and 1934 as Elton increased the income of the BAP (£1,620, £2,480, £2,770) and doubled the number of sources of finance (5, 10, and 9). Consequently the University, which after 1932 provided overhead costs only, gave moral support to the BAP by encouraging it to carry on for a provisional period of three and a half years from January 1935 using external funds. It was made plain to Elton that, if he wished to increase the BAP's annual expenses from £2,450 to £4,500 and acquire a new building for £20,000, he

---

[47] Elton, *Voles*, 184; Cox, 'Elton', 149–60; Madison Grant (1865–1937), *DAB*, president of the New York Zoological Society 1925–37, was interested in conservation and, as a vehement supporter of restricted immigration into the USA, in population dynamics. The Society ran the New York Zoological Park, of which William Reid Blair (1875–1949) was director 1926–40.

[48] *OUG* 62 (1931–2), 212; *HCP* 150 (1931), 167–8; *Bureau of Animal Population Oxford 1932* (Oxford, 1933); *Bureau of Animal Population Oxford University 1933, 1934* (Oxford, 1934, 1935).

would have to raise the money himself without any additional liability being imposed on the University.[49]

By autumn 1935 some of the BAP's grants had run out, including the Leverhulme one which paid Elton's salary as director, leaving him from August 1935 with just £100 p.a. from his university demonstratorship. Some of the outside bodies which had supported the BAP conspicuously (Royal Society £2,000 all told, ICI £1,900, Ministry of Agriculture £1,300, Leverhulme Trustees £1,245, Agricultural Research Council £450, Department of Scientific and Industrial Research £300, Forestry Commission £250) felt it was high time that the University should begin to make a regular financial contribution. At this perilous time Goodrich and Huxley gave such strong backing officially to the BAP and Elton that the University paid him £600 from November 1935 to July 1936 as salary, the Christopher Welch Trust having paid him £250 August–November 1935. The University also set up a committee to advise it on the scientific importance of the BAP, which did no undergraduate teaching. It took advice from H. S. Jennings and Carr-Saunders, both of whom reported enthusiastically about the BAP which they agreed was a research institute for animal ecology, with the increase or decrease of species as its centre of interest.[50] In December 1935 the University agreed to make Elton a reader and to provide no more than £500 p.a. to the expenses of the BAP as bait to external bodies. Early in 1936 Elton was elected a senior research fellow at Corpus at £300 p.a. + £30 superannuation on condition that sufficient external guarantees could be found. In spring 1936 the Agricultural Research Council (ARC) announced its decision to give a five-year guarantee from October 1936 of £300 in the first year and £200 in each of the four following years for running expenses; the MRC gave £400 p.a. (renewable annually for three years) from June 1936 for vole pathology research; and the University included an estimate for the BAP of £507 p.a. for five years in its statement to the University Grants Committee for the next quinquennial grant. The UGC responded generously so that the University set aside £850 p.a. for five years from August 1936 to help to pay running expenses and Elton's salary.[51] So from autumn 1936 the BAP under Elton as reader was given a five-year life as a university institution with regular income of £1,400 p.a. (£850 from the University, £330 from Corpus, £220 on average from the ARC), the remaining half of its budget being raised externally in the form of short-term grants.

Veale was mainly responsible for inducing the University and Corpus to support Elton and the BAP. As registrar Veale had become convinced of the

---

[49] *HCP* 158 (1934), 212–13.
[50] *University of Oxford, Bureau of Animal Population: Annual Report 1935–6* (Oxford, 1936), 6–7; *HCP* 162 (1935), 25–6, 157–61; Herbert Spencer Jennings (1868–1947), professor of zoology, Johns Hopkins University, 1906–38.
[51] *BAP Report 1935–6*, 4–5; *HCP* 162 (1935), 187, 197; 163 (1936), 118; 164 (1936), 93, 237.

utility of the BAP and never lost his belief in Elton as the pioneer in England of animal ecology. Realizing that Goodrich had no interest in live animals, Veale thought that such research as pursued by Elton should be promoted as an investment in the future. Veale processed many of Elton's grant applications and guided him through what Elton called the 'diabolical complications of University and government administration'. As a fellow of Corpus, Veale no doubt persuaded Richard Livingstone, its president and a classicist, to consider making Elton its third scientific fellow.[52] Veale also promoted the BAP to Lindsay, the vice-chancellor, who in any event was keen to gain an Oxford 'first' by supporting a research institute in a new scientific subject in which only Moscow rivalled Oxford.

Though 'population' was a key word of the 1930s, Elton spent a great deal of time raising money even after 1936. It involved what he regarded as a fantastic waste of energy: indeed the labour and anxiety of raising short-term grants by hand-to-mouth means interfered with research and slowed down publication. In 1937–8, for example, the BAP was financed from no fewer than thirteen different sources after umpteen grant applications.[53] Research on rats was not possible because no external body would fund it; ironically it required the war with Germany and the consequent war on waste pursued by the Agricultural Research Council to launch sustained research in late 1939 on the control of rats and mice.[54] Sometimes Elton was a victim of his own honesty: in 1939 the Forestry Commission, while continuing to give the BAP access and facilities, withdrew its support of the vole research because Elton had confessed that a 'very complex series of factors' was responsible for vole numbers and that therefore the problem of controlling voles was no nearer solution.[55] On one occasion he was a victim of his own success. A. D. Middleton had become so expert in research on game, paid for by ICI which made small arms ammunition, that in 1937 ICI recruited him to run game research at its Jealott's Hill Research Station and as a consequence cut its grant to the BAP from £600 to £100 p.a.[56] Within the University few science departments helped the BAP. Tansley, a scientific hero for Elton, provided moral encouragement but not institutional collaboration. The Linacre department carried the overhead costs of accommodation for the BAP and generally Goodrich was encouraging in public about it. Privately, however,

---

[52] Veale memoranda, 21 Nov. 1938, BS/R/1/5, 19 Mar. 1938, OUA, Bureau of Animal Population, UR/SF/Z/2D; Crowcroft, *Ecologists*, 17 (quotation); interview by author with Elton, 18 Nov. 1987; Livingstone to Elton, 17 Feb. 1936, Elton Papers, A4; Sir Richard Winn Livingstone (1880–1960), *DNB*, president of Corpus Christi 1933–50.

[53] Elton, *Voles*, 160–1; *BAP Report 1937–8*, 10–12.

[54] Crowcroft, *Ecologists*, 28–46; D. Chitty and H. N. Southern (eds.), *Control of Rats and Mice*, 3 vols. (Oxford, 1954).

[55] R. Robinson to Elton, 17 Apr. 1939, UR/SF/Z/2D.

[56] *BAP Report 1936–7*, 20; *1937–8 Report*, 21; H. G. Eley to Elton, 12 Apr. 1937, UR/SF/Z/2D; Major H. Gerard Eley (1887–1970), who ran ICI's game research, was Elton's contact with ICI.

he feared by 1938 that the BAP might become a veritable cuckoo in his department's nest so he opposed Elton's plan for a new animal house.[57] The most effective help came from Florey with whom Elton had shared a tent on the Oxford University Arctic expedition of 1924. In 1935 at Florey's suggestion, Elton acquired the services of P. H. Leslie, a pathologist who had been working on whooping cough.[58] Leslie became the BAP's resident biomathematician and demographer until its demise in 1967. In 1936 Elton was indebted to Florey for permitting him to have built in the grounds of the pathology department an isolation laboratory for pathological research on wild animals (paid for mainly by the ARC) and for providing a grant of £400 p.a. for three years from the MRC for A. Q. Wells to work in the Dunn department on the epidemiology of voles. In 1938 Wells discovered tuberculosis in wild voles, thus disturbing the general view that the disease was confined to humans, birds, and captive animals.

Until the outbreak of the Second World War the BAP existed under a system of short-term trial runs, the last of which was that authorized and in part funded by the University in 1936. Outside bodies still thought that the University should be responsible for long-term research. The result was that in spring 1939 Elton was flummoxed about maintaining the vole field-work; and the research on damage caused by rabbits, a problem on which a select committee of the House of Lords had reported voluminously in 1937, almost expired because of lack of funds. Even so the BAP was a notable innovation. It developed a new kind of team-work in zoology as a result of its novel aims and method. Its broad aim remained the study of fluctuations in the population of wild animals, with special reference to disease. Its methods were many and varied: the study of past records; an intelligence system which embraced north America and the USSR; the breeding of laboratory stocks for experimental work; research on reproduction and mortality; the pathology of diseases; the development of better census methods, which included Dennis Chitty's pioneering use of live-trapping and ringing of mammals to measure movements and density of population; and the exploitation of statistical and biomathematical methods.[59] In its combination of field and laboratory work and statistics, Elton's BAP was like Ford's ecological genetics. It was different in that it developed a co-operative operations-research type of approach in which there was a progressive co-ordination of different interests and specialisms into 'single

---

[57] Goodrich to Veale, 30 Nov. 1938, Veale to Goodrich, 1 Dec. 1938, UR/SF/Z/2D.
[58] *BAP Report 1936–7*, 5, 8–9, 15; Crowcroft, *Ecologists*, 18–22; Elton, *Voles*, 197–200; Patrick Holt Leslie (1900–72), second in physiology 1921; Arthur Quinton Wells (1896–1956), second in physiology 1920, came to the BAP from St Bartholomew's Hospital where he lectured in bacteriology and from 1939 to 1956 worked for the MRC.
[59] For Chitty's methods, *BAP Report 1935–6*, 10–11; *1936–7 Report*, 3; Dennis Hubert Chitty (b. 1912), a Toronto graduate who worked at the BAP 1935–61, returned to Canada to become professor of zoology, University of British Columbia, Vancouver, 1961–78.

unified population research'.[60] That was possible because all the staff of the BAP except Elton had no teaching responsibilities in the Linacre department where it was housed: they were full-time or part-time researchers, usually no more than half a dozen at a time, recruited by Elton for a specific purpose and paid by a specific grant. Two key laboratory assistants, Middleton and Ranson, came from Stowe School where they had been laboratory technicians. Two researchers, Chitty and his wife Helen, came from Canada and stayed until 1961. Francis Evans, a Rhodes scholar from the USA, was the BAP's solitary D. Phil. student in the 1930s. The majority of Elton's staff were Oxford graduates, including Leslie and H. N. Southern, both of whom stayed at the BAP until it was abolished in 1967 when Elton retired.[61]

Though Elton claimed he was not a committee man, he was a powerful figure in the British Ecological Society. In 1931 he drew its attention to the increasing number of papers on animal ecology for which there was no obvious outlet. In response the Society asked Elton and Tansley, then editor of its *Journal of Ecology*, to investigate the matter. Their recommendation of a separate *Journal of Animal Ecology* was accepted and Elton was made first editor with Middleton as assistant.[62] This journal, first published in 1932, the year of the foundation of the BAP, provided useful publicity for Elton's pioneering institutional venture in terrestrial animal ecology. As editor until 1951 Elton did not use the normal system of anonymous referees: he made up his own mind but when appropriate took advice from a colleague. Thus as editor as well as director of the BAP Elton encouraged team-work in a framework and under leadership provided by himself.

7.6 Birds

This emphasis on co-operative research was just as apparent in the attempts made at Oxford to promote scientific ornithology as an alternative to the then growing concerns with 'twitching' and protection. These endeavours were so successful that they launched modern British ornithology.[63] The prime movers were Bernard Tucker, an Oxford zoology graduate who became demonstrator in zoology in 1926, and Max Nicholson, an undergraduate historian who had written three books about ornithology by the time he left Oxford. As skilled creators and maintainers of institutions, they were primarily responsible (solo or in harness) for founding the Oxford

---

[60] *BAP Report 1937-8*, 8.

[61] Henry Neville Southern (1908–86), third in Greats 1931, worked for a publisher while maintaining an interest in natural history, took a first in zoology 1938, and immediately joined the BAP where he and Chitty were Elton's leading researchers; obituary in *Journal of Animal Ecology*, 56 (1987), 715–17.

[62] Sheail, *British Ecological Society*, 94–7.

[63] D. E. Allen, *The Naturalist in Britain: A Social History* (Harmondsworth, 1976), 252–8.

Ornithological Society (1921), the Oxford Bird Census (1927), the British Trust for Ornithology (1932), and the Edward Grey Institute of Field Ornithology at the University of Oxford (1938). Though the Grey Institute (henceforth EGI) was initially a one-man band under its director W. B. Alexander, it pre-dated the Cambridge Ornithological Field Station by a dozen years and flourished after the war under David Lack.[64]

The origins of the EGI go back to 1921 when Tucker, an undergraduate, founded the Oxford Ornithological Society of which he was the sustaining secretary. He drew in F. C. R. Jourdain, a well-known ornithologist who was rector of Appleton, conveniently near Oxford. Together they saw to it that the Society concentrated on studying birds in the field, a focus encouraged by Huxley; and in their annual bird reports for Oxfordshire, Berkshire, and Buckinghamshire appreciably raised the level of accuracy in published records. The Society was of national importance because it inspired many local ornithological societies which were stimulated by it and took its annual reports as a model.[65]

In autumn 1926 there was a fortunate coincidence. Wynne-Edwards, a final-year undergraduate zoologist, was secretary of the Society; after a short absence Tucker returned to Oxford as demonstrator in zoology; and Nicholson, who knew virtually nothing about the Society, arrived in Oxford already well known as a writer on ornithology with his propagandist book on *Birds in England*. Tucker had in mind the notion of studying birds in the aggregate using teams of observers working together who would make censuses. He and Wynne-Edwards approached Nicholson to organize the new venture of the Oxford Bird Census begun in 1927. It soon established the first British co-operative trapping station, set up on Christ Church meadow by Nicholson who exploited the pre-war ringing scheme organized by *British Birds*. In 1928 Nicholson organized a national census of heronries, which involved a mass count of a single species by 400 observers, using *British Birds* as a sponsor and as a medium of communication. A glutton for work, Tucker took responsibility for four counties.[66] This enterprise

---

[64] On Tucker see E. M. Nicholson and J. D. Wood, *British Birds*, 44 (1951), 41–6; A. C. Hardy, *OM* 69 (1950–1), 192–4; D. Lack, *Ibis*, 93 (1951), 300–5; Wilfrid Backhouse Alexander (1885–1965), obituaries in *Bird Study*, 13 (1966), 1–4, *Ibis*, 108 (1966), 288–9, a Cambridge graduate who had made a botanical and ornithological career in Australia 1912–26, superintendent of the Tees Estuary Survey run by the Marine Biological Association 1929–30, director of Oxford Bird Census 1930–8 and of EGI 1938–45, was an authority on sea-birds: W. B. Alexander, *Birds of the Ocean* (New York, 1928); David Lambert Lack (1910–73), a schoolmaster at Dartington Hall, Devon, 1933–40, director of EGI 1945–73, on whom see W. H. Thorpe, *BM* 20 (1974), 271–94.

[65] *Oxford Ornithological Society: Miscellaneous Papers, 1921–45* (Bodleian Library, GA Oxon b. 166); Allen, *Naturalist*, 253–4; B. W. Tucker, 'Ornithology in Oxford', *OM* 57 (1938–9), 304–6, 341–3, characteristically underplayed his own role.

[66] Nicholson to author, n.d. [1987]; M. Nicholson, 'B. W. Tucker and the Trust', *The British Trust for Ornithology: Seventeenth Annual Report 1950*, 4–7; Allen, *Naturalist*, 254–6; M.

convinced Nicholson that the day of the individualist bird observer was past, and that Britain needed a central directing body to guide a large corps of observers and digest their data. He envisaged by 1929 an institute of economic ornithology and an endowed university institute to secure these aims. He had tested the market for economic ornithology by launching in 1927 a survey of rookeries in the Oxford district, which showed the effectiveness of rooks in ridding farms of invertebrate pests.[67] Armed with this favourable result, Tucker approached Goodrich, head of the department of zoology, and persuaded the veteran morphologist to accommodate the Bird Census in his department in 1930 and to secure in June 1931 its renaming as Oxford University Research in Economic Ornithology. Goodrich accepted that the description and classification of birds were not the be-all and end-all of ornithology and that co-operative study of their daily lives was a valid venture in academic zoology. To secure funding for the Census, Nicholson and Tucker stressed its economic importance to the Ministry of Agriculture (where they were helped by John Fryer) and to the Empire Marketing Board who provided funds for three years from October 1930 for the employment of Alexander as full-time director of the Oxford Bird Census which would initially continue the research on rooks.[68]

By early 1932 Nicholson was dissatisfied with the 'precarious and unwholesome reliance' of Alexander's outfit on economic ornithology so he proposed that the Oxford scheme be expanded by abandoning its local Oxford character and by remodelling it to become a national co-ordinating centre for ornithology as a whole, a secondary consideration being the continuation of the Oxford scheme with university recognition.[69] Nicholson's concern and drive led in May 1932 to the first steps towards forming a British Trust for Ornithology (henceforth BTO), provisionally established in autumn 1932 and formally so in spring 1933 with Nicholson as the secretary and Tucker the treasurer. In parallel discussions at Oxford Tucker primed

Nicholson, *Birds in England: An Account of the State of our Bird-Life and a Criticism of Bird Protection* (London, 1926); id., 'Report on the "British Birds" Census of Heronries, 1928', *British Birds*, 22 (1928–9), 270–323, 354–72.

[67] M. Nicholson, *The Study of Birds: An Introduction to Ornithology* (London, 1929), 70–2; E. M. and B. D. Nicholson, 'The Rookeries of the Oxford District: A Preliminary Report', *Journal of Ecology*, 18 (1930), 51–66.

[68] Goodrich to vice-chancellor, 25 May 1931, BS/R/1/2; 9 June 1931, BS/M/1/1; *HCP* 149 (1931), 139–40; M. Nicholson, 'The Trust: Origins and Early Days', in R. Hickling (ed.), *Enjoying Ornithology: A Celebration of Fifty Years of the British Trust for Ornithology 1933–1983* (Calton, 1983), 15–28, esp. 16–20; John Claud Fortescue Fryer (1886–1948), FRS 1948, a keen ornithologist, then director of the plant pathology laboratory, Ministry of Agriculture.

[69] Nicholson, 'Trust', 16–17; Proposed institute of field and economic ornithology at Oxford, n.d. [Feb. 1932], Tucker to Veale, 5 and 24 Jan., 15 Feb. 1932, OUA, Edward Grey Institute, UR/SF/Z/2B; Goodrich memorandum, *HCP* 152 (June 1932), 189–92; *OUG* 63 (1932–3), 56; and especially box labelled 'Oxford Bird Census. Papers 1927–30. Papers concerning British Trust for Ornithology and Edward Grey Institute Initiation. University Committee for Ornithology', Alexander Library, Department of Zoology, Oxford.

Goodrich to argue for a national centre in Oxford for field ornithology and to stress that the BTO hoped to raise £8,000 from May 1933 to July 1938 to finance an institute for ornithology if the University would authorize it. The University agreed and the institute came into being in May 1933 on provisional basis until summer 1938 with a token benefaction of £10 p.a. from the University. Otherwise the BTO was responsible for raising all the funding for the institute, including the salary of Alexander, its solitary member of staff, who continued to run the Oxford Bird Census from his temporary accommodation in Goodrich's department while taking on with great reluctance new responsibility as director of field research for the BTO. Though the connection between the University and the BTO was unpopular in the BTO and was to bring pain and grief to Nicholson as the man-in-the-middle, its birth and growth were feasible only through the University which gave some facilities, a research climate, and the use of its prestige in raising money. From the start of their connection there was a basic difference of opinion: the University saw the BTO as existing for the sole purpose of financing the national centre in Oxford, whereas the BTO from its inception pursued its own concerns, only one of which was paying for the Oxford institute.

The BTO failed to raise anything like its stated target of £8,000. It received a donation of £1,400 from Witherby who had been impressed by the accuracy of the co-operative research on heronries obtained by having one observer for every octet of occupied nests, and £400 from the Pilgrim Trust; but generally the slump took its toll.[70] Fortunately for the BTO Tucker had suggested that its general appeal be signed by Grey, the chancellor of the University and a keen field ornithologist.[71] It turned out to be one of his last public acts. When he died in September 1933, Nicholson on behalf of the BTO soon aroused the Grey family's interest in financing the proposed Oxford institute as a memorial to him and naming it after him. This suggestion was apposite as well as opportune. Grey was an active observer, especially of water-fowl, and a best-selling ornithological author. His last speech in the House of Lords was not on fascism but on the adverse effect of oil pollution on sea-birds. Nicholson's suggestion was also piquant: the institute would be named after a great statesman and good ornithologist who as an undergraduate had been sent down from Oxford for idleness and returned to it as chancellor. Initially the Grey family favoured an EGI as a suitable memorial but then the organizers of the Grey Memorial Fund

---

[70] *British Trust for Ornithology: First to Fifth Reports, 1935–9*; Harry Forbes Witherby (1873–1943), founder of *British Birds* 1907 and editor, president of the British Ornithologists' Union 1933–8.

[71] For Grey matters, vice-chancellor to Grey, 15 Nov. 1932, Tucker to Veale, 8 Oct. 1933, Veale memorandum, 14 Feb. 1935, UR/SF/Z/2B; *HCP* 156 (1933), 167–9; F. Pember, 'The National Memorial to Viscount Grey of Fallodon', *Oxford*, 2 (1936), 42–7; E. Grey, *The Charm*

disagreed about the most appropriate form of commemorating him so that its launch was delayed and it lost impetus rapidly. Only in late 1935 did they decide on a tripartite memorial: for Grey the statesman a portrait of some kind in London; for the country-lover a hilltop in Northumberland vested in the National Trust; and for the ornithologist an Edward Grey Institute of bird studies at Oxford. The Grey Fund, which aimed at £16,000, almost all of which was earmarked for the Oxford institute, was launched in January 1936. By summer it had raised only £3,500, of which £500 was destined for a medallion outside the Foreign Office, but the prospect of the remaining £3,000 induced the BTO to think about proper accommodation for the EGI preferably on a site provided by the University.

Early in 1937 St John's offered a building and an acre of land near Bagley Wood for the EGI.[72] The college was prepared to sell at a below-market price, preferably to the University, but it was thought the site was too far from central Oxford and the Chest judged that the proposed EGI did not have sufficient income for its permanent maintenance so it rejected St John's generous offer. The BTO thought the site and building reasonably suitable, provided the latter was revamped. After initial euphoria it turned out that the cost of conversion would be almost £3,000, a sum well beyond the financial capacity of the BTO which had not then laid hands on the Grey money. At this juncture relations between the University and BTO had degenerated into personal frictions and administrative misunderstandings. The University, still conscious of the Owen affair, feared enmeshment and was cautious in the extreme in its dealings with the BTO which it regarded, wrongly, as devoted only to financing ornithology at Oxford. For its part the BTO felt exploited. In 1937–8 the University gave Alexander free accommodation and contributed £10 as annual expenses whereas the BTO gave £550. Moreover the BTO always had wider views than just promoting an Oxford institute: its aim was to advance in quantity and quality the observation and recording of birds. It therefore set up many extensive co-operative ventures supervised by non-Oxonians such as David Lack who directed bands of amateurs to attack problems which no isolated individual could cope with. For example, in June 1937, when its relations with the University were tense, the BTO took over the national bird-ringing scheme from *British Birds* and organized it from the British Museum (Natural History) in London. This choice of location was a palpable hit at Alexander, its director of

of Birds (London, 1927) went through ten editions by 1937; id., *Fallodon Papers* (London, 1926); S. P. Gordon, *Edward Grey of Fallodon and his Birds* (London, 1937); G. M. Trevelyan, *Grey of Fallodon: Being the Life of Sir Edward Grey, afterwards Viscount Grey of Fallodon* (London, 1937), 362.

[72] On the St John's offer and its consequences, *British Trust for Ornithology: Third Report, Summer 1937*, 23; HCP 166 (1937), 133–4; 167 (1937), 273–4; 168 (1937), 175–6; exchanges involving president and bursar of St John's, Nicholson, Veale, Tucker, vice-chancellor, secretary to Chest, president of Magdalen, 4 Jan.–14 June 1937, UR/SF/Z/2B; Nicholson, 'Trust', 20.

field research, who was seen by the BTO as not capable of organizing the ringing scheme and by Nicholson as dragging his feet.

The impasse about a home for the EGI was solved by Tucker who suggested to the registrar that the University might provide rooms rent free in one of its own buildings as an interim measure.[73] Though the organizers of the Grey Fund initially thought that three converted rooms giving only temporary accommodation hardly constituted a dignified and enduring monument to the great statesman, they were won over as was the BTO by the University's move in 1938 of offering rent-free accommodation and an annual grant of £100 p.a. The BTO, keen to avoid indefinite postponement, agreed to temporary premises as better than nothing and, in order to convince the University and the organizers of the Grey Fund that it could deliver financially, it agreed to pay £700 p.a. to the EGI in 1938-9, knowing that £3,000 from the Grey Fund would soon be available. Thus after tense three-way negotiations the EGI was formally established by the University in 1938. Initially it was to be known as the EGI of Ornithology; but in response to the organizers of the Grey Fund who had stressed the importance of field-work compared with the study of the anatomy of birds and to the BTO who wished the emphasis on the live bird to continue, it was changed to the EGI of Field Ornithology. The first director was Alexander, the man on the spot in Oxford, but his appointment rankled with the BTO because as its field director he had not done enough in organizing co-operative ventures and his concerns with the rare and the new were regarded as decidedly *passé*. The alliance made in 1938 between the EGI and the BTO was in any case uneasy: the BTO was primarily concerned with the harnessing of amateurs in national co-operative research projects whereas the EGI was a national research institute for scientific ornithologists.

As the EGI was devoted to research on the numbers, distribution, movements, habits, and economic status of birds, its aims were akin to those of the BAP. Financially each existed mainly on the basis of external funding, the provision of which during the slump was a particularly hard grind. In the case of the EGI the dithering by the organizers of the Grey Fund was particularly unfortunate. In each case the University was cautious, setting up each project on a provisional basis so that it could be scrapped if unviable financially and intellectually, and giving its name but little money. Indeed Nicholson and Tucker did not approach the University for funds for the Oxford Bird Census and the BTO: not wishing to court a rebuff, they used the University's prestige to secure funds from elsewhere.

There were, however, differences between the EGI and the BAP. Each had a director, but Elton was a university reader whereas Alexander had no university post. The EGI was a one-man band whereas the BAP planned and

---

[73] For the establishment of the EGI, *HCP* 170 (1938), 39–41, 215–16; Veale memoranda, 16 Oct. 1937, 13 and 15 Oct. 1938, UR/SF/Z/2B.

executed co-operative research. Elton led the BAP and lobbied for it, whereas Alexander let others such as Tucker and Nicholson take initiatives. Tucker could have followed Elton's practice of cumul by combining the directorship of the EGI with a university teaching post but there is no evidence that he thought of doing so. Temperamentally he was shy, modest, unobtrusive, and unassuming so he preferred to work behind the scenes, patiently and tactfully bridging divergences of interest. Unlike many of his colleagues in the Linacre department he was not a prolific publisher and therefore not as visible: characteristically his major publication involved contributing to a large collaborative venture, the revised edition of Witherby's practical handbook to British birds. As Allen has noted, Tucker possessed a glittering selection of virtues and talents; but he left to Nicholson the work of capturing public attention and of campaigning for a new 'network' approach to ornithology.[74] It was entirely characteristic of Tucker that one of his last acts was to take a large part in the negotiations which led in 1947 to the EGI and the BAP becoming the nucleus of a new department of zoological field studies with the new Linacre professor, Alister Hardy, long their advocate, as its head. While retaining their separate identities, the EGI and the BAP at last secured assured status and income.

[74] Allen, *Naturalist*, 253; Tucker contributed the sections on habitat, field characters, habits, voice, and display to H. F. Witherby, *The Handbook of British Birds*, 5 vols. (London, 1938–41).

8
The Big Battalions

Next to physiology, mathematics and chemistry, with about thirty and forty-five graduates a year respectively by the 1930s, were the largest final honours schools in science. Though there were slightly fewer college fellows in mathematics than chemistry, there were four chairs in mathematics, three inherited from the nineteenth century (the Savilian, Waynflete, and Sedley) and one established in 1928 (the Rouse Ball). In contrast chemistry enjoyed two, the Waynflete and the Dr Lee's, the latter created from an old endowment in 1919. But their internal and external reputations were inversely proportional to the relative number of chairs. Though mathematics was an ancient Hellenistic and gentlemanly subject, its position was ambiguous: it was tolerated but not warmly encouraged by the colleges and the University, an attitude which was reinforced by the feeling of inferiority to Cambridge. There was a tendency among college fellows in mathematics to preserve the status quo. When change occurred two imported Cambridge-trained professors were usually responsible. G. H. Hardy, Savilian professor of geometry 1920–31, who established Oxford's first flourishing research school in mathematics, made Oxford mathematics more visibly nationally and internationally, and launched a campaign which in 1934 secured a small mathematical institute housed in the extension to the Radcliffe Science Library. E. A. Milne came to Oxford in 1929 as the first occupant of the Rouse Ball chair of mathematics. He stayed in Oxford, making it with Plaskett's help into a well-known centre for astrophysics in the 1930s. Thus mathematics was like engineering science and botany in that the importing of established Cantabrigians as professors led to challenges to Oxford's in-built academic conservatism.

In chemistry the college fellows were more prominent than their counterparts in mathematics. Their doyens were Hinshelwood, a Nobel prize-winner in physical chemistry in 1956, and Sidgwick, perhaps the foremost systematizer of chemical data and expositor of chemical theory of his time. Organic chemistry flourished in the Dyson Perrins Laboratory, opened in 1916, under Perkin and Robinson, two successive and imported Waynflete professors who used staff and postgraduates recruited from outside Oxford. The research practices which they brought to Oxford from Manchester were so fruitful that in 1947 Robinson was awarded a Nobel prize. In comparison to physical and organic chemistry, inorganic chemistry at Oxford was unspectacular partly because Soddy, appointed in 1919 as the first occupant of the Dr Lee's chair, was soon at odds with the colleges and the college fellows and failed to live up to the expectations aroused by his previous record and by the award of a Nobel prize in 1921. Even so, chemistry was *the* subject in which Oxford felt decidedly superior to Cambridge, where the subject drifted in the 1930s under the senior professor W. J. Pope, who was an ill man from 1927. Though Cambridge did not produce more than thirty finalists per year in chemistry, in two ways it had the edge over Oxford. It enjoyed more chairs, four new ones being created between 1920 and 1931, in physical chemistry (1920), colloid science, theoretical chemistry, and metallurgy (all 1931). Their occupants were by no means negligible, one of them, Norrish, being eventually a Nobel prize-winner in 1967.[1] Cambridge was also better endowed between the wars. Pope attracted £210,000 from British oil companies to extend and equip his department; and almost £20,000 for metallurgy from the Goldsmiths' Company. Pope had overall charge of the laboratories so that when his chemical Papacy lost impetus in the 1930s it was not easy to recharge his department as a whole. At Oxford, however, in the early 1920s there were five teaching laboratories, run by six colleges, in addition to those belonging to the University. One of them, the Balliol–Trinity laboratory, became an important centre of research in physical chemistry under Harold Hartley and his pupil Hinshelwood, who supervised mainly Oxford graduates whom they had often grabbed via part two of the undergraduate chemistry degree. Three of these college laboratories were still open in 1939, two of them (Balliol–Trinity and Christ Church) closing in 1941 when the new university physical chemistry laboratory was opened. It was only in 1947 that Jesus closed its laboratory which had been mainly devoted to physical chemistry. The long history of the college chemical laboratories showed the sustained commitment of the colleges to chemistry, ensuring that in 1939 the Oxford school was the largest in any British university, a position it was to retain for three decades. One consequence was that, in the early 1960s, of the 76 FRSs who were chemists no fewer than 24 were Oxford graduates in chemistry.

[1] There is nothing on Cambridge chemistry between the wars analogous to G. K. Roberts, 'The Liberally-Educated Chemist: Chemistry in the Cambridge Natural Sciences Tripos, 1851–1914', *Historical Studies in the Physical Sciences*, 11 (1980), 158–83. For an overview see W. H. Mills, 'Schools of Chemistry in Great Britain and Ireland, vi: The University of Cambridge', *Journal of the Royal Institute of Chemistry*, 77 (1953), 423–31, 467–73. William Jackson Pope (1870–1939), professor of chemistry, Cambridge, 1908–39, obituaries in *Journal of the Chemical Society* (1941), 697–715 and *ON* 3 (1939–41), 291–324; Thomas Martin Lowry (1874–1936), first professor of physical chemistry, Cambridge, 1920–36, obituary in *Journal of the Chemical Society* (1937), 701–5; John Edward Lennard-Jones (1894–1954), first professor of theoretical chemistry, Cambridge, 1932–53, obituary in *BM* 1 (1955), 175–84; Eric Keighley Rideal (1890–1974), first professor of colloid science, Cambridge, 1931–46, obituary in *BM* 22 (1976), 381–414; Robert Salmon Hutton (1876–1970), first professor of metallurgy, Cambridge, 1931–42, obituaries in *The Times*, 7, 12, 18 Aug. 1970; Ronald George Wreyford Norrish (1897–1978), professor of physical chemistry, Cambridge, 1937–65, obituary in *BM* 27 (1981), 379–424.

8.1 Mathematics: Locals and Imports

Of the dozen or so college fellows in mathematics appointed between the wars, all but one were Oxford graduates, often scholars or exhibitioners, who had inevitably gained a first-class honours degree and usually university prizes or scholarships. Though a few endured a period as schoolmasters or university teachers elsewhere, the majority were elected to fellowships fairly quickly after graduation, several having served an apprenticeship by undertaking *ad hoc* college teaching previously. They were appointed on their successes in examinations, not on their publications, and to teach undergraduates in the colleges and not to supervise postgraduate students in the University. That situation began to change in the 1930s with the appointment within three years of three college fellows (Phillips, Magdalen, 1933; Haslam-Jones, Queen's, 1936; J. H. C. Thompson, Wadham, 1936) who were Oxford graduates who had stayed on to take a D.Phil.[2] Only six of the college fellows in mathematics had enjoyed some academic experience outside Oxford. Naturally those who had not had little reason to change a system which had rewarded them; and naturally they believed that only Oxford graduates could teach and examine the Oxford course properly. They often prided themselves on their wide erudition and not on focused specialist publication. Though some were not inactive nationally, they tended to be satisfied with local careers. At Hertford Ferrar was a loyal college tutor, acting as bursar and finally as principal. The author of many standard textbooks, he decided in 1937 that research could no longer be combined with arduous teaching and administration so he stopped publishing papers. At New College, Jesus, Merton and Queen's, Poole, Hodgkinson, Newboult, and C. H. Thompson put their teaching first, two of them showing their wide scholarship also in textbooks.[3] Though Poole had taken a D.Phil, he saw his prime function as using his encyclopedic knowledge and facility in several foreign languages to portray mathematics as a whole to his pupils. His research output was restricted by his extreme conscientiousness but was

---

[2] Eric George Phillips (1909–84), first in mathematics 1930, D.Phil. 1932, senior scholar at Christ Church 1930–3, fellow of Magdalen 1933–9; Ughtred Shuttleworth Haslam-Jones (1903–62), first in mathematics 1925, D.Phil. 1927, lecturer at Liverpool University 1927–36, fellow of Queen's 1936–62, obituary in *Nature*, 196 (1962), 413–14; John Harold Crossley Thompson (1909–75), first in maths 1930, D.Phil. 1932, junior research fellow at Merton 1933–6, fellow of Wadham 1936–75, obituary in *The Times*, 25 June 1975.

[3] Edgar Girard Croker Poole (1891–1940), first in mathematics 1915, fellow of New College 1920–40, obituaries in *OM* 59 (1940–1), 10–11, *Journal of the London Mathematical Society*, 16 (1941), 125–30; E. G. C. Poole, *Introduction to the Theory of Linear Differential Equations* (Oxford, 1936); Jonathan Hodgkinson (1886–1940), first 1908, fellow of Jesus 1922–40, obituaries in *OM* 59 (1940–1), 40 and *Journal of the London Mathematical Society*, 15 (1940), 236–40; Harold Oliver Newboult (1897–1949), first 1922, fellow of Merton 1926–49, obituary in *Nature*, 164 (1949), 474–5; H. O. Newboult, *Analytical Method in Dynamics* (Oxford, 1946); Charles Henry Thompson (1865–1948), first 1886, fellow of Queen's 1897–1936, obituary in *OM* 67 (1948–9), 260.

characteristically wide-ranging: he was the only mathematician who collaborated with an engineer. Hodgkinson was a Jesus man to the core, returning to it as lecturer and fellow having been a scholar there. He was first and foremost a teacher, with a continuous succession of firsts in finals. He loved research but regarded it as his own private affair and pleasure. In contrast teaching and examining were for him contractual duties solemnly undertaken and requiring total commitment. At Merton Newboult was more a scholar than a researcher, happy to produce a regular stream of firsts in finals. C. H. Thompson, a Queen's undergraduate, was mathematics fellow there for almost forty years until 1936 when he retired aged 71. Though he published little he had rapport with his pupils who presented him with his portrait and silver when he retired. An exception to the emphasis on teaching and on scholarship at the expense of specialized research was Chaundy at Christ Church who combined all three effectively.[4] The emphasis on local career-building via general scholarship, teaching, administrating, textbook writing, and in some cases editing had two consequences. One was that the college fellows were generally not office-holders in the London Mathematical Society, the leading national society for the subject, though several served on the Council. The exceptions were J. E. Campbell, fellow of Hertford, and his successor Ferrar, who were president and secretary respectively.[5] The other consequence was that, with the exception of J. H. C. Whitehead at Balliol, they did not appreciate the arrival in Britain of German refugee mathematicians as an opportunity to augment Oxford mathematics. Only Chaundy advised Helen Darbishire, principal of Somerville and a Wordsworth scholar who knew nothing about mathematics, about how to secure the services of the renowned Emmy Noether who went to Bryn Mawr College, Philadelphia. None of them was involved in the case of Richard Rado, who was interviewed in Berlin by Lindemann who secured £300 p.a. for two years from Sir Robert Mond to enable Rado to study not at Oxford but at Cambridge under G. H. Hardy for a second doctorate. Generally Oxford did not find niches for refugee mathematicians, whereas at Cambridge Hardy managed to create eighteen posts for them.[6]

[4] Obituary of Chaundy, *Journal of the London Mathematical Society*, 41 (1966), 755–6; T. W. Chaundy, *The Differential Calculus* (Oxford, 1935).

[5] J. E. Campbell, FRS 1905, was president 1918–20, obituary in *Proceedings of the Royal Society of London A*, 107 (1925), pp. ix–xii, and Ferrar secretary 1933–8.

[6] John Henry Constantine Whitehead (1904–60), first in mathematics, 1926, fellow of Balliol 1933–47, Waynflete professor of mathematics 1947–60, obituaries in *Journal of the London Mathematical Society*, 37 (1962), 257–73, *OM* 78 (1959–60), 308–9, and *BM* 7 (1961) 349–63; Helen Darbishire (1881–1961), principal of Somerville 1931–45; Emmy Noether (1882–1935), professor of algebra, Göttingen University, 1922–33; J. W. Brewer and M. K. Smith (eds.), *Emmy Noether: A Tribute to her Life and Work* (New York, 1981), 30–1; for Darbishire's negotiations, SPSL, 532/3; Richard Rado (1906–89), obituary in *BM* 37 (1991), 413–26, ended his career as professor at Reading University 1954–71; Robert Ludwig Mond (1867–1938), obituary in *ON* 2 (1939), 627–32; R. E. Rider, 'Alarm and Opportunity: Emigration of

Nicholson, Pidduck, Whitehead, Ferrar, Haslam-Jones, and J. H. C. Thompson were the six college fellows who had worked outside Oxford. The behaviour of the first two did not reinforce the view that it had been an advantage for them to have done so. By 1930 Nicholson was indeed a warning against making an outsider a college fellow. He arrived at Balliol in 1921 with glowing credentials. He had excelled in the Cambridge mathematics tripos, researched in physics in the Cavendish Laboratory, moved to a chair at King's College, London, and been elected FRS in 1917. Unfortunately his heavy drinking soon became a problem. He was widely known as Boozy Bill who had one good hour a day: before 11 a.m. he was in bed and after 12 he was drunk. He smoked in academic dress and while lecturing removed his painful false teeth and placed them on the lectern. In 1930 he suffered a nervous breakdown, lost his fellowship, and was soon permanently confined to an asylum. Like Nicholson, Pidduck was an eccentric mathematical physicist who did not make the expected mark in Oxford. An Oxford graduate in mathematics and physics who studied at the Technische Hochschule, Charlottenburg, Berlin, Pidduck was first a research fellow of Queen's who demonstrated in the electricity laboratory run by Townsend. In 1921 he became a fellow of Corpus Christi and in 1927 university reader in applied mathematics from which he resigned in 1934. An extensive publisher on the mathematical theory of electricity, Pidduck was antisocial, reclusive, and dwelled on grievances. As such he did not fulfil the expectation that, as Townsend's theorist, he would be Oxford's leading promoter of advanced work in applied mathematics and mathematical physics. But he was an athletic legend: on four carefully chosen occasions he dived off the centre of the concrete bridge in the Parks into the river Cherwell. The third exception was Henry Whitehead, a Balliol mathematician who, after two lucrative years as a financier in the City, returned in 1928–9 to Oxford where he met Oswald Veblen, then a visiting professor from Princeton. Next year Whitehead went to Princeton to study under Veblen who initiated him into the then new and intriguing field of topology and collaborated with him in an important textbook. Becoming a fellow of Balliol in 1933 Whitehead was a leading publisher in algebraic topology as well as cricketer, talker, host, and mountaineer. The most innovative researcher among Oxford's college fellows in the 1930s and so sociable that for him two mathematicians made a congress, he was rewarded with the Waynflete chair of mathematics in 1947. If Whitehead benefited from his period abroad, Ferrar and Haslam-Jones were merely treading water as provincial university teachers waiting for the call to return to Oxford. The case of J. H. C. Thompson revealed the continuing lure of college and university administration. A scholar of New

---

Mathematicians and Physicists to Britain and the United States, 1933–45', *Historical Studies in the Physical Sciences*, 15 (1984), 107–76 (159).

College, he took a first in 1930, won two university scholarships, took his D.Phil. under Milne, and was awarded a junior research fellowship at Merton in 1933. Having had the wit to go to Cambridge to study and publish with Max Born, a refugee mathematician from Göttingen, he was elected mathematics fellow at Wadham in 1936. After this propitious start he contributed little to mathematics, preferring to become right-hand man to Maurice Bowra, warden of Wadham, and in the 1960s becoming the first non-head of a college to act as pro-vice-chancellor of the University.[7]

Only about half the men's colleges had a fellow in mathematics. In 1930 ten colleges lacked a mathematics fellow; in 1939 eight were still bereft. As if to make a telling point against its Cambridge equivalent, a hotbed of mathematics, Trinity had no mathematics fellow until 1961 when J. M. Hammersley was appointed—even then only as a research but not tutorial fellow. Between the wars no women's college had a mathematics fellow. Compared with Cambridge, Oxford mathematics was inferior in number of scholarships, prizes, lecturerships, and fellowships; and the size of its final honours school was about a third of that of the famous Cambridge mathematics tripos, the products of which monopolized the more important chairs in English universities. In a climate still dominated by the classics, only two colleges (Magdalen and Wadham) found it possible in the 1930s to elect their first tutorial fellow in mathematics. Those that did not thought that, as long as there were no more than about thirty finalists a year, it made more sense for them to appoint temporary lecturers, to farm out their undergraduates to a reliable fellow of another college, or occasionally to use their fellow in physics to tutor in mathematics as well

If Oxonians dominated college fellowships in mathematics, they were in a minority at the professorial level where, with three chairs until 1928 and then four, Oxford was on a par with Cambridge mathematics and very well off compared with every science subject at Oxford. Oxford men monopolized the Waynflete chair of mathematics. From 1892 to 1922 its first occupant was E. B. Elliott, an Oxonian who had previously been a college fellow for eighteen years and a specialist on algebra, whose best work was done before 1914. An academic conservative who had no notion of a research school or of postgraduate training, he retired in 1922 aged 71. He was succeeded by

---

[7] John William Nicholson (1881–1955), fellow of Balliol 1921–30, obituary in *BM* 2 (1956), 209–14; F. B. Pidduck, *A Treatise on Electricity* (Cambridge, 1916) was planned with E. W. B. Gill, a demonstrator in physics, and indebted to Townsend; F. B. Pidduck, *Lectures on the Mathematical Theory of Electricity* (Oxford, 1937) was based on Oxford lectures to mathematics undergraduates; Oswald Veblen (1880–1960), professor of mathematics, Princeton University, 1905–32; J. H. C. Whitehead, *The Foundations of Differential Geometry* (Cambridge, 1932); Max Born (1882–1970), obituary in *BM* 17 (1971), 17–52; M. Born and J. H. C. Thompson, 'A Note on the Spectrum of the Frequencies of a Polar Crystal Lattice', *Proceedings of the Royal Society of London A*, 147 (1934), 594–9; Cecil Maurice Bowra (1898–1971), warden of Wadham 1938–70.

A. L. Dixon, then aged 55, who had been a fellow of Merton for over thirty years.[8] Though Dixon had been elected FRS in 1912, he had stopped publishing in 1910. As professor Dixon resumed publication, over half his papers being co-authored with Ferrar, fellow in mathematics at Hertford. Dixon possessed old-world courtesy and charm in abundance, and was a connoisseur of games, languages, and music. He was a popular president of the London Mathematical Society in the 1920s. But his gentle quiescence did not permit him to be a leader or to induce young people to embark on research. Incapable of pushing for mathematics in the University, he withdrew into himself when faced with controversy. Keen undergraduates regarded him at best as unoriginal and at worst as a dud *qua* mathematician, but he soldiered on until 1945 when he retired aged 78. Both Dixon and Elliott never left Oxford and with their interest in Victorian algebra looked longingly backwards.

The Sedley chair of natural philosophy was treated as one devoted more to mathematics than physics by Love who occupied it from 1898 until 1940 when he died in post aged 77.[9] His best work, which dealt with the theory of deformable media and theoretical geophysics, had been published in the 1890s. He was yet another link with Victorian mathematics, teaching classical applied mathematics as found in late nineteenth-century books. By the 1920s Love was revered in Oxford as an endearingly whimsical veteran who was still being honoured externally for his pre-war contributions. But he was so suspicious of the new quantum mechanics that he was no help to young physicists, he had almost no research students between the wars, and in the 1930s, though his faculties remained undimmed, he was physically frail. Even the arrival in Oxford in 1920 of G. H. Hardy, who as a young man had been inspired by Love to turn to mathematical analysis, did not animate the ageing Sedley professor.

G. H. Hardy succeeded William Esson as Savilian professor of geometry. Esson's research had fizzled out before he became a professor aged 59: his best-known work, pursued in the 1860s with A. G. V. Harcourt, a fellow Oxonian, was on chemical reaction kinetics. When Esson died in 1916 aged 78 the chair was suspended and in 1919 resuscitated. In December Hardy was elected to take office the next month. He came from Cambridge to Oxford not so much because he felt its allure but owing to his outrage at

---

[8] Edwin Bailey Elliott (1851–1937), first 1873, fellow of Queen's 1874–92, obituary in *ON* 2 (1938), 425–31; Arthur Lee Dixon (1867–1955), first 1888, fellow of Merton 1891–1922, Waynflete professor 1922–45, obituaries in *Journal of the London Mathematical Society*, 31 (1956), 126–8 and *BM* 1 (1955), 33–6.

[9] E. A. Milne, 'Love', *ON* 3 (1939–41), 467–82 and his reverential tribute in *Journal of the London Mathematical Society*, 16 (1941), 69–80; A. E. H. Love, *Theoretical Mechanics: An Introductory Treatise on the Principles of Dynamics: With Applications and Numerous Examples* (Cambridge, 1897); id., *A Treatise on the Mathematical Theory of Elasticity* (Cambridge, 1892–3).

the dismissal in 1916 of Bertrand Russell by Trinity College, Cambridge, for his pacifism. In some ways Hardy was the last person Oxford wanted. He was an anticlerical atheist who would not enter a religious building such as a college chapel and always referred to God as his personal enemy. He prided himself on his radical political views so in the mid-1920s he was happy to be president of the National Union of Scientific Workers for whose interests he fought publicly. He attacked Greats at Oxford and the mathematics tripos at Cambridge as so fundamentally vicious in their obsessive concern with honours that they should be abolished and not merely reformed.[10] For Hardy an examination could do little harm provided its standard was low. He also attacked the colleges who, greedy for firsts, encouraged tutors to stunt mathematical education by absorption in examination technique, and to tire themselves out in trying to turn a comfortable second into a marginal first. He thought most Oxford tutors did twice as much teaching as any active mathematician should be asked to undertake.

If these beliefs were threatening, his personality and attitude to mathematics were not. In Oxford common-rooms he was a sparkling and welcome conversationalist. Oxford's philosophers liked his occasional lectures on the philosophy of mathematics. Moreover Hardy often took open pride in proclaiming that his kind of mathematics was useless. He regarded some branches of applied science as repulsively ugly and intolerably dull and the physical world as turbid and confused compared with the bright, clear, exquisite, and beautiful world of pure mathematics. In his inaugural lecture he pleased those who were opposed to utilitarian and vocational subjects: he refused to follow chemists and engineers in expounding 'with justly prophetic fervour, the benefits conferred on civilisation by gas-engines, oil, and explosives.... A pure mathematician must leave to happier colleagues the great task of alleviating the sufferings of humanity.' The internal result of such views was that during his reign mathematics on the one hand and applied mathematics and experimental science on the other went their separate ways. That was palpably shown in Hardy's total indifference to the vexed question of the mathematics moderations examination which was a

---

[10] William Esson (1838–1916), professor of geometry 1897–1916, obituary by E. B. Elliott, *Proceedings of the Royal Society of London A*, 93 (1917), pp. liv–lvii; Augustus George Vernon Harcourt (1834–1919), obituary by H. B. Dixon, ibid. 97 (1920), pp. vii–xi; M. C. King, 'The Course of Chemical Change: The Life and Times of Augustus G. Vernon Harcourt, 1834–1919', *Ambix*, 31 (1984), 16–31. On Hardy obituary by E. C. Titchmarsh, *ON* 6 (1949), 447–61; G. H. Hardy, *A Mathematician's Apology with a Foreword by C. P. Snow* (Cambridge, 1967); id., *Bertrand Russell and Trinity: A College Controversy of the Last War* (Cambridge, 1942); id., *Some Famous Problems of the Theory of Numbers and in Particular Waring's Problem: An Inaugural Lecture Delivered before the University of Oxford* (Oxford, 1920), 4; id., 'The Case against the Mathematical Tripos', *Mathematical Gazette*, 13 (1926), 61–71; *Nature*, 114 (1924), 869; G. H. Hardy, *Collected Papers*, 7 vols. (Oxford, 1966–79). When Bertrand Russell (1872–1970) was deprived of his lectureship in 1916, Hardy protested and became a leading campaigner for his reinstatement in 1919 (agreed Nov., effective 12 Dec. 1919).

preparation for finals in both mathematics and physics. Taken at the end of the first year and designed for mathematicians, it did not suit physicists who received no cover of recent mathematical physics and had to make the difficult jump from the pure and abstract world of mathematics moderations to the intense experimental life of finals physics. It was only after Hardy had left Oxford that a new mathematics moderations, aimed at most physicists, many engineers, and some chemists, was contemplated and introduced in 1932 though with limited success.[11]

Until Hardy arrived postgraduates in mathematics were few and inferior; but, as an inexhaustible and inspiring inventor of research projects, he attracted them from abroad as well as Britain to form Oxford's first research school in mathematics which was focused on analysis. Hardy brought internationalism to Oxford mathematics not only in content but also by campaigning against the boycott of German and Hungarian mathematicians from international congresses after the first war. He was conspicuously active in the London Mathematical Society, serving as both secretary and president in the 1920s and being chiefly responsible for founding its journal in 1926. He scored a notable first for an Oxford mathematician when in 1928–9 he visited the USA to savour the intellectual riches of Princeton and the California Institute of Technology, Veblen coming from Princeton to Oxford in his place. Though Hardy was a confirmed bachelor, he supported the higher education of women, going out of his way to help female researchers, then a depressed class, the most important beneficiary being Mary Cartwright.[12] Hardy was sympathetic to unusual students such as Edward Wright who came to Oxford with a London external degree in mathematics obtained while a schoolmaster, attended Hardy's lectures and weekly undergraduate seminar, graduated with a first in 1929, and asked Hardy to supervise his D.Phil., which he gained in 1931. He soon co-authored with Hardy a standard work on the theory of numbers and was elected professor of mathematics at the University of Aberdeen, of which he eventually became vice-chancellor.[13] Hardy also promoted *esprit de corps* among Oxford mathematicians in two ways. First, when Glaisher's *Quarterly Journal of Pure and Applied Mathematics* and his *Messenger of Mathematics* ceased publication in 1927 and 1929, Hardy was the leading spirit in replacing them in 1930 with the Oxford series of the *Quarterly Journal of Mathematics* published by the Clarendon Press and edited initially by three

---

[11] On mathematics moderations, Chaundy memorandum, n.d. [Apr. 1932], PS/R/1/3; *OUG* 62 (1931–2), 639–40.
[12] Dame Mary Lucy Cartwright (b. 1900), first in maths 1923, D.Phil. 1930, FRS 1947 (first woman mathematician), mistress of Girton College, Cambridge, 1949–68, *Who's Who*, letters to author, 5, 8, 10 Apr. 1988.
[13] Sir Edward Maitland Wright (b. 1906), professor of mathematics, University of Aberdeen, 1936–62, vice-chancellor 1962–76, letter to author, 26 Jan. 1989; G. H. Hardy and E. M. Wright, *An Introduction to the Theory of Numbers* (Oxford, 1938).

Oxonians, Chaundy, Ferrar, and Poole. It appeared regularly and, as most contributors were Oxonians, it helped to make Oxford mathematicians visible nationally.[14] Secondly, Hardy initiated a campaign in 1930 for a mathematics institute at Oxford to reflect the improved status of mathematicians and the enhanced prestige of their subject in the world at large.[15] Initially Hardy's aim was to secure such a visible centre in the Bodleian Library extension scheme to which the Rockefeller Foundation contributed heavily; but after he left Oxford it was established in 1934 as a concessionary gesture in the enlarged Radcliffe Science Library where six rooms were occupied until 1952. It was hardly an Oxonian answer to the institutes at Princeton and Göttingen which Hardy admired; and its claims were neglected by the University in the Appeal of 1937 though recognized *in potentia* in the planning of the Science Area. With a common-room, a seminar room, and rooms for Dixon, Titchmarsh, Milne, and Love, it fell far below what the mathematicians wanted in 1930 and 1937; and, given their predilections, it did not double up as a school of theoretical physics.

Hardy returned to Cambridge in 1931 not because he was unhappy in Oxford but for professional and domestic reasons: he wanted to occupy the senior mathematical chair in Britain and at Cambridge he would be able to live in college for the rest of his life. He was succeeded at Oxford by his pupil, protégé, collaborator, and secretary Titchmarsh, an Oxonian who was presumably expected to continue Hardy's approach.[16] In some ways he did. He maintained his mentor's emphasis on pure mathematics, having no personal contacts with the physicists and engineers in the 1930s, even though his research on Fourier integrals and on eigen-functions was not uninteresting to them. His chief research interest, aroused by Hardy, was analysis. Though he supervised nearly all the D.Phil. students in mathematics in the 1930s, he did not maintain Hardy's school of analysis mainly because he was such a quiet, shy, unambitious, and withdrawn man who lacked Hardy's sense of conviviality, his wide range as a mathematician, his internationalism, and his entrepreneurial capacity to innovate. Unlike Hardy, Titchmarsh did not complain about the ambiguous position of mathematics at Oxford. As an

---

[14] Chaundy, Poole, and Ferrar (replaced in 1933 by Hodgkinson) were assisted in 1930 by seven Oxonians, Hardy, Dixon, Elliott, Love, Milne, Pidduck, and Titchmarsh; James Whitbread Lee Glaisher (1848–1928), *DNB*.

[15] G. H. Hardy, 'Mathematics', *OM* 48 (1929–30), 819–21; *HCP* 145 (1930), 187–90; Report of Committee on Needs of Physical Sciences Faculty, 4 Mar. 1931, PS/R/1/2; draft agreement of 2 May 1934, PS/R/1/4; Newboult to Plant, 18 May 1937, UR/SF/MU/2; 12 May 1931, PS/M/1/1.

[16] Edward Charles Titchmarsh (1899–1963), first in maths 1922, professor of pure mathematics, University of Liverpool, 1929–31, FRS 1931, Savilian professor of geometry 1931–63, obituaries in *OM* 81 (1962–3), 212 and by M. L. Cartwright in *BM* 10 (1964), 305–24 and *Journal of the London Mathematical Society*, 39 (1964), 544–65; E. C. Titchmarsh, *The Theory of Functions* (Oxford, 1932), the text of which was extensively scrutinized by Hardy; id., *Introduction to the Theory of Fourier Integrals* (Oxford, 1937) dismissed their application to the mundane world of heat.

administrator he seems never to have overcome the worries he endured as professor at Liverpool. Intimate with few colleagues and never feeling the need to move far afield intellectually and institutionally, Titchmarsh was a brilliant mathematician who needed quiet seclusion but not the stimulus of equal minds.

Like most of the college fellows, Elliott, Dixon, Love, Hardy, and Titchmarsh pursued kinds of mathematics which were unhelpful to physicists and engineers. That also turned out to be the case with E. A. Milne, the first occupant of the Rouse Ball chair of mathematics, who was not quite the sort of mathematical physicist that Oxford's scientists yearned for. The chair was established by the will of Rouse Ball, a mathematician and historian of mathematics of Trinity College, Cambridge, who in 1925 left £70,000 to Cambridge (£25,000 for a chair of mathematics, £25,000 for one in modern English law, and £10,000 each to the libraries of the University and Trinity) and £25,000 to Oxford for a chair of mathematics which he hoped would not neglect its historical and philosophical aspects. Oxford's scientists immediately saw the benefaction as an opportunity to secure a first-rate mathematical physicist as some compensation for the dominance of pure mathematics in the other chairs. When the chair was advertised in March 1928 mathematical physics was declared to be its focus and in June 1928 Milne, the only applicant, was elected unanimously to it effective from January 1929. Hardy and Lindemann, who were electors, were keen to land Milne in Oxford, the former because Milne had been one of his favourite pupils at Trinity, Cambridge, and the latter because he was desperate to secure a mathematician useful to physicists. Milne seemed a perfect choice: he was young, a brilliant mathematician, a solar physicist who had lectured on wave mechanics at Cambridge, and was currently Baeyer professor of applied mathematics at the University of Manchester where he had continued his research on the atmospheres of stars. Only his election drew him to Oxford: having been urged to be a candidate he had declined because he had just bought, decorated, and furnished a house and wanted to stay longer in Manchester. He was surprised at his election: he was not *au fait* with modern quantum theory and thought Fowler and C. G. Darwin more appropriate men than himself.[17] But he was lured to Oxford by its superiority to Manchester

---

[17] Edward Arthur Milne (1896–1950), Baeyer professor of applied mathematics, University of Manchester, 1924–8, FRS 1926, Rouse Ball professor 1929–50, obituary in *OM* 69 (1950–1), 12–14; W. H. McCrea, *ON* 7 (1950–1), 421–43; D. G. Kendall, *Quarterly Journal of the Royal Astronomical Society*, 25 (1984), 147–56; S. Chandrasekhar, 'Edward Arthur Milne; His Part in the Development of Modern Astrophysics', in S. Chandrasekhar, *Truth and Beauty: Aesthetics and Motivations in Science* (Chicago, 1987), 74–91. Poole was the solitary Oxford mathematician who collaborated with an experimentalist, the engineer Binnie: A. M. Binnie and E. G. C. Poole, 'The Theory of the Single-Pass Cross-flow Heat Interchanger', *Proceedings of the Cambridge Philosophical Society*, 33 (1937), 403–11. For the Rouse Ball chair and Milne's election, *HCP* 132 (1925), 192; 133 (1926), 126–7; *OUG* 58 (1927–8), 405, 677; 2 June 1928, UDC/M/41/1; Milne to A. V. Hill, 6 June 1928, Milne Papers; Milne to Lindemann, 15 June 1928, LP, D160;

apropos freedom from administration, an allegedly healthier climate, and above all by the opportunity of using the chair to study theoretical astrophysics as much as he liked.

His inaugural lecture was disappointing to experimentalists. Milne made it clear that for him mathematical physics dealt with 'idealised models' of what nature was conceived to be, models which could never be discovered by merely seeking to account for observations. In his own research he had just turned to developing an alternative to Eddington's notions about stellar structure and was beginning to show a penchant for prescribing a few general physical considerations and then making deductions from them. His lack of constant recourse to specific physical data and his lack of interest in solving particular problems of mathematical physics were revealed in his inaugural when he suggested that the most profitable mode of mathematical astrophysics was to generate general theorems which formed the context in which specific theories might be tested. Milne also gave short shrift to recent quantum mechanics which he thought had dangerously discarded as non-existent the unobservable and the unobserved, and had exploited mathematics of dubious value in order to gain new results about possible relations between measurable quantities. For Milne mathematical physics had to be rigorous mathematics and he deplored the use of sloppy or *ad hoc* mathematical methods to secure results of interest to physicists.[18]

True to his apologia Milne was not useful to experimentalists and Lindemann soon became disappointed with him. Indeed from 1932 Milne became even more theoretical as he developed his theory of kinematic relativity, which he regarded as his main work, as an alternative to Einstein's theory of general relativity. Milne's theory, which originated in problems concerning the structure and expansion of the universe, was given a generally adverse or indifferent reception not least because it had no novel physical consequences, a characteristic which led one Oxford wag to describe it as cosmythology. Though Milne's tendency to build up a system of theory irritated those who wanted him to solve particular problems, his reputation flourished nationally and internationally. Like Hardy he was not only president of the London Mathematical Society (1937–9) but also internationally sought after, lecturing in Berlin in 1931 at Schrödinger's invitation and spending autumn 1932 in the Potsdam astrophysics laboratory with Einstein and

---

Walter William Rouse Ball (1850–1925), obituary in *Nature*, 115 (1925), 808–9; Ralph Howard Fowler (1889–1944), *DNB*, FRS 1925, first Plummer professor of mathematical physics, University of Cambridge, 1932–44; Charles Galton Darwin (1887–1962), *DNB*, FRS 1922, Tait professor of natural philosophy, University of Edinburgh, 1924–36.

[18] E. A. Milne, *The Aims of Mathematical Physics: An Inaugural Lecture Delivered before the University of Oxford on 19 November 1929* (Oxford, 1929); id., 'Kinematical Relativity', *Journal of the London Mathematical Society*, 15 (1940), 44–80 shows the consistency of his views.

Freundlich. Though handicapped by suffering from epidemic encephalitis which first struck in 1923, Milne immediately attracted research students, most of whom were Oxonians who worked in his fields. In their hands his methods were fruitful, leading sometimes to conclusions at odds with his own. Thus his first D.Phil. pupil, T. G. Cowling, who worked on stellar models, made it clear, at a meeting of the Royal Astronomical Society held in 1931 to evaluate Milne's ideas on stellar structure, that he supported Eddington and not his supervisor. Though he thought Cowling disloyal, Milne remained a model for him: Cowling remembered Milne's advice that as a university teacher he should find time for research every day, even if for only half an hour; and Cowling spent his entire career in mathematics departments but his main research interest was theoretical astrophysics. Perhaps the clearest indicator of Milne's status was that Chandrasekhar came from Cambridge in 1933 to work as a guest for a term. Subsequently his theory of white dwarf stars was attacked by both Eddington and Milne but eventually it gained a Nobel prize for him.[19]

Milne was not an institution-builder. His plan of 1930 for an institute of cosmical physics came to nought partly because it became an element in the Radcliffe Observatory affair and partly because Milne never tried to push the scheme through the university bureaucracy: after all he had left Manchester in part to escape from administration. But his seminar was interesting and up to date; and in league with Plaskett, whom he had helped to lure to Oxford, he made Oxford in the 1930s a centre for astrophysics through their colloquium held at the Observatory and through their research students. After his premature death in 1950 he was replaced by C. A. Coulson who delivered what the University had wanted from Milne but not received, namely, the application of mathematics to key problems in physical science.[20] In the year he assumed the Rouse Ball chair, Coulson published his famous book on theoretical chemistry, *Valence*, which was comprehensible to chemists with no mathematical attainments and exploited pictorial terms. Up in the Elysian fields Milne no doubt deplored Coulson's approach: for Milne applied mathematics had to possess the same formal rigour as pure mathematics.

[19] On Milne's research and pupils, McCrea, 'Milne'; G. J. Whitrow (first in maths 1933, D.Phil. 1938 under Milne) in *DSB* and letter to author, 28 May 1987; P. Kerszberg, *The Invented Universe: The Einstein–De Sitter Controversy (1916–17) and the Rise of Relativistic Cosmology* (Oxford, 1989), 354–74; E. A. Milne, *Relativity, Gravity and World Structure* (Oxford, 1935); id., *White Dwarf Stars* (Oxford, 1932); Thomas George Cowling (1906–90), first in maths 1927, D.Phil. 1931, FRS 1947, ended his career as professor of applied mathematics, University of Leeds, 1948–70, obituary by L. Mestel, *BM* 37 (1991), 105–25; T. G. Cowling, 'Astronomer by Accident', *Annual Review of Astronomy and Astrophysics*, 23 (1985), 1–18; Cowling to author, 13 Nov. 1986; K. C. Wali, *Chandra: A Biography of S. Chandrasekhar* (Chicago, 1991), 96, 105–27, 132–3; Erwin Schrödinger (1887–1961).

[20] Charles Alfred Coulson (1910–74), Rouse Ball professor 1952–74, on whom see S. L. Altmann and E. J. Bowen, *BM* 20 (1974), 75–134; C. A. Coulson, *Valence* (Oxford, 1952).

8.2 The College Laboratories

When the first war began the University had its own chemical laboratory on the south side of the University Museum in a building modelled on the Abbot's kitchen at Glastonbury Abbey. The accommodation was so limited and the equipment so outmoded that W. H. Perkin, who had replaced Odling as Waynflete professor in 1913, immediately decided that a new laboratory was needed and by 1916 he had secured the appropriate endowment from Dyson Perrins after whom it was named. The Glastonbury laboratory had sunk into torpor under Odling, a literary chemist who had retired from the Waynflete chair in 1912 aged 83 after occupying it for forty years. A cultivated Nestor of chemistry, Odling was a connoisseur of engravings but uninterested in work at the laboratory bench. Preferring the philosophical and speculative aspects of chemistry, he was not the slave of his laboratory, which he thought it a breach of etiquette for a professor to enter. His last research dated from 1876 so he had founded no research school. One of his obituarists was reduced to claiming on his behalf that 'it was not that Odling *discouraged* research'.[21]

By 1914 there were five college chemistry laboratories which were used not just for teaching but also for research. Their very existence and the founding of two of them in the twentieth century implied that there were perceived deficiencies in Odling's regime. The oldest college laboratory was at Christ Church which had built the Dr Lee's laboratory in 1767 for the teaching of anatomy. When the anatomical material was moved in the early 1860s to the University Museum, the new Dr Lee's reader in chemistry, Vernon Harcourt, redirected the laboratory to chemistry and made it the home of distinguished work in physical chemistry, especially kinetics. In 1903 a third storey was added to satisfy the demand for more laboratory accommodation from H. B. Baker, the Dr Lee's reader, who left Oxford in 1912 for the chair of chemistry at Imperial College, London. Baker was succeeded in 1913 as director of the Lee's laboratory by Andrea Angel, lecturer in chemistry at Christ Church where he had graduated.

Two of the laboratories, that at Magdalen and that jointly run by Balliol and Trinity, dated from the Victorian era. In 1848 Charles Daubeny, professor of chemistry, botany, and agriculture, moved out of the Ashmolean building and at his own expense built a block of laboratories opposite Magdalen and adjacent to the Botanic Garden. Used mainly for chemistry, the Daubeny Laboratory was run from 1888 to 1923 by J. J. Manley, its curator. As at Christ Church, expansion took place early in the twentieth century, with two new laboratories being added in 1902 at a cost to Magdalen

---

[21] William Odling (1829–1921), Waynflete professor 1872–1912, obituaries by J. E. Marsh, *Journal of the Chemical Society*, 119 (1921), 553–64 and H. B. Dixon, *Proceedings of the Royal Society A*, 100 (1922), pp. i–vii (quotation at iii).

of £1,746. The Daubeny Laboratory was used not only by Gunther, Magdalen's science fellow, but by college lecturers in chemistry such as Sidgwick, fellow of Lincoln, and T. S. Moore, who left in 1914 for the chair of chemistry at Royal Holloway College, London. The other Victorian college innovation was the suite of rooms known as the Balliol–Trinity laboratory. The first was built by Balliol in 1853 and a second in 1879. In the latter year Trinity dropped its usual suspicion of its next-door neighbour and successfully proposed a co-operative scheme of science teaching. In 1897 Trinity built its own teaching laboratory which was connected to the Balliol one of 1879. From 1900 to 1920 the Balliol–Trinity laboratory was run by Nagel, an Oxonian who was fellow in chemistry at Trinity where he had been an undergraduate. A good organizer who did no research, Nagel was generous in advising and helping workers in it, even those from other colleges who attended the course in physical chemistry which he planned. In 1901 Nagel acquired an assistant in the form of Harold Hartley, the new fellow in chemistry at Balliol, who was much more research-minded.

The first of two twentieth-century laboratories was erected by Queen's in 1900. Initially it was run by G. B. Cronshaw, fellow of Queen's in chemistry, but when he became more involved in college administration he was replaced as head of the laboratory by F. D. Chattaway, an Oxonian who was already an FRS. Though not made a fellow of Queen's until 1919, Chattaway was a distinguished researcher and effective teacher who quickly made the Queen's laboratory into Oxford's centre for organic chemistry and one which was not eclipsed by the opening of the Dyson Perrins Laboratory. The last of the college laboratories was that erected by Jesus in 1907, purpose-built on three storeys and devoted particularly to physical chemistry. Jesus built it because it disliked the inconvenience and disadvantages of a centralized laboratory. It wanted close and constant oversight of its undergraduate chemists and built its own chemistry laboratory as a part of its tutorial system. The laboratory was run by D. L. Chapman, an Oxonian who was elected a fellow of Jesus in 1907. Though Chapman's later eccentricities made him a figure of fun and his work on chemical kinetics was overshadowed in the 1920s by that of Hinshelwood, his research on photochemical reactions led to his election as FRS in 1913.[22]

---

[22] On the college laboratories see H. Hartley, 'The Contribution of the College Laboratories to the Oxford School of Chemistry', *Chemistry in Britain*, 1 (1965), 521–4, repr. in H. Hartley, *Studies in the History of Chemistry* (Oxford, 1971), 223–32; K. J. Laidler's excellent 'Chemical Kinetics and the Oxford College Laboratories', *Archive for History of Exact Sciences*, 38 (1988), 197–283; Gunther, *Daubeny Laboratory* and *Daubeny Laboratory Registers* are rambling; R. Hutchins, 'Charles Daubeny 1795–1867... Including a History... of the Staff and Work of the Daubeny Laboratory, 1848–1923', typescript, Magdalen College Library, is a good revisionist account; E. J. Bowen, 'The Balliol–Trinity Laboratories Oxford', *Notes and Records of the Royal Society of London*, 25 (1970), 227–36, like Hartley wrote from a college perspective; T. Smith, 'The Balliol–Trinity Laboratories', in J. Prest (ed.), *Balliol Studies* (London, 1982), 185–224; D. A. Long, 'The Sir Leoline Jenkins Laboratories', *Jesus College Record* (1989), 17–20 and

The five college laboratories had worked independently until 1904 when Baker successfully proposed a scheme of co-operation between them. Magdalen was to concentrate on analytical chemistry, Queen's on organic, Christ Church on inorganic, and Balliol–Trinity on physical. Subsequently Jesus joined the scheme and strengthened the commitment to physical chemistry. Baker's initiative gave greater access to the laboratories which previously had tended to be available only to students from the six colleges. It also enabled the laboratories to focus on a particular branch of practical work and their occupants to develop techniques which could be useful in research.

After the first war the flood of undergraduates combined with inflation led the University to approach the government for an emergency grant. As a quid pro quo the Asquith Commission was appointed and in summer 1920 it took evidence from Oxford's scientists, including Frederick Soddy, the newly appointed Dr Lee's professor, who was responsible for inorganic and physical chemistry. An Oxonian who was best known for clarifying the concept of isotopes and for his work with Rutherford on atomic disintegration and radioactivity, Soddy quickly publicized his ambitions for a big school of chemistry based on a new university laboratory for inorganic and physical chemistry costing £100,000, a step on which all heads of science departments were then agreed.[23] He attacked the college laboratories as dissipating effort and resources. By late 1919 he had learned that there was to be no new building for him and submitted to the registrar his first complaint that the University had broken faith with him. Next year in his evidence to the Asquith Commission he called again for a big university laboratory for physical chemistry and launched a vehement attack on the colleges who, he claimed, appointed teachers to the detriment of the University which required people absorbed in experimental research. His call for

(1995–6), 46–57; Herbert Brereton Baker (1862–1935), Dr Lee's reader in chemistry 1903–12, professor at Imperial College, London, 1912–32, obituary in *Journal of the Chemical Society* (1935), 1893–6; Charles Giles Bridle Daubeny (1795–1867), professor of chemistry, Oxford, 1822–54; John Job Manley (1863–1946), curator of the Daubeny Laboratory 1888–1923, obituary in *Proceedings of the Physical Society*, 58 (1946), 332–3; Thomas Sidney Moore (1881–1966), lecturer in chemistry at Magdalen 1907–14, professor at Royal Holloway College, London, 1914–46, obituary in *Chemistry in Britain*, 3 (1967), 494; David Henry Nagel (1862–1920), first in chemistry 1886, fellow of Trinity 1890–1920, obituaries in *OM* 39 (1920–1), 27 and *Journal of the Chemical Society* (1921), 551–3; George Bernard Cronshaw (1872–1928), first in chemistry, Queen's, 1894, fellow of Queen's 1902–28, obituary in *The Times*, 21 Dec. 1928; Frederick Daniel Chattaway (1860–1944), first in chemistry 1891, FRS 1907, demonstrator in Queen's laboratory 1908 and soon head, fellow of Queen's 1919–34, obituary in *ON* 4 (1942–4), 713–16; David Leonard Chapman (1869–1958), first in chemistry 1893, second in physics 1894, fellow of Jesus 1907–44, obituaries by Bowen, *BM* 4 (1958), 35–44 and Hammick, *Proceedings of the Chemical Society* (1959), 101–3.

[23] For the terms of Soddy's appointment, *OUG* 49 (1918–19), 283, 408. For Soddy's campaign 1919–20, F. Soddy, *Accommodation for Inorganic and Physical Chemistry: University of Oxford*, 2 pp., 21 Nov. 1919, in GA Oxon, b. 141 (111), Bodleian Library, Oxford; Soddy, evidence to Asquith Commission, 29 Sept. 1920; *HCP* 114 (1919), 185–6; 115 (1919), 15; 15 June 1920, NS/R/1/2; 15 June 1920, NS/M/1/3; *Asquith Report*, 117.

central university provision for physical chemistry was supported by the Natural Sciences Board, which also acknowledged that such a proposal made the future of the college laboratories vexed, and by the other professor of chemistry, Perkin, who regarded the college laboratories as wasteful and unsatisfactory. The only dissidents from this view were representatives of Queen's, Jesus, and Magdalen, who argued that the college laboratories enabled the tutorial system in science to be retained and remedied the deficiencies of the university laboratories. There is no record on Balliol, Trinity, and Christ Church deviating publicly in 1920 from the line on the college laboratories taken by Soddy, Perkin, the Natural Sciences Board, and eventually by the Asquith Commission which suggested that the college laboratories could be used as temporary homes for developing new subjects or exclusively for research.

By autumn 1923 it had become abundantly clear that Soddy, still a young man in his mid-forties, would not acquire a university laboratory for physical chemistry, that the college laboratories would continue to be thorns in his flesh, and that he and most of the college tutors were at loggerheads. In one respect he was unfortunate. When he arrived in 1919, the wily Perkin had already launched a campaign to secure an extension to his Dyson Perrins Laboratory. It was completed in 1922 at an unexpectedly high cost to the University of £39,000. Some members of the University thought that Perkin was a financial blackmailer and suspected that all chemistry professors were racketeers. Once Perkin had acquired his palace of organic chemistry, it was impossible for Soddy to make comparable claims for physical chemistry: having doubled its borrowing, the University was retrenching financially and chemistry had already been generously treated. Furthermore there was no possibility of Soddy securing external endowment of a new university physical chemistry laboratory as long as the Balliol–Trinity and Jesus laboratories existed. In any case, unlike Perkin, he had taken a stand by 1920 against the dependence of research on commercial interests.

Soddy was statutorily responsible for physical and inorganic chemistry throughout the University; but, given the existence of the college laboratories and of independent college fellows, he was not master in his own house. Two of the leading chemistry fellows had little time for Soddy, perhaps because they had coveted the Dr Lee's chair. Sidgwick ostensibly disliked and denigrated Soddy, but Hartley was more subtle. He saw to it that Soddy had only one success in placing a protégé in a college fellowship when A. S. Russell, a non-Oxonian who had studied radioactivity under Soddy at Glasgow, was elected Dr Lee's reader in chemistry at Christ Church in 1919 and fellow the next year.[24] In 1921 Hinshelwood, a pupil

---

[24] Alexander Smith Russell, obituaries in *The Times*, 17 Mar. 1972, *Nature*, 237 (1972), 120–1, and entry in *Who Was Who*.

of Hartley, was elected chemistry fellow at Trinity and became Hartley's partner in directing the Balliol–Trinity laboratory. In the very year in which Soddy became the first Oxford scientist to win a Nobel prize, they immediately arranged for Hinshelwood to have research accommodation in converted lavatories in Trinity, an act which no doubt infuriated Soddy who had no space and no equipment in his buildings to teach physical chemistry (see Fig. 4). By 1922 the Balliol–Trinity laboratory was a large enterprise: it had five staff, including four college fellows, all of whom were Balliol men (Hartley, Hinshelwood, Raikes, Bowen), twelve rooms, fifty-five laboratory places, and an annual expenditure of just over £2,000 which made it easily the most costly of the college laboratories (Table 8.1). It received informally from Soddy's departmental grant a subsidy equal to £2 per term for every non-Balliol–Trinity student. All told in 1922 Soddy was supposed to pay to the college laboratories £1,000 which was earmarked in his departmental grant. Particularly furious at his lack of control over the Balliol–Trinity laboratory, in summer 1922 he ceased to pay them. By next summer the University decided to pay £600 directly to the Balliol–Trinity laboratory in 1923–4 for teaching physical chemistry on behalf of the University. In autumn 1924 the University went further. It began to pay the laboratory fees of all students in the Balliol–Trinity, Jesus, and Queen's laboratories, thus in principle increasing the subsidy to the Balliol–Trinity laboratory, but Hartley agreed informally not to accept more than £600 p.a.[25]

Soddy felt let down by the University which had not fulfilled its promise to provide him with a new building, had deprived him of his statutory responsibility for physical chemistry leaving him with inorganic chemistry, and in subsidizing the college laboratories out of the university grant to his department had in his view legalized robbery. No wonder that by the mid-1920s he had become intransigent, unyielding, obstinate, and aggrieved, characteristics he had not displayed strongly at Aberdeen. Contrary to the views implied or expressed by Hartley and Bowen, both of whom were devoted to the Balliol–Trinity laboratory, he was perhaps as much sinned against as sinning.[26] He rightly saw chemistry as anomalous compared with other

[25] For polemics 1922–4 between Soddy and the University, *HCP* 121 (1922), 87–93; 123 (1922), 59, 119; 124 (1923), 57; 125 (1923), 157–62; *OUG* 54 (1923–4), 221; 55 (1924–5), 124; Soddy to vice-chancellor, 3 Jan. 1924, OUA, MR/7/2/7; OUA, Balliol–Trinity Laboratory. Correspondence re Accounts 1920–3 and other College Laboratories 1922, MR/6/3/34; OUA, Correspondence re Chemical Laboratories 1923, MR/7/2/7. For details of the University's interim scheme of 1924 for the teaching of chemistry, *HCP* 127 (1924), 193–6.

[26] Bowen, note of 29 Jan. 1974, Soddy Papers B 1; H. Hartley, 'Schools of Chemistry in Great Britain and Ireland, xvi: The University of Oxford', *Journal of the Royal Institute of Chemistry*, 79 (1955), 116–27, 176–84, wrote from local knowledge except on pp. 126–7 where his material on Soddy was derived from F. M. Brewer to Hartley, 3 Jan. 1953, History of Science Museum, Oxford, MS 121. Generally hostile accounts of Soddy's difficulties at Oxford are Fleck, *BM* 3 (1957), 203–16, Fleck in *DNB*, and A. D. Cruickshank, 'Soddy at Oxford', *British Journal for the History of Science*, 12 (1979), 277–88. More sympathetic are: F. A. Paneth, 'A Tribute to

FIG. 4 The exterior of the chemistry laboratory, Dolphin Yard, Trinity College, 1920s. This squalid building, a converted lavatory, was occupied from 1921 to 1929 by C. N. Hinshelwood, who began there the research in chemical kinetics for which he won a Nobel prize. Reproduced by permission of the president and fellows of Trinity College, Oxford.

TABLE 8.1. *College chemistry laboratories 1922–1923*

	Balliol–Trinity	Christ Church	Jesus	Magdalen	Queen's	Total
Undergraduates	94	14	41	23	16	188
Researchers of all kinds	6	2	11	1	9	29
Staff	5	1	2	1	1	10
Annual cost (£)	2,170	560	940	630	650	4,950
Number of rooms	12	4	7	3	3	29
Number of ordinary places	44	23	32	16	13	128
Number of research places	11	5	14	3	10	43

Source: Document of 20 Apr. 1923, MR/6/3/34.

Frederick Soddy', *Nature*, 180 (1957), 1085–7; and A. S. Russell, *Chemistry and Industry* (1956), 1420–1.

sciences at Oxford. He continued to regard the college laboratories as an unfortunate legacy from the past and the tutorial system as outmoded. With the insight of a Namier he characterized Oxford chemistry as 'a scattered collection of teachers and laboratories interconnected respectively by a tangled network of personal relationships and private ownerships'.[27]

Not surprisingly Soddy wished to retain control of his shrunken empire especially through appointments but even on that matter he found himself at odds with the college tutors. In late 1923, when the Hebdomadal Council launched an inquiry into the teaching of chemistry, Soddy thought that two of his demonstrators, Applebey and Lambert, both college fellows, should either not co-operate with it or resign their demonstratorships.[28] Applebey and Lambert refused to resign and, when Soddy threatened to sack them, they so resented this threat to their positions and income that they secured the support of the vice-chancellor who quickly poured emollient on troubled waters after Hartley had told him that Soddy was very unpopular because of his attempt to coerce his demonstrators. In 1927 Lambert was involved in another conflict which this time reached the Privy Council (see Fig. 5). In 1920 the Natural Sciences Board had acted statutorily in appointing Lambert as Aldrichian demonstrator for seven years in Soddy's department to teach physical and inorganic chemistry, without consulting Soddy. In 1927 the Physical Sciences Board renewed Lambert's appointment, again without consulting Soddy who protested to the registrar and vice-chancellor. Garrod, the Regius professor of medicine, supported Soddy and told the vice-chancellor that the exclusion of Soddy from the Board reflected 'the old struggle between the [college] tutors and the professors'. Refusing to compromise or withdraw, Soddy petitioned the Privy Council against Lambert's appointment, the University formally opposed him, and in June 1928 the Privy Council dismissed his petition.[29]

Secure in the Dyson Perrins Laboratory, Perkin could afford to tolerate the peculiarities of chemistry at Oxford, especially the continuing existence of the college laboratories only one of which (Queen's) covered organic chemistry and did so in a style different from his and unthreatening. Perkin agreed with Soddy about centralizing teaching and research in physical chemistry but left him to fight a solitary campaign. The college chemistry fellows, who jealously guarded their statutory independence, resented Soddy's accusations that they could not cover in tutorials the whole of chemistry and that they were feeble researchers. Using their legitimate power in the University's boards, Council, and Congregation, they were able to repel

[27] Soddy to vice-chancellor, 26 May 1923, OUA, MR/7/2/7.
[28] OUA, 1923–4, MR/7/2/7.
[29] Soddy, Petition to the King's Most Excellent Majesty in Council, 19 Sept. 1927; University of Oxford, Reply to the Petition..., 21 Mar. 1928, both in Soddy Papers, B18; *HCP* 141 (1928), 13; *OUG* 50 (1919–20), 781; 51 (1920–1), 179; 57 (1926–7), 602, 739; Garrod to vice-chancellor, 1 Aug. 1927, OUA, Dr Lee's Chair of Chemistry: Aldrichian Demonstrator, 1927, U Sol./6/2.

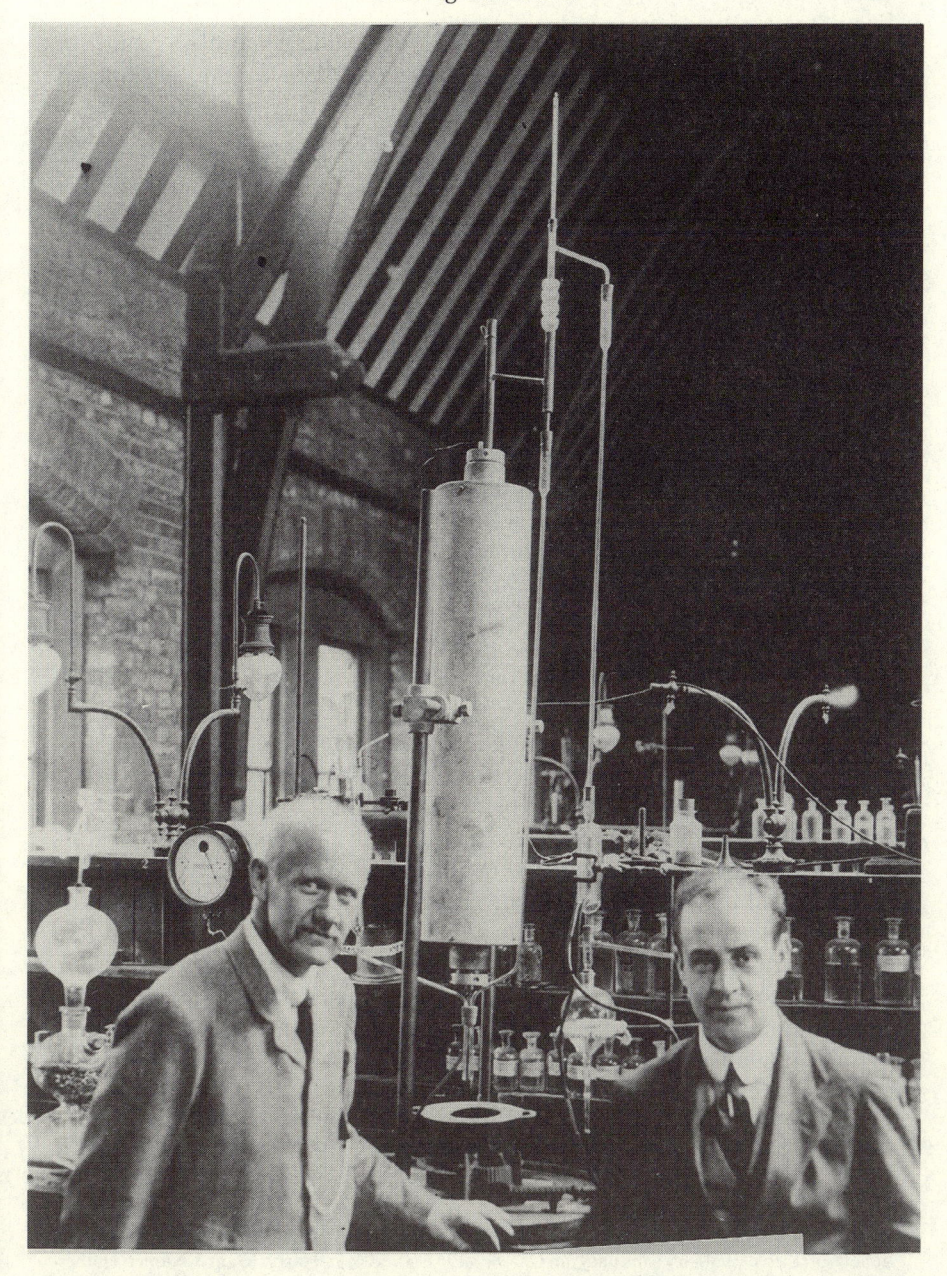

FIG. 5 N. V. Sidgwick and B. Lambert, two long-serving and sturdily independent college fellows, in the 1878 laboratory of the Old Chemistry department, 1926. Reproduced by permission of the Museum of the History of Science, Oxford.

Soddy's attempts at subjecting them and their colleges and to retain the college laboratory system. Once Soddy had made his aims clear, they quickly jettisoned their previous view in favour of a university physical chemistry laboratory because they feared Soddy would rule it as a gauleiter. Only when his retirement was imminent in 1936 did they begin to favour in public a proposal they had liked in 1920 but then quickly dropped.

Two of the college laboratories, Magdalen and Queen's, were closed in 1923 and 1934 respectively on the initiative of the two colleges. Magdalen, which was devoted to quantitative inorganic chemistry in its teaching and research, had the smallest number of researchers and of bench spaces of the five college laboratories in the early 1920s. Given the recommendation of the Asquith Commission in 1922 against college laboratories, given the extra bench spaces made available by the completion of the Dyson Perrins Laboratory that year, given the opposition of Perkin (a fellow of Magdalen) to college laboratories, and given the dire needs of botany whose professor was also attached to Magdalen, it made sense for the college to do a deal with the University in 1923 to promote botany by converting the chemistry laboratory to botanical purposes.

At Queen's there was no such discussion about how the college could serve the University better: it was simply that Chattaway, the laboratory director, intimated to the college in 1932 that he wished to retire in 1934 aged 74. Chattaway had fulfilled his impeccable pedigree. After graduating at Oxford he had taken his Ph.D. under Bamberger and Baeyer at Munich and subsequently studied with Georg Bredig at Heidelberg. An active publisher, Chattaway specialized in the preparation and purification of organic substances. He did not lead a research school; but he had research collaborators and produced a small cohort of pupils who were indebted to him for experimental ingenuity. He was an exacting teacher who insisted that practical organic chemistry was a clean subject which could be conducted in a drawing-room in evening dress. For many years he generated a good spirit in his laboratory which attracted undergraduates mainly from Queen's, Keble, and St Edmund Hall. In his final year he was assisted in the laboratory by a formidable trio of Queen's men, Parkes, Irving, and Eric James.[30]

---

[30] On Chattaway's regime, E. M. Walker to Physical Sciences Board, 11 May 1932, PS/R/1/3; *OM* 62 (1943–4), 188; three Queen's Laboratory Books, 1900–34, History of Science Museum, Oxford, MS 49; E. P. Abraham to author, 13 July 1987. Adolf von Baeyer (1835–1917) was professor at Munich 1875–1917; Eugen Bamberger (1857–1932) ended his career as professor at Zurich Technische Hochschule; Georg Bredig (1868–1944) was head of physical chemistry at Heidelberg 1901–10; George David Parkes (1899–1967), first at Queen's 1922, D.Phil. 1928, fellow of Keble 1930–65, obituary in *OM* 86 (1967–8), 106–7; Harry Munro Napier Hetherington Irving (1905–93), first at Queen's 1928, D.Phil. 1930, fellow in science at St Edmund Hall 1938–51 and vice-principal 1951–61, professor of inorganic and structural chemistry, University of Leeds, 1961–71; Eric John Francis James (1909–1992), first in chemistry, Queen's, 1931, D.Phil. 1933, knighted 1956, high master of Manchester Grammar School 1945–62, first vice-chancellor, University of York, 1962–73, obituary in *Independent*, 21 May 1992.

Chattaway and Queen's gave the University two years' notice of his retirement and of the college's desire to use the laboratory, which had twelve undergraduate and eight postgraduate bench places, for other purposes. As with the closure of the Magdalen laboratory, the demise of the Queen's laboratory involved a college decision in co-operation with the University. That was also the case with Christ Church and Balliol–Trinity whose laboratories were closed in 1941 when the new university physical chemistry laboratory was opened, but not with Jesus whose laboratory was closed in autumn 1947.

At Christ Church from 1920 Russell taught inorganic chemistry to engineering students for a time and continued his own research on radioactivity, especially on the chemistry of protoactinium, until the third storey of the laboratory was removed in 1930. The Dr Lee's laboratory was small compared with that at Jesus which, with seven rooms and almost fifty bench spaces, cost the college almost £1,000 p.a. The University paid nothing to Jesus for teaching physical chemistry on its behalf, but it did give Chapman £200 a year as director of the laboratory. Jesus took its chemical responsibilities so seriously that in 1919 it elected H. J. George as research fellow and lecturer in chemistry and subsequently promoted him to a tutorial fellowship. Jesus's teaching and Chapman's research were devoted to physical chemistry, but with an intellectual range less than that of the Balliol–Trinity laboratory. Towards the end of his long career Chapman's eccentricities became more pronounced but in the mid-1920s he and his wife were still pursuing fruitful research in their favourite field of photochemical reactions. In 1926 they pioneered the use of a rotating slotted disc to produce alternating short periods of light and dark, a technique that was subsequently widely used. In spring 1939 George died and Jesus once again showed its commitment to chemistry by quickly appointing a replacement, L. A. Woodward, an Oxonian who had found his research field by studying Raman spectra with Debye at Leipzig.[31] When Chapman retired in 1944 aged 75, Woodward ran the laboratory until it closed in 1947. It was kept alive during the Second World War because it played its part in the British contribution to the atomic bomb project: it was used for research on the separation of isotopes by gaseous diffusion. It survived after the war because it was one of Jesus's characteristic and attractive features which had ensured that one-seventh of all Jesus undergraduates read chemistry. It was closed only after much controversy about its fate.

[31] On the Jesus lab, Laidler, 'Chemical Kinetics'; Long, 'Jenkins Lab'; HCP 164 (1936), 141–4; J. N. L. Baker, *Jesus College, Oxford 1571–1971* (Oxford, 1971), 128–9, 135; Herbert John George (1893–1939), first in chemistry, Jesus, 1914, fellow of Jesus 1923–39, obituaries in *Nature*, 143 (1939), 752 by A. S. Russell and in *Journal of the Chemical Society* (1939), 1640 by Chapman; Leonard Ary Woodward (1903–76), fellow of Jesus 1939–70, took his Ph.D. at Leipzig under Petrus Josephus Wilhelmus Debye (1884–1966), professor of physics, Leipzig, 1927–35, Nobel prize-winner chemistry 1936.

8.3 The Balliol–Trinity Laboratory

The Balliol–Trinity laboratory gave the lie to Soddy's accusation that college fellows were not interested in research at the laboratory bench. The accommodation there was neither palatial nor purpose-built, but rich in improvisation. The laboratory was bizarre in that in the 1920s and 1930s it occupied a changing suite of converted cellars, washrooms, and lavatories in the two colleges. Hinshelwood was the main beneficiary of two conversions, occupying from 1921 to 1929 a former lavatory whose fittings were still visible and from 1929 to 1941 a former bath-house (see Fig. 6). In these undistinguished buildings, one of which was amazingly described by a German visitor as an institute, Hinshelwood did much of the research that led to his Nobel prize. He was content with his accommodation. As a wag commented, 'That is the worst of being brought up in a cellar; a converted bathroom is then positive luxury.'[32] Balliol and Trinity provided research accommodation not only for their own fellows but for two fellows of other colleges (Bowen, University; Raikes and his successor Wolfenden, Exeter), all Balliol men who helped with teaching. The financing of the laboratory was complicated but there is little doubt that it was subsidized by the two colleges who paid for conversion and running expenses and by the demonstrators. In return for teaching on behalf of the University all honours undergraduates who wished to do practical physical chemistry, the laboratory received from the early 1920s £600 p.a. of which £350 was given to Bowen and £150 to Raikes and then Wolfenden. The University paid Hinshelwood £200 p.a. as head of the laboratory in the 1930s, but Bell conducted practical classes for nothing.[33] All five of them received much less than if they had been college fellows who were demonstrators in a normal Oxford laboratory such as the Dyson Perrins: they were penalized financially because they were college fellows who taught in a college laboratory peopled predominantly by non-Balliol–Trinity undergraduates.

Space was so tight in the Balliol–Trinity laboratory that undergraduates and researchers often worked on adjacent bench spaces, which helped to generate a friendly family atmosphere.[34] Facilities were slender. There was a lab-boy/man but no technical assistance, no secretary, no workshop, and no glass-blowers. Everything possible was home-made in the laboratory. The

---

[32] L. A. Sutton to R. Robinson, 25 Mar. 1934, Sidgwick Papers, 79.

[33] *HCP* 164 (1936), 141–4; Hinshelwood to Margoliouth, 22 Feb. 1937, PS/R/1/5; John Hulton Wolfenden (1902–89), first in chemistry, Balliol, 1923, lecturer in chemistry at Balliol 1925–7, fellow of Exeter 1928–47, obituary in *Exeter College Association Register* (1990), 7–8; Ronald Percy Bell (1907–96), first in chemistry, Balliol, 1928, Bedford lecturer in physical chemistry at Balliol 1932–3, fellow of Balliol 1933–67, obituary in *Independent*, 18 Jan. 1996.

[34] For the facilities of the Balliol–Trinity laboratory, Bowen, 'Balliol–Trinity Laboratories'; R. P. Bell, 'Bowen', *BM* 27 (1986), 83–101 (87–8); E. F. Caldin to author, 21 Aug. 1987; C. J. Danby to author, 20 Sept. 1988; D. Murray-Rust to author, 28 Aug. 1987.

FIG. 6 Interior of the chemistry laboratory, Millard building, Dolphin Yard, Trinity College, Oxford, 1937. This converted bath-house was used by Hinshelwood from 1929 to 1941. From left to right are two undergraduates, J. E. Hobbs (Trinity) and R. M. Lewis (Magdalen), and E. Tommila, a researcher from Finland. Reproduced by permission of Professor K. J. Laidler.

apparatus was simple and light, being mainly glassware plus electrical equipment such as galvanometers. There was no large, heavy, commercially made apparatus. Bowen was perhaps the most inventive researcher. In his research on photochemistry his first source of light was an old street light from Oxford City lighting department; and in that on fluorescence he used old biscuit tins painted grey and a set of false teeth which even Hartley scoffed at. Hinshelwood used immersed electric light bulbs to heat constant-temperature baths. Even undergraduates often designed and assembled the apparatus from scratch, which gave knowledge of how apparatus worked as well as a sense of adventure. Such skills were highly prized by Hartley who at a tutors' meeting in Balliol praised one of his pupils for his brilliance because he was the laboratory's best glass-blower. It could be argued that this do-it-yourself approach, which was partly a result of shortage of money, was first-class training in research and developed initiative, which was useful for those undergraduates who aspired to membership of national élites.

Before the war Nagel, who was a non-researcher, ran the Balliol–Trinity laboratory, leaving research to Hartley. But from 1921, when Hinshelwood succeeded Nagel at Trinity and joined forces with Hartley, the new director of the laboratory, research received greater emphasis. Though Hartley had studied in Munich under Willstätter and Groth, his greatest debt was to T. W. Richards of Harvard who became his role model. Richards, with whom Hartley worked for an inspiring fortnight in 1902, was not a profound theorist but an experimentalist who focused his research on the exact determination of chemical constants and the testing of accepted or provisional generalizations. Obsessed with methods of purification, Richards gave to chemistry new standards of precision measurement and as a meticulous director of research encouraged a wide range of investigations in his laboratory from which issued a constant stream of men trained in his methods and imbued with his ideals. In the Balliol–Trinity laboratory Hartley tried to follow Richards in all these respects. In the 1920s, for example, his main research was on precision measurements concerning the electrochemistry of non-aqueous solutions. Using standardized procedures in accordance with Richards's approach, by 1930 Hartley and his researchers had tested but not contributed directly to the Debye–Hückel theory of strong electrolytes. Hartley's emphasis on procedures and precision was easily adapted to part two of the chemistry degree. As the Balliol–Trinity laboratory techniques were highly standardized, they could be learned and deliver useful results in less than a year. At the same time problems which required more technical development for their solution could be studied by postgraduates and, with a skilful supervisor such as Hinshelwood, by post-doctoral researchers.

Hartley was a busy and efficient man. Through his summer schools in physical chemistry, then neglected in schools, which he ran in the 1920s for schoolteachers, and through his activities for the Science Masters' Association, he directed able students to Oxford and Balliol and away from Cambridge. In the 1920s the Balliol–Trinity laboratory was very much a Hartley enterprise, the four senior demonstrators being Hartley and three of his former pupils (Hinshelwood; Bowen; Raikes followed by Wolfenden). At Balliol he was succeeded as chemistry fellow by two more pupils, Gatty and Bell. He left Balliol in 1930 because the war had given him a taste for high command in industrial chemistry, a taste which could not be exercised at Oxford. Instead he pursued his interests in applied science through consultancies and especially a directorship of the Gas, Light, & Coke Company. He was therefore persuaded by Lord Stamp to join the London, Midland, & Scottish Railway, then in the throes of internal struggles, to organize its research.[35] As a man of action who after the First World War appeared in Oxford still wearing his army uniform and styled himself General Hartley, he wanted a large and new field of operation. Having acquired an FRS in 1926 and a knighthood in 1928, Hartley probably realized that he could go no further as an academic chemist.

Hartley was adept after the first war in placing his Balliol pupils as fellows in other colleges, his successes being at Exeter with Raikes (1919), at Trinity with Hinshelwood (1921), at University with Bowen (1921), and again at Exeter with Wolfenden (1927). This Hartley mafia was consolidated by Hinshelwood who placed his own Trinity pupils as fellows in other colleges besides ensuring that one of them, J. D. Lambert, replaced him at Trinity in 1937: in 1930 Harold Thompson succeeded Applebey at St John's and in 1939 Staveley followed Walden at New College. In both cases a physical chemist replaced a general or inorganic chemist. Thus by 1939 members of the Balliol–Trinity axis, focused on physical chemistry, occupied not just the Dr Lee's chair of chemistry but also six college fellowships. Only Sidgwick at Lincoln rivalled Hartley and Hinshelwood in placing his protégés in college fellowships with Hammick at Oriel (1921), Sutton at Magdalen (1936), and Woodward at Jesus (1939).

Like Richards Hartley did not attempt to pull the whole of his laboratory behind him to work on his pet research field. Instead he encouraged work on kinetics by Hinshelwood, on photochemistry by Bowen, and on acid-base catalysis by Bell. In kinetics Hinshelwood was continuing the Oxford

---

[35] On Hartley, A. G. Ogston, *BM* 19 (1973), 349–73; Ogston interview with author 9 Sept. 1987; E. J. Bowen, *OM* 91 (1972–3), 7–8; R. P. Bell to author, 16 July 1987, interview 30 July 1987; *Nature* 109 (1922), 56–7; 119 (1927), 102–3; H. Hartley, 'The Theodore William Richards Memorial Lecture' (1929), in *Memorial Lectures Delivered before the Chemical Society 1914–1932* (London, 1933), 131–63; Richard Willstätter (1872–1942) worked under Baeyer at Munich 1896–1905; Theodore William Richards (1868–1928), professor of chemistry, Harvard University, 1901–28, Nobel prize 1915, on whom see S. J. Kopperl, 'T. W. Richards' Role in American Graduate Education in Chemistry', *Ambix*, 28 (1976), 165–74; Oliver Gatty (1907–40), first in chemistry, Balliol, 1930, fellow of Balliol 1931–3; Josiah Charles Stamp (1880–1941), *DNB*, chairman of London, Midland, & Scottish Railway, 1926–41.

tradition begun by Harcourt. Apart from military service in the First World War, which aroused his interest in explosive chemical reactions, his whole career was spent at Oxford as Balliol undergraduate, Balliol research fellow 1920–1, fellow of Trinity 1921–37, and Dr Lee's professor of chemistry 1937–64. Much of the work on kinetics for which he received a Nobel prize was done between the wars. This bald statement ignores the key to Hinshelwood's success which was his fertility in developing six subfields in the space of sixteen years. In the 1920s he launched research on surface reactions (1921), unimolecular reactions (1924), and gaseous explosions (1927) which he explained with his important notion of branching chains. In 1930 he began work on reactions in solutions and on the mechanism of composite reactions. Just before the second war (1937) he embarked on the difficult task of examining the kinetics of bacterial growth. In much of this research he was assisted by Oxford undergraduates who worked with him for part two of their chemistry degree on manageable topics which fitted into his large purposes. Of his 48 research collaborators between the wars, 22 were Trinity undergraduates, with 2 from Balliol and 10 from other colleges. Hinshelwood was able to draw publishable work even from two undergraduates who obtained thirds in finals. Thus he exploited a unique local feature (the research year in the chemistry degree) in an unusual site (a college laboratory) to produce work of international calibre. He attracted to this locale six Rhodes scholars, including two with Ph.D.s, who took a D.Phil. or B.Sc. under him. On a couple of specific topics, he enjoyed the services of post-doctoral workers, Moelwyn-Hughes and Arthur Williamson, who worked on solution reactions and gaseous explosions respectively. He took on very few European collaborators, the chief one being Clusius.[36] Hinshelwood received many applications from foreigners to work with him but his limited accommodation in converted lavatories and bath-houses gave him the perfect excuse for rejecting most of them. In any case Hinshelwood

---

[36] For Hinshelwood's research, Laidler, 'Chemical Kinetics', esp. 252; C. N. Hinshelwood, *The Kinetics of Chemical Change in Gaseous Systems* (Oxford, 1926; 2nd edn. 1929, 3rd edn. 1933); id., *The Kinetics of Chemical Change* (Oxford, 1940); id. and A. T. Williamson, *The Reaction between Hydrogen and Oxygen* (Oxford, 1934). His Rhodes scholars were: Robert Emmett Burk, D.Phil. 1926, later polymer chemist with du Pont; Forrest Fairbrother Musgrave, D.Phil. 1933, later manager, overseas department of Albright & Wilson; Frank Henry Verhoek, Wisconsin Ph.D., D.Phil. 1935, later professor of chemistry, Ohio State University; Edgar William Trim (1915–42), B.Sc. 1938; John Gordon Davoud, D.Phil. 1941, later with Courtaulds and the Firestone Company, USA; Carl Arthur Winkler (1909–78), McGill Ph.D., D.Phil. 1939, later professor of chemistry, McGill University, Montreal. Emyr Alun Moelwyn-Hughes (1905–78), obituary in *Nature*, 277 (1979), 334, was in Oxford 1930–3 as an 1851 Exhibition senior student and made his career in Cambridge from 1934. His *The Kinetics of Reactions in Solution* (Oxford, 1933) reveals his debt to Hinshelwood. Arthur Williamson studied with Hinshelwood as an international research fellow and returned to Canada to work in industry. Klaus Clusius (1903–63), Ph.D. 1928 from Breslau Technische Hochschule, in Oxford 1929–30 as a Rockefeller research fellow, worked at the universities of Munich and Zurich. Hutchison, *Autobiography*, 32–42, describes his unforgettable experience as a part two undergraduate with Hinshelwood 1925–6.

was a chemist who believed that the individual was the source of the greatest scientific innovations, so he preferred working with a small group of congenial collaborators: large-scale machine research was anathema to him. Temperamentally he was reserved, reticent, and private, a lone wolf who did not communicate freely and widely with his scientific colleagues at Oxford. He did not attend colloquia or lectures by visiting scientists. Generally he avoided office-holding like the plague before 1939 (though after 1945 he succumbed) and did not covet a chair at Oxford. He was a reluctant candidate for the Dr Lee's chair, only accepting it in January 1937 to prevent Norrish of Cambridge from moving to Oxford. In the 1920s and 1930s he preferred to make his mark in Oxford by his research publications rather than through wide personal contacts. A bachelor, he was comfortable in his Trinity fellowship and laboratory, even continuing as Dr Lee's professor to live in Trinity, and not Exeter to which his chair was attached, on the grounds that he needed to be near the Balliol–Trinity laboratory and that his cats would not tolerate the short move from Trinity to Exeter.

Hinshelwood also started the tradition of Oxford college fellows who were physical chemists writing synthetic works which summarized their research and, as a delegate of Oxford University Press from 1934, he enabled it to flourish. He showed the way in 1926 with his first book on the kinetics of chemical change which deliberately avoided encyclopedic compendiousness by focusing on the clear exposition of general principles illustrated by selected examples of different types of reaction. He was followed by Thompson on spectroscopy (1938), Bell on acid-base catalysis (1941), and Bowen on photochemistry (1942). None of them, however, had such a wide range as Hinshelwood whose penchant for synthesis reached its culmination in his book on *The Structure of Physical Chemistry* in which he made strong claims about physical chemistry as a liberal education, and about science as a humanistic enterprise, in ways which were congenial to his Oxford milieu. As a firm believer in 'a liberal occupation with wide studies', Hinshelwood emphasized the structure and continuity of the whole of physical chemistry and examined its various parts, not as a series of specialized topics, but in their relation to one another. That task in his view involved artistic judgement. In a parallel way Hinshelwood argued that, because science was basically the attempt by the human mind to order facts into satisfying patterns, the imposition of design on nature was an act of artistic creation. For Hinshelwood this view of science as 'a construction of the human mind' was totally compatible with his own experimentalism in the Balliol–Trinity laboratory because some parts of the construction were closely related to 'things of direct experience and observation'.[37]

[37] H. W. Thompson, *A Course in Chemical Spectroscopy* (Oxford, 1938); R. P. Bell, *Acid-Base Catalysis* (Oxford, 1941); E. J. Bowen, *The Chemical Aspects of Light* (Oxford, 1942); C. N. Hinshelwood, *The Structure of Physical Chemistry* (Oxford, 1951), pp. v, 3, 110; J. H.

For almost two-thirds of the twentieth century Balliol enjoyed the services of just two chemistry fellows, Hartley and his pupil Bell. Like Hinshelwood Bell was a beneficiary of Hartley's flexible attitude to the research fields adopted and pursued by his Balliol pupils. For his part two Bell worked on interfaces, an interest which led Hartley to suggest that he should go to study at Copenhagen under Brönsted whose set-up had been praised by Wynne-Jones, a previous Hartley pupil who had spent a year there. Having met Brönsted in Oxford at a Faraday Society meeting, Bell then spent four years in Copenhagen where Brönsted fostered his interest in acid-base catalysis and Bohr's colloquia held in the adjoining Institute for Theoretical Physics launched his interest in quantum theory and its application to chemical kinetics.[38] Bell was funded by a university studentship and by the Goldsmiths' Company with which Hartley had close contacts. Bell's return to Balliol in 1932 as Bedford lecturer in physical chemistry was arranged by Hartley. On Gatty's resignation as chemistry fellow in 1933, Bell was on hand to replace him.

Three of Hinshelwood's pupils who became college fellows (Thompson, J. D. Lambert, and Staveley) followed Bell in finding their research feet through postgraduate experience abroad or outside Oxford. Thompson's initial research under Hinshelwood on kinetics was so promising that, before he graduated in summer 1929, St John's elected him to a research fellowship for 1929–30 to enable him to study in Berlin under Fritz Haber and to be a paying guest with Max Planck, to gain a Ph.D., and to return in 1930 as tutorial fellow. Initially he continued his work on reaction kinetics but soon changed to visible and ultraviolet spectroscopy. This move was encouraged by Hinshelwood and supported by a Royal Society grant but was opposed by Soddy in whose old chemistry department Thompson worked. To learn the new techniques Thompson drove twice a week to Imperial College, London, to work with Alfred Fowler. By the late 1930s Thompson was also working on infra-red spectroscopy and, with the aid of a Leverhulme travelling fellowship, spent a year at the California Institute of Technology learning the latest techniques from R. M. Badger.[39] In finding spectroscopy

Wolfenden, *Numerical Problems in Advanced Physical Chemistry* (London, 1938), which focused on useful procedures, was not a synoptic survey of a research field though it was based on research papers.

[38] Bell (FRS 1944) to author, 16 July 1987, interview 30 July 1987; R. P. Bell, lecture, University of Canterbury, typescript, 1974; Bell, 'Chemical Anecdotes', typescript, n.d; Bell repayed his debt to Brönsted by translating and adapting his *Physical Chemistry* (London, 1937); Johannes Nicolaus Brönsted (1879–1947), professor of physical and inorganic chemistry, University of Copenhagen, 1908–47; William Francis Kenrick Wynne-Jones (1903–82), B.Sc. Oxon. 1925, ended his career as professor of chemistry, University of Newcastle upon Tyne, 1947–68.

[39] Harold Warris Thompson (1908–83), fellow of St John's 1930–75, reader in infra-red spectroscopy 1954–64, professor of chemistry 1964–75, obituary by R. Richards, *BM* 31 (1985), 573–610; Fritz Haber (1868–1934), Nobel prize-winner 1919, director of Kaiser Wilhelm

as his enduring research field and in achieving his FRS in 1946, Thompson was greatly indebted to experience in London and Pasadena. J. D. Lambert, who succeeded Hinshelwood as chemistry fellow at Trinity in 1938, graduated in 1934, worked for two years as an assistant to Hinshelwood paid for by a Royal Society grant, and with Hinshelwood's help then spent 1936–7 on a university studentship working at Göttingen under Eucken who introduced him to the then new field of molecular energy transfer in gases studied by using ultrasonic sound, which became his enduring research interest.[40] On his return he was made lecturer in chemistry at Trinity for a year and then fellow. Another Hinshelwood pupil, L. A. K. Staveley, who graduated in 1936, worked for a year with Hinshelwood who encouraged him to launch out into a line of research not then pursued in Oxford.[41] Financed by a grant from the Goldsmiths' Company, Staveley spent 1937–8 at Munich studying under Klaus Clusius, who had spent a year in Oxford with Hinshelwood in the late 1920s, from whom he learned the techniques of experimental chemical thermodynamics at low temperatures, which remained his lifelong research field. On his return to Oxford Staveley was made lecturer in chemistry at New College for a year and then in 1939 tutorial fellow. The cases of Lambert and Staveley are instructive. Both found it possible and advantageous to visit Nazi Germany to gain essential techniques in an aspect of physical chemistry then ignored in Oxford, and at a time when Jewish physicists and mathematicians were leaving the Third Reich in droves.

The Hartley–Hinshelwood tradition of encouraging prospective college fellows to study abroad was also adopted by Sidgwick who used his friendship with Debye to send to Leipzig in autumn 1928 two of his Lincoln pupils, Woodward and Sutton. Woodward went to Leipzig on a DSIR senior research grant.[42] Having gained his doctorate in 1931, Woodward took three jobs outside Oxford before returning to it in 1939 as chemistry fellow at Jesus where he pursued Raman spectroscopy as his main interest. A similar interest in structure led Sutton in 1928–9 to spend six months in Leipzig on a DSIR studentship learning from Debye the techniques of measuring dipole moments which revealed the detailed structure of molecules. On his return to Oxford Sutton worked for his D.Phil. and ran Sidgwick's research group

Institut für Physikalische Chemie und Electrochemie, Berlin, 1911–33; Max Carl Ernst Ludwig Planck (1858–1947), Nobel prize-winner 1918, had resigned in 1928 from the chair of theoretical physics, University of Berlin; Alfred Fowler (1868–1940), *DSB*, professor of astrophysics, Imperial College, London, 1915–34; Richard McLean Badger (1896–1974), assistant, associate, full professor of chemistry, Caltech, 1929–66.

[40] James Dewe Lambert (b. 1912), first in chemistry, Trinity, 1934, fellow of Trinity 1938–76; Lambert to author, 10 May 1988; Arnold Thomas Eucken (1884–1950), professor of physical chemistry, Göttingen University, 1930–50.

[41] Lionel Alfred Kilby Staveley (1914–96), first, Trinity, 1936, fellow of New College 1939–82; Staveley to author, 6 Apr. 15 May 1988.

[42] On Leonard Ary Woodward (1903–76), first 1926, see A. J. Downs, D. A. Long, and L. A. K. Staveley, *Essays in Structural Chemistry* (London, 1971).

when he was in the USA in 1931. With the backing of Sidgwick and Robinson, he was elected a fellow by examination in organic chemistry by Magdalen in 1932. Again supported by Sidgwick and Robinson, Sutton spent 1933–4 on a Rockefeller fellowship working in Linus Pauling's group at the California Institute of Technology where he learned the then novel techniques of electron diffraction from Lawrence Brockway.[43] On his return he pursued work on molecular structure, using the two techniques of dipole moments and electron diffraction, both of which he had learned abroad. In 1936 he was promoted to chemistry fellow at Magdalen. The careers of Bell, Thompson, Lambert, Staveley, Woodward, and Sutton make it abundantly clear that it is wrong to assert, on the basis of the admittedly spectacular case of Hinshelwood, that physical chemistry at Oxford was entirely indigenous. That was not so. Most Oxford graduates who became fellows of colleges in the 1930s learned about or discovered their enduring field of research and the associated techniques while undertaking postgraduate study under acknowledged experts in London, in Denmark, in Germany even during the Third Reich, and in the USA at the California Institute of Technology. In sharp contrast to the general insularity of mathematics at Oxford, physical chemistry was constantly diversified by topics and techniques imported mainly from abroad.

It was the success of physical chemistry which led to the redefinition of the Dr Lee's chair in 1936–7 once Soddy had intimated his resignation effective at the end of 1936. That redefinition reflected in part the power of the Balliol–Trinity and Lincoln axes, controlled by Hinshelwood and Sidgwick who put their men, mainly physical chemists, into college fellowships. But it was more than just a masonic operation at a time when fellowships were rarely advertised. In the Oxford context physical chemistry enjoyed perceived advantages over inorganic chemistry, advantages which also kept at bay challenges from organic chemistry. Physical chemistry had wide scope: it often dealt with general processes involving many substances, discovered general empirical laws, and created general theories. In contrast inorganic and organic chemistry were limited: they were piecemeal, descriptive, and often merely specific; they could be dismissed as cookery and smells. Consequently physical chemistry was more amenable to comprehensive generalization and intellectual synthesis. Oxford's physical chemists wrote books as

---

[43] Leslie Ernest Sutton (1906–92), first 1928, fellow of Magdalen 1936–73, FRS 1950, obituary by D. H. Whiffen, *BM* 40 (1994), 369–82; Sutton to author, 27 Nov. 1987; L. E. Sutton, 'The Earlier Studies in Great Britain of the Structure of Molecules in Gases and Vapours by Electron Diffraction, with an Epilogue', in P. Goodman (ed.), *Fifty Years of Electron Diffraction* (Dordrecht, 1981), 92–100; Sutton, account of life, Sutton Papers, A2; Lawrence Olin Brockway (1907–79), post-doctoral researcher, Caltech, 1933–7, Guggenheim fellow 1937–8 spent at Oxford and Royal Institution, London, was professor of chemistry, University of Michigan, 1938–79; Linus Pauling (1901–94), professor of chemistry, Caltech, 1931–64, Nobel prize chemistry 1954.

well as papers, whereas their colleagues in inorganic and organic chemistry merely published papers. Oxford's physical chemists in some ways were what Harwood has called comprehensive mandarins who strove for wide conceptual syntheses.[44] As such they were more appropriate agents of the Oxonian aims of a liberal education which would produce a national élite and so was their subject. In terms of college teaching it was claimed with some justice that physical chemists could teach inorganic and organic chemistry but that inorganic and organic chemists could not teach physical chemistry. As long as comprehensive teaching was seen as a primary function of college fellows, that was a significant difference. Physical chemistry was often mathematical in its language and deductive in its arguments. As such it was a better mind-trainer as part of a liberal education than inorganic and organic chemistry which were often merely factual and required only a good memory. In the laboratory, too, physical chemistry could claim an advantage. At Oxford between the wars there was a strong emphasis on precision measurement which the organic chemists eschewed and the inorganic chemists gave less attention to. Through its concerns with instrumentation, measurement, and purification of materials, physical chemistry trained practical chemists effectively. In all these ways physical chemistry was, or was represented as, the most appropriate and flexible of the three branches of chemistry for producing liberally educated chemists who would become leaders in universities, schools, and industry.

As soon as Soddy had given notice of his impending resignation of the Dr Lee's chair, the University took the advice of the Board of Physical Sciences, the subfaculty of chemistry, and an advisory committee of Sidgwick, Hinshelwood, and Chapman. It was quickly agreed that in the new appointment the importance of physical chemistry should be emphasized by appointing to the chair a physical chemist who would be responsible for both physical and inorganic chemistry. Hinshelwood was clearly the man the electors had in mind and he was elected in January 1937. The only dissident was Soddy who objected publicly to the restriction of the chair: as an inorganic chemist he thought inorganic chemistry was the basis of chemistry in all its forms, including industry, a point strenuously rebutted by Sidgwick. Soddy also stressed that appointing a physical chemist to the chair would not solve the problems of control and of accommodation in physical chemistry: the new Dr Lee's professor would have no university laboratory to run and could only secure control by encroaching on the Balliol–Trinity and Jesus laboratories.[45]

---

[44] J. Harwood, *Styles of Scientific Thought: The German Genetics Community 1900–1933* (Chicago, 1993).
[45] For the revamping of the Dr Lee's chair, *HCP* 164 (1936), 141–4; 165 (1936), 125–7; Soddy document, 13 Nov. 1936, UR/SF/CHE/1B; UDC/M/41/1, 20 Nov. 1936, 16 Jan. 1937; Sidgwick in *OM* 55 (1936–7), 218, Congregation debate reported 244–5, Soddy's objections 307–8.

The University's response was twofold. First, Hinshelwood was elected Dr Lee's professor under *ad hoc* arrangements. Though technically a professorial fellow of Exeter, he remained the senior science fellow of Balliol and Trinity, had the same authority in their laboratory as before, and was to retain his rooms in Trinity as long as he was in charge of it. Secondly, as soon as Soddy was out of the way, the proposal for a new university physical chemistry laboratory was revived on the alleged grounds that 'while the fertilising inspiration of a new subject does not depend upon the wealth of material equipment, the less sensational development of it does'.[46] In a remarkable volte-face, the college physical chemistry laboratories were suddenly seen as makeshift, poorly endowed, and limited in equipment and accommodation. By late 1937 Lord Nuffield had been persuaded to divert £100,000 to pay for a new physical chemistry laboratory and ICI had promised £10,000 for maintenance. In spite of the difficulties caused by the run-up to the war and the war itself, a university physical chemistry laboratory, for which Soddy had campaigned so fruitlessly, was ready for occupation in the darkest year of the war, 1941, and the Balliol–Trinity laboratory was closed but not forgotten. Its contribution to physical chemistry at Oxford was commemorated by the arms of the two colleges carved on the front of the new building.[47]

8.4 The Mancunian Inheritance

The arrival in Oxford of W. H. Perkin, junior, as Waynflete professor of chemistry in 1913 was a decisive event in the University's history. For the first time Oxford had head-hunted a provincial university professor of science, with no previous connection with Oxford, who was renowned as a researcher himself, as a research school leader, as a fund-raiser for new laboratories which he helped to design and plan, and as a disciple of Adolf von Baeyer with whom he had worked in Munich from 1882 to 1886. As the leading British organic chemist of his generation, he was lured to Oxford from Manchester specifically to boost research and was so successful that his two successors, Robert Robinson and Ewart Jones, were his chemical son and godson respectively.[48] Given Oxford's inbuilt conservatism, this policy of poaching established non-Oxonian researchers from provincial universities, though subsequently so commonplace as to be notorious, was not only novel before the First World War but also prospectively disruptive.

It was first implemented through Perkin for several reasons. First in 1900 the University had shown that it was prepared to outflank, though not

---

[46] Vice-chancellor's oration autumn 1937, *OUG* 68 (1937–8), 27.
[47] Barrow and Danby, *Physical Chemistry Laboratory*, 76.
[48] Sir Ewart Ray Herbert Jones (b. 1911), Waynflete professor of chemistry 1955–78.

to replace or dismiss, an embarrassingly weak science professor when Townsend, first occupant of the new Wykeham chair of experimental physics, was appointed to remedy the failings of Clifton, the professor of experimental philosophy. Secondly, the bringing together of the college chemistry laboratories in 1904 under Baker's scheme was a public confession of the deficiencies of the University's chemistry laboratory run by Odling, the ageing Waynflete professor of chemistry. Thirdly, in 1908–9 the Hebdomadal Council was so alarmed by Odling's performance that it took the unprecedented step of consulting metropolitan and provincial university professors about how to organize a subject spread between one university laboratory and five college ones. Though the replies were not always harmonious, most of the advisers were trenchant about Oxford's weakness in research. From Manchester Perkin's professorial colleague Dixon advocated the Mancunian model of two professors, a centralized department, and the aim of advanced research. From the Central Technical College, London, H. E. Armstrong denounced as a national disgrace Oxford's failure to count as 'a school of chemical research'. From University College London, Ramsay (a Nobel prize-winner) agreed with Dixon that a professor of organic chemistry was required.[49] Having consulted Oxford's college fellows, who praised the Balliol–Trinity laboratory to the skies, the Council decided that when Odling retired his chair should be filled by an organic chemist, that at some time in the future a chair for inorganic chemistry should be established from the ancient endowment of Dr Lee, and that a new university laboratory, which would work harmoniously with the college laboratories, was required. Meanwhile Poulton had been urging that Oxford should give more attention to being an imperial and international university via research and in 1909 Lord Curzon, chancellor of the University, called on it to give greater attention to research and to postgraduate training. The result of such sustained pressure was that when Odling retired in 1912 the University quickly found £15,000 and a site for a new chemistry laboratory and within a week appointed Perkin, an organic chemist whose forte was research on natural products, as his successor.[50]

It would be misleading to suggest that, when Perkin assumed his Oxford chair in 1913, organic chemistry there was negligible. He inherited several

---

[49] Chemistry. Proposed Reorganisation of Department. 1908–9, OUA, UR/SF/CHE/2; Curzon, *University Reform*, 186, 214; E. B. Poulton, 'The Reform of Oxford University', *Nature*, 80 (1909), 311–12; Harold Baily Dixon (1852–1930), professor of chemistry, University of Manchester, 1886–1922; Henry Edward Armstrong (1848–1937), professor, Central Technical College of the City and Guilds of London Institute, 1884–1911; William Ramsay (1852–1916), Nobel prize-winner 1904, professor, University College London, 1887–1912.

[50] On Perkin, J. Greenaway, J. F. Thorpe, and R. Robinson, *The Life and Work of Professor William Henry Perkin* (London, 1932); R. Robinson, *Journal of the Society of Chemical Industry*, 48 (1929), 1008–12; J. B. Morrell, 'W. H. Perkin, jr., at Manchester and Oxford: From Irwell to Isis', *Osiris*, 8 (1993), 104–26; W. H. Perkin, 'Baeyer Memorial Lecture', in *Memorial Lectures Delivered before the Chemical Society 1914–1932*, 47–73.

demonstrators, whose job was to lecture to undergraduates and to run the laboratory classes for them. Three of these demonstrators were competent organic chemists, namely, James Ernest Marsh, Frederick Chattaway, and Nevil Sidgwick, all of whom had enjoyed postgraduate experience in Germany.[51] In the 1890s Marsh, a pupil of Kekulé's at Bonn, had worked on camphor and the related group of terpenes; he produced a structural formula for camphene which soon suffered eclipse, though he continued to advocate it. Most chemists supported the formula for camphor advanced in 1893 by Julius Bredt, whose views were soon confirmed by the research of Perkin and Thorpe at Manchester on camphoronic and camphoric acids. Thus when Perkin arrived in Oxford, Marsh was eclipsed as the expert on terpenes, and in order to save his credibility he soon moved from laboratory demonstration to teaching the medical and historical aspects of chemistry. Sidgwick, a fellow of Lincoln, preferred interpreting other people's results to experimenting at the bench on a single group of organic compounds such as terpenes or sugars—the kind of procedure that could provide a basis for a research school. His forte was weaving the threads of others' discoveries into an ordered and harmonious pattern, his first major effort in this genre being his *Organic Chemistry of Nitrogen* (1910), which created an accessible and interesting subject out of what had previously been a jumble. Indeed, Sidgwick deprecated Perkin's style of organic chemistry as limited, narrow, and old-fashioned, while Perkin thought that Sidgwick's physical approach to organic chemistry was bogus. Thus, as J. C. Smith stressed, when Perkin arrived in Oxford in 1913, its organic chemistry was elegant and intelligent but not dominating—and quite unlike its counterpart in Manchester under Perkin.

Perkin accepted the Waynflete chair, which he occupied until his death, on the understanding that his research time would not be interrupted by low-level teaching or shredded in other ways, his aim being to build up a school of research as rapidly as possible. For Perkin research, not teaching, was the basis of the reputation of any university school of science. Unlike Odling, he set a grand personal example, appearing and working in the laboratory six or seven hours a day and maintaining a flow of important publications. In 1916 his famous paper on the alkaloids cryptopine and protopine, which ran to 214 pages, took up a whole number of the *Journal of the Chemical Society*. Even his undergraduate lectures were research-

---

[51] James Ernest Marsh (1860–1938); F. Soddy, 'Marsh', *ON* 2 (1938), 549–56; Greenaway *et al.*, *Perkin*, 48–62; Nevil Vincent Sidgwick (1873–1952); H. T. Tizard, 'Sidgwick', *ON* 9 (1954), 237–58; L. E. Sutton, 'Sidgwick', *Proceedings of the Chemical Society* (1958), 310–19; N. V. Sidgwick, *The Organic Chemistry of Nitrogen* (Oxford, 1910) assumed *contra* Perkin that organic chemistry could not be adequately treated without reference to those aspects of physical chemistry which it involves; J. C. Smith, 'The Development of Organic Chemistry at Oxford', 2 vols., unpublished typescript, no date, i. 14. Copies of this valuable work are available in the Dyson Perrins Laboratory, Oxford, and the Royal Society of Chemistry, London.

oriented, with their stress on original sources, and intended mainly for research students; but his top priorities remained research at the laboratory bench and building up a research school. Well aware of the danger of a British university professor becoming 'an academic fossil and unproductive', he emphasized the centrality of 'output of research', and soon Oxford's publications in organic chemistry began to surge at Manchester's expense.[52]

Perkin's research orientation was not always happily received at Oxford. In 1921 his chemical colleagues defeated his and Hope's proposal that there be no lectures between 10 and 11 a.m. in order to give three hours of uninterrupted laboratory work in the morning; and Oxford's tradition of using the time between lunch and tea for various forms of muscular recreation on river and pitch meant that he never implemented for undergraduates a Baeyer regime of ten continuous laboratory hours from 8 a.m. to 6 p.m. But in other ways he was successful in pursuing his vision of research as 'the thing'.[53] He led the successful campaign which induced the University in 1916 to add a fourth year of research (part two) to the existing three-year degree course (part one) in chemistry. Believing that the German emphasis on research at the doctoral level was worth copying, he was influential in connection with the introduction in May 1917 of the D.Phil. degree and the associated creation of the category of advanced student. Thus by summer 1917 Perkin had available *in potentia* a route and rewards for aspiring young Oxford organic chemists: they could do a year's research for the award of a bachelor's degree and, if promising, continue with research for two to three years for a D.Phil. It was characteristic of Perkin, whose marriage was childless, that in his will he made provision for the bulk of his estate (£43,000 gross) to revert after his wife's death to Magdalen College, to which his chair was attached, for research studentships to promote postgraduate research in organic chemistry at Oxford by male graduates from Commonwealth universities and to enable Oxford graduates to visit Commonwealth universities.

[52] Greenaway et al., *Perkin*, 28; Perkin, evidence to the Asquith Commission, 28 Sept. 1920, MS Top. Oxon. b.109; W. H. Perkin, 'Cryptopine and Protopine', *Journal of the Chemical Society*, 109 (1916), 815–1028; id., 'The Position of the Organic Chemical Industry' (Presidential Address, Mar. 1915), *Journal of the Chemical Society*, 107 (1915), 557–78 (561); W. P. Wynne, 'Universities as Centres of Research', *Journal of the Chemical Society*, 127 (1925), 936–54 (939), which used the following major categories: Oxford, Cambridge, Manchester, Imperial College, London, provincial universities other than Manchester, and university colleges; Hutchison, *Autobiography*, 28.

[53] Perkin, 'Organic Industry', 569; Chemistry Sub-faculty Minute Book 1910–27, 14 Nov. 1921, Museum of History of Science, Oxford, MS Museum 136; W. Baker recalls that, in the letter Perkin wrote to him in 1926 offering a demonstratorship at Oxford, Perkin added, 'Research is the thing' (personal communication, 15 Jan. 1987); Edward Hope (1886–1953), obituary in *Journal of the Chemical Society* (1953), 3730–2, came to Oxford in 1916 to join British Dyes team, elected fellow of Magdalen 1919, ill from 1925.

The war also enabled Perkin to introduce into Oxford industrial research, an activity that some of his chemical colleagues regarded as a temporary necessity but one he had long advocated and practised. His views were clearly set out in March 1915 in his second presidential address to the Chemical Society. He attacked the decadence of the British organic chemical industry for its rule-of-thumb methods, its assumption that the dyestuffs industry was like the large-scale heavy alkali industry, its neglect of the importance of research in organic chemistry, and its obsession with short-term commercial interests. At a time of national emergency the answer was to establish contact and co-operation on the German model between industry and universities to mutual advantage, distasteful though that might be to some British academics.

During the war Perkin worked for the Department of Explosives Supplies on making acetone from alcohol, for the Air Board on non-inflammable rubber coating for airships, and on mustard gas; but his chief contribution was to encourage in his laboratory industrial research done by workers employed by chemical firms, especially British Dyes, with which Perkin was strongly connected. Not surprisingly, British Dyes carried out industrial research at Oxford longer (from 1916 to 1925) than Boake, Roberts & Company (eight years) or W. J. Bush & Company (six years). When the Oxford colony began to run down in 1922, Perkin maintained his industrial research interests by being made in 1923 adviser to the research staff of the British Dyestuffs Corporation at Blackley, Manchester, which he visited almost weekly for two years. In a university notoriously suspicious of manufacturing, Perkin had shown that academic work and industrial research could occur side by side to mutual advantage. Yet the success of British Dyes' Oxford colony relied far more on young chemists recruited from outside Oxford than on Oxford graduates. Of the leading or enduring members of the colony, only F. A. Mason was an Oxonian. William Kermack came from Aberdeen University; George Clemo from a Cornish school; Joseph Kenyon from Blackburn Technical College; and Edward Hope from Manchester University. Only Hope found permanent employment at Oxford, the remainder becoming heads of department and professors elsewhere.[54]

[54] For the first use of 'industrial research' in a published Oxford document see W. H. Perkin, 'Report of the Waynflete Professor of Chemistry, 1916', *OUG* 47 (1916–17), 556–7; id., 'Organic Industry', 563; Greenaway *et al.*, *Perkin*, 32–3; Reader, *ICI*, i. 266–75; William Ogilvy Kermack (1898–1970) ended his career as professor of biochemistry at the University of Aberdeen; J. N. Davidson, F. Yates, and W. H. McCrea, 'Kermack', *BM*, 17 (1971), 399–430; Joseph Kenyon (1885–1961) ended as head of chemistry at Battersea Polytechnic, London; E. E. Turner, 'Kenyon', *BM* 8 (1962), 49–66; George Roger Clemo (1889–1983) ended as professor of organic chemistry and head of department at Armstrong College, Newcastle upon Tyne, now the University; B. Lythgoe and G. A. Swan, 'Clemo', *BM* 31 (1985), 165–86; Frederick Alfred Mason (1888–1947), first 1909, gained a Munich doctorate in 1912 under Dimroth in Baeyer's department, worked for British Dyestuffs Corporation 1916–26, lectured at Manchester College

As soon as Perkin appeared in Oxford in January 1913, he attended to the question of the promised new laboratory, which was to be devoted entirely to organic chemistry. He convinced the University that the architect should be Paul Waterhouse, the trusted designer of the 1909 Morley Laboratories at Manchester: he would be cheaper than anyone else because the various measurements made for the Morley Laboratories could be used in Oxford. Perkin even invited Oxonians to Manchester in February 1913 to gain a 'pretty accurate idea of what the suggested accommodation really amounts to'. As before, Perkin collaborated with his architect on the design. The new Oxford laboratory copied the Morley Laboratories in several respects: in height, in the professorial eyrie on the top floor from which Perkin could look down into the main teaching laboratories, and in the public-lavatory style of walls, lined with hard-glazed brown and cream bricks, which Perkin regarded as a necessity, not a luxury.

As at Manchester, Perkin was able to attract external funding for the new building. By summer 1915 he had secured from C. W. Dyson Perrins, after whom the Oxford laboratory was named, a total of £30,000 to supplement the £15,000 available from the University. Dyson Perrins provided £5,000 for the building, £5,000 for equipment, and a permanent endowment of £20,000, the interest on which was to be used for promoting research under the direction of the Waynflete professor only for as long as organic chemistry remained the chief subject of his chair. The Dyson Perrins Trust, used to cover maintenance costs, was of great value to Perkin because it made his laboratory less dependent for income on the University and on laboratory fees. Dyson Perrins, an Oxford graduate in law and partner in the famous Lea & Perrins Worcester Sauce firm, was a very wealthy philanthropist who was presumably impressed by Perkin's general approach and in particular by his research on the alkaloid berberine, of which Dyson Perrins's father had established the empirical formula in 1862.[55]

The Dyson Perrins Laboratory was built in two stages, the first ending with the completion of the central block and western wing in 1916, which enabled Perkin and his co-workers to move from the old Abbot's kitchen laboratory. The second stage was completed in 1922 at a final cost almost twice that of the first, which had required £20,000 for the building and

of Technology 1926–31, and ended his career as one of His Majesty's Inspectors of schools and technical colleges in chemistry and its industrial applications.

[55] Chemistry and Lee's Professorships, OUA, MR/7/2/7; New Chemistry Laboratory Papers, OUA, UM/F/4/15, quotation from Perkin to Poulton, 23 Feb. 1913; Smith, 'Organic Chemistry at Oxford', i. 15–22; *OUG* 45 (1914–15), 781, 804–5; 'Statement of the Financial Arrangements for the Completion of the Organic Chemistry Laboratory', *HCP* 118 (1921), 61–3; 'Report of the Committee on the Finance of the Departments of Chemistry', *HCP* 121 (1922), 87–93; Charles William Dyson Perrins (1864–1958), *DNB* and *Berrow's Worcester Journal* 31 Jan. 1958; Jenkinson to Curzon, 19 May 1920 (quotation), 4 Feb., 4 Mar. 1921, Brasenose College, Oxford, Archives; Paul Waterhouse (1861–1924).

£5,000 for apparatus. Perkin pressurized the University to such an extent that much of the total cost of the second stage (£45,600) was met by it, the rest (£7,500) being secured by Perkin as gifts (mainly £5,000 from the British Dyestuffs Corporation and £1,000 from Barclays Bank). The University resorted to two financial devices. First, in a remarkable act of financial obfuscation it borrowed the capital of £20,000 from the Dyson Perrins Trust with his permission and undertook to replace it with annual instalments of £1,000 payable for twenty years, without affecting at all the interest payable to the laboratory! This loan doubled, at one stroke, the University's total borrowing. Second, it borrowed £19,000 from the Special Reserve Fund. It was forced into these desperate financial moves by Perkin who was keen to alleviate the congestion in his palace of chemistry produced by the post-war increase in research students. His methods provoked dismay and wrath. He gave the University optimistic estimates about his power to raise money from chemical firms, persuaded the University to commit itself to the ambitious scheme for extending the Dyson Perrins Laboratory, and when external money failed to materialize he openly advocated that the University should have no financial reserves, borrow what it could, and 'in forme pauperis' trust to appeals. With relentless opportunism Perkin exploited the administrative confusion between the University Chest and the Hebdomadal Council.

Not until the First World War had ended did vacancies arise or new posts appear in college fellowships in chemistry. Perkin had one early success in securing the election of a Perkinian organic chemist to a college fellowship. Though Queen's predictably rewarded the loyal Chattaway with a fellowship in 1919, Magdalen elected Edward Hope to a tutorial fellowship the same year, thus breaking the monopoly of physical and inorganic Oxonian chemists. Perkin used his own position as a professorial fellow of Magdalen in favour of Hope, one of his Manchester graduates and a member of the British Dyes team. Subsequently Perkin had no success in placing his protégés in fellowships, some of which went to Oxonian physical organic chemists who worked in the Dyson Perrins Laboratory and as college fellows exerted their independence even in the laboratory for which Perkin was responsible overall. Perkin did not like physical chemistry but tolerated it if applied to organic compounds. Not surprisingly, the physical chemists tried to disguise their work by making it smell like classical organic chemistry: Hammick, for example, a demonstrator in the Dyson Perrins Laboratory, used to let 'the Old Man' have a nauseous whiff of pyridine from his bench in order to deceive him into thinking that proper organic research was being done. It is significant that Sidgwick, a long-serving demonstrator, relinquished this post in 1923; he regarded Perkin's exclusive concern with the analysis and synthesis of natural products as narrow and passé.

Another disappointment for Perkin was that the introduction of part two into the undergraduate degree course increased the grip of the college tutors on their more ambitious charges whom they usually steered into physical chemical topics for their research year. The scheme, on which Perkin laid such store, thus backfired: it did not produce a large crop of graduate organic chemists agog to take D.Phil.s in organic chemistry as Perkin protégés. After the war Perkin found only two of his organic demonstrators and researchers from among recent Oxford graduates, namely, Sydney Plant and Harry Ing. Plant had graduated under war-time conditions in 1918 from St John's and Magdalen. In 1919 he became a demonstrator while working for his D.Phil. under Hope (also appointed a demonstrator that year) and Perkin; after Hope was taken ill in 1925, Plant became the general factotum in the Dyson Perrins Laboratory though never a college fellow. In 1921 Ing began a five-year stint of demonstrating while an undergraduate. He took his D.Phil. under Perkin, left to be a research fellow at Manchester under Robinson, and eventually returned to Oxford as reader in chemical pharmacology.[56]

The majority of demonstrators in organic chemistry came from outside Oxford and often had a connection with Robinson or Manchester, John Gulland, R. D. Haworth, and Wilson Baker being cases in point. Gulland, a demonstrator in the Dyson Perrins Laboratory from 1924 to 1931, was an Edinburgh graduate who did research on alkaloids under Robinson at St Andrews and then at Manchester (1921–4). Haworth studied with Lapworth, Perkin's brother-in-law, at Manchester, where he took his B.Sc. in 1919 and his Ph.D. in 1922. Baker, a Manchester B.Sc. (1921) and Ph.D. (1924), had collaborated with Lapworth and Robinson there before coming to Oxford as a demonstrator in 1927, the result of Robinson's having a word with Perkin over dinner. The arrival of Haworth in 1921 as an 1851 Exhibition research scholar permitted Perkin to initiate at last, in 1922, a loose and intermittent lieutenant system in which some of his colleagues gave detailed supervision of research. Perkin suggested the main, though flexible, lines often as small parts of a large project concerned with synthesizing a substance such as morphine. Of his four subalterns, Hope, Clemo, Plant, and Haworth, only Plant was entirely an Oxford product. Clemo was never a demonstrator but part of the British Dyestuffs Corporation team, while Haworth's five years at Oxford were supported mainly by research studentships. Haworth and Clemo found it expedient to leave Oxford in the mid-1920s, being followed by Gulland in 1931; as distinguished professors of organic chemistry at Sheffield, Newcastle, and Nottingham, respectively,

---

[56] Sydney Glenn Preston Plant (1896–1955), obituary in *Journal of the Chemical Society* (1956), 1920; Harry Raymond Ing (1899–1974); H. O. Schild and F. L. Rose, 'Ing', *BM* 22 (1976), 239–55.

they spread the Perkin gospel. Of the imported demonstrators only Hope and Baker secured college fellowships, and Baker did not replace Chattaway as fellow in chemistry at Queen's until 1936, after Perkin's death.[57]

The college structure, the dominance of the physical chemical college fellows, and their control of their pupils' research through the part two examination ensured that in the 1920s relatively few Oxford graduates embarked upon research for a D.Phil. in organic chemistry. There was no equivalent in Perkin's Oxford period to the way in which he had nurtured the research careers of promising young undergraduates at Manchester, such as Robinson and W. N. Haworth, both of whom were to win Nobel prizes. Consequently, as with his British Dyes team, his demonstrators, and his research lieutenants, Perkin recruited many of his graduate students from outside Oxford. Among the subsequently well-known migrants were Thomas Stevens from Glasgow, who eventually followed Haworth as professor at Sheffield; William Davies from Manchester, who as professor of organic chemistry developed laboratory research at Melbourne; Osman Achmatowicz from Poland, who became a professor at Łódź Polytechnic and eventually an official in the Polish Ministry of Higher Education; Louis Fieser and Koepfli, from the United States, who became professors at Harvard and Caltech; and V. M. Trikojus, who became professor of biochemistry at Melbourne. Other graduate students came from as far as Sweden, India, and Japan.[58]

Perkin was a persuasive and efficient administrator who ruled his laboratory without putting pen to paper. In his dealings with the University he showed unscrupulous determination mediated by worldly wisdom. Perkin accepted with some grace the peculiarities of the Oxford system, which gave a professor charge of a university laboratory but which in part staffed it with college fellows who were statutorily independent of him and had as great a say as he had in the subfaculty of chemistry and in the Natural Sciences Faculty Board. Accepting the clear limitations of these arrangements, Perkin persuaded the University and Dyson Perrins to provide for him a laboratory

---

[57] John Masson Gulland (1898–1947), demonstrator in Dyson Perrins Laboratory 1924–31, professor of chemistry, Nottingham, 1936–47; R. D. Haworth, 'Gulland', *ON* 6 (1948), 67–72; Robert Downs Haworth (1898–1990), demonstrator in Dyson Perrins 1925–6, FRS 1944, professor of chemistry at Sheffield University 1939–73, worked in Oxford 1921–5 as a researcher supported by an 1851 Exhibition scholarship and an 1851 Exhibition senior studentship; E. R. H. Jones, 'Haworth', *BM* 37 (1991), 265–76; Wilson Baker (b. 1900), FRS 1946, ended his career as professor of organic chemistry at the University of Bristol 1945–65.

[58] Thomas Stevens Stevens (b. 1900), professor at Sheffield 1963–6; William Davies (1895–1966); Osman Achmatowicz (b. 1899), who worked in the Dyson Perrins 1928–30, was professor at Łódź Polytechnic 1945–53 and under-secretary of state in the Ministry of Higher Education, Poland, 1953–9; Louis Frederick Fieser (1899–1977), having taken a Harvard Ph.D. in 1924, spent 1924–5 at Frankfurt and Oxford; Joseph Blake Koepfli (b. 1904), D.Phil. 1928, researcher at California Institute of Technology 1932–74; Walter Norman Haworth (1882–1950), Nobel prize 1937, professor of chemistry, University of Birmingham, 1925–48.

that was the first major step towards the creation of the Science Area at Oxford. One secret of Perkin's persuasiveness was noted by one of his Magdalen contacts: Perkin 'often gave the impression of being only imperfectly acclimatised and of maintaining a good natured suspicion of those who professed to be bound by statutes or regulations.... He could not have fitted better into an Oxford college with its widely different associations if he had been a member of such a body for the whole of his life.'[59] Perkin's adaptability was based partly on his character, which was amiable and endearing, and partly on his wide interests in music, horticulture, hospitality, and travel, all of which mollified the suspicion that his Germanic emphasis on research engendered in some quarters. At Manchester he was such an accomplished pianist that he played duets with the violinist Adolf Brodsky, the leader of the Hallé Orchestra 1895–6 and principal of the Royal Manchester College of Music 1895–1929, who gave the first performance of Tchaikovsky's violin concerto. At his Oxford home Perkin soon removed a partition wall in order to create a long room for chamber music performances. To the end of his life he practised at his piano every day before breakfast. He believed that like organic chemistry a Beethoven sonata needed to be worked at. For diversion on a train journey he used to read the score of a string quartet. In his horticultural work he also attained as a devotee a high professional standard, specializing in flowering plants, of which many were donated to the University Parks. He kept a good cellar and was extremely hospitable, especially at Magdalen, where he gave many lunch and dinner parties. In the long vacation he and his wife regularly visited the Swiss and Italian lakes. These wide interests, allied to his pleasant, polite, and considerate character, helped to make Perkin acceptable; he was the ideal man to introduce Germanic research practices into Oxford.

Perkin was succeeded in 1930 as Waynflete professor by his favourite chemical son, Robert Robinson, who was proud to be a leading member of what he called the Perkin family of organic chemists.[60] As an undergraduate at Manchester he was inspired by Perkin's lectures to become an organic chemist in the Perkin mould. Having spent no fewer than seven postgraduate years in Perkin's department at Manchester, where they began their long collaboration, Robinson embarked on an academic odyssey that

---

[59] P. V. M. Benecke, 'Laurie Magnus: "Herbert Warren of Magdalen"', Magdalen College MS 407, p. 67. I owe this reference to Dr Brian Harrison. Perkin's affability has been stressed by Drs H. J. Stern and F. W. Stoyle, who worked with Perkin in the early 1920s (personal communications).

[60] On Robert Robinson (1886–1975), Waynflete professor of chemistry 1930–55, A. R. Todd and J. W. Cornforth, *BM* 22 (1976), 415–527 is incisive and comprehensive; R. Robinson, *Memoirs of a Minor Prophet: Seventy Years of Organic Chemistry*, i (London, 1976); T. I. Williams, *Robert Robinson: Chemist Extraordinary* (Oxford, 1990); R. Robinson, 'The Perkin Family of Organic Chemists', *Endeavour*, 15 (1956), 92–102; Lord Todd to author, 17 Nov. 1986; M. Tomlinson to author, 23 Oct. 1988; Smith, 'Organic Chemistry at Oxford'.

took him to chairs at Sydney, Liverpool, St Andrews, Manchester, and University College, London, before he settled in Oxford in 1930. In 1947 Robinson was awarded the Nobel prize for his work on plant products, especially alkaloids, a topic that he first attacked with Perkin in Manchester. Robinson's debt to Perkin may be gauged from his publications: of his first thirty papers twenty-four were written with Perkin; all told they published sixty-four joint papers. Not surprisingly as Waynflete professor Robinson was expected to continue Perkin's modes of work: it was realized in Oxford that Robinson was very much like Perkin, his great friend and teacher, in his mastery of experiment and his enthusiasm for bench work.

In several ways Robinson continued the Perkin regime, but he was more intense, more mercurial, more innovative, and less equable than his chemical father. Like Perkin's his research was focused on elucidating the structure of naturally occurring substances, especially alkaloids and colouring matters, using the established procedures of degradation and synthesis. The former involved breaking down a substance into fragments which could be identified; synthesis involved making the substance from chemical compounds of known composition and structure by a series of controlled reactions whose course was indisputable. Like Perkin he did not employ new physical methods of investigating structures. He maintained Perkin's emphasis on research as the prime academic function, but spent less time at the laboratory bench because he had more research students (about thirty at any one time) than Perkin. In the 1930s his wife, an organic chemist whom he had met at Manchester, spent more time at the bench than her husband who managed only short spells of practical work himself.[61] He confined himself to exploratory experiments in test-tubes, leaving the follow-up to a collaborator. As a more innovative researcher than Perkin, he ignored the university terms contumaciously. Whereas Perkin's forte was the rapid exploitation of new reactions and reagents discovered by others, such as Arthur Michael (1887) and Victor Grignard (1900), Robinson's was the devising of methods of synthesizing such compounds as steroids. One favourite technique involved the synthesis of a substance containing four benzene rings and then modifying it. Another was to obtain a three-ring structure containing the chemical groups which would allow a fourth ring to be added.

In synthetic work Robinson was not as persistent as Perkin. On occasion he abandoned synthetic routes (of which he was often a brilliant deviser with his chess-player's ability to see several moves ahead) which were later shown by others to be practicable. His restless volatility led to his devising a large number of synthetic reactions and procedures but completing a smaller number of actual syntheses. Again in contrast with Perkin, Robinson

[61] Gertrude Maud Robinson (1886–1954), obituary in *Journal of Chemical Society* (1954), 2667–8.

worked in theoretical chemistry and was responsible in 1926 for an electronic theory of organic chemistry.[62] But he had not the patience to consolidate and popularize it. His first general account of it appeared in 1932 in an inaccessible form and then he dropped the topic for seventeen years, being far more interested in the steroids research began in 1932. Meanwhile Ingold had developed a rival treatment from 1926, became wholly absorbed in the study of reaction mechanisms in organic chemistry, kept his ideas continuously on display, and explained the application of his ideas in detail to numerous specific reactions.[62] Robinson's mercuriality of temperament allowed Ingold to be widely seen as the originator and doyen of the electronic theory of organic chemistry and led to poisonous relations between the two men because Robinson, who was legendary for his impatience and irascibility with opposition, accused Ingold of plagiarism. There was also a less obvious source for Robinson's unconcealed hostility to Ingold: whereas Robinson was faithful to Perkin's dismissal of physical methods, Ingold was the leading early exponent in Britain of the new chemical physics.

Robinson inherited from Perkin the capacious and well-endowed Dyson Perrins Laboratory. In the 1930s it received a major extension only once, in 1939–40, at a cost of £29,000 which was raised by Robinson from the Rockefeller Foundation (£23,000) and ICI (£6,000).[63] Robinson had not pushed his case in the University Appeal of 1937 because he thought the prime need was for a new university physical chemistry laboratory; and in 1938 he deliberately went outside the University for money, trusting successfully to Rockefeller's interest in the biological implications of organic chemistry (especially the synthesis of proteins) and to his strong connections with ICI. The need for expanded accommodation had been generated by Robinson's reputation, his productivity, the presence by the end of the decade of sixty research students in the laboratory, and the closure of the Queen's laboratory in 1934. Robinson's high stature attracted post-doctoral researchers of the calibre of Alexander Todd, a Nobel prize-winner in 1957, and Norman Rydon, whose work with Robinson on oestrogens led to the future development of the 'pill'. From Australia he drew J. W. Cornforth, Nobel prize-winner in 1975, and Arthur Birch, deviser of the Birch reduction which was rapidly used by pharmaceutical companies to produce steroids.[64] Like Perkin, Robinson secured few of his researchers

---

[62] W. H. Brock, *The Fontana History of Chemistry* (London, 1992), 522–48; Christopher Kelk Ingold (1893–1970), obituary in *BM* 18 (1972), 349–71, professor of organic chemistry, University of Leeds, 1924–30, professor of chemistry, University College London, 1930–61; G. K. Roberts, 'C. K. Ingold at University College London: Educator and Department Head', *British Journal for History of Science*, 29 (1996), 65–82.

[63] Dyson Perrins Laboratory, 1929–47, OUA, UR/SF/CHE/5.

[64] A. Todd, *A Time to Remember: The Autobiography of a Chemist* (Cambridge, 1983), 23–8; Alexander Robertus Todd, *Who's Who*, and Henry Norman Rydon (1912–91), professor of chemistry, University of Exeter, 1957–77, obituary in *Independent*, 20 Sept. 1991, went to

from Oxford graduates in chemistry: they came mainly from elsewhere in Britain and from abroad. It was Robinson who was responsible for the international composition of the researchers in the laboratory: in Todd's short Oxford period he was struck by the presence of workers from Australia, New Zealand, Sweden, Japan, and Switzerland. As a supervisor Robinson, known to his worshippers as the Waynflete Wonder, was inspiring but not frequent in making contact with his students, from some of whom he received the sobriquet of visiting professor. Robinson's own productivity was extraordinary. In the years 1931–8 inclusive he and his collaborators published no fewer than 226 papers, i.e. a paper per fortnight maintained for eight years. No wonder that by force of example he inspired workers in the Dyson Perrins to produce a maximum in the 1930s of 84 papers in 1934–5 and in the late 1930s he employed Springall, formerly one of his D.Phil. students, as a scientific gentleman's gentleman to help in publishing papers.[65]

In the 1930s Robinson secured only two new university demonstrators, F. E. King and J. C. Smith. Both came from outside Oxford, from London and Manchester respectively where they were known to Robinson, to join the existing quintet of Hammick, Plant, Taylor, Baker, and the ill Hope. Again like Perkin, Robinson was reluctant to use his demonstrators as research lieutenants. He was often possessive and dictatorial with his research students, though when flooded with them he would on occasion use a trusted demonstrator: in the mid-1930s King helped Robinson with the supervision of researchers working with Rockefeller money. Again like Perkin he was unsuccessful in placing his protégés into college fellowships in the 1930s, the solitary exception being Baker who was elected at Queen's in 1936 to replace another organic chemist, Chattaway. One result was that King, who served as departmental and university demonstrator for seventeen years but was never a college fellow, left in chagrin for a chair at Nottingham. Robinson resented a feature that Perkin accepted with some grace, i.e. that he had no control of college fellows who were physical chemists doing research in his laboratory: they enjoyed independence. Hammick, egged on by Sidgwick, was their leader, the others being Taylor and Sutton. Robinson resented the way in which Hammick, a tutor at Oriel, Corpus Christi, and Wadham, sometimes had eight part two students working in the Dyson Perrins Laboratory while his own demonstrators, who were not college fellows, had none or very few. In retaliation Hammick told Robinson that he and

---

Oxford in 1931 and 1937 as 1851 Exhibition senior students; John Warcup Cornforth (b. 1917), *Who's Who*, and Arthur John Birch (1915–95), *Who's Who*, professor of organic chemistry, University of Manchester, 1955–67, and Australian National University 1967–80, arrived in Oxford in 1939 and 1938 as 1851 Exhibition overseas scholars.

[65] Harold Douglas Springall (1910–82), first in chemistry 1934, D.Phil. 1936, ended as professor of chemistry, University of Keele, 1950–75.

the colleges would decide with whom his pupils worked and he won.[66] Furthermore the college fellows in the laboratory, like everyone else except undergraduates, were not controlled by Robinson through rationing of common apparatus and chemicals. Until after the Second World War there was no system for requisitioning and issuing apparatus and chemicals. Like Perkin, Robinson was proud of his paperless administration based on trust.[67] He was too impatient to try to force his views on the college fellows through staff meetings: he held only three in twenty-five years.

Robinson also continued Perkin's enthusiasm for industrial research done in the Dyson Perrins Laboratory and for commercial consultancy.[68] Robinson had worked as an industrial research chemist for a year when he was director of research at British Dyestuffs Corporation, Huddersfield, 1919–20. This experience gave him an understanding of industrial research, useful contacts, and a comprehensive knowledge of dyestuff chemistry. After British Dyestuffs had been absorbed by ICI, Robinson did research for ICI and gave advice in exchange for money, information, and chemicals, both stock and those specially made for him. He was a powerful figure from 1929 in ICI's Dyestuffs Group Research Committee, for which research was done in Oxford, and in the creation of the pharmaceutical division of ICI via the medical section of ICI's Dyestuffs Group at Blackley near Manchester. At Manchester Robinson had taken out patents and he continued to do so in the 1930s at Oxford, registering nine, sometimes in collaboration with ICI.

The legacies of the Perkin–Robinson approach to organic chemistry were a continuing insistence on Baeyer-like degradative and synthetic methods and a marked suspicion of both physical methods and theoretical chemistry. Perkin and Robinson were good bench chemists, proud of their ability to induce a reaction to 'go' and gums to crystallize. They relished their ability to *make* many substances, both naturally occurring and not, and were proud that the latter were useful in dyestuffs, medicine, agriculture, and

---

[66] Frederick Ernest King (b. 1905), *Who's Who*, FRS 1954, departmental demonstrator 1931–4, D.Phil. 1933, university demonstrator 1934–48, professor of chemistry, University of Nottingham, 1948–55, ended as scientific adviser, British Petroleum, 1959–71; John Charles Smith (1900–84), Ph.D. Manchester, first came to Oxford in 1928 as a researcher, university demonstrator 1931–55, reader in organic chemistry 1955–66, obituary in *The Times*, 7 May 1984; Dalziel Llewellyn Hammick (1887–1966), FRS 1952, obituary by E. J. Bowen, *BM* 13 (1967), 107–24; Thomas Weston Johns Taylor (1895–1953), *DNB*, first in chemistry 1920, fellow of Brasenose 1920–46, principal of University College of West Indies, 1946–52, principal of University College of South West [Exeter] 1952–3, knighted 1952, obituaries in *Journal of Chemical Society* (1954), 767–8 and *Brazen Nose*, 9 (1953), 340–2; Robinson to Margoliouth, 9 Nov. 1937, PS/R/1/6, on King as lieutenant; Smith, 'Organic Chemistry at Oxford', ii. 29, on Hammick versus Robinson.

[67] Robinson, Proposals for Improvements in the Technical Administration of the Dyson Perrins Laboratory, 18 Sept. 1945, OUA, UR/SF/CHE/5.

[68] Todd and Cornforth, 'Robinson', 421, 526–7; Reader, *ICI*, ii. 350; Robinson Papers, A35, B67–79, B82–6, esp. B76, Reports from the Dyson Perrins Laboratory to ICI Dyestuffs Group Research Committee, which lists thirteen Oxford researchers.

petrochemicals. In contrast physical chemists, non-makers and often non-applied in their research, seemed merely effete. In the 1930s Robinson was cautious about using infra-red and ultraviolet spectroscopy, X-ray crystallography, and measurements of dipole moments in his own research. Having withdrawn in 1932 from work on reaction mechanisms in organic chemistry, Robinson was suspicious of Ingold's work on this subject which was characterized in Oxford as an Ingoldsby legend.

At Oxford in the 1930s Robinson's approach to the problems of molecular structure was supplemented by that of Leslie Sutton, who worked in the Dyson Perrins Laboratory, and Dorothy Hodgkin, who researched in the mineralogy department. Though both were helped by Robinson at crucial stages of their careers, he and his collaborators did not adopt Sutton's techniques of dipole moments and electron diffraction or Hodgkin's use of X-ray crystallography. In his year at the California Institute of Technology Sutton was impressed by Pauling's team research, which he regarded as professional compared with Oxford's amateur approach, which was satisfactory for doing research but not for research training. Pauling employed mathematical physicists and experimenters in several fields, acting as a link figure who provided ideas. For Sutton Pauling's great strength was his rare combination of chemical instinct and mathematical and physical technique. Sutton wanted to found at Oxford a group devoted to molecular structure using physical methods à la Pauling but found Robinson unresponsive: Robinson proclaimed that he was an organic chemist, that it was up to Sutton to mobilize other physical chemists in promoting his scheme, and that the long-term solution for Sutton was a new university physical chemistry laboratory.[69] In Hodgkin's case it appears she was content with her location in the mineralogy department in the 1930s and made no attempt to transfer to the Dyson Perrins Laboratory. During the war, however, she joined the team working at Oxford under Robinson on the structure of penicillin. By 1945 she had beaten Robinson in determining its structure and shown that Robinson's preferred structure was inadequate. Similarly by 1956 she had beaten Todd in elucidating the structure of vitamin B12. Thus the classic degradative and synthetic methods employed by Robinson and his pupil Todd began to be outflanked in Oxford by the new physical method of X-ray crystallography. Elsewhere the Perkin–Robinson style of research was lethally challenged in the 1950s at Harvard by R. B. Woodward who combined their methods with molecular orbital theory and new physical resources (such as infra-red and nuclear magnetic resonance spectroscopy) to synthesize vitamin B12, chlorophyll, strychnine, cortisone, and cholesterol.[70]

---

[69] Sutton to Sidgwick, 14 Jan., 5 June 1934, and Sutton to Robinson, 25 Mar. 1934, Sidgwick Papers, V. 79.
[70] On Robert Burns Woodward (1917–79), on Harvard staff 1937–79, A. Todd and J. Cornforth, *BM* 27 (1981), 629–95; contrast Todd and Cornforth, 'Robinson'.

8.5 Old Chemistry

Inorganic chemistry, the responsibility of Soddy and then Hinshelwood, was pursued in the 'Glastonbury kitchen' building attached to the Museum. It was physically and mentally separate from physical chemistry, which was located in two college laboratories, and from organic and physical organic chemistry which were in the Dyson Perrins Laboratory. In the 1920s the inorganic chemistry laboratory was informally known as the Old Chemistry department. In Sidgwick's view this odd name was exactly right because that was what Soddy lectured on—old chemistry! Between the wars the Old Chemistry department at Oxford did not make a big national or international impact, the exceptions being Hume-Rothery in metallurgy and Harold Thompson in infra-red spectroscopy. The most curious feature, at least *prima facie*, was that Soddy, appointed to the Dr Lee's chair at Oxford in 1919 and in 1921 the first Oxford professor to be awarded a Nobel prize, did not set up and lead the expected school of radiochemistry there.

The reasons for Soddy's failure to do so are not as simple as some critics have made out. It is well known that Soddy did not have a wide circle of friends among fellow scientists at Oxford, that he was not *persona grata* there, that as a catfish among the cod he was averse to compromise, and that he was at loggerheads with the colleges, the college tutors in chemistry, and the University. In the 1920s he was at odds with two of his senior demonstrators, Lambert and Applebey, who thought him perverse and tactless. These general features of Soddy's regime do not, however, explain why his early attempt to found a school of radiochemistry and his own published research in that subject had foundered by the mid-1920s. Part of the explanation lies in Soddy being a solo worker without a research assistant, collaborator, or lieutenant. In autumn 1919 when A. S. Russell, who had done important radiochemical research with Soddy in Glasgow, was appointed Dr Lee's reader in chemistry with Soddy's warm support it seemed that the core staff of a school had been assembled.[71] Yet Russell never became a demonstrator or informal collaborator in Soddy's department, lectured there only once on radiochemistry, pursued his own research independently it seems in his Christ Church laboratory, was swamped by teaching (fifty-five hours a week at worst), and by the 1930s had turned away from the little research in radioactivity he had managed to do to the study of intermetallic compounds in mercury. Soddy and Russell remained friendly but as researchers they went their separate ways.

It has also gone unnoticed that Soddy's attempts to secure radioactive materials for research, from the Imperial and Foreign Corporation of London in 1921, were doused with cold water by Farnell, the vice-chancellor, so

---

[71] Testimonials, including Soddy, 12 Feb. 1919, and c.v., Russell Papers, DP xx.c.l, Christ Church Archives, Oxford. I owe these references to Dr J. Hughes.

a rare opportunity was lost.[72] In autumn 1921 the Corporation, to which Soddy was scientific adviser, projected a laboratory in England to house radium. Soddy wanted this laboratory to be located in Oxford and, given his unsuitable accommodation, in a separate building for which the Corporation would partly pay. Soddy was agog at the unique opportunity of working with the 2 grammes of radium, worth £70,000, which he had carried from Czechoslovakia to London where it was lodged in the Czech Embassy. In his two years at Oxford he had had no more than 30 milligrammes of radium bromide to work with. He looked forward to having a new radiochemical laboratory as a university institute, with facilities and staff paid for by the Corporation, and to training people there. By October 1921 a draft agreement between the University and the Corporation had been drawn up by Soddy. It permitted Soddy to use the Corporation's equipment and materials for its scientific and technical work for which he would be paid, even though he would be working in university premises. As a quid pro quo the University insisted without prior warning that any scientific research done by Soddy using the Corporation's equipment and materials should be published and should not become private knowledge owned by the Corporation. On hearing of this insistence, prompted by Farnell, that there be no bar on publication, the Corporation dropped the Oxford scheme, leaving Soddy furious. Not for the first time the University, it seemed to him, had let him down. More than that, the University's insistence on there being no restraint on publication had scuppered his ambitious scheme, yet in the Dyson Perrins Laboratory no similar insistence applied to commercial firms doing research there or to Perkin's own industrial research; there seemed to be one university law for Perkin and another for Soddy. And, of course, Soddy was frustrated at the loss of a golden opportunity to investigate the unworked residues from the mines of St Joachimsthal: in his view they constituted an El Dorado for research.

Soddy's chagrin with the University for denying him the patronage of what he called the Czech Radium Corporation and his institutional separation from Russell did not facilitate Soddy's ambitions for research on the purification, extraction, and analysis of radioactive elements and their compounds. His hostility to the college tutors ensured that in the 1920s he attracted only three Oxford graduates, two of them women, to work with him. But he did draw from Japan Satoyasu Imori who became the father of Japanese radiochemistry, a couple of Rhodes scholars, Miss Hitchins from Glasgow who became his private assistant, Paolo Misciattelli as a Ramsay research fellow to do a D.Phil., and J. K. Marsh from Belfast as a post-

---

[72] *Financial Times* and *The Times*, 26 Sept. 1920; exchanges between Soddy and Farnell, Sept.–Nov. 1920, UR/SF/PHA/3; proposed University–Corporation agreement, n.d. [Oct. 1921], *HCP* 120 (1921), 25–6.

doctoral researcher.[73] In 1927 Hitchins left for Africa, Misciattelli returned to Italy, and Marsh went back to Ulster. Without these collaborators Soddy's publishing career as a radiochemist finally ended, though it had been petering out before then. In compensation Soddy spent a lot of time in his laboratory workshop, devising many mechanical gadgets which he patented, the best known being his continuously variable dividing engine which ruled scales for spectrographs and his reversing centrifugal gear. This talent for design was generously put at the service of the University in the late 1920s when he reconstructed the Old Chemistry department.[74] Acting as his own architect, designer, draughtsman, and engineer, Soddy gave particular attention to bench reconstruction, lighting, drainage, and ventilation. Whenever possible he used his own workshop to make artefacts he had designed, he supervised the work done, and he obtained discounts on materials. He avoided the use of expensive and bungling contractors such as those who had cost the University so much when they built the extension to the Dyson Perrins Laboratory in the early 1920s. Showing no sign of chagrin, Soddy modernized his department in 1929–30 at a cost of just under £7,000, his ingenuity saving the University thousands of pounds. Showing no loss of spirit, he continued to press the University, though without success, for extensions to his empire until just before his resignation.[75] Irritatingly fertile in producing, amending, and adding to his designs, Soddy was literally a department-builder, an achievement which his detractors wrongly ignore.

He was not, however, a discipline-builder in physical or inorganic chemistry. This failure was the result of the difficulties he met at Oxford. It was also caused by Soddy's adoption of the roles of sage, prophet, and critic, which resulted in many publications on social and economic problems in their relation to scientific progress. It has often been said that Soddy went off the rails in 1919, foolishly risking the status of an amateur crank in economics and stupidly forsaking chemistry in which he had special gifts. It makes more sense, however, to see him as one who was so devastated by the waste, hatred, and futility of the First World War that he became a seer and propagandist, albeit a solitary one, concerning the social problems of

[73] Soddy's annual reports, *OUG* (1920–7); M. Tanaka and K. Yamasaki, 'Early Studies of Radioactivity and the Reception of Soddy's Ideas in Japan', in G. B. Kauffman (ed.), *Frederick Soddy (1877–1956): Early Pioneer in Radiochemistry* (Dordrecht, 1986), 141–54; Sotoyasu Imori (1885–1982); the Rhodes scholars, subsequently distinguished, were John Hamilton Mennie, B.Sc. 1923, professor of chemistry, McGill University, Montreal, 1940–68, and Milan Wayne Garrett, first in physics 1924, D.Phil. 1926, assistant, associate, full professor of chemistry, Swarthmore College, Swarthmore, Pa., 1927–66; Paolo Misciattelli, D.Phil. 1928; Joseph Kenneth Marsh, D.Sc. Belfast 1923, later a chemical consultant specializing in the rare earths which he had studied with Soddy.

[74] *OUG* 60 (1929–30), 650; Lee's Chair and Old Chemistry Department, 1927–33, OUA, UC/FF/288.

[75] Soddy to Veale, 19 Mar. 1936, Department of Inorganic Chemistry, 1930–47, OUA, UR/SF/CHE/6.

science. He wrestled in particular with the question of how it was that the prospective benefits of science were often vitiated. He concluded that the main villains were the banks, so from the early 1920s he devoted much of his energy and time to writing and speaking about monetary reform. As a chemist who pondered about the relations between science and life, he had concluded by 1919 that the chief problems facing humankind were not physical but moral. Accordingly he concerned himself not only with monetary reform but with the social responsibility of scientists. He was particularly active in 1935. Though he had no leanings towards communism or socialism, he associated himself with Bernal, Blackett, and J. G. Crowther in deploring the misapplication and frustrations to which science was subject. He also led a well-publicized attack on the Royal Society of London which he depicted as a private bureau run by divine right by powerful officers and a self-selecting Council. He was supported by ten of his fellow scientists at Oxford, including Robinson and Chapman, but not by either Sidgwick or Hinshelwood. This was perhaps the only occasion on which he was not isolated while at Oxford. Usually his chemical colleagues regarded his behaviour as abrasive, awkward, and inconsistent, and they deplored his assumption of the role of seer of science as a sterile aberration. Soddy, however, saw himself as 'the pioneer and bearer of a new evangel' and not as a crank or impostor who had strayed from the path of pure science.[76]

The Old Chemistry department under Soddy was different from the Dyson Perrins under Perkin and Robinson in that the professor and senior or university demonstrators were not leaders of research groups or innovative solo researchers. These roles were filled in the 1930s by H. W. Thompson, fellow of St John's and a departmental demonstrator, and by William Hume-Rothery, a metallurgist who was a guest in the department. Soddy's enduring senior demonstrators were Bertram Lambert, Ernald Hartley, and Allan Walden, while Applebey left in 1928 to be succeeded by Freddie

---

[76] T. J. Trenn, 'The Central Role of Energy in Soddy's Holistic and Critical Approach to Nuclear Science, Economics, and Social Responsibility', and H. E. Daly, 'The Economic Thought of Frederick Soddy', in Kauffman, *Soddy*, 179–98, 199–218; M. Davies, 'Frederick Soddy: The Scientist as Prophet', *Annals of Science*, 49 (1992), 351–67; L. Merricks, 'An Invisible Man: On Writing Biography', *History Workshop Journal* (1994), 194–204, a reference I owe to Dr Jeffrey Hughes, is wild on Soddy's radiochemistry but useful on Soddy's membership of the New Britain Group; F. Soddy, *Science and Life: Aberdeen Addresses* (London, 1920), 172; id., foreword, in A. D. Hall *et al.*, *The Frustration of Science* (London, 1935), 5–7 (p. 5 quotation); for Soddy's polemics with the Royal Society, Soddy Papers, B10. Soddy's major works on economics published 1919–36 were: *The Inversion of Science and a Scheme of Scientific Reformation* (London, 1924); *Money versus Man: A Statement of the World Problem from the Standpoint of the New Economics* (London, 1931); *The Role of Money: What it Should Be, Contrasted with What it Has Become* (London, 1934); *Wealth, Virtual Wealth and Debt: The Solution of the Economic Paradox* (London, 1926). Patrick Maynard Stuart Blackett (1897–1974), Nobel prize 1948, president of Royal Society 1965–70, obituary in *BM* 21 (1975), 1–115; James Gerald Crowther (1899–1983) was the first science correspondent of the *Manchester Guardian* 1928–48.

Brewer. All were Oxonians and knew that local esteem could be gained by routes other than those of research and research-leadership. After bitter conflict with Soddy in the 1920s Lambert assumed the role of mediator and peacemaker between Soddy and his staff. He continued his war research on the absorption of gases by solids but was a slow publisher. After Soddy's retirement he received his reward when he was made administrative head of the Old Chemistry department. Applebey also sparred with Soddy in the 1920s and was lured in 1928 to be research manager at ICI's Billingham plant where he joined two of his pupils, Kenneth Gordon, the works manager, and Walter d'Leny, the technical director, to form an agreeable St John's cohort. Walden never undertook original work, preferring to be renowned as a college teacher. Ernald Hartley and Brewer, however, had both begun promisingly as researchers but as demonstrators did not maintain that role. Hartley was indeed famous for the research he had done for thirteen years before the first war with Lord Berkeley who had invited him to work in his country-house laboratory at Foxcombe Hall, 5 miles outside Oxford, on the osmotic pressure of solutions. After war service with Lambert devising gas respirators, he led a quiet life as a demonstrator, publishing little and enjoying country life at his Frilford House home, 8 miles from Oxford. Apparently the time he spent travelling limited the hours he could spend in the laboratory. The fruits of his university labours, in contrast with those done with Lord Berkeley, were the men he helped rather than the papers published. Brewer, the only one of the five main demonstrators to graduate after the first war, was a pupil of Sidgwick who used his American contacts to send Brewer for two years to work at Cornell University as a Commonwealth Fund fellow under L. M. Dennis, which gave him an interest in the chemistry of germanium. On his return to England he trod water for a year at Reading University before becoming a key man in the teaching and organization of inorganic chemistry at Oxford from 1928. Brewer was not a great researcher and did little work at the bench but he gained the rewards of a readership in 1955 for his administration and of being mayor of Oxford 1959–60 as the culmination of his civic labours.[77]

Though all these demonstrators had part two undergraduates to supervise, and Lambert, Walden, and Applebey had the additional advantage of

---

[77] On B. Lambert, first in chemistry 1903, *OM* 82 (1963–4), 60 and *Nature*, 199 (1963), 1136–7; Ernald George Justinian Hartley (1875–1947), first in chemistry 1897, obituary in *Journal of the Chemical Society* (1948), 899–901; on Walden, first in chemistry 1895, *OM* 75 (1956–7), 132–6; on Applebey, first in chemistry 1906, *Proceedings of the Chemical Society* (1957), 214–15; on Frederick Mason Brewer (1902–63), first in chemistry 1924, *OM* 81 (1962–3), 260 and *Proceedings of the Chemical Society* (1964), 381–2; Kenneth Gordon (1897–1955), first in chemistry, St John's, 1921, obituary in *The Times*, 30 Nov., 2 Dec. 1955; Walter d'Leny (b. 1902), second in chemistry, St John's, 1925; Randal Mowbray Thomas Rawdon Berkeley, 8th Earl (1865–1942), obituary by H. Hartley, *ON* 4 (1942–4), 167–82; Louis Munroe Dennis (1863–1936), professor of inorganic chemistry, Cornell University, 1900–33.

being college fellows, none of them used part two of the chemistry degree as Hinshelwood did, as a source of research pupils and of publications. The first demonstrator in Soddy's reign to do this was H. W. Thompson. He enjoyed several advantages. As a pupil of Hinshelwood at Trinity, he spent two long vacations researching on kinetics with his tutor and completed his part two under him. With several publications with Hinshelwood to his name as an undergraduate, St John's offered him a fellowship before his graduation in 1929. Feeling that he needed further research experience, Thompson induced St John's to give him a year abroad as a research fellow before electing him to a tutorial fellowship in summer 1930. In 1929–30 Thompson went to Berlin to the Kaiser Wilhelm Research Institute to work under Fritz Haber for a doctorate which he gained in 1930. A college fellow by the age of 22, the blunt and ambitious Thompson used the part two system to swell his own publications and to push his St John's pupils into print: all four finalists from St John's in 1933 co-published with him. Not surprisingly he taught subsequently distinguished pupils who in the 1930s included Christopher Kearton, Frederick Dainton, and Jack Linnett. Supported by Hinshelwood and Lindemann, but opposed by Soddy, Thompson cleverly shifted his research field from kinetics, where he would have been under Hinshelwood's shadow, to ultraviolet and then infra-red spectroscopy. Thompson's awareness of the importance of studying abroad was shown in the encouragement he gave to Linnett, his favourite research collaborator in the 1930s. Linnett was a St John's graduate tutored by Thompson who supervised his research for a D.Phil. 1935–7. Then Linnett was awarded a Henry fellowship which enabled him to work for a year on infra-red and Raman spectroscopy at Harvard University under Kistiakowsky and Bright Wilson, from whom he also learned much about the application of quantum mechanics to chemistry. On his return to Oxford in autumn 1938 Linnett's career prospered as a research fellow, departmental and university demonstrator, college fellow, and university reader until in 1965 he left for the chair of physical chemistry at Cambridge.[78]

The pioneering work in the science of metallurgy undertaken by the totally deaf William Hume-Rothery was kept alive at Oxford until 1938 by his loyalty to the University, by his own pocket, and by the prescience of

[78] R. Richards, 'Thompson', *BM* 31 (1985), 573–610; Christopher Frank Keaton (1911–92), first in chemistry 1933, chairman of Courtaulds 1964–75, obituary in *Independent*, 29 July 1992; Frederick Sydney Dainton, first in chemistry 1937, vice-chancellor of University of Nottingham 1965–70, Dr Lee's professor of chemistry, Oxford, 1970–3, chairman of University Grants Committee 1973–8; John Wilfrid Linnett (1913–75), junior research fellow at Balliol 1939–45, fellow of Queen's 1945–65, professor of physical chemistry, University of Cambridge, 1965–75, obituary in *BM* 23 (1977), 311–43; George Bogdan Kistiakowsky (1900–82) and Edgar Bright Wilson (b. 1908). Of 12 co-publishers with Thompson in the 1930s, 11 were Oxonians of whom 9 were from St John's; there was a solitary Rhodes scholar, Milton F. Meissner, D.Phil. 1938.

various external funding bodies. Only in 1938, after he had been elected FRS in 1937, was he made a university lecturer in metallurgical chemistry and a fellow of Magdalen by special election. In the post-war years the founding of a department of metallurgy and an undergraduate degree occurred in the mid- to late 1950s owing to Hume-Rothery's efforts, abetted by external pressure. With some justice he became the University's first professor of metallurgy in 1958 and a professorial fellow of St Edmund Hall. His long slog to recognition began in 1922 when, having gained a first in chemistry at Oxford, he was advised to take up metallography by Soddy who was keen to develop metallurgy at Oxford. Soddy arranged for Hume-Rothery to do a Ph.D. under Sir Harold Carpenter at the Royal School of Mines in London, after which it was intended by Soddy that Hume-Rothery should return to the Old Chemistry department to promote the subject. While Hume-Rothery was in London, he developed an interest in intermetallic compounds which did not conform to the rules of valency then current in inorganic chemistry and he began to speculate about the role of free electrons in the solid state. On his return to Oxford in 1925, supported for four years by a senior demiship (i.e. graduate scholarship) at Magdalen, his undergraduate college, he discovered that Soddy, piqued by his disputes about the Balliol–Trinity laboratory, would not accommodate him. Desperate to settle in Oxford, Hume-Rothery secured from Perkin a temporary home for almost four years in the Dyson Perrins Laboratory topped up by occasional visits to the laboratory of engineering science. Then in autumn 1929 he moved to the Old Chemistry department, where he was to stay for many years, through the influence of Lambert, who had been his part two supervisor. It was Lambert who enabled Hume-Rothery to retain his habitation in the Old Chemistry department, even though he was not a member of staff but a mere guest: Lambert smoothed out the innumerable difficulties which Soddy created. It was external funding which enabled Hume-Rothery to maintain his research once his Magdalen post had ended in 1929. For three years he held a research fellowship awarded by the Armourers' and Braziers' Company. In this period he published in 1931 his book on the metallic state in which he used the lattice theory of Lindemann, a supporter and confidant, as a point of departure from which he began to work out empirical rules of alloy formation. As he thought that their meaning would have to be left 'to some Cambridge wave mechanic' Hume-Rothery was well aware that he was so far the Kepler but not the Newton of physical metallurgy. As such he was critical of the insular approaches usually taken by metallurgists and physicists to the features of metals and alloys. The former merely amassed commercially useful data, whereas the latter had investigated underlying principles but their work was vitiated by lack of chemical knowledge, by the use of unsuitable specimens, and by ignorance of practical metallurgy.

Hume-Rothery's research at Oxford would have ended in 1932 but for the lucky accident of the establishment that year of the Gordon Warren research fund of the Royal Society of London for research in engineering, chemistry, physics, and metallurgy. Against hot competition—J. D. Bernal, U. R. Evans, P. M. S. Blackett, J. Chadwick, J. D. Cockroft, C. D. Ellis, W. Sucksmith, E. L. Hirst, and G. M. Dobson—Hume-Rothery and A. J. Bradley, both metallurgists, were elected in June 1932 to the first two fellowships, tenable for four years, renewable on application by the fellow for a further three years, at a salary of £700 p.a. Hume-Rothery probably owed his election to two Oxonians, Sidgwick and Egerton, who were members of the Warren Fund Committee. Sidgwick was interested in Hume-Rothery's early work and had introduced him to Lindemann's lattice theory, while Egerton, then reader in thermodynamics at Oxford, was interested in chemical technology and, as one who had been thrown out of the Old Chemistry department by Soddy in 1921, he had special sympathy with Hume-Rothery's situation there. The Warren fellowship gave Hume-Rothery security, status, and salary not just to 1939 but with one short break to 1955, even though officially it was a limited-term appointment. Supported by considerable grants from the Royal Society for apparatus and materials, which were topped up by Hume-Rothery from his own pocket, on occasion rising to £70 p.a., he built up from 1933 a small research group, recruiting mainly Oxford chemists via the part two chemistry degree, and the odd Oxonian physicist to work with him, sometimes for a D.Phil. In his X-ray work on the determination of lattice and atomic constants, he was greatly helped by Tiny Powell from the nearby department of mineralogy. His growing reputation was confirmed in 1936 with his second book which dealt with the structure of metals and alloys. With its accounts of atomic structure and the theory of the metallic state, this work which was inspired and published by the Institute of Metals gave the first general and accessible account of the principles of structural metallurgy for industrial metallurgists as opposed to physicists. After Tizard had proclaimed as part of the University Appeal of 1937 that special encouragement should be given to the science of metallurgy and after Hume-Rothery had been elected FRS that year, his position at Oxford was regularized: in 1938 he became a fellow of Magdalen, though only for five years, and a university lecturer. He even secured a departmental demonstratorship in 1937 for his favourite pupil and collaborator G. V. Raynor. Though his accommodation in the Old Chemistry department was not palatial, being restricted before the war to one room and a covered yard, Hume-Rothery saw the immediate future of metallurgy as best pursued there. His career to then supplies an extreme example of the importance of external funding. It enabled him in the 1930s to join Carpenter, Desch, Rosenhain, Mott, and Jones as a leading figure in British scientific metallurgy. Hume-Rothery's career also shows that the lack of formal recognition at Oxford until 1938 was in part compensated by the informal help he received at crucial points in his career from Oxonians such as Soddy, Perkin, Lambert, Lindemann, Sidgwick, Egerton, Powell, and Tizard.[79]

8.6 The Chemical Synthesizer

Oxford chemistry was tripartite in terms of sites. Physical chemistry prospered in two college laboratories, organic chemistry burgeoned in the Dyson Perrins Laboratory, and inorganic chemistry was housed in the Old Chemistry department. The same point applied to the chemists themselves. Some physical chemists and physical organic chemists worked in the Dyson Perrins. That happened because there was nowhere else for them: they and their methods were not warmly welcomed by either Perkin or Robinson. If there was a unifying force it was provided by Sidgwick, fellow of Lincoln. His synoptic vision, which embraced the main three branches of chemistry, was revealed in his publications, his contributions to colloquia and seminars,

---

[79] On William Hume-Rothery (1899–1968), G. V. Raynor, *BM* 15 (1969), 109–39; W. Hume-Rothery, 'The Development of the Theory of Alloys', in C. S. Smith (ed.), *The Sorby Centennial Symposium on the History of Metallurgy* (New York, 1965), 331–46; id., *The Metallic State: Electrical Properties and Theories* (Oxford, 1931); id., *The Structure of Metals and Alloys* (London, 1936); Hume-Rothery to Lindemann, 29 Dec. 1928, 6 July 1936, LP, D92; 'Department of Metallurgy', *OM* 78 (1959–60), 306–7; Hume-Rothery to E. J. Bowen, 19 Aug. 1963, Bowen Papers, B 23; Hume-Rothery to G. W. Hedley, 14 Apr. 1924, 9, 26 May 1930, Hume-Rothery Papers; Royal Society of London, Council Minute Book 24: Warren Research Fund Committee, 12, 19 May, 16 June 1932; Royal Society Council minutes show that between 1931 and 1939 Hume-Rothery received £780 from its Dewrance Fund and £395 from its parliamentary grant; S. T. Keith and P. K. Hoch, 'Formation of a Research School: Theoretical Solid State Physics at Bristol 1930–54', *British Journal for the History of Science*, 19 (1986), 19–44 (28–31); H. T. Tizard, 'The Needs of Oxford Science', *Oxford: Special Number* (1937), 52–6. Henry Cort Harold Carpenter (1875–1940), *DNB*, professor of metallurgy, Imperial College, London, 1913–40; Ulick Richardson Evans (1889–1980), FRS 1949, expert on metal corrosion, *BM* 27 (1981), 235–54; James Chadwick (1891–1974), *DNB*, FRS 1927, Nobel prize-winner 1935; John Douglas Cockroft (1897–1967), *DNB*, FRS 1936; Charles Drummond Ellis (1895–1980), FRS 1929, *BM* 27 (1981), 199–234; Willie Sucksmith (1896–1981), FRS 1940, *BM* 28 (1982), 575–88; Edmund Langley Hirst (1898–1975), *DNB*, FRS 1934; Albert James Bradley (1899–1972), FRS 1939, in 1932 world expert on X-ray powder photography in W. L. Bragg's physics department, University of Manchester, Warren research fellow 1932–8, obituary in *BM* 19 (1973), 117–28; Cecil Henry Desch (1874–1958), professor of metallurgy, University of Sheffield, 1920–31, superintendent of metallurgy department, National Physical Laboratory, 1932–9, obituary in *BM* 5 (1959), 49–68; Walter Rosenhain (1875–1934), superintendent of metallurgy department, NPL, 1906–31, obituary in *ON* 1 (1932–5), 353–9; Nevill Francis Mott (1905–96), professor of theoretical physics, University of Bristol, 1933–48, Cavendish professor of experimental physics, University of Cambridge, 1954–71; Harry Jones (1905–86), professor of mathematics, Imperial College, London, 1946–72, obituary in *BM* 33 (1987), 325–42. Hume-Rothery's chief co-researchers before 1939 were: Geoffrey Vincent Raynor (1913–83), first in chemistry 1936, departmental demonstrator 1937–45, professor of metallurgy, University of Birmingham, 1954–81, obituary in *BM* 30 (1984), 547–63; and Peter William Reynolds (b. 1913), first in chemistry 1935, subsequently with ICI.

and his dominance of both the Alembic Club and the Dyson Perrins tea club. It was rightly said after his death that he was a strict but not severe father of the whole Oxford school of chemistry.

Sidgwick was a late developer. An Oxford graduate, he spent three years in Germany learning physical chemistry from Ostwald at Leipzig and organic chemistry from von Pechmann at Tübingen. On his return he was elected in 1901 tutorial fellow at Lincoln, a post he held to 1948. The security conferred by his fellowship enabled Sidgwick to find his chemical feet slowly and to take his time when researching and writing synoptic works. Initially there was little direction to his research. He was not an ardent experimenter at the bench. He was poor at selecting research topics for his pupils and advising on research methods. He was unhappy in conversation about experimental and manipulative procedures. By 1910, with the publication of his first book on the organic chemistry of nitrogen, he had discovered at the age of 37 that his special gift was his ability to bring to chemistry 'not so much the burning and single-minded zeal of the discoverer as the panoramic learning of the scholar'. Sidgwick's first work of armchair or literary chemistry brought physical, inorganic, and organic chemistry together and it disproved, by precept and example, the view held by organic chemists such as Perkin that physical chemistry was all very well but it did not apply to organic substances. The book, which grew out of an Oxford lecture course, established his international reputation.

It was extended in 1927 by his famous work on the electronic theory of valency which eventually sold over 10,000 copies. The immediate stimulus for it was the visit in 1923 of G. N. Lewis, of the University of California, to Oxford where he stayed with Sidgwick. From their discussions and the publication of Lewis's *Valence and the Structure of Atoms and Molecules* later that year, Sidgwick developed an electronic theory of valency which applied to the whole of chemistry. The most original part of the work dealt with inorganic co-ordination compounds: Sidgwick showed that the so-called co-ordinate link was the same as the covalency of carbon, which he explained electronically, thus once again bridging the gap between inorganic, organic, and physical chemistry. As his book presented a much more comprehensive analysis of the uses of the electron in chemistry than had Lewis, it transformed inorganic chemistry from a jumble of unrelated facts into a subject with intelligible principles because it showed how the number of electrons outside the nucleus of an atom of a given element controlled its chemistry.

In 1933 yet another work of synthesis appeared with the publication of his book on the covalent link in chemistry. Like the 1927 book this one took the whole of chemistry for its province. It grew out of his invitation from L. M. Dennis to visit the USA in 1931 as George Fisher Baker lecturer in chemistry at Cornell University, for which he was paid $5,000 salary. About a third of the book was concerned with dipole moments, the importance of which had been communicated to him in 1928 by Debye while his guest in Oxford. In 1937 the second edition of his *Organic Chemistry of Nitrogen* appeared. It was revised and rewritten by T. W. J. Taylor, fellow of Brasenose, and Wilson Baker, demonstrator in organic chemistry, to whom Sidgwick had entrusted the task, having written four draft chapters himself. As eleven colleagues, past and present, collaborated as drafters and as commentators on drafts, the book was very much an Oxford joint-stock effort and a tribute to Sidgwick's wide view of chemistry. Shortly before the Second World War he embarked on another great synoptic work, on no less than the chemical elements and their compounds, which finally appeared in 1950 when he was 77. As a college fellow Sidgwick could afford to bide his time: he published little of major importance until early middle age. As a college fellow in chemistry and the solitary science tutor in a college which had no mathematics fellow, Sidgwick was forced to take a broad view of his subject which helped him to compose four major synthetic and synoptic books.

In the 1930s he was active on the national stage as chairman of the chemistry research board of the DSIR (1932–5), president of the Faraday Society (1932–4), vice-president of the Royal Society (1931–3), and president of the Chemical Society (1935–7), where he dispatched business with bewildering celerity. He had previously been for seven years the first specially appointed chairman of the Publication Committee set up by the Chemical Society to referee papers submitted to the Society's journal. But he was also an Oxford character. Secure in his college fellowship, he was free from professional rivalry and personal vanity. He was the leading representative of science in the delegacy of the University Press. He was famous for biting people verbally, referring once to Wadham men as gutta percha from the neck upwards. As a bachelor don he was a renowned host. For instance, when J. B. Conant of Harvard University visited Oxford, Sidgwick gave a dinner party in Lincoln for him, Milne, Lindemann, Einstein, and Sutton.[80]

---

[80] On Sidgwick, Sidgwick Papers; H. T. Tizard, *ON* 9 (1954), 237–58; L. E. Sutton, *Proceedings of the Chemical Society* (1958), 310–19; C. N. Hinshelwood, *OM* 70 (1951–2), 284–6 (284 quotation); Moore and Philip, *Chemical Society*, 124–5, 157–9; T. S. Moore, *Nature*, 169 (1952), 732–3; A. G. Ogston interview, 9 Sept. 1987; L. E. Sutton interview, 27 Nov. 1984. His major books were: *The Organic Chemistry of Nitrogen* (Oxford, 1910); *The Electronic Theory of Valency* (London, 1927); *Some Physical Properties of the Covalent Link in Chemistry* (Ithaca, NY, 1933); *The Organic Chemistry of Nitrogen: New Edition Revised and Rewritten by T. W. J. Taylor and W. Baker* (Oxford, 1937); *The Chemical Elements and their Compounds* (Oxford, 1950). On the Alembic Club, colloquia, and the tea club, R. P. Bell, Lecture, University of Kent at Canterbury, typescript, 1974, pp. 17–18; Smith, 'Organic Chemistry at Oxford', i. 44–5; Oxford University Alembic Club Minute Books, 1913–18, 1918–23, 1923–30, 1930–8, Museum of the History of Science, Oxford, MSS 143–6. On Gilbert Newton Lewis (1875–1946), professor of chemistry, Berkeley, University of California, 1912–46, and the relation of his work to Sidgwick's, Brock, *History of Chemistry*, 465–83, 591–7, 609–10. James Bryant Conant (1893–1978) was professor of chemistry, Harvard University, 1919–33, and president 1933–53.

An outstanding chemist and an outspoken verbal fencer, Sidgwick exploited his powers of badinage in promoting chemistry socially at Oxford. He was proud of Oxford's chemistry, averring that if someone in Cambridge lit a Bunsen burner it was national news while if someone in Oxford isolated a new element it would be ignored by the press. He was therefore the ever-present life and soul of the Alembic Club, the University's chemical society, between the wars. He kept it going during the first war and afterwards made it a very useful institution for both junior and senior members. It met once a week, with about two distinguished visitors a term, the remaining speakers being Oxonians from all the laboratories. There was a strong emphasis on research in progress and on research tactics and strategy. When a speaker was ill, Sidgwick was happy to step into the breach, sometimes mischievously so. In 1936 when Rideal was unwell Sidgwick spoke on the resonance theory of organic chemistry and glossed the ideas about the structure of benzene of Ingold, Robinson's *bête noire*. As a discussant he was at home with all branches of chemistry. While president of the Chemical Society Sidgwick made some of the Club's meetings joint affairs with the Chemical Society and the Royal Institute of Chemistry. The Club often met in the lecture-room at Jesus and would then migrate after the formal proceedings to Sidgwick's rooms in Lincoln, a stone-throw's distance, where discussion would continue informally.

Sidgwick was also prominent in the joint chemistry–physics colloquia where he revealed his instinct for applying mathematics and physics fruitfully to chemical problems. He loved jousting with the destructive Lindemann and relished correcting the errors of specialists. When he tapped on his bald head with talon-like hands, the audience was agog for the chemical vulture to pounce. Though 'Sidger' was acerbic, he was sociable: whenever possible he presided over the tea club of the Dyson Perrins Laboratory, demolishing the sloppy English of those present, delivering impromptu expositions of chemical topics of all kinds, and attacking cuttingly the electronic theory of organic chemistry held by Robinson, the head of the laboratory where Sidgwick poured out the tea. It was this sort of sociability, allied to his synoptic view of chemistry, that made Sidgwick the dominant figure in Oxford chemistry as a whole between the wars, and not Hinshelwood, Robinson, or Soddy.

8.7 The Lure of Industry

Though Sidgwick did not enthuse about organized applied science, Oxford chemistry graduates were prominent in filling directorships and managerial positions in the chemical industry. In the 1920s taking a D.Phil. was a fall-back position for Oxford graduates when they had failed to gain an Oxford fellowship, an academic post elsewhere, or a good post in industry. Though the University kept applied science as a subject at bay between the wars, Oxford's graduates in chemistry were, and continued to be, powerful as researchers, managers, and directors in the larger chemical firms. No other science school at Oxford remotely matched chemistry in placing its graduates in industry. Several graduates of the 1920s and early 1930s became chairmen of important industrial enterprises: Christopher Kearton of Courtaulds, Peter Allen of ICI, Sydney Barratt of Albright & Wilson, and Michael Perrin of the Wellcome Foundation.[81] Some became research directors for a whole company: witness Geoffrey Gaut at Plessey, Bryan Topley at Albright & Wilson, and at ICI J. D. Rose who followed yet another Oxonian, Wallace Akers. The case of Akers, like that of Ernest Walls, managing director of Lever Brothers, reveals that even before the first war Oxford chemistry graduates were not spurning a career in industry.[82]

Their suitability had been particularly appreciated from 1907 by Brunner Mond whose chief chemist and then research manager at its Winnington plant in Cheshire, Francis Freeth, assiduously and snobbishly collected good researchers from Oxford and made Winnington into a notable research department in the British chemical industry between the wars. Brunner Mond was attractive to Oxford graduates who were aware that it was a prosperous firm which paid good salaries and offered alluring prospects of advancement. At the Winnington plant there was the added inducement of membership of the Winnington Hall Club which in some ways was an extension of Oxford college life into the fertile Cheshire plains. The Club, which excluded non-graduates and commercial staff, provided mechanisms for developing chemical ideas and social relations. With its dining-rooms, bars, croquet lawns, guest rooms, and cordial informality, the Club offered remarkable opportunities for vertical and horizontal communication in a pleasant building with a Tudor wing. The Club, to which members were elected on the basis of their scientific and social qualifications, helped to generate corporate pride and loyalty among the scientifically trained staff of Brunner Mond. Winnington and its Club were dominated by Oxford chemists, a feature which was continued when Winnington became the Alkali

---

[81] For Kearton's centrality to Courtaulds, D. C. Coleman, *Courtaulds: An Economic and Social History*, iii: *Crisis and Change 1940–1965* (Oxford, 1980); Peter Christopher Allen (1905–93), deputy chairman of ICI 1963–8, chairman 1968–71, obituary in *Independent*, 1 Feb. 1993; Sidney Barratt (1898–1975), chairman of Albright & Wilson 1958–67, *Who Was Who*; Michael Perrin (1905–88), obituary in *The Times*, 22 Aug. 1988, chairman of the Wellcome Foundation 1953–70.

[82] Geoffrey Gaut (1909–92), obituary in *Independent*, 3 Sept. 1992, first graduate employee of Plessey as chief chemist 1934–63, director of research 1963–85; Bryan Topley (1901–86) was a pupil of Hinshelwood; John Donald Rose (1911–76), director of research, ICI, 1966–72, obituary in *BM* 23 (1977), 449–63; Wallace Alan Akers (1888–1954), first in chemistry 1909, director of research, ICI, 1944–51, obituary in *BM* 1 (1955), 1–4; Ernest Walls (1881–1961), first in chemistry 1902, managing director of Lever Brothers 1920s, and then chairman of North British Artificial Silk & Rayon Company, Jedburgh.

Group of ICI in 1926 when Brunner Mond, Nobel Industries, British Dyestuffs Corporation, and the United Alkali Company merged. At Winnington there was a small contingent of Cambridge engineers but the neighbouring universities of Liverpool and Manchester were sparsely represented. In 1919, with the development of Brunner Mond's synthetic ammonia plant at Billingham, further opportunities arose for Oxford graduates. Though the demands of high-pressure technology there led to the employment of more engineers than at Winnington, Billingham had its own club for its graduate scientists and managers at Norton Hall, modelled on that at Winnington.

The formation of ICI brought no change in recruitment practices for Winnington and Billingham, except that the visits to the Oxford laboratories were made not by Freeth but by Gordon from Billingham and by H. E. Cocksedge, an Oxford chemist who replaced Freeth in 1926 as research manager at Winnington. Sometimes they were joined by yet another Oxford chemist, W. H. Demuth, who was development director at Winnington. These men recruited fellow Oxonians not just on a mafia basis: they were keen to grab young chemists who had had a year's useful research experience for part two of their degree and in some cases were already co-publishers of papers with their tutors. This was so with Peter Allen, tutored by Hinshelwood, and Christopher Kearton, tutored by Thompson; Allen went to Winnington, Kearton to Billingham.

ICI in general and its Winnington and Billingham plants in particular were attractive to Oxford chemistry graduates who presumably knew that some of their leading teachers, such as Hinshelwood, Perkin, and Robinson, were consultants for ICI. The move of two Oxford chemistry dons, Applebey and Hartley, to research posts in industry in 1928 and 1930 respectively sent out a firm signal to undergraduate chemists that a career in industry was worthwhile. No wonder that between the wars seventeen Jesus men, mainly chemists, joined ICI or its predecessors. The Alkali Group based at Winnington was in a strong position within ICI. Its starting salaries were the highest in any group of the company and its promotion prospects best as a result of its being the most profitable group. Oxonians did well at Winnington. Peter Allen reached £1,000 p.a. eleven years after going there. Lincoln Steel received £1,400 p.a. before he was 30. William Lutyens became chairman of the Alkali Group when just 40, and was succeeded by Digby Lawson and Steel. Though the Alkali Group employed in the 1930s about a fifth of the number of chemists to be found in Dyestuffs, about a third of those in General Chemicals, and about half of those in Fertilisers (Billingham) and Explosives, it was disproportionately well represented and powerful in ICI directors appointed from 1926 to 1952. Of the twenty-nine executive directors, nine came from Winnington and six from Billingham. Only one came from Dyestuffs, which employed about 40 per cent of ICI's chemists. As Winnington discriminated so heavily in its recruitment in favour of Oxford and other groups of ICI cast their nets more widely, that meant that Oxford chemists were prominent as directors of ICI. There was a wave appointed in the early 1940s (Akers, Lutyens, Lawson, and Steel) followed by another in the early 1950s (Allen and Prichard). In 1947 and 1951 they were joined by two Oxford-trained lawyers (Killery and Bingen). If Winnington gave opportunities to Oxonians for managerial advancement, it also looked after researchers content with their lot. For, example C. W. Bunn was recruited to Winnington in 1927 by Cocksedge who sent him back to Oxford to study crystal morphology under Spiller. An expert crystallographer who joined Dorothy Hodgkin in Oxford during the war in the successful attack on the structure of penicillin, Bunn stayed with ICI as a researcher for the whole of his career. Though Winnington was the most attractive destination for Oxford chemists, there was a colony of them, mainly from St John's, at Billingham in the 1930s: Akers was chairman, Gordon general manager, d'Leny technical director, Applebey research manager, and Kearton a rising star. Indeed of about sixty Oxbridge graduates employed at Billingham, six were from St John's.[83]

It was at Winnington, the home of the Alkali Group of ICI, that polythene, ICI's major discovery before the second war, was taken through from laboratory experiments to large-scale production. Several Oxonians played an important role. First, in 1931 Robinson, who was a consultant to ICI's Dyestuffs Group, suggested that several reactions be tried under very high pressures without catalysts. One reaction in his list was that between ethylene and benzaldehyde. The work he outlined was done at Winnington by R. O. Gibson and E. W. Fawcett who studied some fifty reactions. In 1933 they subjected ethylene and benzaldehyde to 2,000 atmospheres' pressure and produced a small amount of polythene as the accidental consequence of looking for the chemical result of a specific reaction. Subsequent experiments on polythene were not reproducible and explosions so common that Dyestuffs Group withdrew its support of the work late in 1933.

Meanwhile Michael Perrin, an Oxford graduate in chemistry (1928), had been recruited to Winnington by Freeth who placed him as an ICI employee in the Amsterdam laboratory of Michels, an expert on high-pressure

---

[83] On ICI's attractions, anon., *A Hundred Years of Alkali in Cheshire* (London, 1973), esp. 108–12; Reader, *ICI*, i. 91–2, 219, ii. 11, 70–80, 93; Baron Kearton to author, 7 Apr. 1988; Sir Peter Allen, telephone conversation with author, 9 Feb. 1987; Baker, *Jesus College*, 131–2; Francis Arthur Freeth (1884–1970), research manager at Winnington 1910–27, obituary in *BM* 22 (1976), 105–18; Herbert Edwin Cocksedge (1884–1962), first in chemistry 1907; William Henry Horner Demuth (b. 1898), chemistry part one 1922, chairman of Dyestuffs Division 1940s; Joseph Lincoln Spedding Steel (1900–85), chairman of Alkali Group 1943–5, *Who Was Who*; William F. Lutyens (1891–1971), third in chemistry 1913, chairman of Alkali Group 1931–9; Digby R. Lawson (d. 1947), short course 1919, chairman of Alkali Group 1939–43; Charles Ross Prichard (1903–76), first in chemistry 1925; Valentine St John Killery (d. 1949), third in law 1922; Eric Albert Bingen (1898–1972); Charles William Bunn (1905–90), first in chemistry 1927, obituary in *BM* 37 (1991), 71–83.

techniques. In 1932 while on holiday in England, Perrin and J. C. Swallow, deputy director of research at Winnington, wrote a report on the desirability of studying the mechanisms of reactions at 20,000 atmospheres. In late 1933 Perrin returned to Winnington where he did high-pressure work under Swallow. Influenced and advised by Hinshelwood, Perrin began to study the effect of high pressure on reactions in solution, initially with Gibson and Fawcett, as a general phenomenon of physical chemistry. The interest in high pressures led him in late 1935 to repeat the 1933 experiment but without benzaldehyde. He succeeded in synthesizing polythene from ethylene and in defining the reaction conditions which enabled the polymer to be made reproducibly and without explosions. Given the Oxford input into the discovery and rediscovery of polythene, it was entirely appropriate that in 1939 Bunn, another Oxonian, revealed its structure using X-ray methods.[84]

[84] Reader, *ICI*, ii. 349–58; D. G. H. Ballard, 'The Discovery of Polyethylene and its Effect on the Evolution of Polymer Science', in R. B. Seymour and T. Cheng (eds.), *History of Polyolefins* (Dordrecht, 1976), 9–53; M. W. Perrin, 'The Story of Polythene', *Research*, 6 (1953), 111–18; R. O. Gibson, E. W. Fawcett, and M. W. Perrin, 'The Effect of Pressure on Reactions in Solution', *Proceedings of the Royal Society of London A*, 150 (1935), 223–40, communicated by Hinshelwood who gave helpful criticism; E. G. Williams, M. W. Perrin, and R. O. Gibson, 'The Effect of Pressure up to 12,000 kg/cm^2 on Reactions in Solution', ibid. 154 (1936), 684–703, received 17 Dec. 1935 and communicated by Hinshelwood who was again thanked for helpful criticism; C. W. Bunn, 'The Crystal Structure of Long-Chain Normal Paraffin Hydrocarbons: The Shape of the CH$_2$ Group', *Transactions of the Faraday Society*, 35 (1939), 482–91.

9
Refugee Scientists

In spring 1933 Jewish intellectuals in Germany suddenly found themselves in a desperate situation as a result of actions taken by the national socialists who had come to power in January of that year. In March the reign of terror launched by the Nazis against the freedoms of speech, press, and assembly, and against the communists, led Einstein to declare publicly that he would not return to Germany because of the Nazis' policies and to resign from the Prussian Academy of Sciences. In April the Nazis moved rapidly and decisively against non-Aryans, many of whom were Jews. Within five days the civil service law was passed, memoranda about the dismissal of Jews were sent to universities, and a person was deemed non-Aryan even if just one of his or her grandparents was Jewish. By the end of the month James Franck and Fritz Haber, two Nobel prize-winners whose service in the first war exempted them from the decrees, had resigned in protest against anti-Semitism. At that time neither intended to emigrate.

Early in May, 20,000 volumes, including some by Einstein, were burned at the Unter den Linden Strasse in Berlin. After this decisive event all German university scientists who were Jews, or Christians with Jewish relatives, had to face a future of persecution, dismissal, expulsion, or emigration. In physics and in mathematics, in which Jews were prominent, the effect of forced migration was pronounced: in the first two years of Nazi rule, about 25 per cent of all physicists and 20 per cent of all mathematicians in German universities were dismissed. For many German scientists who migrated, Britain was a temporary refuge before they moved elsewhere, preferably to the USA, for a permanent haven and employment.

The main organized British response to the plight of German scholars was the voluntary Academic Assistance Council (AAC) formed in late May 1933 with Rutherford as its president. It was funded by donations from British university teachers and the Central British Fund for German Jewry. It acted as an intermediary between possible employers and the refugees, tending to favour those aged 30 to 40 who had an established reputation but were not too senior to compete with the British for junior university posts in the United Kingdom. The AAC, renamed the Society for the Protection of Science and Learning (SPSL) in 1936, was always aware that creating a new post specifically for a German refugee might arouse jealousy among the British who would then look less favourably on the plight of German scientists in general. The AAC was mainly an academic labour-exchange but it also gave temporary grants to refugees in financial distress. Though the AAC was undoubtedly important, a few individuals acted independently of it in trying to place refugee scientists in temporary and, in a few cases, permanent posts. One of the most prominent British activists was Lindemann, Dr Lee's professor of experimental philosophy at Oxford. He was first contacted by the AAC on 24 July 1933. By then he had secured for Britain, and especially Oxford, the services of several German physicists and chemists through his own independent initiatives taken systematically from May 1933 just when the AAC was being formed.[1]

This chapter examines the differential reception of refugee scientists, mainly from Germany, but also from Spain, Czechoslovakia, and Austria, in Oxford between 1933 and 1939. It considers in detail the ways in which Lindemann used the *émigré* physicists to transform the Clarendon Laboratory for which he was responsible. His entrepreneurship makes a striking contrast with the indifference shown by Townsend, Wykeham professor of experimental physics, who ran the other physics laboratory at Oxford, the electrical laboratory, in which no refugee was to be found before the second war. Some writing on the refugee scientists looks at them primarily from their point of view; that is understandable given their desperate situation. My approach is somewhat different in that I use the notion of congruent or mutual interests to explain why a particular refugee was accommodated or not in a particular science department at Oxford. If the local activist and the refugee had or somehow came to have compatible or symbiotic interests, then a deal could be struck; if their interests were incompatible or opposed, then no deal could be done. In those cases where institutional approval or external funding was involved, the interests of the institution or funder had to be congruent with those of the other two parties. If that congruence could not be achieved, then again no deal could be made. The term interest is used here in a large sense to mean not just the selfish pursuit of individual advantage but mainly being concerned with or in something, or affected by

---

[1] Useful general surveys are: A. J. Sherman, *Britain and Refugees from the Third Reich 1933–1939* (London, 1973); A. D. Beyerchen, *Scientists under Hitler: Politics and the Physics Community in the Third Reich* (New Haven, 1977); N. Bentwich, *The Rescue and Achievement of Refugee Scholars: The Story of Displaced Scholars and Scientists 1933–1952* (The Hague, 1953); W. H. Beveridge, *A Defence of Free Learning* (London, 1959); W. E. Mosse (ed.), *Second Chance: Two Centuries of German-Speaking Jews in the United Kingdom* (Tübingen, 1991), esp. P. K. Hoch, 'Some Contributions to Physics by German-Jewish Emigrés in Britain and Elsewhere', 229–41, and P. Weindling, 'The Contribution of Central European Jews to Medical Science and Practice in Britain, the 1930s to 1950s', 243–54; R. E. Rider, 'Alarm and Opportunity: Emigration of Mathematicians and Physicists to Britain and the United States, 1933–1945', *Historical Studies in the Physical Sciences*, 15 (1984), 107–76; P. Hoch, 'The Reception of Central European Refugee Physicists of the 1930s: USSR, UK, USA', *Annals of Science*, 40 (1983), 206–46. James Franck (1882–1964), professor of experimental physics, Göttingen University, 1921–33.

it, in respect of benefit or detriment. This notion of interest is usefully protean: it includes the purely intellectual and the avowedly entrepreneurial.

9.1 The University, Colleges, and Departments

There was no concerted university response to the persecution and dismissal of academic scientists in Germany. That became clear in 1933. In May the Hebdomadal Council set up a committee to examine the obligations and opportunities which the refugees provided. Next month the Council merely expressed the hope that colleges and departments would try to give just temporary accommodation to displaced academics. In so doing it acknowledged that it had no statutory power over the colleges and that it would not impose a staffing policy on the science departments. In any event planning of the science departments and of the Science Area was at that time in its infancy. The Council insisted that whatever was done for refugees should not involve any financial charge on the University. That apparently ungenerous response was the result of the Owen affair which left the University between 1931 and 1938 liable for £750,000 in the associated court case. The official Oxford line was made clear in late June 1933 at a meeting of the Committee of Vice-Chancellors and Principals of British Universities: no refugee was to be given even temporary accommodation if he or she would displace a British subject, any post made available should be so specialized that there would be no competition with the British already in post, and hospitality should be given for no more than three years. The result of such academic protectionism was that colleges and departments were left to act as they thought fit, subject to the restrictions imposed by the University. Even Veale, registrar and master planner, did not try to create a positive university policy: if he was prodded by a concerned science professor, he would lobby the Home Office about the family difficulties of a particular refugee, as he did for Chain at Florey's request. When the Owen affair was settled in 1938, the University took a slightly more favourable attitude to refugees. In February 1939 it made the Sheldonian Theatre available for a large SPSL meeting presided over by Gordon, the vice-chancellor, and strongly supported by Lindsay. Next month the sum of money available to pay refugees for lectures was raised from £70 to £200 p.a. In July the University responded to Lindsay's pressure by permitting Czech students to be admitted without paying college fees, a concession which was extended to all displaced students in October. Otherwise the University's corporate response was limited to setting up in February 1939 a Bureau of Information for Refugee Scholars whose functions were to elicit particulars from colleges of the support they were prepared to give, to recommend appropriate colleges for refugees, and to raise money from the colleges. Preference was to be given to refugees with prospects of re-migration. The Bureau was an attempt to fill the gap caused by the lack of a central university fund. The SPSL opposed the establishment of the Bureau as redundant and suggested that an active liaison officer in Oxford was needed. Ronnie Bell, chemistry fellow of Balliol, was appointed and rapidly became the driving force of the Bureau which formally expired before the start of the war.[2]

By May 1934 Oxford was not in the van in offering sanctuary to displaced scholars. London University had taken 67, Cambridge 31, Oxford 17, and Manchester 16.[3] The hugeness of London, with its federated colleges and institutes, gave it unique advantages. Cambridge, of similar size to Oxford, was more generous partly because it was less anarchic in its organization and partly because two of its leading scientists, Hopkins and Rutherford, promoted the cause of the refugees. By 1935 Hopkins had six in his department of biochemistry. Fearing that a concentration of refugees in his highly visible Cavendish Laboratory might arouse an anti-Semitic backlash, Rutherford helped them mainly through his active presidency of the AAC from its inception. It was easier to spread out mathematicians so G. H. Hardy had no qualms about finding or creating room for more than a dozen of them.[4] At Oxford most of the leading activists on behalf of refugees were arts men such as H. A. L. Fisher, Lindsay, and Sadler, who were heads of colleges, and Gilbert Murray, a professor who was devoted to internationalism. They were reinforced in 1937 by Beveridge, then secretary of the SPSL, who came to Oxford as master of University College.[5] Only two Oxford scientists, Robinson and Sherrington, signed the famous first circular issued by the AAC in May 1933. Robinson was to practise what he urged others to do but Sherrington, then aged 75, was too ill to act as a focus of aid in Oxford in the life sciences.

Given the lack of a corporate response by the University it was left to individual activists in colleges and departments to offer temporary sanctuary

[2] P. J. Weindling, 'The Impact of Central European Medical Scientists on British Medicine: A Case-study of Oxford, 1933–45', in M. G. Ash and A. Söllner (eds.), *Forced Migration and Scientific Change: Émigré German-Speaking Scientists and Scholars after 1933* (Cambridge, 1996), 86–114 is invaluable; OUA, Refugee Committee. 1933–43, UR/SF/RC/1, minutes of Committee of Vice-Chancellors' meeting, 24 June 1933; OUA, Information Bureau for Refugee Scholars, UDC/M/31/1; OUA, Society for the Protection of Science and Learning. 1938–46, UR/SF/PSL/1; OUA, SPSL. Privileges for Refugee Students. 1939–52, UR/SF/PSL/2; OUA, Correspondence between vice-chancellor and Committee, 1938–9, UR/SF/PSL/3; OUA, Correspondence concerning individual scholars. 1938–9, UR/SF/PSL/4; SPSL, 130/3; *HCP* 172 (1939), 135–6; George Stuart Gordon (1881–1942), vice-chancellor 1938–41.

[3] E. Rutherford, 'The Wandering Scholars: Exiles in British Sanctuary', *The Times*, 3 May 1934.

[4] Weatherall and Kamminga, *Dynamic Science*, 60–2; Hoch, 'Physics', 240–1; Rider, 'Alarm', 159.

[5] Michael Ernest Sadler (1861–1943), master of University College 1923–34; George Gilbert Aimé Murray (1866–1957), Regius professor of Greek 1908–36; William Henry Beveridge (1879–1963), master of University College 1937–44.

to refugee scientists. There was no possibility at Oxford of creating research institutes which they would adorn. The University's experience of Owen's nefarious doings at the Agricultural Engineering Research Institute generated a general suspicion of research institutes as uncontrollable both intellectually and financially. The various science faculties at Oxford enjoyed little power and never took initiatives to help refugees. The men's colleges were small: by the late 1930s the average size of their fellowship hovered around fifteen. The women's colleges were even smaller. In such a situation and given the general paucity of tutorial fellowships in science, a few colleges at best gave grants or some form of affiliation to refugee scientists. The response of science departments was also varied owing to the varying interests and resources of prospective helpers and hosts of refugees.

In arts subjects refugees did not need laboratory facilities so it was easier for colleges to help them. It seems that the only refugee scientist to be aided exclusively by a college was Stefan Jellinek, an expert on injuries caused by high-voltage electric shocks.[6] Dismissed by the Nazis from his post as electropathologist at the University of Vienna in 1938 when he was 67 years old, in August 1939 he arrived in Oxford where he was supported by grants from the SPSL, from individuals, and from Queen's which made him a research student and gave him £50 p.a. He landed at Queen's because his son, an undergraduate there, lobbied the provost, R. H. Hodgkin, who was not without a social conscience. Jellinek did not press for laboratory facilities: he was content to write up in book form the results of years of previous research.

Some departments and institutions did nothing to help refugee scientists. In the cases of zoology, forestry, agriculture, experimental psychology, the Old Chemistry department, the college chemistry laboratories, the electrical laboratory, geology, mineralogy, history of science, and history of technology, no interest was taken in refugees. The reasons are clear: most of them were small enterprises, fighting for survival, living off a small budget, and enduring limited accommodation. Some were intellectually insulated or content with their research traditions. Forestry and agriculture were consumed with their own administrative problems. The only recorded case of a refugee applying to one of these indifferent departments concerns Kirchheimer, a palaeobotanist from the University of Giessen whose request to the Registry was passed to the aged Sollas, professor of geology. He realized that 'brains in Germany seem to be going cheap and we have no tariff for them in England'. Accordingly he wrote a discouraging response to the Registry which told Kirchheimer there was no place for him at Oxford.[7] In

---

[6] SPSL, 404/8; Stefan Jellinek (1871–1968) stayed in Britain, producing *Dying, Apparent Death and Resuscitation* (London, 1947).

[7] Sollas to Registry, n.d. [late 1933/early 1934], UR/SF/RC/1; Franz Waldemar Kirchheimer (b. 1911) had just taken his Ph.D. at Giessen.

astronomy and in contrast with Cambridge, Oxford had no resources to offer a refugee. It is clear that Plaskett, the Oxford professor of astronomy, thought very highly of Hermann Brück who, as a Roman Catholic engaged to be married to a non-Aryan, was dismissed in summer 1936 from his post of assistant at the Astrophysics Laboratory, Potsdam. For the next twelve months he was supported by a grant from the Vatican Observatory, Rome. In early 1937 he came second to Zanstra in the advertised competition for the Radcliffe travelling fellowship. He then secured a job in summer 1937 as an assistant observer at the Solar Physics Laboratory, Cambridge, and ended his career as Astronomer Royal for Scotland and professor of astronomy, University of Edinburgh.[8]

In life sciences the botany department was not inactive. Tansley was a very strong believer in academic freedom so he was keen to help Leo Brauner, a botanist at the University of Jena who had been dismissed in April 1933.[9] Hearing about Brauner's plight, on his own initiative Tansley offered him a gratis place in the botany department and by sustained pressure on Beveridge secured for Brauner, whose salary stopped in September 1933, a grant from the AAC of £250 for 1933–4. Tansley was keen to accommodate a plant physiologist whose research overlapped usefully with that of his colleagues Snow and W. O. James. He was also motivated by charity. Brauner arrived in Oxford in August 1933 but in early October went to Switzerland where he successfully negotiated a chair at the University of Istanbul because he needed a larger income than the AAC grant: his parents were destitute, his father having also been dismissed. By late October the AAC money had not arrived so Tansley gave Brauner an advance of £100 for the journey to Turkey and maintenance there. Brauner was one of many German beneficiaries of the reorganization of the University of Istanbul as part of the Turkish government's occidentation policy. When the new university was opened in November 1933, Brauner was one of its founding professors and stayed there until 1955.

In 1939 the botany department came to the aid of Kurt Wohl, a physical chemist who had been dismissed from the University of Berlin in 1935 and who had worked for IG Farben, the huge German chemical cartel, from 1936 to early 1939 as a researcher on detergents. Then, at the instigation of Franz Simon, a German refugee in Oxford who knew Wohl well at Berlin, Wohl was invited by Bell to lecture in Oxford.[10] He secured a temporary visa, arrived in February 1939 in Oxford where he stayed with the Simons,

---

[8] SPSL 325/4; Hermann Alexander Brück (b. 1905), Astronomer Royal, Scotland, and professor of astronomy, University of Edinburgh, 1957–75.

[9] SPSL 196/5; Tansley memorandum on Brauner, 10 Aug. 1933, UR/SF/RC/1; Leo Brauner (1898–1974).

[10] SPSL, 227/5; SP, 14/1/8C. Kurt Wohl (1896–1962) ended his career as professor of chemical engineering, University of Delaware, Wilmington; Franz Eugen Simon (1893–1956).

and next month began unpaid work in the botany department with James, with whom he had exchanged offprints, on the kinetics of photosynthesis in green plants. From April he was supported by Balliol and Queen's (£100 p.a. each) and the department (£50 p.a.). In June Magdalen gave him £25 for six months as a result of the advocacy of Osborn, whose botany chair was attached to Magdalen. Thus the botany department and three colleges combined to help Wohl, who left Oxford in 1942 and soon secured a post at Princeton University.

In the biomedical sciences, physiology, biochemistry, pathology, pharmacology, and human anatomy, refugees were accommodated with a variety of results. Two of them, Chain and Bülbring, had posts elsewhere in England before arriving in Oxford where they stayed for some time. Having prudently emigrated from Germany in spring 1933, Chain worked for two years in Hopkins's department of biochemistry at Cambridge until Florey recruited him to the pathology department at Oxford in autumn 1935, initially on a one-year appointment at £200 p.a. which he was happy to accept. From summer 1936 Chain was a departmental demonstrator, still on a salary of about £200 p.a., until in 1941 he was promoted to a university demonstratorship. Even though he was a joint Nobel prize-winner in 1945 for his work with Florey on penicillin and a cultivated cosmopolitan man, no college made him a fellow and the University did not confer even a readership on him. Frustrated and disappointed on more than one count, and bitterly at odds with Florey, he left Oxford in 1948 to be the scientific director of the International Research Centre for Chemical Microbiology, Rome. The other biomedical scientist to be recruited indirectly to Oxford was Edith Bülbring who came as a research assistant with J. H. Burn when he assumed the chair of pharmacology in 1937. She arrived in England in 1933 and with the help of H. H. Dale became Burn's assistant in the biological standardization laboratory of the Pharmaceutical Society, London.[11] On her move to Oxford she was paid by the Nuffield Medical Research Fund and soon became a departmental demonstrator but remained very much under Burn's wing until in 1946 she became a university demonstrator and in 1948 launched the research on the physiology of smooth muscle which eventually brought her a readership, an Oxford chair, and an FRS.

In the pre-clinical biomedical sciences temporary refuge was found for several refugees. Though Peters had bungled an opportunity to secure no less than Hans Krebs in 1933 for the biochemistry department, in the late 1930s he accommodated Severo Ochoa who was assistant head of the

---

[11] E. P. Abraham, 'Chain', *BM* 29 (1983), 43–91; T.B. Bolton and A. F. Brading, 'Bülbring', *BM* 38 (1992), 69–95; Edith Bülbring, FRS 1958, reader in pharmacology 1960–7, professor 1967–71; Henry Hallett Dale (1875–1968), Nobel prize-winner 1936, director of National Institute for Medical Research, Hampstead, London, 1928–42, president of Royal Society 1940–5.

Institute for Medical Research at Madrid.[12] When the Spanish civil war began Ochoa gained sanctuary amazingly in Germany and then at the Marine Biological Laboratory, Plymouth. A specialist in the biochemistry of muscle action and metabolism, Ochoa came to Oxford in late 1937 and was paid by the Nuffield Medical Research Fund to work with Peters on the metabolism of vitamin B, Peters's main research interest. By 1941 Ochoa had found a haven at Washington University, St Louis. In 1959 he was awarded a Nobel prize for his work in molecular biology. Peters had therefore a strange track record with refugees: he failed to grab one future Nobel prize-winner (Krebs) and to keep another (Ochoa). In pathology Florey was well aware of the benefits conferred on his department by Chain, but his efforts to recruit Hans Sachs, an eminent immunologist from Heidelberg who came to England in April 1936, were unsuccessful because Florey could not finance him.[13] Sachs was happy in principle to come to Oxford; in practice he stepped westwards in 1936 to Ireland where he died. In physiology Claude Douglas aided Ernst Brieger who stayed in England and Heinrich Schwarz who left for the USA in 1940. Even before Brieger, medical director of the tuberculosis hospital at Breslau, was dismissed effectively in April 1934, J. S. Haldane had arranged for him to work at Oxford under Douglas and Priestley learning useful physiological techniques pertaining to human respiration. From summer 1934 to summer 1935 Brieger worked in Oxford and at the Papworth tuberculosis hospital, Cambridgeshire, being supported by a grant of £150 p.a. from the AAC and a donation from the Papworth Village Settlement which became his long-term employer in summer 1935.[14] Douglas also found laboratory space for Heinrich Schwarz, a Catholic endocrinologist who was dismissed from his post at the University of Vienna and was prohibited from private medical practice in 1938. In desperation he wrote to the SPSL and A. V. Hill who put him in contact with Douglas, who gave him bench space in September 1939, three months after his arrival in England.[15] Schwarz was supported at Oxford by the SPSL, the Catholic Committee for Refugees from Germany, and the Quakers' Germany Emergency Committee, giving him £250 p.a. all told until in May 1940 he left for the USA for a research post in the Philadelphia General Hospital. In human anatomy Clark, who hated fascism, gave help to three refugees. In 1937 he found accommodation for six months for Pio del Rio-Hortega, a famous neurohistologist who was an expert on brain

---

[12] SPSL 418/9; Severo Ochoa (1905–93) made his career in the Medical School, New York University, where he was professor of pharmacology 1946–54 and biochemistry 1954–74.
[13] SPSL 544/3; Hans Sachs (1877–1945).
[14] SPSL 387/3; Ernst Brieger (1891–1969).
[15] SPSL 385/2; Heinrich Schwarz (b. 1899); Archibald Vivian Hill (1886–1977), Nobel prize-winner 1922, Foulerton research professor, Royal Society, London, 1926–51, keenly supported the AAC and SPSL.

tumours and head of the National Cancer Institute, Madrid, which was destroyed in 1937 during the civil war.[16] Through the efforts of Hugh Cairns, professor of surgery, he came to work with Clark and then from May 1938 in Cairns's new laboratory where he was employed as a neuropathologist paid by the Nuffield Medical Research Fund to classify tumours. His wife, a laboratory assistant, was paid from the same source. She returned to Spain in summer 1939 but he went the next year to Argentina. Clark's interest in neuroanatomy made him sympathetic to Benno Schlesinger, a well-known Viennese neuro-surgeon, who applied to the SPSL for help after he was dismissed in March 1938, the month of the Anschluss of Austria. He was employed by Clark as a research assistant funded by the Nuffield Medical Research Fund to work for eight months in 1938–9 on the venous draining of the brain.[17] Schlesinger eventually left England for the USA in June 1940.

In the clinical medical departments the new facilities provided by the Nuffield medical benefaction enabled several refugees to be given a temporary home, some of them being paid as researchers by the Nuffield Medical Research Fund. The leading activist was Cairns. He provided space and salary for the Hortegas from May 1938. He also made arrangements for refugees to work as volunteers in other Nuffield departments or in his own. Having failed to secure a post at the Littlemore Hospital, Oxford, for Eugen Pollak, a Viennese neuropathologist who came to England in January 1939, Cairns landed him a three-month job working in the pathology laboratory of the Radcliffe Infirmary which he left in summer 1939 for a post in the neurosurgery department of Manchester Royal Infirmary where he ended his days.[18] Cairns also placed in the Radcliffe pathology laboratory Gustav Steiner, an Austrian refugee, who worked voluntarily on cerebrospinal fluids.[19] In his own department Cairns accommodated Ludwig Guttmann, a neurologist at the Jewish Hospital, Breslau, who was informed in August 1938 of his dismissal effective in October.[20] He immediately approached Cairns, arrived in Oxford in March 1939, and stayed with Lindsay for three weeks in the Master's Lodge at Balliol. With support from Balliol and the SPSL and Cairns's encouragement he continued his research on sweating, a topic which was related to the diagnosis and treatment of spinal and peripheral nerve injuries. Cairns deplored the messy laboratory habits of Guttmann, who was frustrated because he was excluded from patients. To Cairns's relief, he left Oxford early in 1944 to achieve fame as director of the

---

[16] Fraenkel, *Cairns*, 121, 262–3; *OUG* 69 (1938–9), 367; Pio del Rio-Hortega (1882–1945), *DSB*.
[17] SPSL 424/13; Benno Schlesinger (b. 1900).
[18] SPSL 397/ll; Eugen Pollak (1890–1953).
[19] *OUG* 70 (1939–40), 382.
[20] SPSL 394/8; Fraenkel, *Cairns*, 134; obituary of Ludwig Guttmann (1899–1980), *BM* 29 (1983), 227–44.

National Spinal Injuries Centre at Stoke Mandeville Hospital, Aylesbury. In the department of anaesthetics, Professor Macintosh secured £250 in 1938–9 from the Nuffield Medical Research Fund for Grita Weiler to work in the physiology department on the effect of barbiturates on muscles.[21] In the Nuffield Institute for Medical Research run by Gunn, Ilse Sachs worked for the same time for the same salary from the same source on adrenaline until in 1939 she married Arthur Cooke, a research physicist in the Clarendon Laboratory.[22] Gunn also gave accommodation for six months just after the outbreak of war to Otto Loewi, a Jew in his mid-sixties who had been imprisoned for two months after the Anschluss and dismissed from his chair of pharmacology at Graz.[23] Loewi had shared a Nobel prize in 1936 with H. H. Dale for their work on the chemical transmission of nerve impulses. Having been deprived by the Germans of both his pension and Nobel prize money, Loewi came to England in September 1938 and then spent eight months working in Brussels. On the failure of a plan for Loewi to go in summer 1939 to the Banting Institute, Toronto, he returned to England and, at Dale's instigation, was given accommodation at Oxford in Gunn's Institute until in spring 1940 he went to the USA to be research professor at New York University. In the department of clinical medicine, Witts gave temporary accommodation to three refugees including Gabriele Ehrlich, a Jewish spinster who was dismissed in June 1938 from her post as a researcher in medical chemistry at the Children's Hospital, Vienna.[24] Supported by St Hilda's, she worked with Witts in spring 1939 but was unhappy because she felt cold-shouldered by two taciturn north-country colleagues. In autumn 1939 she left for the USA aided by St Hilda's who raised £65 for her passage.

In the pre-clinical and clinical biomedical sciences many refugees were helped by the Nuffield medical benefaction of the late 1930s. It paid for new facilities in which nine refugees were accommodated as volunteers in 1938–9. Through its Medical Research Fund it provided that session £1,780 in the form of salaries for six refugees (Hortega, Mrs Hortega, Weiler, Sachs, Ochoa, Schlesinger) and £2,230 if the £450 paid to Bülbring is included.[25] There was no equivalent general benefaction and no associated research fund internally available in the physical sciences so that the funding of refugees in those subjects was more difficult than in medical sciences. In mathematics there was the missed opportunity of Emmy Noether and the sad case of Artur Winternitz. In June 1933 Helen Darbishire, principal of Somerville, told Beveridge that her college wanted to offer hospitality to a

[21] *OUG* 70 (1939–40), 377, 387. Weiler stayed until 1950.
[22] Ibid. 384; Ilse Sachs (d. 1973); Arthur Hafford Cooke (1912–87).
[23] SPSL 414/10; Otto Loewi (1873–1961), obituary by Dale, *BM* 8 (1962), 67–89, esp. 81–2.
[24] SPSL 367/2; Gabriele Ehrlich (b. 1899).
[25] *OUG* 70 (1939–40), 377–92.

distinguished German woman scholar.[26] She then heard that Noether had been suspended from her post at Göttingen and began negotiations which were not resolved at the end of September. The Rockefeller Foundation would not fund Noether for just one term, Darbishire saw some teaching as essential but was worried about finding it, and she felt that she had to consult her governing body which was not due to meet until October. She was thinking of paying Noether £24 for a term's lectures and giving her free board and residence in Somerville, but Noether preferred action to inconclusive negotiation and accepted the offer of a post at Bryn Mawr College, Philadelphia, in autumn 1933. Winternitz was peculiar in that he had been born in Oxford so, when he lost his assistant professorship of mathematics at the University of Prague in 1938, he appealed in desperation to Whitehead, fellow in mathematics at Balliol, who helped to secure a grant of £125 for six months from the SPSL for Winternitz who arrived in England destitute in April 1939.[27] He worked unpaid in Oxford in loose association with Whitehead who thought him one of the goofiest men he had ever met and incapable of doing anything except mathematics. Unemployable and unmovable, Winternitz survived the war in Oxford on grants from the SPSL and the Leverhulme Trust.

In engineering science Southwell was initially cool about taking any refugee: his laboratory was already full. But when he heard about Andreas Gemant, who had written to the vice-chancellor, he changed his mind. Gemant was a Hungarian specialist in dielectrics which he had taught and written about as a *privat-docent* at the Berlin Technische Hochschule until he was dismissed in October 1933. Southwell was keen to develop work on dielectrics under Moullin, the reader in electrical engineering. So, too, was Fleming, the director of research and education for Metropolitan-Vickers Electrical Company, Manchester.[28] Through the mediation of Lindemann, Met-Vick agreed to pay Gemant £500 p.a. for two years from January 1934 to research in Southwell's department, the grant being made via the AAC so that Gemant did not seem to be on the staff of either Met-Vick or the University. In December 1935 Met-Vick renewed its grant for all 1936, but on its expiry was not prepared to employ Gemant as a researcher: he was, in its view, not an initiator and was too 'Continental' to be put into a British industrial laboratory. Even so, Met-Vick, through Fleming's influence, kept Gemant on its books at £500 p.a. for all 1937, giving him in March £375 as

---

[26] SPSL 532/23.
[27] SPSL 286/3; Winternitz to Whitehead, 25 Jan. 1939, UR/SF/PSL/4; Artur Winternitz (b. 1893).
[28] SPSL 328/4–5; Southwell to Veale, 5 Aug. 1933, UR/SF/RC/1; Andreas Gemant (b. 1895); A. Gemant, *Electrophysik der Isolierstoffe* (Berlin, 1930) and *Liquid Dielectrics* (New York, 1933); Arthur Percy Morris Fleming (1881–1960), *DNB*, director of research and education, Met-Vick, 1931–54.

advance salary and a big hint that he should job hunt in the USA. He went in May 1937 to the USA where he received five offers, and accepted the best one from the University of Wisconsin for September 1937. Though Gemant was welcomed by Southwell as an active publisher in a field which the latter wished to develop, Gemant became a problem in two ways: he would not teach; and being temperamentally optimistic he found it difficult to accept that his position at Oxford was not permanent and that his grant from Met-Vick was only temporary.

Apart from Lindemann, the leading activist for refugees among physical scientists was Robinson, Waynflete professor of organic chemistry. A signer of the AAC circular of May 1933, he appreciated the great contributions made by Germans to organic chemistry so he quickly gave free accommodation to two organic chemists, Fritz Arndt from Breslau and Arnold Weissberger from Leipzig, both of whom had been dismissed from their posts.[29] They were paid £600 p.a. and £400 p.a. respectively by ICI for two years as beneficiaries of Lindemann's private programme for supporting refugees, and they received expenses of £268 from the Rockefeller Foundation. In 1934 Arndt moved to a chair at Istanbul where he stayed until he retired. Weissberger had greater difficulty in finding a permanent home. Though he published articles and an important book on organic solvents in 1935 and supervised research in the Dyson Perrins laboratory, his ICI grant was not renewed in autumn that year so reluctantly in 1936 he left academia for a post as an industrial researcher in synthetic organic chemistry with Kodak in Rochester, NY. Herbert Appel, another organic chemist from Leipzig, was not officially dismissed though classed as a non-Aryan but he sought sanctuary first in Switzerland and then for a year from October 1934 working with Robinson on anthocyanins before moving on to Birmingham for a year in Norman Haworth's chemistry department.[30] Robinson also helped Gerhard Weiler, a specialist in small-scale organic chemical analysis, a technique then new to the Dyson Perrins Laboratory.[31] He arrived from Berlin in March 1934 with his apparatus, was installed in the Dyson Perrins Laboratory, and quickly established his micro-analytical enterprise as an important resource for many scientists in Oxford and in industry. By the late 1930s he was in commercial partnership with Fritz Strauss selling micro-analytical services from private premises in Oxford. He was naturalized in 1946 and ended his days in Oxford. In late 1938 Robinson tried to help a prospective refugee, Richard Willstätter, an ageing organic chemist and Nobel prize-winner who had resigned from his chair at Munich in 1924 in protest against growing

[29] Robinson note, 23 Nov. 1933, SPSL 24/8; SPSL 208/4, 227/3; Fritz Arndt (1885–1969); Arnold Weissberger (b. 1898) worked for Kodak 1936–75; A. Weissberger and E. S. Proskauer, *Organic Solvents: Physical Constants and Methods of Purification* (Oxford, 1935).
[30] SPSL 208/3; Herbert Hans Karl August Appel (b. 1907).
[31] SPSL 226/5; Gerhard Weiler (1899–1995).

anti-Semitism.[32] Robinson deeply admired Willstätter's work on natural colouring matters and realized that some pretext was needed to persuade the Nazis to allow Willstätter to leave Germany. Robinson officially invited him to Oxford to lecture in 1939, telling the registrar that Oxford should grab him before the Americans fell over each other to do so. In the event Willstätter was offered sanctuary and unique facilities by the Swiss government in January 1939 and migrated in March to Locarno where he died during the war.

The most successful activist in Oxford for securing posts for refugees, mainly in Oxford but also elsewhere, was Lindemann who persuaded ICI in May 1933 to provide temporary research grants for them. Naturally his first concern was the Clarendon Laboratory for which he was responsible. His recruitment of a cohort of low-temperature physicists from Breslau and of a spectroscopist from Göttingen transformed the research output and practices of the laboratory. Lindemann appreciated their contribution so much that he managed to find funding for them up to the outbreak of war. In theoretical and nuclear physics the refugees he recruited were less successful and, like many of their contemporaries, were birds of passage. In order to appreciate the effects of the refugees on the Clarendon, it is necessary to analyse Lindemann's inheritance from Clifton (his predecessor) and the regime at the Clarendon 1919–33, not forgetting Oxford's other physics laboratory, the electrical, where no refugee was to be seen.

For a total of sixty-seven years from its establishment in 1872 to its evacuation in 1939, the original Clarendon Laboratory was directed by only two professors of experimental philosophy, Robert Bellamy Clifton and Lindemann. When Clifton resigned from his chair in 1915 aged 79, having occupied it for no less than half a century, the laboratory was widely regarded as moribund. In 1939 it had been transformed: the research and academic staff had increased from two to about twenty, the technical staff from one to five, and a new Clarendon building was ready for occupation. The laboratory had at last acquired an identity in research. Through the work of a trio of low-temperature physicists whom Lindemann had recruited from the Technische Hochschule at Breslau, namely Franz Simon, Kurt Mendelssohn, and Nicholas Kurti, the Clarendon had gained an international reputation in cryogenics. It had also become a leading laboratory for high-resolution spectroscopy, through the combined efforts of Derek Jackson and Heinrich Kuhn whom Lindemann had invited to Oxford from the University of Göttingen in 1933. In atmospheric physics Gordon Dobson had made Oxford well known as the informal centre of international collaborative research on atmospheric ozone. Under Clifton research was simply

---

[32] SPSL 566/2; Robinson to Veale, 30 Nov. 1938, UR/SF/RC/1; on Richard Willstätter (1872–1942), Nobel prize-winner 1915, see Robinson's adulatory Willstätter memorial lecture, *Journal of the Chemical Society* (1953), 999–1026.

not encouraged; but under Lindemann, group research flourished from 1933 in low-temperature physics, with induction of postgraduates into research taking place via the master–apprentice relation which replaced Lindemann's previous mode of chucking promising young researchers in at the deep end. This change of laboratory practice was made most manifest in publications. Compared with the nuclear physicists in the laboratory, the cryogenicists were steady and prolific publishers: in the last academic session before the war no fewer than 19 of the 22 papers published were in low-temperature physics.[33]

9.2 Lindemann: Inheritances, Aims, Early Recruits

Poor Clifton has had a bad press.[34] Recent historians describe him as *fainéant* and characterize physics at the Clarendon under him as languishing, dull, undistinguished, and neglected. The laboratory in 1915 has been described as moribund, defunct, a sorry legacy, and a depressing inheritance. It has been castigated as quietly sinking into a trough for half a century. Its career under Clifton has been painfully contrasted with that of the Cavendish which rose under Rayleigh and Thomson in those years to a peerless position in British physics. Even the obituaries of Clifton admit that he did not welcome research and that he jealously guarded his apparatus as being too sacred to be used by ham-fisted students: it was to be dusted and cleaned, but not violated by undergraduates. This view of the Clarendon as a mausoleum persisted outside Oxford until well after Clifton's resignation: in 1932 McKerrow of Metropolitan-Vickers Electrical Company, which had sold an atomic hydrogen welding device to the Clarendon, averred that one could not tell whether the laboratory was buying apparatus to use or as a 'museum piece'.[35] Clifton himself had contributed to that idiosyncratic reputation when in 1897 he formally announced the end of his original work in the

---

[33] B. Bleaney, A. H. Cooke, N. Kurti, and K. W. H. Stevens, 'F. A. Lindemann, Viscount Cherwell (1886–1957): Head of the Clarendon Laboratory, University of Oxford (1919–56)', *Physics Bulletin*, 37 (1986), 261–3, is a brilliant and perceptive vignette. For an earlier version of part of this chapter see J. B. Morrell, 'Research in Physics at the Clarendon Laboratory, Oxford, 1919–1939', *Historical Studies in the Physical and Biological Sciences*, 22 (1992), 263–307.

[34] Robert Bellamy Clifton (1836–1921); J. L. Heilbron, *H. G. J. Moseley: The Life and Letters of an English Physicist, 1887–1915* (Berkeley and Los Angeles, 1974), 34–5; P. Forman, J. L. Heilbron, and S. Weart, *Physics circa 1900: Personnel, Funding and Productivity of the Academic Establishments*, Historical Studies in the Physical Sciences 5 (Princeton, 1975), 37; G. Gooday, 'Precision Measurement and the Genesis of Teaching Laboratories in Victorian Britain' (University of Kent at Canterbury, unpublished Ph.D. thesis, 1989), ch. 6, rescues the early but not the late Clifton; F. W. F. Smith, 2nd Earl of Birkenhead, *The Prof in Two Worlds: The Official Life of Professor F. A. Lindemann, Viscount Cherwell* (London, 1961), 88–91; Clark, *Tizard*, 51; Sanderson, *Universities and British Industry*, 38–41; R. T. Glazebrook, 'Clifton', *Proceedings of the Royal Society A*, 99 (1921), pp. vi–ix; *Nature*, 102 (1921), 18–19.

[35] G. McKerrow to J. D. Cockroft, 15 Aug. 1932, Cockroft Papers, Churchill College, Cambridge. I owe this reference to Anna Guagnini.

laboratory, on the grounds that the University's failure to provide a new laboratory for undergraduate teaching of electricity had forced him to do this teaching in his own research laboratory.[36] Clifton was relieved of such teaching in 1900 when Townsend was appointed to the new Wykeham chair of experimental physics endowed by New College and the Drapers' Company. From that time there was a statutory division of labour with Townsend teaching magnetism and electricity, leaving mechanics, heat, light, and sound to the Clarendon.[37] In 1910 the Drapers' Company again came to Oxford's aid when it paid £23,000 for the new electrical laboratory which enabled Townsend to employ six demonstrators at a time and to give research space to Henry Moseley who was in Oxford in a private capacity just before the outbreak of war waiting for Clifton's resignation and his chair.[38] When war broke out it dismembered Townsend's flourishing group which worked on his favourite topic of electrical discharges in gases: it was the electrical laboratory which provided no fewer than eight physicists for employment as boffins in the First World War, the best known subsequently being Henry Tizard, a fellow of Oriel in natural science and demonstrator in the electrical laboratory; in contrast the Clarendon contributed only the solitary I. O. Griffith, an assistant demonstrator, to the war effort.[39] Though Townsend's career and his laboratory work between the wars have attracted caustic comment, there is no doubt that in 1914 the palm went to the electrical laboratory and not to the Clarendon. When Clifton resigned in 1915 after fifty years service, as did his assistant Henry Walter after forty-three years, the University suspended the chair for a year and then in May 1916 suspended it again until the end of hostilities. At the end of the war there was a strong feeling among Oxford's leading scientists that great changes in the Clarendon were required and that a new professor should be appointed quickly because other universities with vacant chairs in physics were snapping up the best people. Accordingly the chair was resuscitated and on 23 April 1919 Lindemann was elected to it, just three weeks after Rutherford had been elected to the Cavendish chair of experimental physics at Cambridge.[40]

Lindemann's career before 1919 was unusual for a future Oxford professor.[41] Though he was English he was born in Germany and his

[36] Janet Howarth, 'Science Education in Late-Victorian Oxford: A Curious Case of Failure', *English Historical Review*, 102 (1987), 334–71, at 340–2.

[37] H. E. Hurst, 'Recollections of the Study of Physics in Oxford at the Beginning of the Twentieth Century', *OM* 88 (1969), 59–60, offers telling vignettes of Clifton and Townsend.

[38] Heilbron, *Moseley*, 35–6, 80, 110; *OUG* 45 (1914–15), 650.

[39] Clark, *Tizard*, 23–48; *OUG* 48 (1917–18), 473; Idwal Owen Griffith (1880–1941), biographies in *Nature*, 148 (1941), 589 and *Brazen Nose*, 7 (1931–44), 165–9.

[40] 11 May 1915, NS/M/1/3; D. H. Nagel and G. B. Cronshaw to registrar, 25 Nov. 1918, NS/R/1/2; *OUG* 46 (1915–16), 540; Birkenhead, *Lindemann*, 81.

[41] The main biographies are: Birkenhead, *Lindemann*; Harrod, *Cherwell*; G. Thomson, 'Lindemann', *BM* 4 (1958), 45–71; Bleaney et al., 'Lindemann'; R. Berman, 'Lindemann in Physics',

higher scientific training was obtained there and not at Oxford, Cambridge, or London universities. His crucial German experience involved spending four to five years working with Walter Nernst at the Physikalisch-Chemisches Institut of the University of Berlin where he obtained his doctorate in 1910. Nernst regarded Lindemann as such a good experimentalist that they collaborated in research on specific heats of solids at very low temperatures attained by using liquid hydrogen; this work led in 1911 to the Nernst–Lindemann formula for specific heats. Apart from co-publication Lindemann was indebted to Nernst in other ways. In 1906 Nernst had enunciated his heat theorem, later called the third law of thermodynamics, which dealt with energy at absolute zero and next year Einstein produced his quantum theory of specific heats which highlighted the diminution of specific heats at low temperatures. Nernst quickly realized that the world of low temperatures promised to be unique in physics because in it the phenomena explicable only by quantum physics would be revealed. Lindemann appeared in Nernst's laboratory in 1908 just when Nernst had begun to work on low-temperature physics, especially specific heats. From Nernst Lindemann derived a deep interest in low-temperature physics pursued in the contexts of the quantum theory and of Nernst's heat theorem. He also admired the way in which Nernst swung the whole laboratory onto this work, thus identifying the laboratory with cryogenics. He also learned that Nernst's strange working hours, which involved late rising, were not a bar to scientific eminence and indeed he copied them. His career was promoted by Nernst who, when he arranged the first Solvay physics conference in Brussels in 1911, ensured that Lindemann was its co-secretary. Through Nernst he made contact with distinguished German theoretical physicists such as Einstein and in the laboratory made friends with two young English researchers, Tizard and Alfred Egerton. It is not surprising that in his Oxford period he was sometimes so consumed with nostalgia for his heady days in Berlin with Nernst that he slipped into German when reminiscing about them. He remained in contact with Nernst, seeing him for the last time in Berlin in 1935. He took great trouble in arranging for an honorary D.Sc. to be conferred on his old supervisor and mentor by the University in 1937. After Nernst's death in 1941 Lindemann paid his last tribute in an affectionate obituary which he wrote with Simon.[42]

When he left Berlin Lindemann could afford to take his time in deciding on his future career because he was very wealthy: for him science was not a

Notes and Records of the Royal Society of London, 41 (1987), 181–9; R. V. Jones, 'Lindemann beyond the Laboratory', *Notes and Records*, 41 (1987), 191–210; T. C. Keeley, 'The Right Honourable Viscount Cherwell', *Year Book of the Physical Society* (1958), 79–82; R. V. Jones, 'Lindemann', *Nature*, 180 (1957), 579–81.

[42] Birkenhead, *Lindemann*, 32–53; K. Mendelssohn, *The World of Walter Nernst: The Rise and Fall of German Science* (London, 1973), 61, 74, 76, 78, 160, 171–4; Hermann Walther Nernst

livelihood but a lifetime's choice. When in 1913 he accepted an invitation from Millikan to lecture at the University of Chicago the fee of $500 was not the chief consideration. Unemployed when war broke out and keen to serve, in 1915 Lindemann joined the Royal Aircraft Factory, Farnborough, where he worked on a variety of problems including the aerodynamics of fighter planes. Though the First World War has often been described as the chemists' war, physicists and engineers were recruited by Farnborough's superintendent, O'Gorman, to try to transform the hazardous art of flying into a reliable science. By 1916 Lindemann was one of half a dozen scientists who took flying courses and became certificated pilots. Next year he undertook his celebrated experiments in which he deliberately put aircraft into a spin in order to test his theory about how to extricate them from it. The chief result of these perilous experiments was a standard manœuvre for pilots to recover from a spin, his procedure being contrary to their instinctive reaction.[43] At Farnborough he met some younger physicists whom he invited to Oxford subsequently. Lindemann had a small physics laboratory in which one of his assistants from 1917 was Thomas Keeley who had recently graduated in physics at Cambridge. Keeley's main role in the future was to be that of day-to-day manager of the Clarendon and loyal assistant to Lindemann. A sister institution of Farnborough was the Central Flying School at Upavon, Wiltshire, where Lindemann was taught to fly. There he met Dobson, a Cambridge physicist who had worked as meteorological adviser to the School from 1913. When war broke out Dobson, the man on the spot, was put in charge of scientific research there, working on bombs and bomb-sights. Dobson was soon joined at Upavon by Tizard, an old friend of Lindemann, and by Griffith, who became a new one. Another Farnborough colleague of Lindemann may have been the spectroscopist Thomas Merton who had enjoyed such an outstanding career as an undergraduate at Balliol that his tutor Harold Hartley had arranged for him to proceed straight to a research degree. From 1910 to 1914 Merton had used his considerable wealth to finance private research in a London house on gas spectra and astrophysics, using the best modern equipment. At the start of the war Merton became the first scientist and inventor to be employed by the British Secret Service.[44]

When Lindemann assumed the chair of experimental philosophy at Oxford on 1 May 1919 at the age of 33 at a salary of £900 p.a., having

(1864–1941); F. A. Lindemann and F. E. Simon, 'Nernst', *ON* 4 (1942), 101–12; J. Mehra, *The Solvay Conferences on Physics: Aspects of the Development of Physics since 1911* (Dordrecht, 1975), 13, 52–3, 67–8.

[43] Birkenhead, *Lindemann*, 53–4, 57–80; Thomson, 'Lindemann', 50–5; Mervyn Joseph Pius O'Gorman (1871–1958).

[44] Clark, *Tizard*, 26–36; Thomas Clews Keeley (1894–1988), obituary in *The Times*, 27 Dec. 1988 and *Independent*, 2 Jan. 1989; Gordon Miller Bourne Dobson (1889–1976); J. T. Houghton and C. D. Walshaw, 'Dobson', *BM* 23 (1977), 41–57; Thomas Ralph Merton (1888–1969); H. Hartley and D. Gabor, 'Merton', *BM* 16 (1970), 421–40.

been greatly aided in his application by Tizard, he knew that the finances of the Clarendon would be a dire problem. Even so, he had high ambitions. He wished primarily to build up and inspire a research school of about a dozen workers. He insisted right from the start that the power to acquire good men to research on promising problems was more important than the possession of big apparatus in magnificent buildings. He knew that some leading figures at Oxford wanted the Clarendon to end its eclipse by the Cavendish. That was more easily said than done not least because the Clarendon was in competition with the other department of physics, the electrical laboratory under Townsend, for limited funds and facilities. After a short honeymoon, this structural division of physics within the University led to bitter rivalry between the two laboratories, with little co-operation between them, until they were effectively merged in 1945. Relations between the departments continued to be so strained that in the early 1930s, when the Clarendon staff heard that the electrical laboratory intended to give a course on high vacua, they thought of mounting a rival called higher vacua. In 1919 Lindemann was a new boy, unfamiliar with Oxford, whereas Townsend was the established professor with more demonstrators and more college fellows under his direction. That staffing situation persisted. When the new category of university demonstrator was established in 1927, Townsend secured five men in that denomination (H. F. Biggs, E. W. B. Gill, R. T. Lattey, S. P. McCallum, and J. H. Morrell) at a cost to the University of £2,450 p.a.; whereas Lindemann had just two (Griffith and Dobson) at a cost of £1,100 p.a. Moreover two of Townsend's demonstrators (Gill, McCallum) were college fellows as against one of Lindemann's (Griffith).[45]

Lindemann arrived in Oxford at a peculiarly unpropitious time. The University was in such a desperate plight financially that after the war like Cambridge it was given an annual emergency grant of £30,000 by the government to save it from financial collapse. Clearly Lindemann could not expect money for research to flow profusely from the University to the Clarendon. In his view the University saw teaching as central and its authorities were either unsympathetic to scientific research or ignorant about it. His early ideal was an external annual grant of over £1,000 for research for about five years which would attract and retain good men who would put the Oxford physics school in such a commanding position that the University would continue the grant. Like his early vision of physical expansion of the

---

[45] Lindemann to R. A. Millikan, 8 Mar. 1919, LP, A21; Joseph Wells, warden of Wadham, to Lindemann, 23 Apr. 1919, LP, A22; R. V. Jones, 'Oxford Physics in Transition', in R. Williamson (ed.), *The Making of Physicists* (Bristol, 1987), 113–26, on 114; *OUG* 57 (1926–7), 543; Henry Francis Biggs (1882–1934), obituary in *OM* 52 (1933–4), 335, was an import from Manchester; Robert Tabor Lattey (b. 1881), first in chemistry 1903; James Herbert Morrell (1892–1965), third in physics 1905, was also a barrister at the Inner Temple; Ernest Walter Brudenell Gill (1883–1959), fellow of Merton 1909–58; Stanley Powell McCallum (1895–1940), fellow of New College 1927–40; Griffith was fellow of Brasenose 1920–41.

Clarendon at a cost of over £30,000 to accommodate 50–60 honours undergraduates and 30 researchers, it was not realized for many years because money was very tight. The limited sum available for the Clarendon from the University was spent on such ordinary matters as improving the water supply, installing mains electricity, and making conversions to the building. The financial constraints were such that in 1922 and 1923 Lindemann himself contributed £100, one-ninth of his annual salary, to the income of his own department (see Table 9.1). Apparatus and materials beyond the ordinary were difficult to acquire in his opening years: in 1922 he agitated successfully for liquid air to be provided, Oxford by then being the only English university without it.[46]

When Lindemann assumed his chair he wanted to avoid the departmental dreariness which would be the inevitable result of retailing other men's results at second hand. He turned quickly to his favourite field of low-temperature physics as the particular means of realizing his general aim of furthering research. By the end of 1919 he had created space in the Clarendon for compressors and for apparatus for the liquefaction of gases, had secured the loan from the Air Ministry of four liquid air containers, and had induced Townsend to give a liquid air machine. A year later what he called 'plant for low-temperature research' was being installed in the old engine room and was used until 1922 by four researchers who worked mainly on the properties of alkali metals at low temperatures. It was at this stage that difficulties appeared. In 1920 Lindemann failed to secure from the Air Ministry some liquid helium which he wanted for studying low-temperature anomalies: to Lindemann's chagrin it was shipped to McLennan at Toronto University. But he did buy a hydrogen liquefier, designed by Nernst and built by Hoenow, Nernst's head mechanic. Even the prototype of this liquefier was temperamental and needed Hoenow's skilled hand to make it function. Lindemann's optimism was soon shown to be baseless: nobody at the Clarendon could make the Nernst–Hoenow liquefier work. Lindemann's first attempt to create a school of Nernstian low-temperature physics was a patent failure by 1922 and it may have convinced him that a research lieutenant or assistant fully conversant with the intricacies, eccentricities, and dangers of liquefiers would be necessary for future success.[47]

---

[46] Lindemann to D. Vickers, 8 Nov. 1919, LP, B4; *Asquith Report*, 53, 114; Lindemann, memorandum on Clarendon, n.d. [1920], Asquith Papers, MS Top. Oxon. b. 107, ff. 122–5; Birkenhead, *Lindemann*, 85–90; F. A. Lindemann, 'Clarendon Laboratory', *HCP* 173 (1939), 225–34, at 228–9, draft in LP, B25, ff. 1–9; NS/R/1/3, 5 May 1922; NS/M/1/3, 16 May 1922; *HCP* 122 (1922), 145.

[47] *OUG* 50 (1919–20), 688–9; 52 (1921–2), 75–6, 679; 53 (1922–3), 671; LP, B4; Mendelssohn, *Nernst*, 70–1, 164–5. The four researchers were: Ivan George Evans who relinquished a research scholarship in Mar. 1920 to take up a post at the Patent Office; Henry Herman Leopold Adolf Brose (1890–1965), whose 1925 D. Phil. taken under Townsend was on the motion of electrons in oxygen; Tielman François Tertius Malherbe (b. 1899) returned to South Africa where he

TABLE 9.1. *Income of Clarendon Laboratory 1919–1939*
(to pay for demonstrators, assistants, apparatus, etc.) (£s)

Year	Balance from last account	University (including government grant)	Lab fees	Sales	Bank interest	Leigh fund	Lindemann donation	Christ Church grant	Lecture fees	Rockefeller grant	Duke of Westminster gift	Oxford Appeal (Higher Studies Fund)	Total
Calendar 1919	567	500	126	559	14	—	—	—	—	—	—	—	1,767
Calendar 1920	273	3,041	333	7	25	—	—	—	—	—	—	—	3,704
Calendar 1921	—	2,850	436	—	3	200	—	—	—	—	—	—	3,513
Calendar 1922	197	2,467	643	—	—	—	100	—	—	—	—	—	3,407
Calendar 1923	47	2,350	428	—	—	300	100	—	—	—	—	—	3,275
Calendar 1924	50	2,367	468	—	—	200	—	—	—	—	—	—	3,126
Jan–July 1925	14	1,600	299	—	—	—	—	—	—	—	—	—	1,940
Session													
1925-6	217	2,300	384	—	—	300	—	90	50	—	—	—	3,353
1926-7	249	2,650	463	13	—	—	—	30	60	—	—	—	3,465
1927-8	454	2,130	398	6	—	—	—	50	81	—	—	—	3,119
1928-9	443	1,950	338	80	—	200	—	50	59	—	—	—	3,118
1929-30	310	1,950	440	—	—	—	—	20	89	—	—	—	2,808
1930-1	127	1,800	557	13	—	—	—	20	113	—	—	—	2,629
1931-2	280	2,200	516	—	—	—	—	20	98	—	—	—	3,118
1932-3	544	2,112	764	—	—	400	—	110	—	—	—	—	3,937
1933-4	456	2,212	788	—	—	250	—	50	—	500	—	—	4,295
1934-5	76	2,212	670	12	—	300	—	50	—	500	500	—	4,320
1935-6	114	2,712	617	4	—	300	—	50	—	—	—	(bank o/d 177)	3,796
1936-7	127	2,762	622	—	—	300	—	50	—	—	—	—	3,987
1937-8	186	2,762	682	—	—	200	—	50	—	—	—	300	4,280
1938-9	402	2,762	693	—	—	200	—	50	—	—	—	300	4,507

Note: The Categories of income are not exhaustive.

Source: University accounts published annually as an appendix in *OUG*. From session 1932–3 the lab fees included lecture fees.

Perhaps bruised by the fiasco of the hydrogen liquefier, Lindemann's own experimental research soon petered out, his last original paper in physics being published in 1924 only five years after his accession to his Oxford chair. Unlike his fellow professors such as W. H. Perkin in organic chemistry and Sherrington in physiology, from the mid-1920s Lindemann did not set a grand personal example of arduous research at the laboratory bench; and he did not develop the lieutenant system by which he could have supervised research vicariously. As Hinshelwood stressed, Lindemann was not the officer among his troops and did not build up in the 1920s a school focused on a topic with which he was publicly associated.[48] During the five years in which he was publishing research papers, his reputation as a researcher and as head of a research laboratory hardly rivalled that of Rutherford and of W. L. Bragg, two Nobel prize-winners, at Cambridge and Manchester respectively. In this period he worked with Keeley on instruments such as a new silica fibre electrometer and a photoelectric radiation pyrometer, which was appropriate occupation for a professor of experimental philosophy. He also played the role of a Socratic oracle, not only with his postgraduates but also with his colleagues such as Dobson and Hinshelwood. Dobson's discovery of ozone in the upper atmosphere depended on Lindemann's idea that it was feasible to use observations of meteors to study the variation of the density of air with height and on Lindemann's explanation of the warm layer at 50 kilometres height which they discovered, namely, that the warmth was caused by the absorption of solar ultraviolet radiation by a layer of ozone.[49] Some of Hinshelwood's most important research in chemical kinetics, that on unimolecular reactions in the gaseous phase, was stimulated by Lindemann's oral contribution to a Faraday Society discussion; characteristically his hypothesis, subsequently confirmed by Hinshelwood, was briefly presented and lacked equations.[50] From the mid-1920s Lindemann was still quick to spot a promising new line of research but he did not consolidate it. For instance, details about the Raman effect were first published in spring 1928; by November of that year, Lindemann, Keeley, and N. R. Hall published from the Clarendon the first British paper on this important phenomenon, their results being avowedly preliminary and

pursued a career in secondary school education; John Gustave Pilley (1899–1968), first in chemistry 1922 and later professor of education, University of Edinburgh, 1951–66.

[48] Birkenhead, *Lindemann*, 114.

[49] G. M. B. Dobson, I. O. Griffith, and D. N. Harrison, *Photographic Photometry: A Study of Methods of Measuring Radiation by Photographic Means* (Oxford, 1926), 5; G. M. B. Dobson, *The Uppermost Regions of the Earth's Atmosphere Being the Halley Lecture Delivered on 5 May 1926* (Oxford, 1926); id., 'Forty Years' Research on Atmospheric Ozone at Oxford: A History', *Applied Optics*, 7 (1968), 387–405, at 387–8.

[50] K. J. Laidler, 'Chemical Kinetics and the Oxford College Laboratories', *Archive for History of Exact Sciences*, 38 (1988), 197–283, at 256–60.

incomplete.[51] There were no further papers from the Clarendon on the Raman effect; and none of the three authors went on to establish a reputation based on studying it.

As Lindemann's experimental research died out so did his appreciation of current trends in theoretical physics. He did not attend any of the postwar Solvay conferences. He resented the arid mathematical formalism of Schrödinger's wave mechanics. He was certain that ordinary experimental physicists ought not to be content with just a set of mathematical equations, however coherent and logically satisfying, even if they complied with experimental results: they ought to ask also for the physical meaning of mathematical formulations, especially what he called 'abnormal mathematical expedients'. Lindemann often referred contemptuously to recent mathematical physics as mere squiggles on paper and, like Milne, he castigated contemporary exponents of quantum mechanics for their timidity in being satisfied 'with a physical world composed of little black marks on a white sheet of paper'. This derogatory attitude ensured that rising young theoretical physicists such as Mott regarded Lindemann as sadly passé.[52]

Given Lindemann's characteristics it is not surprising that in the 1920s and beyond his own postgraduates did not pursue a main line of research. Instead he launched these men on a series of isolated researches in which time and effort were expended in entering a new field which often was not consolidated. Inexperienced researchers became involved in the wearing business of making new and delicate apparatus, so that the main problem was lost sight of. Before 1933 Lindemann's postgraduate researchers worked as prospectors on a variety of problems while he himself did less and less in the laboratory. Lacking any gradual induction into the craft of scientific research via apprenticeship under an acknowledged and active leader, they were thrown in at the deep end and sank or swam. In both his own research and as a research supervisor Lindemann's speciality was the flash of intuition but not the steady flame.[53] In that respect he was different from Arthur Tyndall who in the early 1930s was building up a powerful research department at Bristol. The researches pursued under Tyndall at that time were diverse but consolidated. Tyndall's career as a discipline-builder at Bristol shows that disparate fields could be pursued concurrently and successfully,

[51] D. A. Long, 'Early History of the Raman Effect', *International Reviews in Physical Chemistry*, 7 (1988), 317–49, at 330–1; F. A. Lindemann, T. C. Keeley, and N. R. Hall, 'Frequency Change in Scattered Light', *Nature*, 122 (1928), 921.

[52] F. A. Lindemann, *The Physical Significance of the Quantum Theory* (Oxford, 1932), p. vi; id., 'The Place of Mathematics in the Interpretation of the Universe', *Philosophy*, 8 (1933), 14–29, at 29; Mendelssohn, *Nernst*, 173; N. F. Mott, review of Lindemann, *Quantum Theory*, in *Nature*, 130 (1932), 330–1.

[53] Jones, 'Lindemann'; H. R. Calvert, letter supporting Jones, *Nature*, 180 (1957), 1146; Jones, 'Oxford Physics', 120–2; Thomson, 'Lindemann', 55–6.

in spite of the danger of dispersal of effort.[54] In contrast Lindemann's interests were too widely scattered to be effective and in temperament he was restlessly unpersevering in the face of temporary difficulties.

These features of Lindemann's reign ensured that relatively few foreigners came to work in the laboratory in the 1920s, a decade in which European physicists and their students travelled a great deal in order to have first-hand experience of work being done in such capitals of research as Munich, Leipzig, Göttingen, Leiden, Zurich, Copenhagen, Berlin, and Cambridge. At the Cavendish in 1930, for instance, half the laboratory workers were from overseas. In contrast, until 1933 the Clarendon was not on the regular international itinerary. Even within Britain, Lindemann did not produce young disciples who modelled themselves on him as a renowned practitioner whose reputation was based on a continuing mastery of a particular field; it is not possible to draw up a tree showing convincingly his far-reaching influence on British physics through his research pupils, most of whom were Oxford graduates. Of his research students who became demonstrators in the Clarendon during the 1920s, the longest serving were Atkinson and Bolton King. Though each was helped by Lindemann, neither could be regarded as his disciple *qua* physicist. Atkinson graduated in physics from Hertford in 1922 and with Lindemann's financial aid immediately began a four-year stint as a demonstrator. When Atkinson fell out with his college which had elected him to a research fellowship, Lindemann used his pre-war acquaintance at Berlin with James Franck to secure a two-year Rockefeller travelling fellowship for Atkinson tenable at Göttingen where Franck was professor of experimental physics. After a year as a teaching assistant at the Technische Hochschule, Charlottenburg, Atkinson refused an offer to return in some capacity to the Clarendon because he wanted a permanent ascertainable position and academic home. Having received a more attractive offer from Rutgers University he crossed the Atlantic for an eight-year stay in New Jersey which he left in 1937 to become chief assistant at the Royal Greenwich Observatory until 1964. In the early 1960s his career reached its zenith: he was president of the British Astronomical Association and was awarded the Eddington medal of the Royal Astronomical Society.[55] Like Atkinson, Bolton King ended his career a long way from the Clarendon. Having acquired a first in physics from Balliol in 1923, he was the first recipient in 1924 of the Duke of Westminster research studentship in physics tenable for six years at Christ Church, Oxford. It was typical of Lindemann that he had used friendship with an aristocrat to persuade him to endow a

---

[54] B. Lovell, 'Bristol and Manchester: The Years 1931–9', in Williamson, *Physicists*, 148–60; S. T. Keith, 'Scientists as Entrepreneurs: Arthur Tyndall and the Rise of Bristol Physics', *Annals of Science*, 41 (1984), 335–57.

[55] Robert d'Escourt Atkinson (1898–1982); Lindemann to father, 29 June 1922, LP, A93; Atkinson to Lindemann, 15 June 1927, 23 Oct. 1928, 15 Nov. 1928, 2 Feb. 1930, LP, D12.

research fellowship and that its first recipient should be Bolton King, a protégé and friend. From this base Bolton King launched a career with Lindemann's support as managing director 1930–40 of the Oxford Instrument Company, located just a stone's throw from the Clarendon; it made and sold high-quality photoelectric cells. After successful war-work on proximity fuses, Bolton King ended his career as director of the scientific department of the British Council. Though he regarded Lindemann as the biggest factor in his life from 1922 to 1940, Lindemann did not secure for him an academic post at Oxford or elsewhere.[56] Admittedly there were few vacancies in physics or in mathematics and physics in the colleges between the wars; and Lindemann had only limited success in placing his men in them. He helped to secure a fellowship for Keeley in 1924 at Wadham, where he was a professorial fellow; but in 1927 Townsend no doubt used his pull as a professorial fellow at New College to facilitate the election of his protégé, S. P. McCallum, to a fellowship there. The rivalries between Lindemann and Townsend, and between the Clarendon and electrical laboratories, were exacerbated by the fiercer competition between them for money. The financial fortunes of the Clarendon declined between 1927 and 1931. In those years its annual departmental income, which was used to buy apparatus and pay demonstrators and assistants, dropped from £3,465 to £2,629 mainly because the annual grant from the University was reduced from £2,650 to £1,800 (see Table 9.1). Given such depressing financial circumstances it is not surprising that Lindemann could not create permanent demonstratorships in his own laboratory for his best D.Phil. students.

For senior research workers, Lindemann had no cohort of college tutorial fellows in physics upon whom he could call and rely. He inherited from Clifton just one such fellow, I. O. Griffith, who from 1919 to 1931 acted as his senior demonstrator. Griffith's forte turned out to be university politics, especially in the 1930s when he was chairman of the Physical Sciences Board, vice-chairman of the General Board of the Faculties, and a member of the Hebdomadal Council, in which capacities he was a masterly ally of Lindemann; an emollient and not metallic personality who specialized in telephone negotiations and played golf every day in term, he was not a prolific publisher or a noted researcher. In view of the dearth of available college physicists, Lindemann brought people in from outside, especially talented young physicists whom he had met in Berlin or Farnborough. He avoided the obvious mode of recruitment which would have been to entice to Oxford rising researchers from established centres such as Cambridge or Manchester. Instead he chose men who had been chums and fellow researchers at either Berlin or Farnborough; two of them (Keeley, Dobson) were

---

[56] Edward Bolton King (1900–74), entry in *Balliol College Register*; Birkenhead, *Lindemann*, 122–3.

Cambridge men but they were picked out not for their training at Cambridge but for their proved worth at Farnborough. This meant that until 1933 disparate lines of research were pursued by the senior workers in the Clarendon and therefore there was no hope of swinging the whole laboratory into one field of research as Nernst had done in Berlin when Lindemann worked with him. It was the outsiders, whom Lindemann recruited mainly in 1919 and 1920, who were largely responsible for the increased reputation in the 1920s of the Clarendon. Though Lindemann's own bench research was soon to end, he lured them to Oxford, defended their interests, provided what help he could muster, and generally supported them without interfering. Like Sir Thomas Beecham, he believed in appointing or appropriating those whom he regarded as the best men and then letting them play. In this way in the 1920s atmospheric physics (Dobson), spectroscopy (Merton and Jackson), and applied thermodynamics (Egerton) became the main lines of research, with the general management of the laboratory becoming increasingly the responsibility of a fourth outsider, Keeley.

Keeley met Lindemann in 1917 at Farnborough and was brought to Oxford in autumn 1919 to continue as research assistant, being funded by a grant from the Department of Scientific and Industrial Research. In conjunction with Lindemann he worked mainly on the design of instruments, especially electrometers and pyrometers, the former being produced commercially by the Cambridge Instrument Company and by Spindler & Hoyer in Germany. From 1924 his career prospered. In that year he was elected science fellow at Wadham, Lindemann going to the length of subsidizing his salary for ten years. In 1928 he became a demonstrator in the Clarendon and in autumn 1931 replaced Griffith as senior demonstrator. From the late 1920s Keeley, like Lindemann, was no longer a productive researcher at the bench.[57]

Three of Lindemann's other recruits, Dobson, Merton, and Jackson, were wealthy men who possessed private means. For Lindemann that was a highly desirable qualification which guaranteed devoted commitment to science as a lifetime's vocation and not as a mere livelihood. He felt that one could not be certain that a young physicist without independent means was not researching hard more for the sake of his wife and family than out of pure devotion to physics; but with a physicist of independent means he knew that such mundane considerations could not apply. As a rich man himself, who employed two servants and a chauffeur, he put scientific gentlemen in a category superior to that of the players—before 1933. Men of independent means had no need to ask Lindemann to go cap in hand to the University for money for equipment. Wealthy researchers enabled Lindemann to build up

---

[57] Lindemann memorandum, 5 Feb. 1927, PS/R/1/1; Keeley memorandum, 9 Nov. 1932, PS/R/1/3; Birkenhead, *Lindemann*, 95–9; Wells to Lindemann, 1 Dec. 1924, LP, B121.

research in the Clarendon at low cost to the University. On the other hand, those of private means had no need to work in the Clarendon so that if circumstances became less than happy they could withdraw from it partly or totally. Dobson and Merton were cases in point.

Dobson, a first in the Cambridge natural sciences tripos, began his career as a student assistant 1911–13 at the Kew Observatory, London, working on atmospheric electricity. In 1913 he became meteorological adviser to the Military Flying School, Upavon. For the duration of the war he was director of the experimental department at the Royal Aircraft Factory, Farnborough, where he met Lindemann, who was so impressed by Dobson's work that in 1920 he secured for Dobson a university lectureship in meteorology and a Clarendon demonstratorship. By this time Dobson's wife enjoyed a private income and in 1924 Dobson himself inherited land in the Lake District from his prosperous father. Comfortably placed financially, Dobson was in a position to distance himself from the Clarendon and to pursue his research in his country home on Boar's Hill, Oxford. In the early 1920s he collaborated with Lindemann in discovering the earth's ozone layer. His early work on photographic spectrophotometry was aided by a Lindemann–Keeley electrometer and by one of Merton's spectrographic gadgets. Later in the decade Dobson designed a photoelectric spectrophotometer, using a photocell made by Keeley. Thus in the 1920s Dobson was indebted to three Clarendon colleagues either for theoretical promptings or instrumental ingenuity. But his research was done at his home where he studied the variation in the amount of ozone in the atmosphere relative to latitude and season. Though he was made reader in meteorology in 1927, no college fellowship came his way until 1937 when Merton elected him to one and for years he remained a mainly solo worker in Oxford with very few students all of whom, bar one, were Oxford graduates.

In sharp contrast to his isolation in Oxford, his private meteorological laboratory became by 1926 the centre of an international network of research on atmospheric ozone. With the aid of a Royal Society grant, Dobson persuaded observatories in Ireland, Shetland, Sweden, Germany, and Switzerland to make daily routine measurements of ozone in the atmosphere using instruments all made and calibrated in his home. Supported by a grant from the Department of Scientific and Industrial Research, by 1927 Dobson and an assistant had derived sufficient information from the photographic plates sent from these laboratories to be able to correlate the European distribution of ozone with atmospheric pressure systems. By 1929 the scheme had been extended by Dobson to the USA, Egypt, India, and New Zealand. The instrument used for this collaborative work was Dobson's variant of Féry's spectrograph which had two faults: by the time the plates had reached Oxford they were out of date; and the instrument was not much use on cloudy days. To remedy these defects Dobson began work in 1927 on a photoelectric spectrophotometer of which a prototype was produced by Beck for £500 provided by the Royal Society of London. In the 1930s Dobson spent much time at his home, making, adjusting, and calibrating the Beck instruments which remained standard for several decades. They were first tested in 1932 not in Britain but at Arosa, Switzerland, at the invitation of F. W. P. Götz in a project concerned with the vertical distribution of ozone. As the 1930s passed Dobson did not move nearer the Clarendon but as a researcher became more detached from it and from Lindemann of whom he saw little. Even when Dobson taught in the Clarendon in the morning, he cycled for lunch to his home-cum-laboratory. In 1937 he improved his private facilities when he moved to a 10-acre site on Shotover Hill, Oxford, where he built a new laboratory, a workshop, and a paved observing platform from which the instruments could 'see' the sun all day unobstructed and on which they did not have to be moved as had been the case at Boar's Hill. From 1934 to 1950 Dobson served as chairman of the Atmospheric Pollution Committee of the Department of Scientific and Industrial Research, which took up much of his time, the biggest project being undertaken from 1937 to 1939, with his former pupil A. R. Meetham, on atmospheric pollution in Leicester. Though Dobson was the only person who was continuously a demonstrator in the Clarendon from 1920 to 1939, his research life was based on his home, his private means, and his ability to extract grants for apparatus and assistance from the Royal Society and the Department of Scientific and Industrial Research. By the late 1930s his work was more fully recognized financially beyond England than by his own University: other countries had given £5,000 to support what he called his 'international research on the upper atmosphere', while the University was then giving him a mere £100 p.a. for apparatus. It was Dobson's internationalism which led to his friendship with Victor Conrad, a geophysicist at the University of Vienna who was dismissed in 1938. Dobson tried to arrange finance for Conrad from the Royal Meteorological Society, Balliol, and Oxford University Press, and offered him accommodation, but in the event Conrad went straight to the USA in spring 1939 without staying in Oxford.[58]

Dobson may well have found the model of the country-gentleman scientist in T. R. Merton who started spectroscopic research at Oxford in 1919 but left in 1924 for a rural house in Herefordshire, where for twenty-four years he lived the life of an affluent gentleman, working in his private laboratory,

[58] Houghton and Walshaw, 'Dobson'; Dobson, 'Ozone'; Dobson memorandum, 26 Nov. 1938, PS/R/1/7; Friedrich Wilhelm Paul Götz (1891–1954), head of the meteorological observatory, Arosa, Switzerland, from 1921; Dobson's main collaborators were Alfred Roger Meetham (b. 1910), who ended as a principal scientific officer at the National Physical Laboratory; and Douglas Neill Harrison (1901–87) as head of upper air instruments at Meteorological Office Development branch, Harrow. Dobson's Merton fellowship, tenable from 1937 to 1944 in the first instance, brought him £300 per year. On Victor Anton Conrad (1876–1962), SPSL 325/6.

taking out patents, and collecting Italian Renaissance pictures worth almost £2 million when he died. By spring 1919, before Lindemann was elected to his chair, Merton's appointments as stipendless reader in spectroscopy from October 1919 and as a research fellow of Balliol had been made public. Initially he worked in the department of inorganic chemistry under Soddy but in August 1920 he moved to the Clarendon just two months after he had been elevated to a titular chair. Presumably Soddy kicked Merton out and Lindemann seized the opportunity to recruit a brilliant researcher, recently elected FRS, who was capable of attracting from Japan Professor Takamine of Tokyo, who was one of just three physicists sent to Europe by the Japanese government. Lindemann could offer Merton only a dark room in 1920 and in 1921 two rooms and a workshop which restricted his research students to two. Even so, using mainly three men from his own college, some of whom were undergraduates, Merton's little research group dominated the Clarendon's somewhat meagre publications in 1922. Within two years he exchanged the cramped conditions of the Clarendon for the untrammelled delights of a country house, and he went on to act as treasurer to the Royal Society for seventeen years and to be knighted in 1944 for his inventions used in the Second World War. For Merton, reverting to private research with the added attractions of good salmon fishing and collecting Botticellis was a more alluring proposition than enduring the physical constraints of the Clarendon Laboratory.[59]

After a gap of a few years, research in spectroscopy was resuscitated from 1927 by D. A. Jackson who had graduated with a second in physics at Cambridge that year. Rutherford wanted him to work in the Cavendish on nuclear physics but Jackson, stimulated by his Cambridge tutor H. W. B. Skinner, wished to work in the new field of the hyperfine structure of spectral lines. As a very rich man Jackson defied Rutherford but accepted Lindemann's offer of space at the Clarendon where he appeared as a researcher with all his apparatus bought by himself. He turned out to be a most promising spectroscopic yearling: by 1933 he was a self-taught expert on hyperfine structures, nuclear spins, and the magnetic moments of the alkali and alkaline earth elements. Though 'fearless, wild-tempered and wickedly amusing', he was also by summer 1933 the most prolific solo publisher in the Clarendon over the previous two and a half years. Neither his temperament nor his style of life permitted him to take on research students. He moved in artistic and aristocratic circles, the first of his six wives being the daughter of the painter Augustus John. Part owner of the *News of the World*, Jackson

---

[59] Hartley and Gabor, 'Merton'; *OUG* 49 (1918–19), 288, 368; 50 (1919–20), 717; 52 (1921–2), 75, 679; 4 Feb. 1919, 9 Mar. 1920, NS/M/1/3; *HCP* 116 (1920), 109–10; 119 (1921), 155–6. Merton's research collaborators were: Sidney Barratt; Raynor Carey Johnson (1901–87), lectured in physics at Belfast and King's College, London, before being master of Queen's College, Melbourne, 1934–64; and D. N. Harrison.

also owned horses which he rode when fox-hunting and competing at important race meetings. Once he was fined £10 by the stewards at Sandown Park for rough riding; when he appeared before them, he pulled out a £100 note and asked for the change.[60]

Before the Germans arrived in 1933, the most productive and sustained research group in the Clarendon was that run by Egerton. After 1933 his frustrations increased so much that in 1936 he left Oxford for the chair of chemical technology at Imperial College, London. It is significant that Lindemann could not retain Egerton, even though Egerton's dedication to research was highly congenial. Egerton enjoyed considerable success at Oxford: he had a coherent programme of research and infused his research students, brilliant or otherwise, with his own enthusiasm; but eventually his career there foundered whereas those of Simon and Mendelssohn, which slightly overlapped it, did not. That contrast deserves analysis.

When Egerton arrived in Oxford in 1919 he had enjoyed no previous contact with the University. His first mentor was Lord Rayleigh who inspired the young pupil at Eton College, who had founded its Scientific Society, to defy his parents' preference for a naval career and to adopt science instead. Rayleigh also advised him to read chemistry from 1904 to 1908 at University College London under Ramsay, who received in 1904 with Rayleigh a Nobel prize for his discovery of the inert gases in the atmosphere. Ramsay became a second mentor to Egerton and imbued in him the notion that a university's chief function was research. After four years as an instructor at the Royal Military Academy, Woolwich, London, where he began research on explosives, he enjoyed research training under Ramsay through whom he met Nernst who was in England with Lindemann. At Nernst's invitation, Egerton and his wife Ruth went in 1913 to Berlin where he worked for a year on the chemical analysis of gases after the passage of electricity through them, began research on the vapour pressure of metals, and became friendly with Lindemann. After work as a boffin in the first world war on explosives, he was attracted to Oxford in 1919 by the combined efforts of Lindemann and Tizard, who had returned to Oxford in spring 1919. Initially Egerton endured a strange existence: spiritually he was in Lindemann's department, but physically he worked in Soddy's chemistry department, researching on the vapour pressure of metals, supported by a research grant from the Department of Scientific and Industrial Research. This unstable situation changed in November 1920 when the readership in thermodynamics, to which Tizard had been elected in February 1920 and from which he had resigned in the summer, was advertised. In January 1921

---

[60] Derek Ainslie Jackson (1906–82); H. G. Kuhn and C. Hartley, 'Jackson', *BM* 29 (1983), 269–96; Bleaney *et al.*, 'Lindemann', 262; Herbert Wakefield Banks Skinner (1900–60) researched in the Cavendish Laboratory 1922–7 and was a Coutts Trotter research student at Trinity College, where Jackson was an undergraduate.

Egerton was elected to this readership at a salary of £300 p.a. By autumn 1921 Soddy had ejected Egerton who then began a fifteen-year stint in improvised accommodation in the attics and cellars of the Clarendon.[61]

Egerton's research was based on the assumption, not widely cherished at Oxford, that there was no Jordan between applied and pure science. His three main fields of work reflected that conviction. First, in response to a request from the British Electrical Research Association and using facilities at Imperial College, he updated the research of H. L. Callendar on thermodynamic data pertaining to steam. Secondly, from 1919 to 1935 he continued the work done in Nernst's laboratory on the vapour pressure of metals where, as in the steam work, his preoccupation was to establish data. Thirdly, in 1924 he began to investigate various aspects of combustion, a field initially suggested to him by Tizard who just after the war had launched from his Oxford base some notable research for the Asiatic Petroleum Company on knocking as the main limiting factor in the performance of petrol engines, one result of which was the modern system of rating fuels by their toluene or octane number. Egerton's research on engine knock soon toppled Tizard's assumptions that knocking in internal combustion engines was detonation and that the effect of anti-knocks might be to delay combustion. The remainder of Egerton's time at Oxford was spent on tracing the sequence of events occurring in an engine cylinder during a working cycle, especially the progress of oxidation which involved very difficult quantitative analyses of gas mixtures. By the time he left Oxford Egerton had shown that knock was a special type of combustion involving a branch chain reaction which in turn involved the production of peroxides. For the work on vapour pressure of metals (which had led to his FRS in 1925) and on combustion, Egerton built up a research group. As a supervisor he was the opposite of Lindemann: Egerton was patient and spared no pains in helping his young researchers, both chemists and physicists by training, to overcome their difficulties in work on intriguing and difficult hybrid topics. Though not a college tutor, he attracted to his group almost a researcher a year, initially from Oxford but increasingly from elsewhere, the best-known example of the latter category being Llewellyn Smith, a Manchester graduate who was later to become managing director and chairman of Rolls-Royce Motor Cars. Such was the eminence of Egerton's research group that from 1927–8 it was processed separately as a subdepartment in the university annual accounts. In the 1920s it received £250 p.a. from the University, and in the 1930s until Egerton left usually £288 p.a. from the same source. Significantly no formal external endowment was ever recorded in Egerton's accounts.

---

[61] Alfred Charles Glyn Egerton (1886–1959); D. M. Newitt, 'Egerton', *BM* 8 (1960), 39–64; Egerton, *Egerton*; *OUG* 50 (1919–20), 689, 691; 51 (1920–1), 225, 337; 52 (1921–2), 78, 681; *HCP* 119 (1921), 155. John William Strutt, Lord Rayleigh (1842–1918).

In spite of his burgeoning reputation, Egerton's accommodation remained limited: his problem was not so much lack of apparatus but lack of space and split space at that because he occupied one room on the top floor and two rooms in the cellar. He was compelled to waste time in making continual peregrinations between regions that were widely separated. As the cellars had been used by C. V. Boys for determining the gravitational constant of the earth they gave a firm foundation for delicate instruments but suffered from two disadvantages: they were populated by vicious mosquitoes in summer and formed part of the natural ventilation system of the lecture theatre above them. As Ubbelohde gleefully recalled, chemical vapours from Egerton's researches found their way through florally shaped ventilation holes in the steps of the theatre and assailed the nostrils of lecturers including Lindemann. In autumn 1931 the problem of space was partly solved by decanting some of Egerton's research into the engineering science laboratory where Southwell was interested in engineering chemistry, calorimetry, and heat engines. In autumn 1931 Egerton and Llewellyn Smith began work on detonation in petrol engines and the use of lead tetra-ethyl in fuel as 'anti-knock dope'. Two years later Llewellyn Smith, who had gained his D.Phil., was replaced by Ubbelohde in work showing that aldehydes and peroxides were produced during the compression stroke of a petrol engine. From 1934 this research was extended to diesel engines and fuels by Egerton and another D.Phil. student, J. W. Drinkwater, a Manchester graduate in engineering. This combustion research done in the engineering science laboratory came to an end in 1936 when Egerton moved to Imperial College and Drinkwater to a lectureship in mechanical engineering at the University of Birmingham.[62]

It was probably the experience of supervising research in split locations and of devoting consultancy fees to buying apparatus that led Egerton to propose in 1934 the establishment of a separate new laboratory or institute

[62] Egerton memorandum, Nov. 1931, PS/R/1/3; A. R. Ubbelohde, 'Egerton', *Fuel*, 39 (1960), 100–3; *HCP* 159 (1934), 47; *OUG* 62 (1931–2), 664; 64 (1933–4), 203; 65 (1934–5), 198; 66 (1935–6), 211; 67 (1936–7), 225. Frederick Llewellyn Smith (1909–88), obituary in *The Times*, 24 Aug. 1988; Alfred René Jean Paul Ubbelohde (1907–88); F. J. Weinberg, 'Ubbelohde', *BM* 35 (1990), 383–402. Egerton's chief postgraduate collaborators who were Oxford graduates were: William Bell Lee (b. 1886), fourth in physics 1922; Stanley Frederick Gates (1897–1972), second in chemistry 1921, later ICI research, plant, and works manager, Lancashire; William Edmondson (b. 1902), second in chemistry 1925, later research chemist, Coxeter Ltd.; A. R. Ubbelohde, first in chemistry 1930; John Mylne Mullaly (1901–25), second in chemistry 1924; Michael Milford (1905–78), second in physics 1928, left the Clarendon in 1935 to spend his career as physics and senior master, Repton School; and Frank Allan Cunnold (b. 1912), first in physics 1933, D.Phil. 1935, made his career as a civil servant in the War, Defence, Supply, and Air ministries. Those from other universities were: Llewellyn Smith, John Wilson Drinkwater (both firsts in engineering from Manchester), F. F. Coleman, R. J. Bracey, and L. M. Pidgeon, later a researcher, National Research Council, Canada. Hugh Longbourne Callendar (1865–1930), *DNB*, professor of physics, Royal College of Science, London, 1902–7, Imperial College, London, 1907–30; Charles Vernon Boys (1855–1944) visited the Clarendon in the early 1890s.

for thermodynamics at Oxford. Nothing came of this proposal partly because the University was undecided about the relation of thermodynamics to physics and partly because Mendelssohn and especially Simon had not only exacerbated the problem of congestion in the Clarendon but had shown that Egerton's brand of thermodynamics, with its emphasis on combustion research, was not the only way of pursuing the subject profitably. Moreover in 1935 Ubbelohde, Egerton's most promising and productive researcher, left Oxford for the Royal Institution, London, where he was Dewar fellow under W. H. Bragg who wanted to exploit Ubbelohde's expertise in thermodynamics, gained not only at Oxford but also in 1931–2 under Eucken at Göttingen. Next year Egerton was invited by his old friend Tizard, rector of Imperial College, to occupy the chair of chemical technology there in succession to W. A. Bone. He accepted because in contrast to Oxford he would be in sole charge of a department with good workshops and laboratories in which he would be able to develop postgraduate training and where he could increase the department's already formidable reputation in fuel technology and combustion chemistry. In addition to these increased disciplinary opportunities and the removal of accumulated frustrations, Egerton had served actively for several years on various London committees and was a rising figure in the Royal Society. His move to London enabled him to become even more active in such work which led eventually to his becoming a secretary to the Royal Society 1938–48 and to a knighthood in 1943.[63]

While Dobson, Jackson, and Egerton were pursuing productive research in the late 1920s, Lindemann lost interest in personal laboratory experimentation and his credibility began to suffer. To some he seemed idle, truculent, and tiresome. To the genial Rutherford he was the most detestable of men. To others he became the butt of spiteful jokes because he spent so much time moving in polite, aristocratic, and political society. It was once asked why he was like a channel steamer, the venomous answer being that he ran from pier to pier. Lindemann was also on close personal terms with some leading industrialists and from 1921 with Winston Churchill to whom he acted as a scientific adviser on a wide range of matters including the new water garden at Chartwell in 1928.[64] While pursuing the life of a socialite, Lindemann jousted a good deal in defending and promoting science at Oxford. He developed a chip on his shoulder about the status of science at Oxford and sometimes was paranoiac about its relation to classics. He became a publicist and even a prima donna, fighting both real and mock battles on behalf of science. As a substitute for the loss of his personal experimental research drive, he gave even more attention to polemics.

[63] *HCP* 159 (1934), 173–4; Ubbelohde to author, 1 July 1987; William Arthur Bone (1871–1938), professor of chemical technology, Imperial College, London, 1912–36.
[64] Birkenhead, *Lindemann*, 121, 124, 127–31; R. V. Jones, 'Churchill', *BM* 12 (1966), 35–105, on 66–9; D. Wilson, *Rutherford: Simple Genius* (London, 1983), 434.

In his early years at Oxford he suffered some defeats and not a little bruising. One of his *bêtes noires* was H. W. B. Joseph, New College philosophy tutor, who was renowned as a ruthless pricker of half-truths and pretentious verbiage: in 1919 at a meeting of the Jowett Society he ridiculed Lindemann's defence of the theory of relativity. In 1923 Joseph was the chief extinguisher of a scheme supported by Lindemann for Science Greats which would have combined the study of philosophy with the history and philosophy of science. Joseph represented complacency and the vested interests of classicists and philosophers, both of which Lindemann deplored. When he expressed his misgivings about the low place of science at Oxford to the wife of the warden of All Souls, she assured him that he should not worry because a man who had got a first-class degree in classics and philosophy could get up science in a fortnight. Not surprisingly Lindemann was keen for scientists to hold more college fellowships because they carried prestige and power while laboratory demonstrators without fellowships were regarded as helots. That is why he made such a row about the refusal of Christ Church to make the Duke of Westminster research fellow in physics a member of its governing body. He always tried to ensure that Oxford did not lose any useful scientific resource. Thus from 1930 to 1934 he led the opposition against the proposal of the Radcliffe Trustees that their Observatory, hitherto in Oxford, should be moved to South Africa. Again in 1930 he was active in successfully opposing Christ Church's proposal to close its own laboratory. Characteristically he argued that the existence of college chemistry laboratories generated mutual competition and criticism which ensured that Oxford's chemistry was lively in research.[65]

9.3 The Clarendon and the New Diaspora

Also in 1930 Lindemann's interest in low-temperature physics revived to such an extent that he turned to Franz Simon at the University of Berlin for help in producing liquid hydrogen in the Clarendon. There were several reasons why Lindemann approached Simon and not another cryogenic expert such as W. H. Keesom of Leiden who had visited Oxford in 1926 to talk about the liquefaction of helium and his own success in being the first to solidify it.[66] Simon had taken his Ph.D. under Nernst at the University of Berlin just as Lindemann had. The subject of his Ph.D., obtained in 1921, was dear to Lindemann's heart and mind: it concerned specific heats at the low temperatures that could be procured with liquid hydrogen. The

---

[65] Harrod, *Cherwell*, 15–27, 52–3, 55, 58, 116–17, 145–59; Lindemann to Dean of Christ Church, 28 Feb. 1930, LP, B127.

[66] *OUG* 57 (1926–7), 75; Willem Hendrick Keesom (1876–1956), director of the Kamerlingh Onnes Laboratory, Leiden, 1923–45.

framework of Simon's thesis was Nernst's heat theorem, again a subject which Lindemann cared about. Lindemann knew Simon and his work from the mid-1920s and presumably appreciated that in 1930 Simon had built a new hydrogen liquefier, which was an improved version of the old temperamental Nernst–Hoenow instrument which no one at Oxford had been able to use satisfactorily. Simon had risen at Berlin by 1927 to be an associate professor and was the chief maintainer of work in cryogenics once Nernst had left in 1922. By 1931 Simon had built up such an effective research school that in April he left Berlin for the Breslau Technische Hochschule where he had been appointed full professor and was joined that summer by Kurt Mendelssohn, who had been his chief assistant in Berlin from 1929, and Nicholas Kurti, a Hungarian research student at Berlin who gained his Ph.D. in 1931. At Breslau Simon soon attracted a new Ph.D. student, Heinz London, who worked on superconductivity.[67]

Wishing to avoid the fiasco of the early 1920s Lindemann sent Keeley on two visits to Berlin, in December 1930 and spring 1931, accompanying him on the latter, to see Simon's hydrogen liquefier in action with a view to buying one for the Clarendon. By April 1931 Simon had left for Breslau but Mendelssohn had not, so it fell to Mendelssohn to demonstrate the apparatus to Lindemann and Keeley. Next month the hydrogen liquefier was duly working in the Clarendon to the delight of Lindemann who, having imported a successful instrument, tried to lure Mendelssohn to work at Oxford in 1931–2. Mendelssohn declined this invitation because in June 1931 he was going to Breslau where he would be in charge of research during Simon's leave of absence in the USA for the first six months of 1932; but he floated the possibility of visiting the Clarendon in 1932–3.[68]

By early spring 1932 the hydrogen liquefier in the Clarendon was working so well that Lindemann decided to extend his low-temperature resources by securing a helium liquefier like the one he had seen working at Berlin in Simon's department. As Simon was away from Breslau on study leave in the USA, Mendelssohn was contacted by Keeley in February 1932 about one of these miniature liquefiers. The model sought could have been the desorption one developed by Simon in the mid-1920s, which produced cooling by desorption of helium previously adsorbed by charcoal at the temperature

[67] N. Kurti, 'Simon', *BM* 4 (1958), 225–56; Nancy Arms, *A Prophet in Two Countries: The Life of F. E. Simon* (Oxford, 1966); Kurt Alfred Georg Mendelssohn (1906–80); D. Shoenberg, 'Mendelssohn', *BM* 29 (1983), 361–98; Nicholas Kurti (b. 1908); 'Kurti', *McGraw-Hill Modern Scientists and Engineers*, ii: *H–Q* (New York, 1980), 195–6; N. Kurti, 'Undergraduate in Paris 1926–8: Graduate Student and D.Phil. in Berlin 1928–31', in Williamson, *Physicists*, 86–90; Heinz London (1907–70); D. Shoenberg, 'H. London', *BM* 17 (1971), 441–61; Lindemann to Simon, 6 June 1924, SP, 14/3/L.

[68] Keeley to Simon, 2 Dec. 1930, 7, 12 Mar. 1931, SP, 14/1/2; Mendelssohn to Simon, 22 Oct. 1932, SP, 14/1/1; K. A. G. Mendelssohn, 'The World of Cryogenics, iv: The Clarendon Laboratory Oxford', *Cryogenics*, 6 (1966), 129–40, at 129–31; Mendelssohn to Lindemann, 22 May 1931, MP, B8.

of liquid or solid hydrogen; or it might have been a small Linde liquefier, based on the Joule–Kelvin effect, which was developed by Ruhemann by 1930 in Simon's laboratory. Though Keeley was well aware that Simon's laboratory at Berlin and then Breslau was one of only four research centres in the world where liquid-helium temperature work was carried out (the others being Leiden, Toronto, and the Physikalisch-Technische Reichsanstalt in Berlin), he did not know that at Berkeley in spring 1932 Simon was developing a new type of helium liquefier which used single-stroke adiabatic expansion of helium gas under pressure at the temperature of liquid hydrogen.[69]

At the time Mendelssohn, who had taken his Ph.D. in 1930 under Simon and had spent a good deal of effort in developing and building improved versions of Simon's miniature liquefiers, was supported financially by an annually renewable grant from the Notgemeinschaft der Deutschen Wissenschaft: he was not a *privat-docent* or a paid assistant to Simon. In June 1932 he told Lindemann that he would be happy to visit Oxford to install a helium liquefier if a Rockefeller Foundation fellowship could be arranged for him for 1932–3. Simon agreed to this arrangement, being quite aware that Lindemann wanted to begin experimental work with liquid helium and was therefore 'interested in transferring the technique perfected in our institute to his laboratory'. The application to the Foundation for 1932–3 was made too late to be processed so with Simon's approval Lindemann applied to it in December 1932 for a fellowship for Mendelssohn at Oxford for 1933–4. This application involved a difficulty, namely, that Mendelssohn did not have an approved type of academic post in Germany to which he could return; and by April 1933 it had become clear that this was an insurmountable obstacle. Meanwhile Lindemann was becoming frustrated about the proposed liquid-helium work because he realized that it was more difficult than that with liquid hydrogen and that Mendelssohn's experience would be vital. By October 1932 Lindemann was hoping that Mendelssohn would visit Oxford in either the next Christmas or Easter vacation to give advice and install a helium liquefier which he would bring with him. On being informed by Mendelssohn about the new expansion liquefier, Lindemann quickly arranged for Mendelssohn to visit Oxford in early January 1933 for a few days with all expenses paid. Lindemann went so far as to tell Mendelssohn by letter in November 1932 that if a Rockefeller grant were to be quickly obtained there would be no need for him to return to Breslau because he could come to Oxford for good.[70]

[69] Keeley to Hoenow, 19 Feb. 1932, Hoenow to Mendelssohn, 23 Feb. 1932, MP, B10; W. H. Keesom, *Helium* (Amsterdam, 1942), 150–80, gives a good account of helium liquefiers and liquid-helium techniques. For low-temperature research in general see R. G. Scurlock (ed.), *History and Origins of Cryogenics* (Oxford, 1992), esp. 322–56 on Oxford by R. Berman.

[70] Mendelssohn to Lindemann, 15, 28 June, 16 Sep., 6 Oct., 21, 25 Nov., 8 Dec. 1932, MP, B8; Lindemann to Mendelssohn, 23 June, 3 Oct., 14, 29 Nov. 1932, MP, B8; Mendelssohn to Simon,

Mendelssohn duly visited Oxford in the Christmas vacation 1932–3, set up a helium liquefier, and within a week had produced without hitch or trouble the first liquid helium in Britain. Lindemann was cock-a-hoop: by importing Mendelssohn, whom he thought a first-class man in ability, industry, and power of work, he had at last managed to upstage Cambridge and wipe the eye of the Cavendish. From 1930 Kapitza had been keen to develop low-temperature work there and, backed by Rutherford, secured £15,000 for a purpose-built laboratory from the Ludwig Mond bequest and a Royal Society Messel professorship for himself. By 1931 the Mond Laboratory had begun to be erected in the courtyard of the Cavendish and it was widely expected that it would be the first in Britain to liquefy helium. In the event the Clarendon secured this coup in January 1933, a month before the Mond Laboratory was formally opened. Kapitza suffered the chagrin of liquefying helium in his expensive new Mond Laboratory in 1934, a year after Mendelssohn at the Clarendon. Lindemann was quick to give publicity to his success in both *The Times* and *Nature*, stressing that the apparatus, of a Simon–Mendelssohn type and grandiloquently referred to as 'plant', was small and at £30 inexpensive, though the compressor and liquefier for hydrogen cost £350. Characteristically Lindemann rubbed salt into Kapitza's wounds by alleging that the efficiency of his pyrex Dewar flasks for storing liquid hydrogen equalled that claimed for the more complicated double metal vessels developed by Kapitza.[71]

In early 1933 Mendelssohn was helped in the Clarendon by Keeley, who worked the hydrogen liquefier, and by Babbitt, a Canadian Rhodes scholar who had graduated in physics in 1932 and become a demonstrator. For further experience Babbitt soon went to work with Mendelssohn in Breslau from which he reported to Keeley, his Oxford supervisor, that he had 'learnt considerable [*sic*] about the work'. By April 1933 Breslau was so afflicted by Nazi violence that Mendelssohn, who was of mixed Jewish–German parentage, decided to seek refuge in Oxford. Lindemann realized that Mendelssohn was a valuable man, well versed in low-temperature physics and also an effective cryogenic engineer, who would quickly develop low-temperature research in the Clarendon if he could be kept there (see Fig. 7). Within a few days of Mendelssohn's arrival in Oxford, Lindemann had induced Imperial Chemical Industries to make him a provisional offer of £400 a year for two years which he was happy to accept. By late July 1933 Mendelssohn had signed a contract with ICI backdated to take effect from 1 May 1933. The

22 Oct. 1932, SP, 14/1/1; Simon to H. M. Miller, 21 July 1932, SP, 14/1/13 (quotation); Simon to W. E. Tisdale, 19, 26 Nov. 1932, SP, 14/1/13; Miller to Simon, 6 Apr. 1933, SP, 14/1/13.

[71] *The Times*, 28 Jan. 1933; F. A. Lindemann and T. C. Keeley, 'Helium Liquefaction Plant at the Clarendon Laboratory, Oxford', *Nature*, 131 (11 Feb. 1933), 191–2; Jones, 'Oxford Physics', 122; Piotr Leonidovich Kapitza (1894–1984); D. Shoenberg, 'Kapitza', *BM* 31 (1985), 325–74, at 341–2; id., 'Royal Society Mond Laboratory, Cambridge', *Nature*, 171 (1953), 458–60.

FIG. 7 K. A. G. Mendelssohn, the first to liquefy helium in Britain and the first refugee from Germany to find sanctuary in the Clarendon Laboratory, with low-temperature apparatus there, 1930s. Reproduced from Mendelssohn Papers, Eng. misc. b.388, by permission of the Bodleian Library, Oxford.

advent of the Third Reich gave Lindemann his opportunity to capture Mendelssohn's expertise, having previously borrowed it for a fortnight. Mendelssohn did not disappoint him: by late September 1933 in a paper on the production of high magnetic fields at low temperatures Mendelssohn had publicly confirmed that the Clarendon's cryogenic research had been launched by him.[72] This success induced Lindemann to contemplate a larger scheme of importing some of Mendelssohn's low-temperature colleagues from Breslau to Oxford where they would be paid neither by the University nor by the colleges but by ICI. After all, it was Mendelssohn's coup with the liquefaction of helium at the Clarendon which had enabled Lindemann at long last to steal a march on the Cavendish which in 1932 had enjoyed an *annus mirabilis* with the discoveries of the splitting of lithium and boron atomic nuclei by Cockroft and Walton and of the neutron by Chadwick. Through a cohort of Breslau low-temperature physicists, who would develop the cryogenic work already begun by Mendelssohn, the Clarendon would have better prospects of outsmarting the Cavendish again; and, at the same time, through them the status of science in Oxford and its reputation elsewhere could be raised. Moreover, there were personal resonances: Mendelssohn was a pupil of and senior assistant to Simon, who like Lindemann had been a Ph.D. pupil of Nernst in Berlin. The success of Mendelssohn's early research in Oxford presumably suggested to Lindemann the idea of creating in Oxford a worthy continuation of Nernst's cryogenic school, which would develop research in which he had been involved as a young man, using Mendelssohn's Breslau colleagues. Emotionally this scheme satisfied Lindemann's long-lasting and deep loyalty to Nernst: it would enable him to build a new world of old memories around himself at the Clarendon. In implementing this dream Lindemann enjoyed a stroke of luck: on a visit to the USSR Kapitza was detained in September 1934, thus permanently removing the Clarendon's chief British rival in the field of cryogenics.

It was the combination of Nernstian loyalties and Mendelssohn's *coup de théâtre* which was responsible for Lindemann's taking the lead in helping German physicists, especially Jews, who found themselves without prospects and jobs in Nazi Germany. With car and chauffeur he toured Germany in spring 1933 in an attempt to provide refuge for German physicists by placing them, even if temporarily, in British universities including of course his own. Exploiting his friendship with Sir Harry McGowan, chairman of ICI, he persuaded the company to finance German scientists for two years, with the

---

[72] Mendelssohn, 'Cryogenics'; Keeley to Mendelssohn, 22 Mar., 2 Apr. 1933 (quotation from PS by Babbitt), MP, B12; Mendelssohn file, SPSL, 335/2; Mendelssohn to J. M. King, ICI, 28 July 1933, MP, B14; Mendelssohn, *Nernst*, 166; id., 'Production of High Magnetic Fields at Low Temperatures', *Nature*, 132 (14 Oct. 1933), 602, dated 28 Sept. 1933; John David Babbitt (1908–1982), a Canadian Rhodes scholar, first in physics 1932, D.Phil. 1934, became a physicist with the National Research Council, Ottawa.

possibility of renewal. His basic aim was to improve Oxford physics by inviting refugee physicists to the Clarendon. To that end he co-operated with the Academic Assistance Council which aimed to help all German scholars irrespective of field; but more importantly he had turned earlier to ICI which was interested in German physical scientists.[73] Mainly through the ICI scheme, Lindemann managed to gain for the Clarendon six experimentalists, namely, four Breslau low-temperature physicists, Mendelssohn, Simon, Kurti, and Heinz London; Heinrich Kuhn, a spectroscopist from Göttingen; and Leo Szilard, a Hungarian nuclear physicist. Of these Simon, Mendelssohn, Kurti, and Kuhn were retained at Oxford until 1939 and indeed spent the rest of their working lives there. Having previously lured Einstein to make a couple of visits to Christ Church, Oxford, Lindemann also recruited from Berlin a couple of distinguished theoretical physicists, Schrödinger and Fritz London, neither of whom stayed long.

It was natural for Lindemann to try to provide for Simon in Oxford, having witnessed for himself the speedy prowess of Mendelssohn. Simon was a more desirable catch than Mendelssohn. He was the senior man. He had an international reputation shown in his proved ability at Berlin and Breslau to attract to his laboratory non-German researchers such as Kurti from Hungary via the Sorbonne, R. C. Swain from the USA, and R. Kaischew from Bulgaria; in contrast Mendelssohn had not had the time or opportunity to attract researchers.[74] Mendelssohn's research fields were or were about to be superconductivity and superfluidity in liquid helium, whereas Simon had already shown himself to be a powerful investigator on two related topics which were particularly dear to Lindemann; these were Nernst's heat theorem and specific heats at low temperatures. A good deal of Simon's work in the 1920s had been concerned not just with defending the theorem but with reformulating and elevating it into the third law of thermodynamics. Simon's research was far more obviously than Mendelssohn's a continuation and extension of Lindemann's own pre-war work. Lastly, Mendelssohn was a solo worker, whereas from 1931 Simon had enjoyed the services of Kurti as a research collaborator and right-hand man. Lindemann's top priority was therefore to secure Simon and Kurti with whom lengthy coded negotiations were conducted, via Mendelssohn as a helpful intermediary, about salaries. In May 1933 Simon paid a brief visit to England at Lindemann's invitation, next month he resigned his post at Breslau, in July he formally accepted the Oxford job about which he and Lindemann had spoken, in August he

---

[73] Birkenhead, *Lindemann*, 101–4; Rider, 'Alarm', 146–51; Hoch, 'Refugee Physicists', 221–31; Lindemann to McGowan, 24 May 1933, LP, D95; Lindemann to F. Haber, 9 Jan. 1934, LP, D86; Lindemann and W. Rintoul to McGowan, 24 July 1934, LP, D96. Harry Duncan McGowan (1874–1961), chairman of ICI 1930–50, was rewarded with an honorary Oxford DCL in 1935.

[74] Arms, *Simon*, fig. 8.

arrived in England bringing with him two helium liquefiers, in September he received an offer from ICI of £800 p.a. for two years from 1 September 1933, and next month accepted it. Kurti arrived after Simon in September 1933, and Heinz London, having completed his Ph.D. at Breslau in late 1933, joined the Breslau colony in Oxford early in 1934, both being supported by ICI grants.[75]

The recruitment of the spectroscopist Heinrich Kuhn was both opportune and opportunist. During a recruiting tour of Germany in which he collected the visiting cards of physicists keen to leave their country and wrote evaluative comments on them, Lindemann saw in Göttingen James Franck, an old friend from before the First World War. One of Franck's most promising pupils was Kuhn, who had taken a Ph.D. in 1926 and by early 1933 was a demonstrator and *privat-docent* in Franck's department. Because of his non-Aryan descent, Kuhn had been immediately dismissed as a result of the Nazi legislation of April 1933. He needed to leave Germany quickly. In early June in Franck's house he met Lindemann for the first time and was invited to Oxford to work with Jackson, who in principle had agreed to the partnership. By 18 June 1933 Kuhn had gladly accepted an ICI grant to enable him to work at Oxford for two years, with a possibility of extension for a third year but not beyond. By mid-August not only had Kuhn and his wife arrived in Oxford but he had met Jackson who reported quickly to Lindemann that he thought the partnership with Kuhn was workable. Thus, as a result of negotiations which took a fortnight and of arranging to leave Germany which took a month, Kuhn's career had been transformed totally: in August 1933 he found himself in the Clarendon under a professor whom he had met only in June and with a research collaborator whom he had not met before.[76]

In 1933 Lindemann was keen to recruit theoretical physicists as well as experimentalists. This was partly the result of frustration with Milne, the Rouse Ball professor of mathematics, and of the inspiration provided by Einstein in his two summer visits made in 1931 and 1932. Milne's chair was intended to be devoted to mathematical physics. In practice Milne showed himself uninterested in solving the particular problems posed by experimentalists. Though Lindemann initially wanted Milne to come to Oxford, he was subsequently disappointed by Milne's approach which he regarded as mere mathematical rehashing of existing findings and he deprecated Milne's going off into cosmological work of little interest to physicists.[77] In contrast to

[75] Lindemann to Simon, 24 Apr., 18 May, 18 June, 24, 29 July 1933; Simon to Lindemann, 12 May, 3 July 1933; Simon to A. Black, ICI, 5 Oct. 1933, all SP, 14/3/L; Black to Simon, 18 Sept. 1933, SP, 14/1/3; Kurti, transcript of interview for Archive for History of Quantum Physics; H. London file, SPSL, 334/5.

[76] Heinrich Gerhard Kuhn (1904–94); Kuhn to author, 23 Jan. 1987; S. Lukes, 'Interview with Heini Kuhn', *Balliol College Record* (1987), 42–53.

[77] Harrod, *Cherwell*, 66; Lindemann memorandum, 19 May 1938, UR/SF/PHE/7.

Milne, Einstein was his beau ideal of a theoretical physicist. Through Lindemann's initiative he first appeared in Oxford in spring 1931 as Rhodes memorial lecturer, his topic being his own theory of relativity. He was accommodated in Christ Church, the calm cloisters of which he relished as much as Oxford relished him. In summer 1931 he was happy to accept the offer of a research studentship at Christ Church for five years at £400 p.a. His last appearance in Oxford was in June 1933 as Herbert Spencer lecturer, his theme being the method of theoretical physics.[78] Einstein's visits presumably convinced Lindemann that it would be useful to induce a couple of German theoretical physicists to seek refuge in Oxford for two to three years. He had in mind Fritz London and Hans Bethe, who had been recommended by Sommerfeld, because unlike Milne each could answer problems and was not 'the more abstract type who would spend his time disputing with the philosophers'. In the event London came but not Bethe. A Jewish *privat-docent* at Berlin and assistant to Schrödinger, Fritz London had been dismissed in May 1933. He was a distinguished theorist, best known for his work in the late 1920s with Heitler on homopolar chemical bonding which interpreted chemical valency in terms of electronic spin. With Mendelssohn's help he arrived in summer 1933 in Oxford where he was supported by an ICI grant for two years initially.[79] The second and even better theoretical catch was a non-Jewish physicist, Erwin Schrödinger, whom Lindemann met in Berlin in spring 1933. Though Lindemann was suspicious of Schrödinger's wave mechanics, he accepted Einstein's high opinion of Schrödinger *qua* theoretical physicist. Like the Breslau physicists, Schrödinger was supported financially mainly by ICI, initially for two years. Unlike them he received a much greater grant (£1,000 p.a.) and his base was not the Clarendon but Magdalen, the result of Lindemann's plan that Schrödinger be elected a non-stipendiary supernumerary fellow there but with a small research grant. In early November 1933 Schrödinger arrived in Oxford and within a few days heard that he had been awarded a Nobel prize. At this stage, with two distinguished theoretical physicists, one of whom was a Nobel prize-winner, installed in Oxford, it seemed possible that theoretical physics would blossom to complement the

[78] Birkenhead, *Lindemann*, 51–2, 159–62; Harrod, *Cherwell*, 47–50, 91; Elsa Einstein to Lindemann, 11 May 1931, Lindemann to Einstein, 29 June 1931, both LP, D56; *OUG* 61 (1930–1), 224; 62 (1931–2), 131; 63 (1932–3), 582; A. Einstein, *Theory of Relativity: Its Formal Content and Present Problems* (Oxford, 1931); id., *On the Method of Theoretical Physics* (Oxford, 1933).

[79] Lindemann to Einstein, 4 May 1933, LP, D57; Fritz Wolfgang London (1900–54); K. Gavroglu, *Fritz London: A Scientific Biography* (Cambridge, 1995), esp. 96–138; F. London file, SPSL 334/4; Hans Albrecht Bethe (b. 1906), Nobel prize-winner in 1967 for his theory about the source of energy in stars, spent 1933–5 at the universities of Manchester and Bristol before leaving for Cornell University, Ithaca, NY; Arnold Sommerfeld (1868–1951), professor of theoretical physics, University of Munich, 1906–31; Walter Heinrich Heitler (1904–81) left Göttingen in 1933 to be a research fellow, University of Bristol, until 1941.

experimental work in cryogenics and spectroscopy which the *émigrés* were promoting in the Clarendon.

That did not happen because of Schrödinger's personality and behaviour and because of the difficulty of continuing the ICI grant for Fritz London. It has been argued in Schrödinger's case that there was no place, literally and intellectually, at Oxford for him. Certainly Schrödinger felt that he was living off munificent generosity without being able to return any true service, which was for him a burden. He did lecture regularly at Oxford on elementary wave mechanics but he seems to have had little involvement with the Clarendonians. He felt he was caught on the prongs of a fork: either he was paid for doing almost nothing or he was treated as a pauper. There is no doubt that he was worried about the lack of a permanent position and felt unappreciated. All these things counted but were less important than his unconventional approach to marriage. For Schrödinger bourgeois marriage, though comfortable, was incompatible with the romantic love for which he yearned and which in his view was destroyed by everyday humdrum living. By late summer 1933 his latest mistress was Hildegunde March, wife of the German physicist Arthur March, and she was pregnant by Schrödinger. In November the Schrödingers and the Marches arrived in Oxford, and Schrödinger made no attempt to conceal that he had a wife and a mistress, sometimes living openly with them in a *ménage à trois*. In May 1934 Schrödinger's child by Mrs March was born. This led to strained relations with Lindemann to whom Schrödinger had pleaded that March should come to Oxford ostensibly as his assistant but in fact to facilitate his own affair with Mrs March. Lindemann's verdict was that that 'bounder' should go, though he modified this verdict in summer 1935 when he failed to persuade the Physical Sciences Faculty Board to support a readership in theoretical physics for Schrödinger. Perhaps Lindemann appreciated Schrödinger's decisive intervention at Magdalen in 1934 when he secured the election to a fellowship of James Griffiths, a protégé of Lindemann. On the whole, however, Schrödinger's sojourn in Oxford was not a success: he stayed there for two and a half years before moving in 1936 to the University of Graz.[80]

Fritz London was a married man of impeccable bourgeois morality who for three years until August 1936 was supported by an ICI grant. Unlike Schrödinger, London did important research in his Oxford period. Stimulated by the work on superconductivity of his younger brother Heinz, Fritz relinquished his work on quantum-mechanical theoretical chemistry and with Heinz produced their well-known paper on the electrodynamics of

[80] Erwin Schrödinger (1887–1961); P. K. Hoch and E. J. Yoxen, 'Schrödinger at Oxford: A Hypothetical National Cultural Synthesis which Failed', *Annals of Science*, 44 (1987), 593–616; W. J. Moore, *Schrödinger: Life and Thought* (Cambridge, 1989), 267–71, 273, 280, 293–8 (298 quotation), 317; Schrödinger to Sidgwick, 1 Mar. 1936, Sidgwick papers, V. 75; PS/M/1/1, 18 June 1935; Dirk ter Haar, 'James Howard Eagle Griffiths 1908–81', *Magdalen College Record* (1982), 38–43.

superconductors. This fruitful partnership, which lasted just two years, ended in 1936 when Heinz left Oxford in February. That spring of 1936 was not a happy time for Fritz: he had lost his scientific collaborator, an experimentalist who had worked in the Clarendon and given Fritz, a theoretician, some contact with it; his interests were so theoretical that an industrial or commercial post was out of the question; he was too senior to apply for junior posts in British universities and senior ones were few and far between; he knew that his ICI grant, which had been extended for one year from 1 August 1935, would expire irrevocably on 1 August 1936; and in May he fell out of a railway carriage and broke his ribs. Compared with other migrants he had great difficulties with the English language. When his ICI grant ended he was supported for two months by a temporary grant from the AAC before he assumed a post in Paris at the Institut Henri Poincaré in October 1936. In late 1938 he held a visiting post for six months at Duke University, Durham, North Carolina, and transformed it into a permanent chair of theoretical physics there in 1939. Presumably Lindemann regretted London's departure from Oxford because in August 1936 he was negotiating for a lecturership at Brasenose for London as part of a larger financial arrangement which he failed to implement.[81]

The migrant experimentalists in the Clarendon also faced problems of funding. Generally from 1933 to 1935 they were on two-year ICI grants which were renewed for one year until the summer of 1936. The ICI programme was intended to benefit the company through advances made in research, to improve the training of university students who would afterwards become university teachers or work for ICI, to raise the general level of long-range British scientific research, and to plug embarrassing intellectual gaps in British university laboratories. The scheme tried to avoid displacing British scientists and those Germans chosen as recipients in 1933 were generally not closely connected with industry because of ICI's wish to avoid jeopardizing its good relations with IG Farben, its German equivalent. For Lindemann the ICI scheme was not only philanthropic: it enabled him to take advantage of a 'golden opportunity of buying up many of the best brains in Germany'. The ineluctable difficulty was that ICI would not fund the scheme indefinitely and wished to bring it to an end in summer 1936.

Various solutions were found to the problem of paying the former recipients of ICI grants, depending on their seniority, the nature of their research, and its relation to Lindemann's and ICI's central preoccupations. In the case of Heinz London, whose field was superconductivity, his lack of manual dexterity as an experimenter was notorious: he was clumsy and untidy. In early 1936 he migrated to Bristol as a research assistant at £200

---

[81] SPSL 34/4; A. H. Cooke, interview with author, Sept. 1986; K. J. Spalding to Lindemann, 8 Aug. 1936, LP, D148; F. and H. London, 'The Electromagnetic Equations of the Supraconductor', *Proceedings of the Royal Society of London A*, 149 (1935), 71–88.

p.a. and maintained his reputation as a physicist whose ideas and knowledge counterbalanced his sloppiness in the laboratory. In contrast, Kuhn managed to remain in Oxford from 1936 when his ICI money ran out: he was supported by two colleges, St John's and Queen's, with grants totalling £300 per year. Simon and Kurti not only stayed at the Clarendon but in 1936 their ICI grants were renewed for five years. One presumes that Lindemann's very high evaluation of Simon's work and the pressure he exerted on ICI, which had an abiding interest in low-temperature physics, ensured that Simon and Kurti belonged to that small band whose grants were renewed beyond 1936.[82]

At Breslau Simon's salary was the equivalent of £2,000 p.a.; but his grant from ICI was initially £800 p.a. for two years from autumn 1933 augmented to £1,000 p.a. in autumn 1935 for three years. Lindemann wanted to keep Simon in Oxford: he was unequalled as a low-temperature physicist, he had drive, organizational prowess, amiability, and a sense of humour; he was an inspiring supervisor of postgraduates, and his wife, who spoke English perfectly, was popular with the ladies of north Oxford. In late spring 1936 Egerton's resignation from the university readership in thermodynamics gave Lindemann his chance to secure a permanent university post for Simon. By June 1936 he was supporting Simon warmly to the Physical Sciences Faculty Board, stressing that he was so eminent that there was no comparable British candidate. His advocacy won the day: Simon was elected to the readership for five years from 1 October 1936 at a salary of £500 p.a., making him one of that select trio of German physicists who secured a quasi-permanent British university post before 1939. Simultaneously ICI reduced its grant to him from £1,000 to £500 p.a. but guaranteed it until 1941. Simon's total emolument of £1,000 p.a. was five-sixths of that then paid to many science professors at Oxford and compared well with £750 p.a. which was the maximum paid to university demonstrators who did not hold a college post. Though no university or college post was found in 1936 for Kurti, Simon's chief collaborator, ICI saw that the partnership was fruitful and gave a five-year grant to him worth £300 p.a..[83] Mendelssohn was also so important to Lindemann's plans that in 1938 he regretted that he could offer no room at Oxford to Lise Meitner, who was soon to be the co-

---

[82] H. Molson to Lindemann, 1 May 1933, LP, 161 (quotation); H. London file, SPSL, 334/5; K. Mendelssohn, 'Pre-war Work on Superconductivity as Seen from Oxford', *Reviews of Modern Physics*, 36 (1964), 7–12, at 8; A. M. Tyndall to Simon, 21 Jan., 26 June 1936, SP, 14/3/T–Z; Simon to D. C. Thomson, 13 June 1939, SP/14/3/K.

[83] Lindemann and Rintoul to McGowan, 24 July 1934, LP, D96; Lindemann to Slade, ICI, 4 Oct. 1936, LP, D156; Lindemann to C. G. Robinson, 5 Mar. 1936, LP, D228; Lindemann memorandum, 6 June 1936, PS/R/1/5; Simon, application for Birmingham physics chair, 29 Feb. 1936, SP, 14/2/67; Lindemann to A. D. Lindsay, 23 Sept. 1938, LP, D154; Lise Meitner (1878–1968). The three German physicists who secured permanent British university posts were: Simon; Max Born, chair, Edinburgh, 1936; and Rudolf Ernst Peierls (1907–95), chair, Birmingham, 1937.

discoverer of nuclear fission, because his top priority was to support Simon and Mendelssohn. In the event, procuring a stipend for Mendelssohn involved protracted and complicated negotiations. Mendelssohn's first contract from ICI was for two years from 1 May 1933 at £400 p.a. It was renewed for three years from May 1935 at £400 p.a. Fearing that he might lose Mendelssohn's services in May 1938, Lindemann persuaded ICI in late 1936 to spread out its grant to Mendelssohn until 30 September 1939. Mendelssohn was to be paid by ICI at £400 p.a. until 30 November 1936, and then at £200 p.a. until September 1939. At the same time Lindemann persuaded Sir Robert Mond to transfer to Mendelssohn for three years the grant of £300 p.a. he had previously given to Richard Rado, a refugee mathematician who had just acquired a permanent post. Thus from December 1936 Mendelssohn received £200 p.a. from ICI and £300 from Sir Robert Mond. Unfortunately Mond died in October 1938 and by May 1939 it was known that his widow had decided not to renew in late 1939 her husband's grant to Mendelssohn who supported not only his own family but also his parents who lived with him. Lindemann's response was to turn in mid-May to the University's recently established Higher Studies Fund and to secure a promise of a grant of £450 p.a. maximum when the grants from ICI and Mond ran out, £450 being the sum stated by Lindemann to the registrar as Mendelssohn's total annual income.[84]

It would be wrong to assume that the migrants who were to remain in the Clarendon were always totally happy with their lot. At some point most of them were anxious about their situation in Oxford and their prospects. Their feeling of insecurity seems to have increased with seniority. Kurti was certain that Oxford and ICI would somehow look after Simon and himself so he did not apply for another post. In contrast Kuhn experienced such worrying uncertainty in 1935–6 that he applied for jobs abroad, though he drew the line at a post at Kharkov in Stalinist Russia. At the same time Mendelssohn applied for the chair of physics at Raffles College, Singapore, and for a lectureship at Rangoon University, having earlier declined to consider a post at Leningrad. Simon was the most restless and ambivalent about his position and facilities at Oxford, a situation which persisted after 1936 when he was elected a university reader; indeed his anxiety was not totally allayed in 1937 when the University finally decided to build a new Clarendon Laboratory. No college elected him to a fellowship; but through his friendship with Cyril Bailey, classics don at Balliol, he was made in 1935 an honorary member of

---

[84] Mendelssohn to J. M. King, ICI, 28 July 1933; Mendelssohn to L. A. Inglis, ICI, 25 Apr. 1935; Inglis to Mendelssohn, 14 Aug. 1935, all MP, B14; H. L. Nathan to Lindemann, 17 Nov. 1936; R. E. Slade, ICI, to Lindemann, 20 Nov. 1936; Lindemann to Nathan, 24 Nov. 1936; Lindemann to Slade, 24 Nov. 1936; Lindemann to R. Mond, 27 Nov. 1936; Mendelssohn to Mond, 27 Nov. 1936; Lindemann to Veale, 19 May 1939, all MP, B15; *OUG* 70 (1939–40), 51; Robert Ludwig Mond (1867–1938); J. Thorpe, 'Mond', *ON* 2 (1939), 627–32.

the Balliol senior common-room with lunch and dinner rights. His anxiety about his situation in Oxford was not reduced by the offers he received of jobs elsewhere and an unsuccessful application he made. No sooner had Simon arrived in Oxford than he was apprised of a new chair of physical chemistry at Istanbul at about £1,100 p.a. salary for five or ten years, an offer which he rejected rather quickly. Two years later the offer was increased to £2,000 p.a. salary for ten years, which Simon refused after considerable deliberation. In 1936 he turned down the offer of a chair of physics at Jerusalem. The one post for which he applied was the chair of physics at Birmingham to which Oliphant was elected in spring 1936. Simon vacillated about whether to apply, partly because he received contrary advice from colleagues elsewhere; Rutherford, Cockroft, and Blackett urged him to apply, but Chadwick discouraged him on the grounds that a German refugee should not apply for a post for which under normal circumstances he would be ineligible. Even after Simon had secured the Oxford readership, a permanent position which enabled him to plan research more effectively, he yearned for a fully equipped and staffed laboratory totally under his own control. In 1937 he appeared in print agreeing with Lindemann that in Britain money for posts and equipment in universities was tighter than in the USA, Germany, and Russia, and stressing the superiority of the equipment of laboratories in the USA which he had just visited. In autumn 1938 Simon was not entirely unresponsive to a scheme proposed by C. F. Squire, an American post-doctoral researcher with Simon, that Carnegie patronage be sought for a new low-temperature centre in the USA to be run by Simon and Kurti.[85]

Further causes of Simon's discontent were focused on his income, the status of himself and of science in Oxford, and his German nationality. It is clear that Simon felt he had been demoted as a result of his forced move from Breslau: his salary at Oxford was at best just half of that paid to him at Breslau; and his move to Oxford at the age of 40 was accompanied by the loss of a permanent full professorship. No longer was Simon Herr Professor: from 1933 to 1936 he existed in Oxford on what he regarded as a charitable basis, his salary coming from ICI and some of his research expenses being covered by a grant of £550 p.a. from the Rockefeller

[85] Kurti, AHQP tape; Kuhn to author, 23 Jan. 1987; Kuhn file, SPSL 333/3; Mendelssohn file, SPSL 335/2; C. M. Skepper to Mendelssohn, 1 Nov. 1934, MP, 4; Arms, *Simon* 69–73; C. Bailey to Simon, 25 Jan. 1935; Simon application for Birmingham chair, 29 Feb. 1936, both SP, 14/2/67; C. S. Gibson to Simon, 28 Sept. 1933, SP, 14/3/G; Simon to A. Black, ICI, 5 Oct. 1933, SP, 14/4/L; J. Chadwick to Simon, 19, 28 Feb. 1936; Simon to Chadwick, 2 Mar. 1936; Simon to J. Cockroft, 26 Feb. 1936; Cockroft to Simon, 2 Feb. 1936, all SP/14/3/CD; Simon, letter to *Daily Telegraph*, 11 June 1937; Simon to Bridgman, 20 May 1936, 11 Feb 1937, SP/14/3/B; Simon to W. F. Giauque, 23 Apr. 1936, SP/14/1/8; Squire to Simon, 16 Sep. 1938; Simon to Squire, 5 Oct. 1938, both in SP/14/3/S; Charles Francis Squire, Johns Hopkins University Ph.D. 1937, was a research fellow in Paris; Mark Oliphant (b. 1901), professor of physics, University of Birmingham, 1936–50; Cyril Bailey (1871–1957), fellow of Balliol 1902–39.

Foundation received in 1933–5. From 1936 his readership gave him a university position which *de jure* was tenable for five years and not permanent. Simon was happy to have what he hoped would be a permanent university post but was not entirely content with his readership which he regarded as a crypto-chair which did not give him control of a department or institute he could call his own. He was also unhappy about the lack of a 'physikalische atmosphäre' in Oxford: it seemed to him that the University despised science and that scientists were treated as plumbers who were called in only when needed. Within his own research group, however, he promoted the desired ethos not least because its migrant members spoke to each other in German up to about 1939.[86]

In applying for funds from British institutions such as the Royal Society of London and the Department of Scientific and Industrial Research, Simon was at a disadvantage until 1938 when he was naturalized. As a German he could not apply directly for grants but had to work through Lindemann. Funds for research at the Clarendon were so low in his opening years that had he not managed to bring with him from Breslau some of his own apparatus, including two helium liquefiers, his research would have been seriously hampered. In order to continue his long-standing interest in the properties of fluids at high pressures, he had to travel to Amsterdam, and in order to develop his research on magnetic cooling begun in 1934, he had to go to Paris. In neither case was the appropriate apparatus available in the Clarendon. The Amsterdam work was done with the co-operation of Michels whose high pressure apparatus enabled Simon to study the melting curve of helium at pressures up to about 5,000 atmospheres.[87] More important in Simon's view was the new research on magnetic cooling which required strong electromagnetic fields. He turned for help to Aimé Cotton, the director of the Laboratoire de l'Électroaimant de Bellevue, near Paris, which housed a huge and powerful magnet (150 tons), then unique in its great adaptability for a wide range of experiments. In return for the use of this facility for seven periods, each of one month, between August 1935 and April 1938, Simon offered the French portable facilities for low-temperature research. He and Kurti took their apparatus, with its built-in expansion helium liquefier, by car from Oxford to Paris where in spring 1936 they produced France's first liquid helium. The work on adiabatic demagnetization done at Bellevue was an international affair involving Simon, Kurti, Squire, Rollin who was an Oxford postgraduate, and Lainé who was Cotton's assistant. The research was important not only for its investigations of the absolute scale of temperature below $1°K$ and of the properties of paramagnetic salts at such low temperatures but also because it provided a

---

[86] Arms, *Simon*, 69–71, 86 (quotation); LP, B15; Kurti, interview with author, Dec. 1986.

[87] Kurti, 'Simon', 236–7; Antonius Mathias Johannes Friedrich Michels (1889–1969), director of the van der Waals physics laboratory, University of Amsterdam, 1929–60.

general resource for experiments done at below 1°K. As befitted an international collaboration in Paris, the work on adiabatic magnetic cooling was published mainly 'en français Kurti' by the Paris Academy of Sciences which supported the Bellevue installation. The Paris visits ended because Simon resented the travelling involved, the loss of holidays, and the relentless intensity with which work at Bellevue had to be carried out. In spring 1938 the research on magnetic cooling in Paris came to an end and the outbreak of war vitiated plans to resume it in Oxford using a big magnet paid for by a Royal Society grant of £1,500 to Simon in 1939.[88]

In some British laboratories the migrants from Germany faced problems of adjustment to the indigenes working on similar research problems. At the Clarendon there was an obvious contrast between spectroscopy and cryogenics; in the former Jackson, a native, was an active researcher but in the latter there was no indigenous researcher working with indigenous apparatus. This difference led to contrasting developments once the migrants had settled in. In spectroscopy there were initial problems of *approchement* between Jackson and Kuhn, who were strangers. The ebullient Englishman, a master of repartee and wicked fantasy, was a very rich man, was mainly self-taught as a researcher, and had never worked under or with anyone. Inevitably there was a period of adjustment by both men to their new situations: it was short because each appreciated the virtues of the other. Primarily an experimentalist who had worked on hyperfine structures, Jackson respected Kuhn's more theoretical approach as epitomized in his book on atomic spectra written by 1933. They quickly established a fruitful partnership, working together on making spectral lines as sharp as possible, a topic on which they published eleven joint papers from 1935 to 1939 inclusive. When Jackson was away from Oxford at race meetings, fox hunts, and dinner parties, Kuhn developed his existing theoretical and experimental interest in a field which did not excite Jackson, namely, the artificial pressure-broadening of spectral lines on which Kuhn published solo. Thus Jackson's periodic absences from the laboratory gave Kuhn time to extend a long-standing interest. The nature of their partnership as well as Jackson's personality precluded their supervising any research students before the war. Nor was their field of high-resolution spectroscopy central to Lindemann's aims: it was a side-show compared with low-temperature and atomic

---

[88] Kurti, 'Simon', 238–40; Bellevue Laboratory, Paris, correspondence, SP/1/8E, including Simon to P Lainé, 17 Apr. 1936 (quotation); Aimé Auguste Cotton (1869–1951), professor of physics at the Sorbonne and director of physics laboratories 1920–41, on whom see T. Shinn, 'The Bellevue Grand Electroaimant, 1900–1940: Birth of a Research-Technology Community', *Historical Studies in the Physical and Biological Sciences*, 24 (1993), 157–87; Bernard Vincent Rollin (1911–69), Oxford first in physics 1933, D.Phil. 1935, was a Commonwealth fellow, University of California, Berkeley, 1937–9, did war-work at Oxford, became a university lecturer at Oxford in 1945 and fellow of Wolfson College in 1967; obituary in *The Times*, 27 June 1969.

physics. Hence Kuhn's financial position was difficult from 1936. In compensation, space and equipment were not problematic. Kuhn and Jackson shared a large laboratory which was sumptuously equipped by Jackson at a cost of £4,000 by 1939. It was not the University, not the colleges, and not an industrial sponsor, but Jackson's own pocket which provided the physical resources for the two main lines of experimental spectroscopy pursued in the Clarendon.[89]

In cryogenics a tense situation quickly developed from the competing interests of Simon and Mendelssohn. Inclined to be awkward and paranoiac, Mendelssohn had launched cryogenics at Oxford, he was the first of the Breslau cohort to settle in Oxford, and he had played an important role in helping Simon and Kurti to migrate from Breslau to Oxford. He then found himself number two to Simon. He resented his subordinate position and soon became jealous of Simon, who he felt had usurped his own position at Oxford as a result of the contingency of the Nazis' accession to power. There were many difficult situations between them, involving conflict about apparatus, materials, and on occasion field of work. An informal agreement was made to avoid abrasive competition: Simon and Kurti were to focus on thermodynamic properties, including the magnetic cooling method, leaving Mendelssohn to cover superconductivity and liquid helium with his own independent research group. Inevitably on occasion one territory overlapped the other because the new cooling method employed by Simon and Kurti led to the discovery of new superconductors below 1K; and in 1937 Rollin, one of Simon's research students, discovered the liquid helium 2 film, sometimes called the Rollin film. Suspicious of what he regarded as Simon's Herr Professor mentality, Mendelssohn resented what seemed to him to be the invasion by Simon of his sacrosanct territories of superconductivity and superfluidity in liquid helium. It was fortunate for the Clarendon's future that Simon's character was genial and he did not retaliate against Mendelssohn's jealousy. It was therefore possible for each of them to build up a productive research group at the Clarendon, working independently on contiguous though sometimes overlapping areas.[90]

As the less senior, Mendelssohn did not attract as many researchers as Simon; but in collaboration with four postgraduates before the war (John Babbitt, Judith Moore, Rex Pontius, and John Daunt, all of whom were Oxford graduates in physics) he made important discoveries about superconductivity and superfluidity in liquid helium, which in the period 1934–9

---

[89] Kuhn and Hartley, 'Jackson', 274–80; Kuhn to author, 23 Jan. 1987; *OUG* 64 (1933–4), 355; 65 (1935–6), 655; H. G. Kuhn, *Atomspectren* (Leipzig, 1934): Lindemann memorandum, 17 Feb. 1939, LP, B25, ff. 3–4; OUA, Spectroscopy. Reader and lecturer in, UR/SF/Spec/1.

[90] Shoenberg, 'Mendelssohn', 381–2; Simon to Bragg, 1 Nov. 1946, SP, 14/3/M; Simon to Rollin, 12 Dec. 1938, SP, 14/3/R.

inclusive were published in no fewer than twenty-five papers. The most spectacular work, done by Mendelssohn and Daunt, concerned the way in which any solid surface in contact with liquid helium 2 became covered with a frictionless thin film of liquid helium (not a distillate) whose flow was analogous to the resistanceless electrical current which flowed in superconductors. Mendelssohn was not only a highly ingenious experimenter and cryogenic engineer who designed in Oxford the first helium liquefiers through which the liquid could be seen. Though he had no research lieutenant, he saw to it that his pupils secured their D.Phil.s and with the Simon group he and his pupils spread the Clarendon's growing reputation in low-temperature physics via an increasing flow of regular publications. But none of his D.Phil. pupils made an academic career in Oxford subsequently. Judith Moore married R. A. Hull, one of Simon's researchers, and stopped research; Pontius and Babbitt returned to the New World where, after the war, Daunt joined them.[91]

The larger group working in the Clarendon was that of Simon and Kurti. Their favourite fields of research before the war were the new one of very low temperatures, i.e. those below 1K, attained by magnetic cooling, and the old one of specific heats in relation to Nernst's heat theorem. They used the demagnetization of paramagnetic salts to study the thermal and magnetic properties of these salts and of other substances as a function of temperature below 1K, once the thermodynamic temperature scale below that temperature had been established. By the late 1930s Oxford had overtaken Leiden and Berkeley as the world's leading centre for magnetic cooling work. The very success of the method violated Nernst's heat theorem so it was natural for Simon to continue his interest in not only defending it from attack but also elevating its status to that of the third law of thermodynamics, equal in importance to that of the first and second laws. In 1927 and 1930 Simon had reformulated the heat theorem in ways to which Nernst objected because Simon had added riders to what Nernst regarded as a universally applicable theorem. In the confused period of the early 1930s, the theorem was strongly attacked by the likes of Fowler and Eucken as limited in application, irrelevant, and useless. To counter such objections, in 1937 Simon revealed his third formulation of Nernst's heat theorem with the statement that the contribution to the entropy of each subsystem which is in internal equilibrium disappears at absolute zero. This formulation covered

---

[91] Shoenberg, 'Mendelssohn', 371–6; J. G. Daunt and K. Mendelssohn, 'Transfer of Helium II on Glass' and 'Transfer Effect in Liquid Helium II,' *Nature*, 142 (1938), 911–12, 475; John Gilbert Daunt (1913–87), first in physics 1935, D.Phil. 1937, ended his career as professor of physics, Queen's University, Kingston, Ontario; Judith Rachel Moore (1911–43), second in physics 1933, D.Phil. 1937, was the first woman to undertake postgraduate research in the Clarendon; Rex Bush Pontius (1909–87), second in physics 1935, D.Phil. 1937, a Rhodes scholar from Idaho, spent most of his career in the Kodak research laboratory, USA.

all cases and by the 1940s was widely accepted as a third general law of thermodynamics, one which was a particularly useful guide in low-temperature physics.[92]

Simon soon began to attract research students not only from Oxford but also from Cambridge and the European continent. The first Oxonian was Rollin, a graduate in 1933, who helped with the magnetic cooling work in Paris. He was soon joined by A. H. Cooke, who was still an undergraduate. Towards the end of his second year Cooke began research with Simon and wrote his first published paper before he graduated in 1935 when he began research for his D.Phil. on calorimetry and on the properties of magnetic materials at very low temperatures. Other Oxford physics graduates were excited by Simon's research and steadily joined his group, namely G. L. Pickard (1935), R. T. Kerslake (1936), B. Bleaney (1937), and H. S. Arms (1938), all of whom worked for D.Phil.s on properties of matter at very low temperatures. In 1936 Simon gained a post-doctoral researcher, R. A. Hull, an Oxford physicist who acquired his D.Phil. in 1936 on the subject of photoelectric phenomena of thin films. Before submitting his thesis Hull had turned to low-temperature physics as his research field. In contrast to Mendelssohn's students who gained doctorates, those of Simon were patently successful in making careers at Oxford. In the 1940s three of them (Hull, Cooke, and Bleaney) became college tutorial fellows in physics at Brasenose, New College, and St John's respectively. Cooke went on to be head of his college while Bleaney succeeded Simon as Dr Lee's professor in 1957. After distinguished work as a boffin, Rollin became a university lecturer in 1945. Pickard left Oxford in the late 1930s for the Royal Aircraft Factory, Farnborough, and eventually landed in Vancouver University as associate director of its physics department. The only one not to make an academic career was the American Arms, who capitalized on his war-work on the atomic bomb project to end his career as chief engineer, Atomic Power Division, English Electric Company, Rugby.[93]

---

[92] Kurti, 'Simon', 233–5, 238–40; F. E. Simon, 'The Third Law of Thermodynamics: An Historical Survey', *Year Book of the Physical Society* (1956), 1–22; M. and B. Ruhemann, *Low Temperature Physics* (Cambridge, 1937), 239–40.

[93] Cooke (1912–87), first in physics 1935, D.Phil. 1938, was fellow (1946–76) and warden of New College (1976–85), obituaries in *Independent*, 1 Aug. 1987, and *The Times*, 1 Aug. 1987; Simon to secretary of Harmsworth Trust, 8 Sept. 1937, SP, 14/3/CD; F. Simon, A. H. Cooke, and H. Pearson, 'Liquefaction of Hydrogen by the Expansion Method', *Proceedings of the Physical Society*, 47 (1935), 678–83; George Lawson Pickard (b. 1913), first in physics 1935, D.Phil. 1937, ended his career as professor and director of the Institute for Oceanography, University of British Columbia, Vancouver; Ralph Trevor Kerslake (b. 1915), first in physics 1936, did not complete his D.Phil. and ended his career as a senior brewer, Guinness, London; Brebis Bleaney (b. 1915), first in physics 1937, D.Phil. 1939, fellow of St John's 1947–57, Dr Lee's professor of experimental philosophy 1957–77; Henry Shull Arms (1914–72), second in physics 1938, D.Phil. 1949, was a Rhodes scholar from Idaho; Richard Albert Hull (1911–49), first in physics 1932, D.Phil. 1936, fellow of Brasenose 1944–9, obituary in *Brazen Nose*, 9 (1949–54), 53–4.

By the late 1930s Simon's eminence was recognized by the Belgian government which was prepared to support the newly begun work of D'Or, professor of chemistry at Liège, on specific heats at very low temperatures. Spurning nearby Leiden, which specialized in large-scale cryogenic techniques, it paid for two post-doctoral assistants to D'Or, Duyckaerts and Désirant, to be trained by Simon in the Clarendon. Another visitor from north-west Europe was Dr B. S. Blaisse from Holland. Nearer home Helen Megaw came from Cambridge in May 1935 to work with Simon until summer 1936. Megaw was a Cambridge graduate who took her Ph.D. there in X-ray crystallography under the youthful and inspiring direction of J. D. Bernal, who advised her in summer 1934 to spend some time in Vienna in Hermann Mark's laboratory and then to study thermodynamics intensively under Simon in Oxford. In the event she did not learn much about thermodynamics *per se* but with the practical help of Kurti and the supervision of Simon learned about the hazards and demands of doing experimental runs lasting five to six hours using the helium liquefier. An expert on ferroelectricity in crystals, and the first woman to be given a staff appointment at the Cavendish Laboratory, Megaw ended her career as fellow of Girton College, Cambridge, and director of studies in natural science there. Before the arrival of the Breslau cohort of physicists, it would have been inconceivable for a Cambridge post-doctoral physicist to have made a rewarding pilgrimage to the Clarendon.[94]

A key element in the success of the Simon group was the work of Kurti as collaborator with Simon and, increasingly as the 1930s passed, as the experienced lieutenant who taught young men and women how to use the equipment. From the mid-1930s Lindemann was rarely in the Clarendon and was often away; the unremitting day-to-day management of the laboratory was in the capable hands of the endlessly patient and gruffly laconic Keeley; and towards the end of the decade Simon was not in the laboratory every day though often around. Mendelssohn taught his own students how to use the apparatus, whereas in Simon's group these essential practical skills were often taught by Kurti. His role cannot be fully appreciated if one follows Birkenhead in assuming that the helium liquefiers were simple in design and easy to use. It is true that the expansion liquefier avoided the large-scale engineering necessary in a separate installation for supplying liquid helium continuously, that it was relatively cheap because of its small scale at a time when the price of helium was high, and that it avoided waste which was anathema to Simon. But each liquefier took three to four weeks to build and, once working, up to eight hours could elapse before the investigation proper could be begun. Such apparatus was not easily constructed: it required a lot

---

[94] D'Or to Simon, 8 Apr. 1938; Simon to D'Or, 27 Apr. 1938, both SP, 14/3/NO; Helen Dick Megaw (b. 1907), Cambridge Ph.D. 1934, later fellow of Girton 1946–70, worked in Oxford from Easter 1935 to summer 1936; Megaw to author, 30 Nov. 1988, 26 Jan. 1989.

of workshop time and skill. Moreover to work it successfully required an appreciation of the subtle problems of heat and liquid flow in a small-scale system as well as caution concerning the obvious dangers of liquid hydrogen leaking and exploding and of releasing helium gas at 150 atmospheres via a valve into glassware. When explosions involving liquid hydrogen occurred, the results could be devastating. In one incident at Berlin Kurti destroyed fourteen window panes, a wall, a door, and twenty glass Dewar vessels. Soon after Mendelssohn's arrival in Oxford he badly damaged a hand while pressure testing a small steel vessel containing hydrogen, ICI gallantly paying £30 of the Radcliffe Infirmary fees of £34 15s. incurred by him. Even by the late 1940s Mendelssohn suffered injury again when a blocked tube caused concentric glass Dewar vessels filled with liquid hydrogen and liquid oxygen to shatter.[95] Thus the liquefaction techniques brought to Oxford by Mendelssohn, Simon, and Kurti were neither easily copied elsewhere nor easily learned in Oxford. The tacit skills accumulated by these three Breslau physicists in designing, maintaining, and working miniature liquefiers meant that their field of research could not be rapidly appropriated by other physics departments: even after Simon, Pickard, and Cooke had demonstrated the attainment without a hitch of 0.12 absolute at a Very Low Temperatures Exhibition held at the Science Museum, London, in May 1936, the University College London triumvirate of Professors Donnan, Ingold, and Andrade still found it necessary to see Simon's apparatus working *in situ* in Oxford and to take his advice before installing liquid hydrogen facilities in London.[96] It cannot be over-stressed that, when Lindemann found a home for Simon, Kurti, and Mendelssohn, they not only imported apparatus but also brought with them the skills of designing and working it which could not be learned from textbooks or exhibitions. For them small-scale cryogenic engineering, i.e. the means of generating extremely low temperatures, played a major role in low-temperature physics (and continued to do so until the mid-1950s). It was that conjunction which made it difficult for other laboratories in the 1930s to copy the Clarendon's techniques and programmes quickly and easily.

In his plans for the new Clarendon Laboratory, completed in autumn 1939, Lindemann gave as much prominence to nuclear as to low-temperature physics. From 1932, when the neutron was discovered, he had begun to give more attention to the former though his own successful experience in the field

[95] Birkenhead, *Lindemann* 105; Shoenberg, 'Mendelssohn', 364, 384; Mendelssohn to W. Rintoul, ICI, 29 Nov. 1935, 13 Jan. 1936, J. M. King, ICI, to Mendelssohn, 16 Jan. 1936, all in MP, B14; Bleaney interview, Oct. 1987; Cooke interview; Megaw to author; Kurti to author, 13 Sept. 1991.
[96] *Nature*, 136 (1936), 939; Andrade to Simon, 3 May 1938, SP, 14/3/A; Donnan to Simon, 10 June 1936, SP, 14/3/CD; Ingold to Simon, 4 May 1937, Simon to Ingold, 26 May 1937, both SP/14/3/IJ; Frederick George Donnan (1870–1956), professor of chemistry, UCL, 1913–37; Edward Nevill da Costa Andrade (1887–1971), Quain professor of physics, UCL, 1928–50.

was meagre. The results obtained over the next seven years, however, made a pointed contrast with those in low-temperature physics. Though the numbers of staff and postgraduates involved were not negligible, nuclear physics did not take off before the war whereas cryophysics did. No major discoveries were made; and in the last academic session before the war, of the 22 published papers emanating from the Clarendon no fewer than 19 were about low-temperature physics: unlike Fermi in Rome, Lindemann in Oxford did not succeed in the 1930s in rivalling established centres.[97]

In a fit of enthusiasm inspired by the *annus mirabilis* of the Cavendish in 1932 and blithely ignoring the difficulties of starting nuclear physics from scratch, Lindemann had installed in the Clarendon in 1933–4 a high-voltage apparatus. Questions remained about who was to use it and for what. Lindemann had available by early 1935 three Oxonian graduates and researchers, namely, Collie, Hurst, and Griffiths. Collie, an Oxford first in chemistry in 1925, had in effect then taken a degree in physics while working for his 1927 B.Sc. on radioactivity partly in the Clarendon and partly in the Christ Church laboratory under Russell. In 1929 he was appointed a lecturer in physics at Christ Church and next year tutorial fellow and Dr Lee's reader in physics there. By the mid-1930s he had some useful publications and one successful D.Phil. student to his credit (O. A. Gratias on radioactive branching). Claude Hurst, an Oxford double first in mathematics and physics from Jesus, had taken his D.Phil. in 1933 on infra-red radiation and when elected in 1934 to a tutorial fellowship in mathematics and physics at Jesus had published just two papers based on his thesis. J. H. E. Griffiths, an Oxford first in physics from Magdalen, had taken his D.Phil. in 1933 on the Kerr cell and its applications, and on his election in October 1934 to a fellowship by examination at Magdalen had published just one paper which Schrödinger valued highly.[98] Given the relative inexperience in nuclear physics of this trio, none of whom had strayed out of Oxford to learn about the subject, Lindemann not unnaturally turned to the device which was already proving successful in cryophysics, i.e. importing a Continental expert or cohort. By June 1935 he had secured on Simon's recommendation the Hungarian Leo Szilard, and an ICI fellowship for him for three years.

[97] F. A. Lindemann, 'Designing a New Physics Laboratory', *Oxford*, 5 (1938), 53–62, at 55; G. Holton, 'Fermi's Group and the Recapture of Italy's Place in Physics', in G. Holton, *The Scientific Imagination: Case Studies* (Cambridge, 1978), 155–98; Enrico Fermi (1901–54), Nobel prize-winner 1938.

[98] *OUG* 64 (1933–4), 200; 65 (1934–5), 194; Carl Howard Collie (1903–91), Dr Lee's reader in physics 1930–71, obituary in *The Times*, 28 Aug. 1991; O. A. Gratias and C. H. Collie, 'The Parent of Proto Actenium', *Proceedings of the Royal Society of London A*, 139 (1933), 567–75; Collie to author, 6 Feb. 1991; Orvald Arthur Gratias (b. 1909), a Canadian Rhodes scholar, D.Phil. 1932, spent much of his career in the Canadian branch of J. C. P. Coats; Claude Hurst (1907–87), fellow of Jesus 1934–75, was teaching up to thirty hours a week in the late 1950s; J. T. Houghton, 'Hurst', *Jesus College Record* (1988), 7–9; ten Haar, 'Griffiths'; Griffiths was undergraduate, postgraduate, fellow, vice-president, and finally president (1968–79) of Magdalen.

Szilard, a Berlin Ph.D. in 1922, had been a *privat-docent* there from 1925 until 1933 when he left for England and with Beveridge formed the Academic Assistance Council which he continued to support. Initially Szilard worked with T. A. Chalmers, a physicist at St Bartholomew's Hospital, London, where in 1934 they evolved a new way of separating radioactive materials isotopically. Having failed to secure facilities at Cambridge and Imperial College, London, Szilard persuaded Lindemann to employ him in the Clarendon to build up nuclear physics, particularly by the study of nuclear chain reactions, in collaboration with Collie and Griffiths. For two years from summer 1935 Szilard and his two new colleagues worked on slow neutrons and their absorption but his character and concerns precluded his achieving in nuclear physics what the Breslau cohort was doing in cryophysics. Szilard was restless, alternating intense work with absence from the laboratory. By summer 1937 he was so worried about the situation in Europe that he told Lindemann that in future he wished to spend at least half his time in the USA and the rest in Oxford. Though Lindemann deplored this scheme, it was formalized on 1 January 1938. By autumn 1938, after the Munich crisis, Szilard had resigned his ICI fellowship and lived thereafter in the USA. A brilliant, mercurial, and exotic bird of passage, Szilard's style was to live in Oxford from two 18-inch suitcases (so it was said) and not even to start the slogging work of inducting research pupils into the skills of making and working apparatus. A quintessential maverick, he was not the easiest man to have as a laboratory colleague: in 1938 the charitable Simon told him that he had exhausted the patience of his Oxford friends, ICI, and the SPSL. On one occasion Szilard wanted two dustbins full of paraffin wax for his experiments on neutrons; in the event he ordered a ton of wax which duly filled the dustbins and many other containers long after his departure.[99]

Even though Lindemann knew that Szilard was looking yearningly to the USA, in autumn 1937 he secured a grant from the University of £300 p.a. for five years for apparatus for accelerating electrons to be developed by Szilard. As an insurance measure Lindemann had lured to Oxford, as a collaborator with Szilard and as substitute for him, J. L. Tuck, a rich man who had worked with Polanyi at Manchester. Supported by a fellowship awarded by the Salters' Institute of Industrial Chemistry, Tuck's work on electron acceleration, done with Oxford graduate W. C. Morgan, was unpublished

[99] Leo Szilard, *His Version of the Facts: Selected Recollections and Correspondence*, ed. Spencer R. Weart and Gertrud Weiss Szilard (Cambridge, Mass., 1978), 18–21, 41–2, 48–52; W. Lanouette, *Genius in the Shadows: A Biography of Leo Szilard: The Man behind the Bomb* (New York, 1992), 139–72; R. M. Cooper (ed.), *Refugee Scholars: Conversations with Tess Simpson* (Leeds, 1992), 29–33, 241–8 on Szilard and AAC; Szilard, SPSL, file, 167/2; Szilard to Lindemann, 30 Mar. 1935, LP, D237; Lindemann to Szilard, 30 July 1937, LP, D237; Simon to Szilard, 23 Aug. 1938, SP, 14/3/S; K. Mendelssohn, 'The Coming of the Refugee Scientists', *New Scientist*, 7 (1960), 1343–4; Leo Szilard (1898–1964).

when it was interrupted by the outbreak of war. Kerst, who devised the betatron in 1940, regarded Tuck's machine as the most promising and technically accomplished precursor. Tuck's technical astuteness led to his being Lindemann's personal assistant in the war and to his involvement in the atomic bomb project at Los Alamos.[100] Thus, for different reasons, the two imported nuclear physicists, Szilard and Tuck, enjoyed limited success in the Clarendon. A third projected import, Lise Meitner, who was under pressure from the Nazis from March 1938, could well have settled at the Clarendon in autumn 1938 as Szilard's successor and promoted nuclear physics; but Lindemann had no facilities for her, his first duty being to keep Simon and Mendelssohn who had the prior claim.

The indigenous nuclear physicists in the Clarendon responded to their situation in different ways. In the mid-1930s Hurst was put in charge of a Cockroft–Walton accelerator, which was inconveniently installed in the lecture theatre with the target in the basement. This was very different work from his previous research on infra-red radiation; Hurst published little in this new field and by early 1938 the expensive high-voltage apparatus for disintegrating nuclei, which was a Cambridge speciality, was described by a caustic critic as 'derelict'. Unlike Hurst, who had a research pupil (W. D. Allen), Griffiths had none; and though his work on neutrons was highly regarded he was not as regular a publisher as Simon, Kurti, and Mendelssohn.[101] Collie was the busiest supervisor in nuclear physics, with half a dozen D.Phil. students under his direct or indirect care in the 1930s (Gratias, Roaf, Touch, Booth, I. R. Jones, Duckworth). As a college tutor at Christ Church he was helpful, interesting, and up to date, for which he was admired and liked by his undergraduate pupils. As a researcher he maintained his interest in radioactivity while taking on the new field of nuclear physics. Indeed in the late 1930s his research was carried out in two separate places; in the Clarendon Laboratory he pursued work on the counting of nuclear particles; in the Christ Church chemical laboratory he worked on radioactivity using materials secured on loan from the Czech government in 1935 for three years by Lindemann. Collie's research interests also included electronics, a subject which he promoted enthusiastically and profitably with two of his Christ Church pupils, E. H. Cooke-Yarborough and Martin Ryle. Characteristically Collie did not try to force such promising pupils into his own favoured research fields. It was Collie who brought them together,

---

[100] Lindemann to Veale, 29 Oct. 1937, UR/SF/PHE/4; Lindemann memo, 6 May 1938, PS/R/1/6; *HCP* 168 (1937), 170–1; D. W. Kerst, 'Historical Development of the Betatron', *Nature*, 157 (1946), 90–5; James Leslie Tuck (1910–80), Manchester B.Sc. 1930 and M.Sc. 1931, ended his career as associate leader, physics division, Los Alamos Laboratory, 1954–73; William Clifford Morgan (b. 1916), second in physics 1938, ended as a director of EMI Electronics.

[101] E. W. B. Gill to Veale, 1 Mar. 1938, UR/SF/PHE/7; William Douglas Allen (b. 1914), an Adelaide graduate 1936, gained his D.Phil. at Oxford 1940 and ended his career as associate divisional head, Rutherford high energy laboratory, Didcot.

which enabled them to set up the university amateur radio station which in turn was useful experience for their war-work on radar countermeasures. With three research interests, two laboratories, his D.Phil. students, and his Christ Church undergraduates occupying his busy days, the urbane Collie had neither the time nor the single-minded temperament to build up a research school focused on one particular aspect of nuclear physics.[102]

The modesty of Oxford's nuclear physics in 1939 was confirmed in 1940 when the British atomic bomb effort was launched under the auspices of the Maud Committee. This urgent work on nuclear physics, both theoretical and practical, was carried out by teams at Birmingham, Cambridge, and Liverpool. The Clarendon Laboratory did, however, make a vital contribution to the work of the Maud Committee through the research done by Simon, Kurti, Kuhn, and Arms (a D.Phil. student of Simon's) on the diffusion method of separating isotopes of uranium. Later in the war Simon, Kurti, Kuhn, and Arms went temporarily to the USA to contribute to the Manhattan project. Only one senior Oxford nuclear physicist was involved in the atomic bomb project: Tuck went to Los Alamos where he did important work on fusing mechanisms using focused shock waves. The others, such as Collie and Griffiths, devoted themselves to radar research in the Clarendon as part of the war effort.[103]

In the 1930s Lindemann was not closely involved in the day-to-day running of the laboratory. Nor was he involved regularly in university politics with its possibilities for intrigue, persuasion, and attrition; this was left to I. O. Griffith. Lindemann's absence was partly the result of his increasing involvement in the application of science to national defence, which by 1934 had become an obsessive concern, one which he shared with Winston Churchill. They worried about the potential threat posed by Hitler's Luftwaffe; and they were critical of the view, recently expressed to Parliament by Stanley Baldwin, Lord President, that the bomber would always get through. Suspicious of the Air Ministry's supine attitude, they persuaded

---

[102] Douglas Roaf (b. 1911), second in physics 1932, D.Phil. 1936, became demonstrator and lecturer in physics in 1946; Arthur Gerald Touch (b. 1911), first in physics 1933, D.Phil. 1937, left Oxford in 1936 to work on radar under Watson-Watt and ended as chief scientist, British government communications headquarters, 1961–71; Eugene Theodore Booth (b. 1912), a Rhodes scholar from Georgia, D.Phil. 1937, became professor of physics at Columbia University after the war; Ivor Rhys Jones (b. 1916), first in physics 1937, made his early career in the Indian Civil Service; John Clifford Duckworth (b. 1916), first in physics 1938, was managing director of National Research and Development Corporation 1959–70 and has been chairman of Lintott Control Equipment Limited from 1980; Birkenhead, *Lindemann*, 110; *OUG* 66 (1935–6), 17; Edmund Harry Cooke-Yarborough (b. 1918), wartime honours physics 1940, ended as head of electronics and applied physics, Atomic Energy Research Establishment, Harwell; Martin Ryle (1918–84), first in physics 1939, Nobel laureate 1974; F. Graham Smith, 'Ryle', *BM* 32 (1986), 497–524, at 499–500; E. H. Cooke-Yarborough to author, 18 Sept., 15 Oct. 1990.

[103] C. H. Collie, 'Oxford Physics and the War', *Oxford*, 9 (1946–7), 54–8; M. Gowing, *Britain and Atomic Energy 1939–1945* (London, 1964), 57–8, 68, 119–20, 219–20; Sanderson, *Universities and British Industry*, 342–4.

Ramsay MacDonald, the Prime Minister, to set up in early 1935 a special air defence research subcommittee of the Committee of Imperial Defence, which by reporting directly to MacDonald would bypass the Air Ministry. Lindemann was convinced that any connection with that Ministry was fatal: it was defeatist and its strategies were limited to counter-bombing and reprisals; and other ministries such as the War Office had a legitimate interest in air defence. Lindemann was therefore deeply suspicious of the Air Ministry's own committee for the scientific survey of air defence, established in late 1934 with Tizard as its chairman. In summer 1935 Lindemann, as Churchill's scientific adviser, agreed to serve on Tizard's committee and began a long feud with Tizard which has become a *locus classicus* for science-policy pundits such as C. P. Snow.[104]

One cause of dispute was the priority given by Tizard and his colleagues to radar, which in Lindemann's view was of dubious value. Instead he promoted two pet schemes of his own, the first being small aerial mines which would be dropped by parachute in front of hostile aircraft. The second was to establish the position of approaching aircraft not with traditional searchlights but by detecting the infra-red radiation emitted by the engines of the aircraft. This scheme was not a pipe-dream because in Oxford there was a ranking expert, R. V. Jones, an Oxford graduate who had completed his D.Phil. in the Clarendon on infra-red detection. Though officially Jones was working in the University Observatory from autumn 1934 as the first Skynner senior student in astronomy, he spent most of his time in 1935 working in the Clarendon on infra-red aircraft detection. Through Lindemann's lobbying of H. E. Wimperis, director of scientific research in the Air Ministry, Jones was employed from January 1936 to spring 1938 by the Air Ministry to research on airborne infra-red detection, initially part-time at £300 p.a. and then full-time from September 1936 at £500 p.a. In 1937 Jones was joined by George Pickard, who had recently completed his D.Phil. in the Clarendon on low-temperature physics. They continued to apply Oxford science to national defence until the end of March 1938 when Pickard left the Clarendon for Farnborough and Jones to work on communications development in the Air Ministry headquarters and then on behalf of the Air Ministry at the Admiralty Research Laboratory at Teddington. Lindemann's promotion of air defence was not, of course, confined to the Clarendon. In 1937 he stood unsuccessfully as the unofficial Conservative candidate in a university by-election, making much of his aeronautical research in the first war as an unchallengeable basis for his expertise in the scientific aspects of air rearmament. On this occasion his concerns about air defence made for him many political enemies in Oxford. He alienated the left

---

[104] Birkenhead, *Lindemann*, 146–56, 172–210: Harrod, *Cherwell*, 163–5, 177–8; Clark, *Tizard*, 105–48; C. P. Snow, *Science and Government* (London, 1961).

and, by splitting the Tory vote, allowed the Independent candidate to win a previously safe Tory seat. His intervention split the Tory scientists at Oxford: the official Tory candidate was no less than Buzzard, Regius professor of medicine.[105]

The spurt of research triggered by the arrival of the *émigrés* provided at last a powerful justification for the erection of a new Clarendon Laboratory which was completed by autumn 1939. In this expensive enterprise Lindemann was unwittingly helped by Townsend simply because by the mid-1930s it had become abundantly clear that the electrical laboratory was unlikely to be a locus of growth, especially in research. Having been appointed in 1900, Townsend was exempt from the post-war rule that professors had to retire when 67 years old, so that his advancing years intensified the characteristics he had already revealed. He had a steady but small group of researchers, often Rhodes scholars, some of whom became important in academic physics in the antipodes and one of whom, van de Graaff, became famous later as the inventor of a well-known high-voltage generator. For Townsend small was beautiful and he was uninterested in expanding his laboratory. Most research students found him very difficult to work with, partly because he showed a general antagonism to new theories in physics: a firm believer in classical statistical mechanics, he was not reconciled even by the 1930s to the early quantum theory of the atom. He was out of touch with recent developments in his subject: he was notorious for not attending conferences, for not relating his research to other branches of physics and electrical engineering, and for his culpable negligence in not reading other authors. Though Townsend delighted in a sporting Hibernian argumentativeness, aspiring young physicists found him a repressive incubus. By the late 1930s he was sadly declining: he lectured slowly and drearily, at times writing with chalk not only on the blackboard but on the desk in front of him; and in fields such as electronics and electromagnetic theory he was out of date or ignorant and not in the same league as Moullin, the reader in electrical engineering in the engineering science department. Not surprisingly, once it had decided to build a new Clarendon Laboratory, the University set up in 1938 a working party to consider the relations between the two physics laboratories and professors. W. L. Bragg was brought in as an external adviser who deprecated overlap of expensive apparatus and leaned towards converting Townsend's chair into one of theoretical physics; but the problems of doing this were so formidable that in late 1938 the University shelved the issue until Townsend retired or died. Townsend soldiered on until 1941 when his

---

[105] Reginald Victor Jones (b. 1911), first in physics 1932, D.Phil. 1934, professor of natural philosophy, Aberdeen, 1946–81; R. V. Jones, *Most Secret War* (London, 1978), 9–44; Lindemann to Wimperis, 11 Dec. 1935, 25 Nov. 1936; Jones to Lindemann, 18, 20 Sept. 1936; Wimperis to Lindemann, 5 Oct. 1936, all LP, D123; Harry Egerton Wimperis (1876–1960), director of scientific research, Air Ministry, 1925–37.

intransigence about helping in the war effort provoked the University to set up a visitatorial board. It found him guilty of grave misconduct and advised that he should be sacked or resign. Townsend, who had been knighted in January, retired in September on condition that the decision of the board would remain confidential.[106]

Lindemann's campaign for a new laboratory involved great effort and some disappointments for him. It began in earnest in 1934 immediately after I. O. Griffith and Southwell produced a plan for what became known as the Science Area of the University. Initially Lindemann argued that, as the Clarendon was completely out of date, hopelessly overcrowded, and a fire risk, then at least an extension or preferably a new laboratory costing £80,000–£100,000 was needed. Two years later he had made significant progress towards the latter aim: the University's submission of May 1936 to the University Grants Committee stressed that a new Clarendon was among the three most needed science buildings; and in his oration of October 1936 the vice-chancellor admitted that the overcrowded Clarendon was 'wholly obsolete' (see Fig. 8). By spring 1937 the University had conceded the case for a new Clarendon building costing £77,000 not just as an acknowledgement of the justness of Lindemann's claim, but also because in the scheme initiated in autumn 1935 for the Science Area a new Clarendon building was the key to the most urgent changes to be made there: geology would move into the old Clarendon, human anatomy would be extended, physical anthropology improved, and both zoology and the Pitt Rivers Museum extended. In order to initiate this sequence of departmental improvement involving gradual but continuous reconstruction of the Science Area, a new Clarendon had to be built; but this argument did not extend to three major new items of equipment which Lindemann wanted, namely, a large magnet costing £5,000 maximum and modelled on that at Paris, an atom smasher at £5,000, and a cyclotron costing £5,000 maximum. Indeed the University made it utterly clear in spring 1937 that it was in no way committed to providing these new pieces of equipment, not least because they implied policy decisions about the future of physics. The University's caution was not surprising: it found the estimated cost of the building by borrowing up to £77,000 from the Ministry of Agriculture and Fisheries by the sale of securities held by the Ministry on behalf of the University, a procedure permitted by the Universities and Colleges' Estates

---

[106] Llewellyn Jones, 'Townsend', *Yearbook of the Physical Society* (1957), 106–10; Victor Albert Bailey (1895–1964), first in engineering 1920, became professor of experimental physics at Sydney; Leonard George Holden Huxley (1902–88), second in physics 1925, became professor of physics at Adelaide; Charles Melbourne Focken (1901–78) became senior lecturer in physics, Otago; Robert Jemison van de Graaff (1901–67); Cooke-Yarborough to author, 18 Sept. 1990; OUA, UR/SF/PHE/7, covered 'Physics. Question of future of department of experimental philosophy and of electrical laboratory'; Townsend inquiry, OUA, MR/7/1/7.

FIG. 8 The Clarendon Laboratory (1870–2) photographed c. 1940 by T. C. Keeley, its devoted administrator. Most of it was demolished in 1946 to make way for a new building for geology. Reproduced by permission of the Museum of the History of Science, Oxford.

Act of 1925. By spring 1938 the University had reaffirmed its view of 1937 that no more than £7,000 would be available for equipment. It was prepared to pay the cost of some new small items and of removing apparatus from the old Clarendon and installing it in the new one, but it excluded the three major items that Lindemann wanted. The University's niggardliness about major apparatus irritated Lindemann because he felt that the new ship was being spoilt for a ha'p'orth of tar; after all the cost of the essential equipment would be only about one-fifth of that of the new building.[107] To his chagrin both Lord Nuffield and ICI joined the University in frustrating his ambitions.

There was rich irony in Nuffield's indifference then, though not later, to the new Clarendon Laboratory. In 1936 another motor-car manufacturer, Herbert Austin, had made a bequest of no less than £250,000 to the Cavendish Laboratory, a gift which apparently so depressed Rutherford that he gladly handed over the planning of the new facilities, including a cyclotron, to Cockroft. In contrast Lord Nuffield, who first met Lindemann as late as 1927, found it difficult to understand the importance of physics which presumably seemed to him difficult, abstract, and remote from practical application. By the end of 1937 he had donated a total of £3,598,000 to the University and colleges, but not one of the pleasant financial shocks he administered galvanized the Clarendon. His preferred objects of philanthropy were the medical school (£2,216,000), the new college named after him (£1,000,000), financially poor colleges (£172,000), the Higher Studies Fund of the University (£100,000), and to Lindemann's irritation £100,000 for a new university physical chemistry laboratory, with £70,000 allocated to the building and £30,000 for equipment. Nuffield's largess reduced Lindemann's chances of fund-raising elsewhere because it conveyed to the world at large the impression that Oxford was so well endowed that it did not need more benefactors. Moreover senior university officials discouraged Lindemann from making a personal approach to Nuffield. In 1938 Nuffield rubbed salt into Lindemann's wounds when he gave £60,000 to the physics department at the University of Birmingham. As for ICI, though it had been helpful in giving fellowships to *émigrés* working in the Clarendon and even though Lindemann was on close personal terms with its chairman, he failed to persuade it not to earmark any donation it might make to the University Appeal launched in 1937: the chemical cartel gave £10,000 specifically for the new physical chemistry laboratory.[108]

[107] HCP 159 (1934), 171–2, 174–5; OUG 67 (1936–7), 25 (quotation); HCP 167 (1937), 31–4, 57; 170 (1938), 33–5; 173 (1939), 15–16; Lindemann, 'New Laboratory', 62; Lindemann to Veale, 15 Mar. 1937, UR/SF/PHE/4.

[108] Crowther, *Cavendish Laboratory*, 230–1; Andrews and Brunner, *Nuffield*, 259–60, 309–12; Lindemann to A. D. Lindsay, 13 Oct. 1937, 26 Mar. 1938, UR/SF/PHE/4; Lindemann to H. McGowan, ICI, 18 Mar. 1937, McGowan to Lindemann 27 Apr. 1937, LP, D99; Birkenhead to Morris, 21 Nov. 1927, LP, B12.

The new Clarendon was well designed, not least because of the inspections made of the laboratories at Eindhoven (Philips), Amsterdam, and Leiden. What it lacked was money for apparatus and especially for permanent posts. In the difficult economic circumstances of the 1930s, the University decided to spend its limited resources on urgent non-repeatable capital projects such as buildings, and not on new recurrent commitments such as posts. It was particularly wary of those which involved an incremental salary scale. This was yet another reason why Lindemann had to fight so hard for university demonstrators. During the 1930s the Clarendon enjoyed the services of just two compared with four in Townsend's electrical laboratory. That meant that in the Clarendon more departmental funds had to be devoted to paying departmental demonstrators than was the case in the electrical laboratory. By the end of the 1930s the University was allocating to the Clarendon only 2 per cent of its total expenditure on university demonstrators.[109]

In these depressing circumstances Lindemann sought external funding for apparatus and posts. By summer 1939 the Royal Society of London was dismayed by the persistence of his applications for research grants made annually since 1935. Believing that brains were more important than bricks and mortar, he spent a considerable amount of time and effort, not always unsuccessfully, trying to find endowments for posts. He objected to casual benefaction because it inhibited long-term planning, had an adverse effect on his researchers' mentality, relegated research to a subordinate position, and had already led to the loss of able researchers such as Rollin in 1937.[110] By 1939 he was so desperate that in spring he raised with the University what was then a novel issue, that of the payment of research workers as opposed to teaching staff. Lindemann revealed that a high proportion of the twenty-three researchers in the Clarendon (including Dobson but excluding the self-financing Jackson) were supported by temporary external grants, many of which would soon expire. In these circumstances he feared that he would occupy the new Clarendon with plenty of space but no men. Of the total salary bill of £7,230, about £1,800 (25 per cent) was contributed by the colleges, £1,400 (19 per cent) by the University, £1,120 (15 per cent) by the Department of Scientific and Industrial Research (DSIR), £1,000 (14 per cent) by ICI, and £1,910 (26 per cent) by various outside sources. Of the nine recipients of college funding, only Keeley, Collie, Hurst, and Griffiths were in permanent college posts, each of them receiving an average of just over

---

[109] *OUG* 62 (1931–2), 22; 67 (1936–7), 23; Lindemann memo about demonstratorship for Hull, 4 May 1939, PS/R/1/7.

[110] F. A. Lindemann, 'Scientific Research in National Life', *OM* 57 (1938–9), 496–7; Royal Society Council minutes, 15 June 1939; Lindemann to Duke of Westminster, 21 Mar. 1937, LP, B17; Lindemann to A. D. Lindsay, 23 June, 18 Oct. 1937, LP, B18; Lindemann to Sir Lionel Faudel-Phillips, 12 Nov. 1937, LP, B19; Lindemann to Leverhulme Trust, Mar. 1938, Lindemann to Lindsay, 11 May 1938, both LP, B22; Lindemann to Veale, 15 Mar. 1937, UR/SF/PHE/4.

£300 from his college. The university contribution of £1,400 was spent mainly on Keeley as a demonstrator (£350) and Simon as a reader (£500), with £350 going into departmental funds to pay for three part-time demonstrators. About two-thirds of the Clarendon staff were dependent on external funding. ICI provided for the trio of Simon, Mendelssohn, and Kurti. The DSIR supported two post-doctoral researchers (Cooke, Hull) and four postgraduates. Outside sources, other than the DSIR and ICI, included the Salters' Company (Tuck, £300), the 1851 Exhibition Commission (Holbourn, £400), the Duke of Westminster (Roaf, £400), Sir Robert Mond (Mendelssohn, £300), and the Rhodes Trust (Arms, Allen). At the end of the academic session or soon afterwards about £1,850 of the outside benefactions would come to an end, none of them being renewable for the individuals concerned. That meant that the low-temperature groups, who made up almost half of the Clarendon's population, were likely to decline or expire because the five most senior cryophysicists (Simon, Mendelssohn, Kurti, Cooke, and Hull) were supported by temporary grants, three of which were due to end in 1939 and two in 1941.[111]

This appalling prospect dismayed the University which in principle acknowledged the justness of Lindemann's case. But it offered no immediate aid: it confined itself to agreeing that additional financial support for the Clarendon's researchers was necessary and that the University Grants Committee should be apprised of the situation. The University had accepted that in the Clarendon an undue proportion of the research done was financed by temporary grants from outside sources, with its attendant dangers for all concerned. The desired peaceful and planned remedies were more college fellowships, more university demonstratorships, and an increased government grant; but, in the event, many of the Clarendon's researchers were saved for physics and for Oxford by the convulsive outbreak of the Second World War and their employment as boffins.

[111] Lindemann memorandum, *HCP* 173 (1939), 225–35, draft in LP, B25, ignored Dobson's £300 p.a. from Merton from 1937. My figures ignore Dobson's college income, because the University responded to the figures as presented by Lindemann. Lindemann's memorandum also ignored Hans Epstein who had arrived in England from Germany in early 1939 and, through Simon's influence, had been given accommodation in the Clarendon in May. The destitute Epstein was funded from June by a temporary grant from the SPSL. Athelstan Hylas Stoughton Holbourn (1907–62), D.Phil. 1936, was laird of Foula, Shetland Islands; on Hans Epstein (b. 1909), SPSL 327/1.

10
Conclusion

In 1929 William Diplock, a disenchanted Oxford graduate chemist who became a famous judge, predicted that the better facilities for science at Cambridge 'by the inexorable law of progress, will remove our incubus of scientists, while Oxford, by the inexorable law of Oxford, will remain, as through ten centuries she has remained, a citadel of the ancient culture'.[1] Diplock believed that it was stupid for Oxford to try to rival Cambridge in science and mathematics. Several features of Cambridge in comparison with Oxford gave credence to his view that Oxford should remain content with its traditional strengths in arts and cede science to Fenland. Mathematics at Cambridge had the same prestige as Greats at Oxford. Cambridge had established a peerless reputation in experimental physics and in physiology in the late nineteenth century and had maintained it. Its biochemistry under Hopkins between the wars was world-famous. It was *the* scientific university in Britain. In the late 1930s it had more staff in science and technology (186) than anywhere else except the University of London.[2] It was followed by Manchester (176), Leeds (142), Oxford (119), and then six English universities (Birmingham 79, Bristol 72, Durham/Newcastle 73, Liverpool 78, Nottingham 70, and Sheffield 83). In Scotland the Royal Technical College, Glasgow (later the University of Strathclyde), mustered 95, with Glasgow (60) and Edinburgh (51) trailing. In arts subjects Oxford enjoyed 457 posts which exceeded all the University of London (364) and Cambridge (233), with only Manchester (100) of the rest attaining a triple figure. Though Bernal's numbers may be semi-quantitative, they do reinforce the old contrast of Cambridge for science and Oxford for arts while also revealing that relative to most other universities Oxford was not impoverished in posts in science and technology.

In student numbers Cambridge was Britain's largest university with 5,840 full-timers in the mid-1930s, with Oxford next (4,830). Only five other British universities (Glasgow, Edinburgh, Manchester, Liverpool, University College London) topped 2,000 (see Table 10.1). Cambridge's total full-time student population was equal to that of Bristol, Sheffield, Leeds,

---

[1] W. J. K. Diplock, *Isis or the Future of Oxford* (London, 1929), 94; William John Kenneth Diplock (1907–85), *DNB*, second in chemistry 1929.
[2] J. D. Bernal, *The Social Function of Science* (London, 1939), 417.

TABLE 10.1. *Student numbers in some British universities 1934–1935 (nearest 10)*

University	Number of full-time students	Number of full-time men students	Number of full-time women students	Number of full-time research students	Number of full-time arts students	Number of full-time pure science students	Number of full-time medical/dental students	Number of full-time technology students	Number of full-time agrarian students
Birmingham	1,520	1,070	450	80	620	250	440	220	0
Bristol	1,010	690	320	60	430	240	270	70	0
Edinburgh	3,350	2,440	910	150	1,660	270	1,210	140	80
Cambridge	5,840	5,330	510	330	3,400	1,250	480	590	120
Durham/Newcastle	1,770	1,350	420	40	780	400	380	160	40
Glasgow	4,390	3,240	1,150	30	2,510	490	1,110	240	40
Leeds	1,620	1,250	370	70	580	280	520	200	40
Liverpool	2,130	1,600	530	30	690	300	780	370	0
London, Imperial College	930	910	20	190	0	410	0	520	0
London, University College	2,400	1,550	850	230	1,100	560	570	250	0
Manchester[a]	2,730	2,090	640	220	1,040	600	590	490	0
Oxford	4,830	3,950	880	250	3,970	590	130	40	90
Reading	610	290	320	20	280	120	0	0	210
Sheffield	770	640	130	50	270	150	170	180	0

[a] Includes University and College of Technology (from 1905 faculty of technology of University).

Source: Tables in *UGC Report for the Period 1929–30 to 1934–35* (London, 1936).

Birmingham, and Imperial College, London, put together. Cambridge had the most full-time research students (330) with only Oxford (250), UCL, and Manchester exceeding 200. Cambridge had as many research students as those of Birmingham, Bristol, Durham/Newcastle, Leeds, Liverpool, and Sheffield aggregated. In terms of full-time students in pure science, Cambridge (1,250) had twice as many as anywhere else, its nearest rivals being Manchester (600), Oxford (590), and UCL (560). In applied science, too, it had the most (590), though strongly challenged by Imperial College (520) and Manchester (500), with Oxford (44) trailing. Disregarding the peculiar case of Reading (210), Cambridge (120) and Oxford (90) led in agrarian students. In number of medical students Cambridge (480), a small market town in the fens, was not in the same league as Glasgow (1,100) and Edinburgh (1,210) but was not far behind Leeds (520), Liverpool (780), UCL (570), and Manchester (590). Oxford (130) trailed lamentably in medical numbers. Oxford (3,970) had the edge on Cambridge (3,400) in arts students, with only Edinburgh, Glasgow, UCL, and Manchester topping 1,000. At the undergraduate level Cambridge had much larger numbers in mathematics and engineering than Oxford: in 1929, when Diplock graduated, Cambridge generated 87 and 95 graduates in mathematics and engineering respectively: the equivalent figures for Oxford were 24 and 10.

At Cambridge the various faculty boards were more widely accepted than at Oxford where their main roles between the wars were to supervise examinations, prescribe curricula, frame lists of lectures, and make recommendations to the General Board of the Faculties and to the Hebdomadal Council. In 1926, when new statutes were introduced in line with the recommendations of the Asquith Commission, a difference of kind was added to that of degree. At Oxford the colleges were permitted to continue to charge for a large part of tuition fees, but at Cambridge that procedure was ended and all formal tuition fees were paid henceforth to the University. The persisting financial power of the colleges at Oxford meant that from 1926 the income of the University from tuition fees was much less than at Cambridge and it was the chief reason why the University's total income was only about two-thirds of Cambridge's. In 1934–5, for example, tuition fees at Oxford generated for the University £46,000, which was 10 per cent of its total income of £452,000; at Cambridge the same fees produced £173,000, which amounted to 27 per cent of the University's total income of £646,000 (see Table 10.2). Thus the way in which Oxford colleges continued to charge tuition fees, whereas those at Cambridge did not, was largely responsible for Cambridge's greater wealth. At Cambridge from 1926 the strength of the faculties, the financial power of the University, and the commitment to science and applied science in staff and student numbers went hand in hand. At Oxford in contrast the strength of the colleges, the relative impoverishment of the University, and the devotion to arts subjects were mutually reinforcing. That said only

TABLE 10.2. Income of some British universities 1934–1935
(in thousands to nearest £1,000)

University	Total income	Endowments	UGC grant	Government depts.	Tuition fees
Birmingham	216	34	76	0	51
Bristol	196	25	57	28	32
Edinburgh	285	53	95	5	73
Cambridge	646	157	107	54	173
Durham/Newcastle	231	22	66	12	58
Glasgow	258	44	88	0	78
Leeds	256	13	71	7	61
Liverpool	248	35	86	4	71
London, Imperial College	216	9	not known	18	46
London, University College	275	51	not known	0	87
Manchester[a]	415	49	95	4	94
Oxford	452	147	97	27	46
Reading	119	14	35	29	22
Sheffield	149	8	50	0	29

[a] Includes University and College of Technology.

Source: Tables in *UGC Report for the Period 1929–30 to 1934–35*.

Manchester (£415,000) enjoyed in the mid-1930s an annual income near that of Oxford. The next seven English universities (Leeds, £256,000; Liverpool, £248,000; UCL, £237,000; Durham/Newcastle, £231,000; Imperial College and Birmingham, £216,000 each; Bristol, £196,000) and the biggest Scottish universities (Edinburgh, £285,000; Glasgow, £258,000) trailed behind Cambridge, Oxford, and Manchester. Cambridge had easily the greatest total income in the 1930s. It possessed the biggest endowments, it received the largest UGC grant, it earned almost twice as much from government departments as anywhere else, and it pocketed tuition fees approaching double those received elsewhere. Financially Cambridge was Britain's premier university until the late 1930s when it was challenged by Oxford which began to benefit from Nuffield's huge medical benefaction.

Diplock's prediction that Oxford would cede science to Cambridge turned out to be false. The preceding chapters have shown in detail that by 1939 Oxford's science had prospered via research but not usually through the expansion of undergraduate numbers. It has also been stressed that Oxford science was not homogeneous. Though each subject faced common structural features and dominant interests, the responses to these varied considerably so that each science was to some extent *sui generis*. Though the importance of local agendas and agents in changing Oxford science has been stressed throughout this book, we must also consider some general themes which have pervaded it. In this concluding chapter I discuss the effects of the First World War, the arrival of planning, the appearance of academic entrepreneurs, the various modes of entrepreneurship, new sources of funding, the purposes of the funders, and the specificities of Oxford science.

The First World War prompted developments which were unthinkable in peacetime. Through the extensive contributions made to the war effort by many Oxford scientists, the importance of science was grudgingly acknowledged in the University. As part of that effort, even industrial research was launched in Oxford. The war was the occasion for the introduction of two research degrees, the fourth year of research (known as part two) in the undergraduate chemistry degree (1916) and the D.Phil. (1917), which acknowledged for the first time at Oxford the importance of training in research. Buoyed up by their war-work Oxford's scientists appealed after the war to government for a temporary grant to enable them to cope with post-war problems. It soon became the regular UGC grant to the University. As a quid pro quo the government set up the Asquith Commission to report on the universities of Oxford and Cambridge, which had escaped detailed scrutiny by the Thomson Committee during the war.[3] The Commission's report revealed some national expectations of Oxford: science, in its pure but not applied form, was viable there and strengthenable by private funding, especially if more of the University Parks could be allocated to science. When new statutes were introduced in 1926 it became clear that these hopes would have to be pursued without disturbing the academic and financial power of the colleges which continued to pocket most of the fees paid for teaching.

The Asquith Report advocated the appointment of an effective registrar to modernize Oxford's anarchic administration. From 1930 Veale, formerly a planner and civil servant in the Ministry of Health, reduced ad hocery and introduced bureaucratic mechanisms which created the Science Area and a procedure for distribution of the UGC grant. Aided by the Owen scandal Veale introduced a measure of centralized financial control in league with McWatters, the new secretary to the University Chest. Though some of their methods were unpopular, they were reluctantly accepted not least because, as Oxford graduates in classics, Veale and McWatters were loyal internal piecemeal reformers and not external intruders. The introduction of bureaucratic efficiency did not automatically aid the cause of science; but it did provide a regular and recognized way in which the concerns and claims of

---

[3] *Report of the Committee Appointed by the Prime Minister to Enquire into the Position of Natural Science in the Educational System of Great Britain*, Parliamentary Papers, ix (1918), 471–556. J. J. Thomson was its chairman.

science could be openly registered, discussed, and judged as a whole. It became less likely that such concerns would be lost or indefinitely postponed in Oxford's polycratic structure. Bilateral deals between a head of a science department and a representative of the University became less prevalent. Veale's innovations affected only the University: the colleges remained anarchies unto themselves.

In comparison with the years before 1914 Oxford's scientists between the wars were more entrepreneurial as builders of departments or disciplines. They took heart from what Perkin and Sherrington, head-hunted before 1914, had achieved by the early 1920s but they were not homogeneous in situation or in outlook. Some were college fellows, usually Oxonians, working within the college structure. Others were professors, often without previous connections with Oxford and recruited as established stars from Cambridge, Manchester, Sheffield, London, Harvard, and even Sydney. Some of these professors accommodated to collegiate values but others did not. Broadly speaking research prospered through college fellows, professors who compromised with Oxford's peculiarities, and professors who did not. It is therefore foolish to portray Oxford science as monolithic and to see entrepreneurship as flourishing only outside the colleges.

Some of Oxford's most prominent scientists helped to remove suspicion of science by shining as connoisseurs in activities highly valued in Oxford. Among college fellows Hinshelwood and Sidgwick stood out. The former's interests included comic verse, food, drink, classics, and the visual arts, as well as reading Dante in the original in his college laboratory. Sidgwick, also a classical scholar, was renowned for pungent wit and generous hospitality. The outstanding leaders of research schools in the professoriate in the 1920s were Perkin and Sherrington who showed that distinction in specialist publication did not exclude personal refinement. Perkin's achievements as a musician, horticulturalist, and host made his Germanic notions about research less unattractive, while Sherrington showed that experimental physiology was quite compatible with bibliophily, history, philosophy, and poetry.

Many entrepreneurs presented their subjects, either through conviction or expediency, as distinguished, small, and non-applied. If a science could be pursued with distinction at the research level and if its graduates were enabled by their education through that subject to join national élites (and not local rank and file), then it fitted in two ways Oxford's image of itself as a national university. If a science was small in size, it was unthreatening to vested interests, not expensive, and therefore did not require non-stop fundraising which tended to shred research time. If a science was presented as non-applied, it slotted easily into Oxford's emphasis on liberal education and its suspicion of vocational, technical, industrial, and commercial subjects. The college fellows in chemistry were particularly clever. They taught their subject in the tutorial hour, using the standard techniques of essay-writing and discussion, as a vehicle of liberal education while not denying that chemistry had industrial applications.

Two Oxford professors, Lindemann and Florey, took their cue from Perkin by trying to make their subject big and good. They made heavy demands on the University to remedy what they regarded as inadequate provision for capital and running costs; simultaneously they sought considerable external funding in order to outflank established structures and interests. By 1939 Lindemann had acquired a new Clarendon Laboratory paid for by the University; of his twenty-three researchers in physics, fourteen were funded externally from seven different sources. Florey's achievement in pathology was even more spectacular. He inherited the splendid Dunn building and endowment, but only a handful of staff. In 1938–9 he attracted thirty researchers, usually working in teams on different facets of a given problem. Most of them were externally funded by bodies whom Florey had robbed as an academic highwayman. The most ingenious gatherer of external funding was Charles Elton who secured grants in one year from thirteen different sources for his Bureau of Animal Population. The likes of Lindemann, Florey, and Elton paid the price of expending time and effort on the gruelling and often frustrating task of raising and administrating temporary grants from external sources. They knew that long-term planning was inhibited, that able researchers, especially those unrecognized by the colleges, could be lured elsewhere, and that vigorous research teams would suddenly expire if their funding ended.

Entrepreneurship also showed itself in the recruiting of refugees from 1933, in the employment of cherished laboratory assistants, and in the growing internationalism which prevented insularity. Most refugees were birds of passage who found temporary refuge in Oxford with a host who looked for someone with a pertinent field of research or technique. Thus Robinson secured organic chemists to work on topics he studied and to introduce the new technique of semi-micro-organic analysis. The first activist at Oxford was Lindemann who was also the most persistent and successful. Though he failed to retain two theoretical physicists and one atomic physicist, he saw that a group of low-temperature physicists from Breslau could enable him to realize some long-held ambitions and he enjoyed sufficient pull with ICI to secure special funds for them at a time when money and jobs were not widely available for indigenous physicists. The persisting congruence of the interests of the Breslau cohort, ICI, and Lindemann himself was responsible for the disproportionately important effect that refugees had in physics at the Clarendon Laboratory.

Laboratory assistants were like college servants—essential but unsung. They provided not only technical skill but social cohesion and continuity. In chemistry there were two sterling examples. In 1916 Perkin appointed Fred

Hall as steward of the Dyson Perrins Laboratory. He was an effective substitute demonstrator and a skilled quantitative analyst for Perkin who, dedicated to research, gave him a free hand. Hall supervised cleaners, ordered and distributed apparatus and chemicals, kept meticulous financial records, and chastised wasteful workers. After he retired in 1955 the University recognized his great contribution by conferring an honorary MA on him in 1957.[4] In the Balliol–Trinity laboratory James Warrell began as a laboratory boy in 1889 and under Hartley and Hinshelwood was valued as general factotum. The laboratory revolved around him not least because he had green fingers in inducing home-made and improvised apparatus to work. His centrality became even clearer in 1941 on the closing of the Balliol–Trinity laboratory; he supervised the move to the new university physical chemistry laboratory where he stayed for seven years full-time.[5] Two laboratory assistants were researchers in their own right. In pharmacology Burn brought with him to Oxford in 1937 Harold Ling, who had been his personal assistant for about fifteen years. Ling became head technician, administrator, and accountant of the department, while co-authoring several papers with Burn. In entomology Poulton appointed Albert Hamm as an assistant in 1897. He amassed vast collections, some of which he donated to the University, and became an expert on the courtship and mating habits of empid flies as examples of evolution. Having retired in 1931 he received an honorary MA in 1942.

Another facet of entrepreneurship was the growing internationalism. It was evident in mathematics through Hardy, Whitehead, and Milne. In the 1920s Hardy opposed the hostility and isolation to which German mathematicians were subjected until about 1928 and he exchanged posts for a year with Veblen of Princeton. Through Veblen Whitehead learned about topology, went to Princeton to study it, and eventually returned to Oxford as first a fellow and then a professor. Milne not only attended international colloquia and worked in foreign observatories. He lured to Oxford as professor of astronomy Harry Plaskett, a Canadian then working at Harvard. In chemistry young Oxford graduates made pilgrimages to Denmark and especially to Germany to learn about new fields and techniques even during the Third Reich. Less well known is the way in which Oxford's chemists were indebted in the 1930s to the USA not for money but for skills and opportunities. For new theoretical and physical methods, Caltech and Harvard were the favourites. The ageing Sidgwick was happy to visit Cornell where he developed and published his views on valency. This debt to the USA was not new in the 1930s: early in the century Hartley had found his scientific model at Harvard in the form of T. W. Richards.

[4] Smith, 'Organic Chemistry at Oxford', i. 69–73; Frederick Charles Hall (1882–1962).
[5] Barrow and Danby, *Physical Chemistry Laboratory*, 77–8.

Between the wars new sources of funding, and old ones which became more important, were tapped by the University which after all was old, famous, and beautiful. Relative to the colleges the University became less impoverished. Between 1922 and 1938 its total income rose from about £180,000 to about £400,000, with the ratio of university income to aggregate college income rising from 1:3 to 2:3. This increase was not due primarily to income from rents and dividends, from fees and dues, or from the Common University Fund derived from taxation of the colleges. All these remained relatively stable. The University became wealthier through increased trust funds, culminating in Nuffield's benefaction, through greater college contributions to some professorial salaries, but mainly through the grant from UGC which rose from £30,000 p.a. in 1919 to £108,000 p.a. just before the second war. From the late 1920s until the Nuffield medical bequest, the UGC grant provided about 30 per cent of the University's income, exceeding examination, graduation, matriculation, and registration fees, which in the 1930s totalled about £80,000 p.a. Oxford's UGC grant was one of the largest by the mid-1930s: it was exceeded by Cambridge (£108,000), with Manchester (£96,000), Edinburgh (£95,000), Glasgow (£88,000), and Liverpool (£86,000), close behind (see Table 10.2).

It was the UGC grant which enabled Oxford to create the new post of university demonstrator in 1927, to give less stingy annual grants to science departments in the 1930s, and by the late 1930s to appoint sixty university demonstrators in science who cost about £26,000 in salaries. The allocation of these university demonstrators ensured that even small departments had a core staff of three to five, which could be supplemented by departmental demonstrators if the department had the money to pay them. In the early 1920s there were about 55 departmental demonstrators; by 1939, 77 of both kinds. The greatest growth was in life sciences, with zoology increasing from 2 to 8 (5 university ones) and botany from 1 to 4 (all university). In biomedical sciences biochemistry shot from 3 to 8, human anatomy 2 to 5, pharmacology 1 to 3, and pathology 2 to 4 in demonstrators of both kinds. In these six subjects there was a threefold increase of demonstrators in less than twenty years, whereas physical sciences did not change much. Sometimes the number of demonstrators in zoology and botany exceeded the number of undergraduate finalists. Such a staff/student ratio permitted personal teaching and left time for research. Not surprisingly zoology became an enclave of excellence and the springboard for new institutional developments in animal ecology and field ornithology.

Other government bodies besides the UGC supported science at Oxford. The Department of Scientific and Industrial Research (founded 1916) and the Medical Research Council (reconstituted 1920) gave arm's-length support; but the Ministry of Agriculture (reconstituted 1919), the Agricultural Research Council (founded 1931), the Forestry Commission (founded 1919), and the Colonial Office were more intrusive. The DSIR promoted neglected or promising fields, assisted the research of established people, and supported research-training of postgraduates. The main beneficiaries in the 1920s were Perkin in organic chemistry, Barker in crystallography, Merton in spectroscopy, Dobson in meteorology, Egerton in thermodynamics, Lindemann in instrument design, Hinshelwood in kinetics, and Huxley in genetics, usually for collaborative work. In the 1930s, though there were useful grants to James (plant physiology), Peters (biochemistry), and Wilkins (mycology), new staff in new fields such as Sutton (dipole moments), H. W. Thompson (spectroscopy), and Hume-Rothery (metallurgy) and new researchers such as Willis Jackson and Hooker (engineering science) in the physical sciences were favoured recipients.[6]

The MRC supported much important work done in biomedical sciences at Oxford.[7] It focused its aid for non-clinical neurology on Sherrington who received all he asked for, and it rejoiced that it helped his researchers to make the important technical breakthrough of making simultaneous records of electrical and mechanical responses in mammalian reflexes. The MRC persistently backed Peters's research on vitamins from the moment he assumed the chair of biochemistry. It supported Dreyer's research enthusiastically and, even after he had blotted his copy-book in 1923, it encouraged the work done by Vollum, his colleague, on tuberculosis. From Florey's first appearance in Oxford, the MRC steadily supported his various research projects in experimental pathology. In 1938–9 no fewer than ten researchers, besides Florey, were aided by the MRC. Once Zuckerman was settled into the human anatomy department, the MRC regularly patronized his research on hormones.

In agriculture the Ministry and the ARC supported the department and two research institutes at Oxford with annual and *ad hoc* grants which rose to £44,000 in 1925–6 and averaged about £22,000 p.a. in the 1930s. Even so government was very dissatisfied with much of its investment in Oxford agriculture and was clearly happier with the achievements of Cambridge with its battery of state-funded institutes (Animal Nutrition founded 1911, Plant Breeding 1912, Agricultural Botany 1919, Small Animal Breeding 1919, Animal Pathology 1923) and research stations (Low Temperature 1922, Horticultural 1923, Potato Virus 1926) topped off by the privately endowed Molteno Institute of Parasitology (1921).[8] In forestry the tables were for once turned in favour of Oxford which, drawing on its long con-

---

[6] Data from annual reports of the DSIR for 1919–20 to 1937–8, all published as Parliamentary Papers.

[7] Data from annual reports of the Medical Research Council for 1919–20 to 1937–8, published as Parliamentary Papers except for gap for much of 1920s when published only separately by HMSO.

[8] Cooke, *Agricultural Research Council*.

nection with India, was the British centre for imperial forestry, Cambridge's department being closed in 1933. The Forestry Commission with annual grants of between £2,500 and £4,000, and the Colonial Office, whose grant always exceeded that from the Commission, ensured that in the relatively small field of forestry research and training Oxford was given special treatment.

Individual benefactors were also important. Among industrialists Lord Nuffield was in a class of his own with gifts totalling by 1939 almost £4 million, the biggest benefaction to any British university between the wars. Roughly £2.7 million went to medicine and £100,000 paid for the new university physical chemistry laboratory opened in 1941. Other industrialists, often with an Oxford connection or contact, operated in the £10,000–£30,000 range and secured eponymous recognition. Perrins, of the Worcester Sauce firm, gave £30,000 for a new chemistry laboratory; Whitley, of Greenall Whitley brewers, produced £10,000 for a biochemistry chair; and Pollock, of British Oxygen, contributed £500 p.a. towards a readership in electrical engineering. Another industrialist, Lewis Evans, a retired paper manufacturer, donated his valuable collection of scientific instruments which, with the help of the Goldsmiths' Company, was the basis of the University's Museum of the History of Science. Of non-industrialists, Rouse Ball, who endowed a chair in mathematics with £30,000, was a well-known Cambridge figure. Mrs Watts and J. E. Crombie were not, yet their financial contributions were essential in establishing experimental psychology and in sustaining seismology.

Having turned away from ameliorating the lot of victims, some philanthropic bodies turned to scientific medicine as a preferred beneficiary of their largess. At Oxford the Dunn Trustees gave £103,000 all told in 1922 for pathology (contrast their £165,000 to Cambridge for biochemistry) and the Rockefeller Foundation £75,000 in 1924 for biochemistry (contrast its £133,000 to Cambridge for pathology). At Oxford between the wars Rockefeller was the third biggest benefactor, its contribution being exceeded only by Lord Nuffield and the British government. Its greatest beneficence was directed in the 1930s to the new Bodleian Library, to which it contributed £616,000. It also gave useful grants for research in low-temperature physics, experimental anatomy, protein synthesis and structure, and brain metabolism. It produced £23,000 for an extension to the crowded Dyson Perrins Laboratory. Through its fellowships it enabled Oxford scientists such as Young and Abraham to study abroad. Rockefeller undoubtedly fuelled the transformation of Oxford biomedical sciences but its masterpiece was its grant in 1928 to Cambridge of £700,000 of which £250,000 went towards a new university library and the rest to science. This Rockefeller grant confirmed that Cambridge was Britain's premier scientific university.

New or newly important national companies, especially chemical ones, played a useful role at Oxford. In the early 1920s British Dyestuffs contributed £5,000 towards the Dyson Perrins Laboratory. ICI, formed in 1926, paid for industrial research there, employed Oxford chemists as consultants, provided apparatus for promising researchers such as Hodgkin, and through McGowan, its chairman, supported low-temperature and theoretical refugee physicists to the tune of almost £4,000 in 1935–6. In response to the 1937 Appeal it earmarked £10,000 for the new physical chemistry laboratory, Shell (which began operations in Britain in 1928) doing likewise with £25,000 for a new geology building. Oxford fared quite well from the Salters' Institute of Industrial Chemistry, founded in 1918 by the Salters' Company to promote education for industrial chemistry, with seven fellows coming to Oxford, mainly to work in the Dyson Perrins or Jesus laboratories, and three Oxford graduates going elsewhere to study chemical engineering.[9] In engineering science various resources were provided by the research-minded Metropolitan-Vickers Company of Manchester but not by Morris Motors of Cowley near Oxford.

It was only in 1937, almost ten years behind Cambridge, that Oxford tapped the wealth of its alumni and well-wishers by launching an Appeal which aimed to gather at least £500,000. Given a flying start by donations of £100,000 each from Rhodes Trustees and Lord Nuffield, it raised £427,000, the greatest part of which was spent on the Bodleian Library extension. But the Appeal provided £80,000 for a new Clarendon Laboratory and allowed a new Higher Studies Fund to be created.

Learned societies, trusts, and commissioners took note of the increasing salience of Oxford's scientists. The Royal Society supported a few promising projects and men. From 1932 it kept metallurgy and Hume-Rothery alive at Oxford and helped the Bureau of Animal Population. In 1936 it gave £2,000 for a Svedberg centrifuge to be erected in the biochemistry department under Peters and Philpot, making Oxford and the Lister Institute the two British centres for work on macro-molecules. The New York Zoological Society was the immediate cause of the establishment of the Bureau of Animal Population in 1932. Two trusts, the British Trust for Ornithology and the Grey Trustees, raised the money to create the Grey Institute of Field Ornithology in 1938. Apart from paying scholars, the Rhodes Trustees spent their funds mainly in South Africa and Oxford, where their greatest largess was reserved for the Appeal and for Rhodes House, completed in 1929 at a cost of £150,000. It housed the Rhodes Library, which was the Commonwealth and American history section of the Bodleian, and cost the Trustees £3,000 p.a. in the 1930s. Science was a minor consideration for them but they did make useful contributions to the endowment of the chair of forestry and

---

[9] *The Salters' Institute of Industrial Chemistry* (London, 1971) lists its fellows. I owe this reference to Dr J. Hughes.

the readership in engineering. For the commissioners of the 1851 Exhibition science was their exclusive focus through their senior studentships and research scholarships. Cambridge was the greatest beneficiary with Oxford trailing, except in organic chemistry which secured 80 per cent of the awards made to Oxford (seven to researchers with Perkin, and thirteen to those with Robinson). Thus the Exhibition commissioners paid for one researcher a year, usually a highly promising one, in organic chemistry at Oxford between the wars.[10]

Why did external funders choose to give money to Oxford? After all they could have directed their money to Cambridge, Manchester, or Imperial College. In the case of the UGC it gave a little more to Cambridge than Oxford by the 1930s but by then it had implemented the recommendation of the Asquith Commission that each university receive a grant of about £100,000 p.a. The UGC acknowledged that Oxford and Cambridge offered the greatest amenities of university life. It welcomed their superiority in endowment income because it was the best guarantee of intellectual freedom and innovation. It lauded their memorable record of progress in new buildings costing £780,000 at Cambridge and *circa* £950,000 at Oxford. In the 1930s the UGC professed liberal education as its ideal, while suspecting practicality and vocationality in universities. It believed in 'education for life', with 'mere acquirement of knowledge' not being sufficient. It saw the exclusive concern with professional training for particular careers as a great mistake. Accordingly it proclaimed that, when universities taught scientists and engineers who would enter industry, they should focus on understanding those principles of science and technology which would facilitate success in the practical and technical work to be done in industry. The UGC clearly saw Oxford and Cambridge as the most successful universities in pursuing education for life and as the inheritors of the Greek tradition of intrepid thinking about the fundamental issues of individual and social life.[11] All these views, so welcome at Oxford, were not unconnected with the presence of Oxonians on the UGC. The first two secretaries (Walter Buchanan-Riddell and Alan Kidd) were Oxford arts graduates, Buchanan-Riddell being a fellow and head of Hertford. The second and third chairmen were Buchanan-Riddell and Walter Moberly, an Oxford classicist who had been a fellow of two Oxford colleges for sixteen years.[12]

---

[10] *Record of the Science Research Scholars of the Royal Commission for the Exhibition of 1851: 1891–1960* (London, 1961). I owe this reference to Dr Hughes.

[11] University Grants Committee, *Report for the Period 1929–30 to 1934–35* (London, 1936), 10, 12, 21, 32, 43.

[12] Walter Robert Buchanan-Riddell (1879–1934), secretary 1919–22 and chairman 1930–4 of UGC, fellow 1903–12 and principal 1922–30 of Hertford; Alan Kidd (d. 1933), secretary of UGC 1922–33; Walter Hamilton Moberly (1881–1974), *DNB*, chairman of UGC 1935–49, fellow of Merton 1904–5, of Lincoln 1906–21. Kidd was succeeded by John Baldwyn Beresford (1888–1940), a Cambridge historian, as secretary 1933–40.

The DSIR and the MRC were government-funded bodies who adopted the arm's-length approach: they gave grants to selected people for selected projects and then gave them their heads. For the DSIR the presence at Oxford of eminent or promising researchers, usually in chemistry and physics, was sufficient guarantee that their postgraduates would benefit. Though Oxonians did not dominate the DSIR's Advisory Council, the presence on it at various times of Lindemann, Egerton, Sidgwick, and Dobson, backed by Tizard's power as a secretary for nine years, ensured that Oxford's researchers were not neglected. For the MRC Sherrington showed that world-class biomedical science was possible at Oxford, an achievement reinforced no doubt by three Oxonians (Sherrington, Garrod, and Dreyer) who sat on the twelve-strong Council 1924–7. In the MRC the influence of Fletcher, its secretary for much of the inter-war period, was decisive and not unfavourable to Oxford. Fletcher was obsessed with Cambridge where he had been undergraduate, Coutts Trotter student, and then fellow at Trinity College. He had been a demonstrator in physiology, his best-known research being that on muscle metabolism done with Hopkins, so he thought highly of Cambridge physiology and biochemistry. He extended these fixations to Oxford. He believed that Oxford and Cambridge were the two eyes of the English mind, that their antiquity gave them a special glamour, and that they attracted the ablest brains from all classes from all parts of Britain. For Fletcher Oxford was the next best vehicle after Cambridge for producing a biomedical élite which would enable Britain to compete effectively in international medical research and would produce from the laboratory knowledge which would improve clinical practice. He believed in freedom of research, in its returns being rarely applicable in the short term, and in the pre-eminence of physiology and biochemistry as pre-clinical sciences. He deplored the frequent separation of pathology and bacteriology from experimental physiology and biochemistry. In his view British pathology and bacteriology were unsatisfactory precisely because commercial and practical concerns handicapped them as sciences. He therefore backed Dreyer, a scientific pathologist whom he saw until 1923 as Oxford's equivalent to Hopkins, poured money into Sherrington's non-clinical approach to neurophysiology, and supported Peters's biochemical research as an extension in Oxford of that on vitamins done at Cambridge by Hopkins, Peters's mentor and Fletcher's hero. It was Fletcher who steered to Oxford big endowments of biochemistry and pathology from the Rockefeller Foundation and the Dunn Trustees in the early 1920s, with even larger ones going to Cambridge. Fletcher was proud of freeing pathology from the destructive effect of commercial work at what he called the two senior universities.[13] His successor Edward Mellanby was equally enthusiastic about experimental patho-

[13] M. Fletcher, *The Bright Countenance: A Personal Biography of Walter Morley Fletcher* (London, 1957), esp. 248; *Report of the MRC for the Year 1923–24* (London, 1924), 14–22.

logy. He played a decisive role in the appointment of Florey to the Oxford chair and then responded favourably to his begging. Mellanby, a Cambridge physiologist who had researched with Hopkins, thought experimental pathology the most neglected of the pre-clinical sciences and he knew from their time together at Sheffield how promising Florey's work was.[14] Though Mellanby's elder brother John followed Sherrington in the Oxford chair of physiology in 1936 and was a member of the MRC 1935-8, it displayed no largess to him. It preferred to support Florey, Zuckerman, and Peters, whose success in giving the first demonstration of vitamin action outside the intact animal it hailed as revolutionary.

The UGC, DSIR, and MRC were apparently content with the results of their grants at Oxford where the recipients appreciated their arm's-length approach. That was not so in agriculture and forestry where direct funding from the Ministry of Agriculture, the ARC, the Colonial Office, and the Forestry Commission led to administrative friction, disenchanted staff, and the notorious Owen case. In agriculture Oxford was saved from eclipse by Cambridge and Reading through the efforts of Sir Daniel Hall, a loyal Oxonian. In forestry, where there was little domestic competition, Oxford became and remained the imperial centre not least because it was in general supported by four loyal Oxonians, Ormsby-Gore, M. J. MacDonald, and especially Furse, at the Colonial Office, and Roy Robinson in the Forestry Commission. As a Rhodes scholar from Australia, Robinson took cognizance of Oxford's imperial connections and promoted them via forestry, though not uncritically.

Of individual benefactors of Oxford Lord Nuffield was unique. A loyal local boy made good, he was concerned about Oxford's townscape and secured eponymous fame by trying to modernize the University, especially through strengthening the links between the academic and practical worlds. Between the wars he gave nothing to engineering science and physics at Oxford presumably because he thought them incorrigibly unapplied: he preferred to support physics at Birmingham. Some benefactors, like Whitley and Dyson Perrins, were Oxford graduates concerned to develop a new subject or provide new facilities. Pollock craved and secured recognition at Oxford through promoting electrical engineering there. Evans, Mrs Watts, and Crombie were close acquaintances of three Oxford savants, Gunther (history of science), Brown (experimental psychology), and Turner (seismology), whom they supported. Rouse Ball was neither an Oxonian nor connected with Oxford: he wished to upgrade Oxford's mathematics by founding a new chair.

The Dunn Trustees and especially the Rockefeller Foundation revered the two old English universities and in medical matters were guided by Fletcher.

[14] *MRC Report 1933-4*, 26; *MRC Report 1937-8*, 15.

A major aim of Rockefeller was to make the peaks higher, that is to support the best or most promising institutions. At both Cambridge and Oxford, Rockefeller contributed substantially to new university libraries, but its patronage of science differed. At Cambridge it upgraded biological and agricultural sciences as a whole. At Oxford it avoided large-scale expenditure on departments, except for biochemistry, preferring to back promising individual researchers.

The attitudes of new or newly important national companies were not uniform. British Dyestuffs and ICI appreciated the quality of the consultancy work done for them by leading physical chemists such as Hartley and Hinshelwood and especially that done by organic chemists such as Perkin and Robinson. ICI saw Oxford chemistry as a good stable from which to recruit graduates whereas it promoted low-temperature physics to develop postgraduate training and to plug an embarrassing gap in British physics. As Oxford's geologists such as Douglas and Harrison had strong links with the Anglo-Persian Oil Company, the gift of £25,000 by Shell in 1937 for a new geology building was not a reward for consultancy work. Presumably Shell, a multinational firm new to Britain, felt that Oxford's facilities for geology needed to be put on a par with those at Cambridge. Metropolitan-Vickers Electrical Company, reconstituted in 1919, pursued a less oblique interest: ignoring the Clarendon and electrical laboratories, it helped engineering science at Oxford because it hoped to benefit from the research of Moullin's group on dielectrics.

Societies, trusts, and commissioners had varied motives. The Royal Society of London wished to promote new subjects, such as metallurgy and animal ecology as pursued by Hume-Rothery and Elton, and a flexible new instrument such as the ultra-centrifuge. The New York Zoological Society supported animal ecology because one of its key figures was deeply impressed by Elton in person. Three trusts (that for Ornithology, Grey, and Rhodes) had very strong Oxford connections, whereas the 1851 commissioners simply saw Perkin and Robinson as leading British organic chemists who were worthy of support.

There were several features of Oxford science between the wars which were peculiar to Oxford. The most obvious was the way in which chemistry and physics were split between several sites. Physical chemistry pursued by Oxonians flourished in two college laboratories; organic chemistry pursued by outsiders prospered in the Dyson Perrins Laboratory; and inorganic chemistry survived in the Old Chemistry department. This triple division was not just a consequence of micro-politics but of the perceived superiority of physical chemistry in the Oxford milieu. In physics the Clarendon and electrical laboratories competed for scarce resources but their amalgamation was shelved. The second feature was that no other British university had a fourth year of research in its undergraduate chemistry degree. This year was a valuable resource for students and, in the hands of Hinshelwood, the basis of his research groups in kinetics and of his Nobel prize.

Rhodes scholars were unique to Oxford. They hastened the recognition of research by taking D.Phil.s and they helped to promote the image of Oxford as *the* university of the English-speaking world. Between the wars those from the USA were mainly non-scientists but about two-thirds of those from the Commonwealth were scientists. Rhodes scholars were not unknown in the Clarendon and electrical laboratories, and in pathology under both Dreyer and Florey, but those who studied physiology under Sherrington formed a cohort which was subsequently particularly distinguished. Three of them, Penfield, Cairns, and Holman, became famous surgeons. Another three, Porritt, Lovelock, and Owen-Smith, became famous athletes. A further trio, Aitken, Davison, and Cluver, shone as medical and university teachers and administrators. A couple, Fulton and Eccles, collaborated with Sherrington for whom they modified or invented crucially important instruments. Fulton followed Sherrington as a historian of medicine and Eccles followed him as a philosopher of mind. Two Australians, Florey and Eccles, followed their mentor in winning Nobel prizes.[15]

When the second war broke out there were only two women science fellows, one of whom (Hodgkin) became a Nobel prize-winner, in the five women's colleges (Somerville, Lady Margaret Hall, St Hugh's, St Hilda's, and the Society of Oxford Home Students which became St Anne's in 1952), even though Oxford had admitted women to full membership of the University in 1920, with Cambridge not following suit until 1948. In 1923 Cambridge limited the numbers of women undergraduates to 500, a restriction which increased Oxford's attraction for women. In 1927 Oxford imposed its own limit of 840 on women undergraduates, one rarely reached until 1945, raised in 1948, and abolished in 1957. In the mid-1930s, for example, Oxford housed more full-time women students of all kinds (880) than any other English university, with only University College London (850) and Manchester (640) totalling over 600 (see Table 10.1). Most of the women undergraduates at Oxford read English, history, and modern languages. In the 1930s the five women's colleges generated no graduate in geology, one in engineering science, and on average less than two graduates per annum in physics, less than three per annum in chemistry, three per annum in physiology, botany, and zoology, and less than four per annum in mathematics. As the women's colleges, all founded between 1879 and 1893,

---

[15] Emile Frederic Holman (1890–1977), first in physiology 1916, head of Stanford University medical school and professor of surgery 1926–55; John Edward Lovelock (1910–49), world mile record 1933, third in physiology 1934; Harold Geoffrey Owen Owen-Smith (1909–90), third in physiology 1933, captain of England rugby team 1937; Eustace Henry Cluver (1894–1982), first in physiology 1916, professor of physiology 1919–26, preventive medicine 1935–59, University of Witwatersrand.

were financially impoverished relative to the older men's colleges, their fellowship was small and, reflecting the subjects chosen by their charges, dominated by arts subjects. Inevitably the women's colleges were reluctant to elect tutorial fellows in science. They preferred to appoint on short-term contracts college lecturers and tutors, mainly Oxford graduates, who usually taught in more than one college. In the 1920s the regular tutors at several or even all the women's colleges were Jane Kirkaldy, a zoologist (science), Elizabeth Farrow (chemistry), and Dorothy Wrinch (mathematics). In the 1930s Kirkaldy, who retired, was replaced by Jean Orr-Ewing (a medical scientist); Farrow, who married and had children, by Margaret Leishman (chemistry); but Wrinch prevailed as the solitary mathematics lecturer to all five colleges almost to the war.[16] In the 1930s two colleges, Somerville and St Hilda's, began to employ tutors who did not teach elsewhere. At Somerville Christine Pilkington, a botanist, was a research fellow for a year, then as Mrs Snow science tutor for a year. She resigned to research at home with her husband but taught botany at Somerville informally to 1948. As tutor/lecturer in science she was replaced for four years by Irene Hilton, a Liverpool graduate in zoology who had taught at Swansea and Edinburgh.[17] She left Oxford to promote women's employment and was replaced by Hodgkin who soon became an Oxford rarity in being wife, mother, researcher, and college fellow. Hodgkin could combine these roles because she lectured only rarely, did not demonstrate in a laboratory, and did not tutor outside Somerville. At St Hilda's Sybil Cooper from the physiology department doubled as research fellow and science lecturer for three years until she resigned in 1934 after her marriage. She was succeeded by Margaret Cattle, a botanist who lasted until 1941 when she resigned to have children after her marriage in 1939.[18] The three women science fellows between the wars presented two familiar role models and one unfamiliar one. Orr-Ewing, elected fellow of Lady Margaret Hall in 1938, was a dedicated single woman who worked in Florey's pathology department. Another spinster, Leishman, elected fellow of St Hugh's in 1932, resigned in 1937 to become a schoolmistress, her departmental demonstratorship in the Old Chemistry depart-

---

[16] C. Dyhouse, *No Distinction of Sex? Women in British Universities, 1870–1939* (London, 1995) argues that Oxford and Cambridge were particularly objectionable. Jane Willis Kirkaldy (d. 1932), first in zoology, Somerville, 1891, a teacher there for thirty-five years, was a translator and textbook writer but not a researcher, obituary in *OM* 51 (1932–3), 17; Elizabeth Monica Openshaw Farrow, second in chemistry, St Hugh's, 1922, taught there 1925–31; Jean Orr-Ewing, second in physiology, Lady Margaret Hall, 1919, lecturer 1929–38 and fellow 1938–44 there; Margaret Augusta Leishman (1903–76), second in chemistry, St Hilda's, 1926, taught at St Hugh's and St Anne's from 1931, fellow of St Hugh's 1932–7.

[17] Irene Alexandra Francis Hilton (b. 1902), tutor at Somerville 1931–4 and lecturer in zoology 1934–5, later secretary to Women's Employment Federation.

[18] Sybil Cooper, research fellow at St Hilda's 1925–34, 1946–68, lecturer 1941–5; Margaret Cattle (b. 1908), first in botany, St Anne's, 1930, D.Phil. 1933, lecturer at St Hilda's 1934–9, tutor 1939–41, fellow 1940–1.

ment having expired in late 1936 when Soddy retired. In contrast Hodgkin showed that it was feasible though difficult to combine a college fellowship with marriage, motherhood, and productive research.

Between the wars changes in the teaching of science were not momentous but in research achievement they were. Given the widespread realization from the 1920s of the importance of undergraduate numbers, a standard pattern of change in British universities was for research in science to ride the back of considerable expansion in undergraduate numbers which generated additional accommodation, equipment, and staff, all of which could be diverted to research. That did not happen at Oxford. Though there was a modest increase in undergraduate scientists, by all major criteria the University remained dominated by arts subjects. That dominance, which was intimately connected with the statutory independence of the colleges, was disturbed by three sorts of entrepreneurial scientists. There were first some college fellows, mainly Oxonians, who recruited postgraduates internally. Then there were two types of professors, those who adopted by conviction or expediency an accommodating policy of promoting their subject as good, small, cheap, and a means of liberal education, and those who outflanked Oxford's established structures and interests by acquiring external funding to enable their subject to be good and big. Research in science at Oxford had prospered by 1939 against all the odds in a University that was notorious for its anarchic inertia. It was a hard school for those scientists who were discipline-builders metaphorically and sometimes literally. They were well prepared to make a notable contribution as boffins during the Second World War, sometimes adapting their specialist skills to new topics and working in co-operative interdisciplinary teams. They also ensured that the big expansion in undergraduate teaching of science which came in the 1950s was based on a strong reputation for research.

INDEX

Abercrombie, M. 278
Aberdeen University 285, 313, 322, 342
Abraham, E. P. 209–11, 443
Academic Assistance Council 369–70, 372, 374, 376, 379–80, 407, 411, 423
academic conservatism 63–7, 80, 82, 307
Achmatowicz, O. 346
Acland, H. W. 22, 163–4, 202, 256
Adam, M. G. 257
Adelaide University 242
administration of University 67–74
admissions to colleges 53–4, 64
Adrian, E. D. 174, 181
Agricultural Economics Research Institute 112, 114–15, 123–30, 132, 136, 140
Agricultural Engineering Research Institute 71–4, 112, 115–18, 125, 129–42, 160
Agricultural Research Council 129, 136–7, 139–40, 219, 295–7, 441–2, 447
agriculture 112–42, 373, 442
Agriculture, Board/Ministry of 71–3, 112–42, 151, 236, 295, 300, 428, 441–2, 447
Ainley-Walker, E. W. 76, 161, 182, 185, 208
Air Ministry 387, 425–6
Aitken, R. S. 175, 449
Akers, W. A. 365, 367
Albright and Wilson 365
Alexander, W. B. 299–304
Allen, D. N. de G. 101
Allen, P. C. 365–7
Allen, W. D. 424, 432
All Souls College 64–5
Amory, C. 294
Amsterdam University 367, 415
anaesthetics 108, 162

anatomy, comparative 269–73, 277–8
anatomy, human 164–6, 194–204, 216, 376–7, 428, 442
Anderson, H. K. 65
Andrade, E. N. da C. 267, 421
Angel, A. 7, 318
Anglo-Persian Oil Company 222, 225, 448
Anson, W. R. 30, 245
anthropology 258–61
anthropology, physical 194–8, 200–2
Appeal, University (1937) 25, 42–5, 68, 108, 221, 267, 314, 349, 430, 444
Appel, H. H. K. A. 380
Applebey, M. P. 49, 60, 324, 331, 353, 356–7, 366–7
applied science 22, 24, 161–4, 236–7, 312, 314, 331, 364–8, 397–400, 425–7, 437–9, 445
Appointments Committee 270
Arkell, W. J. 47–8, 218, 220–5, 275
Armourers' and Braziers' Company 359
Arms, H. S. 419, 425, 432
Armstrong, H. E. 339
Arndt, F. 380
arts dominance 63–7, 77, 80, 82–93, 115, 125–6, 163, 435, 449–50
Ashby, A. 124, 129
Ashmole, E. 264–6
Ashmolean Building (old) 218, 263–7, 318
Ashmolean Museum 243, 263–7
Asquith Commission 17–25, 30, 34, 54, 320–1, 326, 435, 437, 445
Astbury, W. T. 231
astronomy 217, 219, 244–58, 305, 315–17, 374
Atkins, W. R. G. 242
Atkinson, R. d'E. 391

atomic bomb project 425
Austin, H. 430

Babbitt, J. D. 404, 417
Bacteriological Standards
  Laboratory 9, 183, 185–6, 208
Baden-Powell, D. F. W. 221
Badger, R. M. 334
Baeyer, A. von 326, 338, 351
Bagley Wood, Oxford 152, 292, 302
Bagnold, R. A. 223
Bailey, C. 413
Baker, Henry 237
Baker, Herbert Arthur 224
Baker, Herbert Brereton 318, 320, 339
Baker, J. R. 204, 217, 268–9, 273, 278–81, 292
Baker, William George 240, 243
Baker, Wilson 345–6, 350, 363
Baldwin, J. M. 269, 283, 287
Baldwin, S. 425
Balfour, H. 258–61
Balliol College 55, 64, 124–6, 212, 372, 375, 377, 396, 414
Balliol–Trinity chemistry
  laboratory 306, 319–23, 327–39, 440
Bamberger, E. 326
Bancroft, H. 119
Bannister, R. G. 172
Barcroft, J. 190, 214
Barker, T. V. 34, 40, 76, 218, 227–8, 442
Barger, G. 189
Barratt, S. 365
Bates, H. W. 287
Bayzand, C. J. 221–2
Belfield, V. 97–8
Belgium, government of 420
Bell, R. P. 328, 331, 334, 372
Bellamy, E. F. B. 246–7, 254, 257
Bellamy, F. A. 246, 252–3, 257
Berenblum, I. 207
Berkeley, R. M. T. R., eighth Earl 357
Berkeley University 403, 418
Berlin University 334, 358, 374, 384, 391, 401–3

Bernal, J. D. 229–32, 280, 356, 360, 420, 433
Best, C. H. 181
Bethe, H. A. 409
Beveridge, W. H. 372, 374, 423
Biggs, H. F. 386
Bingen, E. A. 367
Binney, F. G. 275, 292
Binnie, A. M. 48, 97–8, 106
biochemistry 31, 34, 164–7, 170, 184, 186–94, 209–11, 216, 375–6, 441–4, 446–7
Biological Sciences Faculty Board 89, 164, 224, 241
Birch, A. J. 349
Birkenhead, first Earl (Smith, F. E.) 250
Birmingham University 51, 201–2, 204, 380, 399, 414, 425, 430, 433–6, 447
Black, A. N. 98, 101
Blackaby, J. H. 140
Blackett, P. M. S. 356, 360, 414
Blackman, F. F. 238
Blackwood, B. M. 194, 196–7, 200–2
Blair, W. R. 294
Blakiston, H. E. D. 17–18
Bleaney, B. 419
Boar's Hill, Oxford 236, 395
Bodleian Library, Oxford 42–3, 45, 220, 255, 261–7, 314, 443–4
Bodley, R. S. 164
Bohr, N. 232, 334
Bolton King, E. 391–2
Bone, W. A. 400
Booth, E. T. 424
Born, M. 310
Botanic Gardens 31, 234–6, 240, 242–3
botany 31, 34–5, 217, 233–44, 270, 326, 374–5, 441, 449
Bourdillon, R. B. 7–8, 11, 47
Bourne, G. C. 13, 268, 270–2
Bourne, R. 153–5
Bowen, E. J. 110, 322, 328, 330–1, 333
Bowman, H. L. 220, 225–7, 230
Bowra, C. M. 310
Boys, C. V. 399
Bradley, A. J. 360

Bradley, J. 248
Bragg, W. H. 229–30, 400
Bragg, W. L. 389, 427
Brasenose College 53, 55, 58, 96, 411, 419
Brauner, L. 374
Bredig, G. 326
Breslau Jewish Hospital 377
Breslau Technische Hochschule/ University 381, 402–4, 406–8, 439
Breslau Tuberculosis Hospital 376
Brewer, F. M. 229, 356–7
Bridges, R. F. 208
Brieger, E. 376
Bristol University 390–1, 411, 433–6
British Association for the Advancement of Science 246–7, 253
British Association for the Advancement of Science, Presidency 26, 166, 290
British Council 392
British Dyes/Dyestuffs Corporation 8, 342, 344–6, 351, 444, 448
British Ecological Society 237, 298
British Empire Cancer Campaign 204, 207
British Museum (Natural History) 272, 281, 288, 302
British Pharmacological Society 213, 216
British School of Archaeology 222, 229
British Petroleum 267
British Trust for Ornithology 219, 299–303, 444, 448
Brockway, L. O. 336
Brodsky, A. 347
Brönsted, J. N. 334
Brown, A. B. 88
Brown, C. A. C. 139–40
Brown, G. L. 164
Brown, J. D. 275
Brown, W. 86–92, 170, 447
Brück, H. A. 374
Brunner Mond 365–6
B.Sc. degree 15, 29, 228, 332
Buchan, J. 280
Buchanan-Riddell, W. R. 445

Building Research Board/Station 97–8
Bülbring, E. 214–15, 375, 378
Bunn, C. W. 367–8
Bunting, A. H. 239
Burdett-Coutts scholarship 222, 228
Burdon-Sanderson, J. S. 167, 182, 212
Bureau of Animal Population 76, 218–19, 242, 269, 280, 291–8, 303–4, 439, 444
Bureau of Information for Refugee Scholars 371–2
Burn, J. H. 166, 211–12, 214–16, 375, 440
Burt, C. 86–7
Burtt-Davy, J. 153
Buxton, L. H. D. 76, 196–7, 200, 202, 258
Buzzard, E. F. 108, 162, 165, 186, 427

Cairns, H. W. B. 108, 162, 165, 175, 178, 208, 278, 377, 449
California Institute of Technology 313, 334, 336, 346, 352, 440
Callendar, H. L. 398
Cambridge University:
  agriculture 112–13, 115–16, 121, 123–4, 141–2, 442
  and Asquith Commission 20–1
  attracts Oxonians 46–8, 170, 223–4, 278
  biochemistry 183–4, 186–7, 189–90, 193–4, 443, 446
  botany 234–6, 238–9, 242
  chemistry 306, 358
  college heads 65
  college size 54
  engineering 66, 94–9, 101–2, 106, 109, 135, 366, 435
  experimental psychology 85, 89
  forestry 145, 150, 443
  income 23, 45, 108, 435–6, 441
  mathematics 305, 308–12, 314–15, 433, 435
  medical school 163
  ornithology 299
  pathology 190, 206, 443

Cambridge University (*cont.*)
  physics 382–6, 389, 391–4, 396, 404, 406, 422–5, 430
  physiology 174, 177, 179, 181, 214, 446
  and refugees 193–4, 372, 374–5
  reputation in science 24–5, 66, 364, 433, 443
  solar physics 256
  student numbers 433–5
  X-ray crystallography 229, 420
Campbell, J. E. 56, 308
Carleton, A. B. 16, 200
Carleton, H. M. 169–70, 175, 271, 278–9
Carnegie Corporation 219, 414
Carpenter, G. D. H. 268, 283–4, 286, 290–1
Carpenter, H. C. H. 359–60
Carr-Saunders, A. M. 48–9, 178–9, 271, 275, 285, 292, 295
Carrington, J. B. 265
Carter, C. W. 62, 165, 191, 193
Cartwright, M. L. 313
Cattle, M. 450
Cecil, H. R. H. G. 120
centrifuge, ultra 192–3, 444, 448
Chadwick, J. 360, 406, 414
Chain, E. B. 26, 50–1, 166, 204–5, 207–11, 215, 371, 375
Chalk, L. 151, 153–4
Champion, H. G. 159
Chandrasekhar, S. 257, 317
Chapman, D. L. 61–2, 319, 327, 337, 356
Chattaway, F. D. 61, 319, 326–7, 340, 344, 346
Chaundy, T. W. 27, 74, 308, 314
Chelmsford, Lord (Thesiger, F. J. N.) 269
Chemical Society of London 26, 342, 363–4
chemistry 187, 305–6, 318–68, 439–40, 448–9
chemistry, Alembic Club 230, 362, 364
chemistry, college laboratories 21–4, 306, 318–39, 373, 401, 448

chemistry, Dyson Perrins
  Laboratory 305, 318–19, 343–6, 349–53, 356
  extension (1922) 321, 326, 355
  accommodates Hume-Rothery 359
  and refugees 380
  and Science Area 30, 38, 40, 42, 346–7
  steward 439–40
chemistry, Dyson Perrins tea club 362, 364
chemistry, inorganic 336–7, 353–61, 448
chemistry, old laboratory 318, 343, 353, 355–60, 373, 448, 450
chemistry, organic 320, 324, 326–7, 336–52, 380–1, 445, 448
chemistry, part two degree 14–15, 366, 437, 448–9
  and promotion of physical chemistry 229, 306, 330, 332, 334, 345–6, 350, 358
chemistry, physical 319–20, 322, 324, 327–38, 344–6, 351–2, 361, 368, 448
chemistry, physical chemistry laboratory 42, 45, 109–11, 306, 320–1, 326–7, 337–8, 349, 352, 430, 440, 443–4
chemistry tutorials 60–2, 93
Chest (University) 34, 69–73, 78, 227, 231, 302, 344, 437
Chicago University 92, 222, 257, 385
Chitty, D. H. 297–8
Chitty, H. 298
Chitty, L. 101
Christ Church 120, 401, 407, 409
Christ Church chemistry laboratory 306, 318, 320–1, 327, 353, 401, 422, 424–5
Christ Church, Duke of Westminster Research Scholarship 391–2, 401, 432
Christopherson, D. G. 29, 103–4
Chrystal, R. N. 153
Church, A. H. 217, 235, 237
Churchill, W. 400, 425
Clapham, A. R. 27, 217, 233, 238–9, 241
Clark, A. J. 214

Clark, W. E. Le G. 26, 47, 50, 166, 194–5, 198–204, 216, 278, 376–7
Clarke, G. R. 122
Clements, F. E. 241, 293
Clemo, G. R. 49, 342, 345
Clifton, R. B. 6, 339, 381–3, 392
Clinton, Lord 144–5, 150
Clusius, K. 332, 335
Cluver, E. H. 449
Cockroft, J. D. 360, 406, 414, 430
Cocksedge, H. E. 366–7
Cohen, R. L. 129
Cole, G. D. H. 110
college fellows 52–9, 63–5, 74, 78, 438–9, 450–1
 and agriculture 123, 126
 biochemistry 165, 193
 chemistry 305–6, 318–22, 324–8, 330–7, 344–6, 350–1, 353–4, 357–8, 361–2, 364
 geology 221
 mathematics 305, 307–10, 312
 pathology 182, 204
 pharmacology 212, 215
 physics 386, 392, 401, 424–5, 431–2
 physiology 174–5
 zoology 268–9, 279
college heads 52, 65, 165
college life 54–9, 93–4
college tutorials 52, 59–63, 93, 321, 324, 337
colleges, autonomy of 21–2, 52–4, 318–38, 371, 373, 435, 437
colleges, finances 20, 23, 53–4, 68, 431–2, 435, 437
Collett, R. 292
Collie, C. H. 422–5
Collingwood, R. G. 84
Colonial Office 34–5, 113, 145–50, 153–60, 442–3, 447
combine-harvester 134
Common University Fund 53–4, 68, 441
Conant, J. B. 363
Congregation xvi, 52, 68–70, 77, 324
 and allowances for heads of science departments 78

forestry 144, 151
Observatory residence 245
old Ashmolean building 266
Science Area 31, 38, 40–1, 46
Science Greats 84
connoisseurship 58, 179–80, 215, 347, 438
Conrad, V. A. 395
Convocation xvi, 16, 18, 31, 144, 146
Cooke, A. H. 378, 419, 421, 432
Cooke-Yarborough, E. H. 424
Cooper, S. 174–6, 179, 450
Copenhagen 334
Cornell University 357, 362, 440
Cornforth, J. W. 349
Corpus Christi College 47, 64, 295–6
Cotton, A. A. 415
Coulson, C. A. 317
Cowley, A. E. 220, 262, 264, 266
Cowling, T. G. 317
Creed, R. S. 175, 177, 179
Cripps, S. 122
Crombie, J. E. 246–7, 443, 447
Cronshaw, G. B. 319
Crowfoot, J. W. 229
Crowther, J. G. 356
Cunliffe, N. 119, 122
Cunningham, J. T. 271
Curzon, G. N., Lord 15, 19, 24, 265, 339
Cushing, H. 174, 178, 180

Dainton, F. S. 26, 358
Dale, H. H. 214, 375, 378
Darbishire, H. 308, 378–9
Dartington Hall 130
Darwin, C. G. 315
Darwin, C. R. 260
Darwinism 269, 272–5, 278–9, 283–7, 290
Daubeny, C. G. B. 318
Daunt, J. G. 417–18
Davies, W. 346
Davison, W. C. 175, 449
Dawes, G. S. 165
Day, W. R. 151, 153–4

De Beer, G. R. 28, 47, 56, 62, 268–70, 273, 275, 277, 279, 284
Debye, P. J. W. 327, 330, 335, 363
degrees, new 14–16, 341
De La Warr, ninth Earl 118, 120–1
demonstrators 74–6, 79, 441
  in anatomy, human 200–1
  biochemistry 193
  botany 233, 237–8
  chemistry 324, 328, 340, 345–6, 350, 353, 356–8, 360, 450–1
  engineering 97
  forestry 151–3
  geology 221, 225
  mineralogy 228
  pathology 204, 207, 375
  pharmacology 212
  physics 383, 386, 391–4, 401, 431–2
  physiology 174–5
  zoology 268, 273
Demuth, W. H. H. 366
Denham, H. J. 135–9
Dennis, L. M. 357, 362
Denny-Brown, D. E. 177, 277
Department of Scientific and Industrial Research 47, 233, 239, 295, 335, 363, 393–5, 397, 415, 431–2, 441–2, 446
Desch, C. H. 360
Development Commission 114–15, 123–6, 129, 131–2
Diplock, W. J. K. 433
Dixey, F. A. 289
Dixon, A. L. 80, 311, 315
Dixon, H. B. 339
d'Leny, W. 357, 367
Dobson, G. M. B. 76, 255, 360, 381, 385–6, 389, 392–5, 431, 442, 446
Donnan, F. G. 421
Dougan, J. L. 139–40
Douglas, C. G. 10, 76, 167, 170–2, 175, 376
Douglas, J. A. 12, 80, 84, 220–5, 448
D.Phil. degree 14–16, 28–9, 57–8, 75–6, 437, 449
  in astronomy 244, 257

biochemistry 193
botany 233, 239
chemistry 230, 332, 335, 341, 345–6, 350, 354, 358, 364
entomology 290
geology 221
mathematics 307, 310, 313–14, 317
physics 418–19, 422, 424–6
physiology 174–5
zoology 268, 298
Drapers' Company 383
Dreyer, G. 9, 182–6, 206, 208, 211, 442, 446, 449
Dreyer, J. L. E. 247
Drinkwater, J. W. 399
Druce, G. C. 233, 241, 243
D.Sc. degree 15, 231
Duckworth, J. C. 424
Duke University 411
Dunn Trustees 31, 166, 182–6, 189, 208, 443, 446–7
Durham/Newcastle University 433–6
Duthie, E. S. 204, 209
Dyson, F. W. 249, 254
Dyson Perrins, C. W. 343, 443, 447

Eccles, J. C. 49, 164, 175–7, 180–1, 277, 449
ecology 218, 236–43, 275–7, 280–1, 285–6, 291–8, 444, 448
Eddington, A. S. 254, 316–17
Edinburgh University 116, 142, 145, 186, 189, 193, 212, 374, 433–6, 441
Education, Board of 270
Edward Grey Institute of Field Ornithology 218–19, 269, 298–304, 444
Egerton, A. C. G. 46–7, 76, 360–1, 384, 393, 397–400, 412, 442, 446
Ehrlich, G. 378
Einstein, A. 254, 316, 363, 369, 384, 407, 409
Elliot, W. E. 121
Elliott, E. B. 310
Ellis, C. D. 360
Ellis, J. C. B. 121

Elmhirst, L. K. 130
Elton, C. S. 27–8, 76, 218–19, 268–9, 273, 275–8, 280, 284–6, 291–8, 303–4, 439, 448
Elton, G. Y. 291, 293
Eltringham, H. 27, 289
embryology 270–1, 274, 279
Empire Marketing Board 128–9, 147–8, 293, 300
engineering science 30, 47–8, 94–106, 136, 141, 308, 314, 379–80, 399, 442–5, 447
Engledow, F. L. 116, 141
Entomological Society of London 27, 289–90
entomology 286–91
Epstein, L. A. 209
Esson, W. 311
Eucken, A. T. 335, 400, 418
Eugenics Society 203
Euler, H. K. A. S. von 211
Evans, A. J. 263–4, 267
Evans, F. C. 280, 298
Evans, L. 219, 258, 263–5, 443, 447
Evans, U. R. 360
Ewens, R. V. G. 209
Ewing, A. C. 88
Ewing, J. A. 95
examination, mathematics moderations 312–13
examination, natural science moderations 270
exclusiveness, social 53
Exeter College 203, 258
Expedition, Arctic (1924) 222, 275, 277, 292
Expedition, British Guiana (1929) 281
Expedition, Faeroes (1937) 280
Expedition, Greenland (1928) 281
Expedition, Merton College Arctic (1923) 275
Expedition, New Hebrides (1933) 279–81
Expedition, Spitsbergen (1921) 275–7, 280–1, 285, 291
Exploration Club 218, 222, 280–1

Fankuchen, I. 232
Faraday Society 363, 389
Farnborough, Royal Aircraft Factory 385, 392–4, 419
Farnell, L. R. 52, 188, 264, 353–4
Farrell, B. 92
Farrow, E. M. O. 450
Fawcett, E. W. 367–8
Federov, E. S. 227
Fermi, E. 422
Ferrar, W. L. 27, 56, 307–9, 311, 314
Fieser, L. F. 346
finances (University) 17–21, 23–4, 321–2, 343–4, 371–2, 428, 430, 435–6, 441–8
Findlay, J. R. 265
Fisher, H. A. L. 17–18, 63, 372
Fisher, J. 281
Fisher, R. A. 281–3
Fisher, R. B. 193
Fisheries, Ministry of 271, 284
Fleming, Alexander 166
Fleming, Arthur Percy Morris 379
Fletcher, C. M. 210
Fletcher, W. M. 183–5, 446–7
Florey, H. W. 26, 50, 166, 182, 204–11, 439, 442, 447
  as doctor-explorer 277
  helps Elton 297
  and honours pathology degree 164
  and Nuffield medical benefaction 162
  as physiologist 165, 179, 216
  and refugees 371, 375–6
  as Rhodes scholar 175, 449
Florey, M. E. 210
Flugel, J. C. 86
Ford, E. B. 28, 65, 77, 218, 268–9, 273–5, 279, 281–4, 292, 297
forestry 31–41, 112–13, 142–60, 242, 373, 442–4, 447
Forestry Commission 14, 34–5, 113, 144–9, 151–8, 295–6, 441, 443, 447
Forestry Institute/Imperial Forestry Institute 34, 113, 143–60
Fotheringham, J. K. 76, 218, 247, 252, 257

Fox, L. 101
Fowler, A. 334
Fowler, R. H. 315, 418
Franck, J. 369, 391, 408
Frank, F. C. 103
Franklin, K. J. 180, 212–15
Fraser, E. R. 182
Fraser, T. R. 212
Freeth, F. A. 365–7
Freundlich, E. F. 254, 317
Fry, S. M. 229–30
Fryer, J. C. F. 300
Fulton, J. F. 174–6, 178–80, 205, 213, 449
Furse, R. D. 147, 155, 158, 447

Galton, F. 202
Gardiner, J. S. 284
Gardner, A. D. 9, 76, 183, 186, 208, 210, 292
Garrod, A. E. 184, 186, 189–90, 262, 324, 446
Garstang, W. 271, 279
Gaskell, W. H. 177
Gatty, O. 48, 331, 334
Gaut, G. 365
Gemant, A. 103, 379–80
General Board of the Faculties xvi, 69–71, 75, 89, 392, 435
genetics 272–5, 279, 281–4
geology 217–18, 220–5, 373, 428, 444, 448
George, H. J. 327
Gibson, A. G. 182, 185, 208
Gibson, R. O. 367–8
Giessen University 373
Gill, E. W. B. 11, 56, 386
Gilmour, J. 118–19
Glasgow, Royal Technical College 433
Glasgow University 94, 433–6, 441
Godber, G. E. 164
Godwin, H. 238, 242
Goldsmiths' Company 219, 265, 306, 334–5, 443
Goodrich, E. S. 13, 27, 84, 151, 263, 268–73, 277–9, 285, 294–7, 300–1

Goodrich, H. P. 13, 278
Gordon, G. S. 371
Gordon, K. 357, 366–7
Gordon, S. P. 275
Gotch, F. 86, 166–7
Göttingen University 335, 379, 381, 391, 400, 408
Götz, F. W. P. 395
Gough, H. J. 102
government grant 17–21, 23–4, 30, 54, 68, 75
  see also University Grants Committee
Granit, R. 164, 176–7
Grant, M. 294
Gratias, O. A. 422, 424
Gray, J. 271
Graz University 378, 410
Greats xvi, 25, 82–3, 86–9, 312, 433
Greek, compulsory 16, 84
Grensted, L. W. 289, 291
Grey, E., Earl 218–19, 251, 301
Grey Memorial Fund/Trust 301–3, 444, 448
Griffith, I. O. 11, 40–2, 383, 385–6, 392–3, 425, 428
Griffith, J. G. 57
Griffiths, J. H. E. 410, 422–5, 431
Groth, P. H. von 226–7, 330
Gulbenkian, C. S. 267
Gulland, J. M. 49, 345
Gunn, J. A. 184, 211–14, 216, 378
Gunther, R. W. T. 76, 218–19, 258, 263–7, 271, 319, 447
Guttmann, L. 377–8

Haber, F. 334, 358, 369
Haeckel, E. H. P. A. 270, 274, 279
Haldane, J. B. S. 48, 84, 170–1, 271
Haldane, J. S. 10, 166–72, 187, 376
Halifax, Lord (Wood, E. F. L.) 109
Hall, A. D. 123–4, 126–7, 132, 135, 243–4, 447
Hall, F. C. 439–40
Hamm, A. H. 288, 440
Hammersley, J. M. 310
Hammick, D. L. 27, 61, 331, 344, 350–1

Hanitsch, K. R. 289, 291
Harcourt, A. G. V. 229, 311, 318
Hardy, A. C. 13, 268–9, 271, 275, 277, 284–5, 304
Hardy, G. H. 47, 55, 84, 305, 308, 311–15, 372, 440
Harley, J. L. 239
Harrison, A. St B. 110
Harrison, J. V. 224–5, 448
Harrod, H. R. F. 250–1
Hartley, E. G. J. 356–7
Hartley, H. B.
  regime in Balliol–Trinity laboratory 306, 319, 321–2, 330–1, 334
  as college fellow 60, 78, 385
  promotes electrical engineering 102
  honours 26
  and industry 49, 366, 448
  promotes history of science 265
  debt to Richards 440
  supports Science Greats 84
  war work 7
Harvard University 48, 88, 174, 253, 330, 346, 352, 358, 438, 440
Haselfoot, C. E. 56
Haslam-Jones, U. S. 307, 309
Haworth, R. D. 49, 345
Haworth, W. N. 346, 380
Hazel, A. E. W. 52
Heatley, N. G. 209–11
Heaton, T. B. 212, 215, 278
Hebdomadal Council xvi, 69–73, 77–8, 344, 435
  and agriculture 113
  biochemistry chair 188
  chemistry 324, 339
  experimental psychology 89
  Imperial Forestry Institute 150
  Museum of History of Science 266
  physics 392
  Radcliffe Observatory 251
  refugees 371
  Science Area 34–5, 38–41
Heitler, W. H. 409
Henchley, D. V. 105

Hertford College 56, 391
Hewitt, C. G. 292
Higher Studies Fund 42, 181, 291, 413, 430, 444
Hiley, W. E. 153
Hill, A. V. 376
Hilton, I. A. F. 450
Hinshelwood, C. N.
  regime in Balliol–Trinity laboratory 306, 321–3, 328–38, 364, 440, 442
  exploits chemistry, part two degree 448–9
  as college fellow 60, 305, 358
  as connoisseur 58, 438
  as Dr Lee's professor 353
  and forestry site 151
  honours 26
  and industry 366, 368, 448
  Lindemann, view of and debt to 389
  and Royal Society reform 356
  war work 8–9
Hirst, E. L. 360
history of astronomy 247–8
history of medicine 180, 213–14, 261–2
history of science 217–20, 258, 261–7, 373
history of technology 217–18, 258–61, 373
Hobby, B. M. 290
Hodgkin, D. C. 26, 218, 226, 228–33, 352, 367, 444, 449–51
Hodgkin, R. H. 230, 373
Hodgkin, T. L. 230
Hodgkinson, J. 307–8
Hogarth, D. G. 264
Holbourn, A. H. S. 432
Holman, E. F. 449
Hooker, S. G. 29, 58, 101, 442
Hope collections (entomology) 286–91
Hope, E. 8, 342, 344–6, 350
Hopkins, F. G. 170, 183, 186, 189–90, 192–4, 372, 375, 433, 446–7
Hopkins, H. J. 105
Hopkinson, B. 100
Hora, F. B. 239

Hornsby, T. 248
Hückel, E. 330
Hudson's Bay Company 292–4
Hughes, J. S. 247, 254, 257
Hull, R. A. 418–19, 432
Hume-Rothery, W. 228, 353, 356, 358–61, 442, 444, 448
Humphrey, G. 92
Hurst, C. 422, 424, 431
Huxley, J. S. 273–7
  as college fellow and tutor 62, 268, 271
  as Darwinian 269
  debt of pupils to him 279–81, 284–5, 291–3, 295, 442
  as editor 27
  leaves Oxford 47
  and Oxford Ornithological Society 299
  supports Science Greats 84

I. G. Farben 374, 411
Imori, S. 354
Imperial Agricultural Bureaux 156
Imperial Chemical Industries:
  and agricultural engineering 136
  attractiveness to Oxford graduates 365–8, 448
  supports Bureau of Animal Population 219, 295
  employs consultants from Oxford 351, 366, 444
  supports Dyson Perrins Laboratory extension 349
  provides apparatus for Hodgkin 230, 444
  lures Oxford academics 49, 236, 296, 357
  supports physical chemistry laboratory 45, 338
  aids refugees 380–1, 404, 406–14, 421–3, 430–2, 439, 444, 448
  research-minded 98
Imperial College, London 24, 46–7, 98, 106, 217, 238–9, 318, 334, 397–400, 423, 434–6

Imperial Forestry Bureau 156
imperialism 142–60, 288, 293, 339, 341, 443, 447
India Office 142–3
Indian Forestry Service 142–4, 154
Industrial Fatigue Research Board 10, 169
industrialism, suspicion of 22, 94–111, 131–2, 138, 140–1, 158, 167–8, 312, 438–9, 445
industry, careers in 364–8
industry, research for 342, 351, 354–5, 437
Ing, H. R. 49, 345
Ingold, C. K. 349, 352, 364, 421
Institute of Agrarian Affairs 130
Institute of Metals 360
International Astronomical Union 246–7
International Geodetic and Geophysical Union 247, 253
International Geographical Union 223
internationalism 313, 350, 440
Irvine, J. C. 149–50, 156, 158
Irving, H. M. N. H. 326
Istanbul University 374, 380, 414

Jackson, D. A. 381, 393, 396–7, 400, 408, 416–17, 431
Jackson, W. 26, 103, 442
James, E. J. F. 326
James, W. O. 27, 217, 233, 238–40, 442
Jellinek, S. 373
Jenkin, C. F. 11–12, 49–50, 95–8
Jenkin, H. C. F. 95
Jenkinson, A. J. 19
Jenkinson, J. W. 6, 270, 274, 279
Jennings, H. S. 295
Jennings, M. A. 210, 211
Jerome, W. J. S. 212
Jervis-Smith, F. 95
Jesus College 422
Jesus College chemistry laboratory 306, 319–23, 327, 337, 364, 366, 444
Joachim, H. H. 84
John Innes Research Institute 243–4

Jolliffe, A. E. 47
Jones, E. R. H. 338
Jones, F. W. 198
Jones, H. 360
Jones, I. R. 424
Jones, R. V. 426
Joseph, H. W. B. 84, 401
Jourdain, F. C. R. 275–7, 284, 299

Kapitza, P. L. 404, 406
Kapteyn, J. C. 245, 248
Kearton, C. F. 26, 358, 365–7
Keble College 57
Keble Road triangle 30, 40–1
Keeble, F. W. 31, 49, 84, 234–6
Keeley, T. C. 385, 389, 392–4, 402–4, 420, 431–2
Keen, B. A. 116
Keesom, W. H. 401
Keith, A. 198
Kelvin, Lord (Thomson, W.) 100
Kendall, D. G. 257
Kenyon, J. 342
Kermack, W. O. 342
Kerslake, R. T. 419
Kidd, A. 445
Killery, V. St J. 367
King, F. E. 350
King, J. 129–30
King's College, London 47, 309
Kirchheimer, F. W. 373
Kirkaldy, J. W. 450
Kistiakowsky, G. B. 358
Knowles, F. H. S. 197
Knox-Shaw, H. 248–52, 254–5
Koepfli, J. B. 346
Krebs, H. A. 193–4, 375–6
Kuhn, H. G. 381, 407–8, 412–13, 416–17, 425
Kurti, N. 381, 402, 407–8, 412–18, 420–1, 424–5, 432

laboratory assistants 439–40
Lack, D. M. 299, 302
Lady Margaret Hall 231, 450
Laidler, K. J. 60

Lamarckism 272–3, 279, 283, 287
Lambert, B. 7, 56, 324–5, 353, 356–9, 361
Lambert, J. D. 331, 335
Langley, J. N. 177
Langmuir, I. 232
Lapworth, A. 345
Last, H. M. 259
Lattey, R. T. 386
Lawson, D. R. 366–7
leavers 46–51
Leeds, E. T. 266
Leeds University 433–6
Leiden University 401, 418, 420
Leipzig University 228, 335, 362, 380
Leishman, M. A. 450
Leslie, P. H. 297–8
Leverhulme Trust 295, 334, 379
Lewis, G. N. 362
liberal education 93–4, 122–3, 159, 161–3, 333, 336–7, 438–9, 445
Liddell, E. G. T. 164, 174–7, 180–1
Lincoln College 206, 361–4
Lindemann, F. A. 381–432
  and Department of Scientific and Industrial Research 442, 446
  as discipline builder 179, 439
  and forestry site 38, 151
  attitude to Milne 315–16
  distanced by Plaskett 257
  and Radcliffe Observatory 219, 248, 250–2, 254–5
  aids Rado 308
  aids refugees 370, 439
  and Science Greats 84
  dines and jousts with Sidgwick 363–4
  encourages Wrinch 232
  unattractiveness to undergraduates 105
  aids X-ray crystallography 230
Lindsay, A. D.:
  on agriculture 141
  likes Bureau of Animal Population 296
  and Imperial Forestry Institute 151, 155, 158

Lindsay, A. D. (*cont.*)
  and John Innes Research
    Institute 244
  and Nuffield College 109–11
  aids refugees 371–2, 377
  and Science Area 35, 38
  alliance with Veale 68
  applauds Zuckerman 201
Ling, H. 215, 440
Linnett, J. W. 358
Literae Humaniores Faculty Board 83, 89
Liverpool University 48–9, 166–7, 170, 174, 176, 186–9, 315, 366, 425, 433–6, 441
Livingstone, R. W. 296
Llewellyn Smith, F. 398–9
Lloyd, A. H. 151, 153
Locke, J. 86, 89
Lodge, T. A. 41–2, 44–5
Loewi, O. 378
London, F. W. 407, 409–11
London, H. 402, 407–8, 410–12
London hospitals 161–3, 203
London Mathematical Society 308, 311, 313, 316
London, Midland, and Scottish Railway 331
London University 371, 433
  see also Imperial College; King's College; Royal Holloway College; University College London
Longstaff, T. G. 275–6, 281
Love, A. E. H. 80, 311, 315
Lovelock, J. E. 449
Lutyens, W. F. 366–7

Mabbott, J. D. 88
MB degree 16
McAtee, W. L. 287
MacBride, E. W. 279, 287, 290
McCallum, S. P. 56, 386, 392
McCarthy, L. 236
MacDonald, J. R. 118, 426
MacDonald, M. J. 158, 447

McDougall, W. 10–11, 48, 86–8, 167, 170
McGowan, H. D. 406, 444
Macintosh, R. R. 162, 378
MacIver, D. R. 195
MacKeith, M. H. 212–13
McKerrow, G. 382
McWatters, A. C. 73, 437
Madan, F. 220, 261–2, 264
Madigan, C. T. 224
Madrid, Institute for Medical Research 375–6
Madrid, National Cancer Institute 377
Maegraith, B. G. 204
Magdalen College:
  aids agriculture 120
  and botany 233–5, 238
  and chemistry 336, 341, 344, 359
  and electrical engineering 102
  relations with Gunther 263–4
  and mathematics 310
  and physics 422
  and physiology 174–5, 277
  presidency 65
  aids refugees 375, 409–10
  tutorials 62
Magdalen College chemistry/Daubeny laboratory 318–19, 326
Manchester University:
  and botany 242
  and chemistry 167, 305, 338–52, 366, 438
  general data 433–6, 441, 449
  and mathematics 315
  and physics 389, 392, 398–9, 423
  aids refugees 372
Manley, J. J. 318
Mann, G. 167
Marett, R. R. 200, 203, 258
March, A. and H. 410
Marsh, J. E. 340
Marsh, J. K. 354–5
Mason, F. A. 342
Matamek Conference 294
mathematics 305, 307–17, 378–9, 440, 450

Index

mathematics tutorials 62–3
Mathematical Institute 46, 232, 305, 314
Maxton, J. P. 130
Medawar, P. B. 26, 28, 51, 62, 204, 268–9, 278
Medical Research Committee/Council 9, 178, 183–6, 193, 204, 206–8, 293, 295, 441–2, 446–7
Medical Sciences Faculty Board 89, 163–4, 187
medicine, clinical 161–3, 169, 172, 205, 377–8
Meetham, A. R. 395
Megaw, H. D. 420
Meitner, L. 412, 424
Mellanby, E. 181, 206–7, 446–7
Mellanby, J. 181, 447
Mendelssohn, K. A. G. 381, 397, 400, 402–7, 412–13, 417–21, 424, 432
Merton, T. R. 50, 385, 393, 395–6, 442
Merton College 56, 394
metallurgy 358–61, 442, 444, 448
Metropolitan-Vickers Electrical Company 98, 103–4, 106, 379–80, 382, 444, 448
Michels, A. M. J. F. 367, 415
Middleton, A. D. 293, 296, 298
Middleton, T. 126
Miers, H. A. 226–8
Milk Marketing Board 129
Millard, T. 95
Millikan, R. A. 385
Milne, E. A. 250–5, 257–8, 305, 310, 314–17, 363, 390, 408, 440
Milne, J. 246–7
mimicry 281–4, 287–90
mineralogy 217–18, 220, 225–33, 373
Mintoff, D. 105
Misciattelli, P. 354–5
Moberly, W. H. 445
Moelwyn-Hughes, E. A. 332
Moir, J. C. 162
Mond, A. M. 49, 236
Mond, R. L. 308, 404, 413, 432
Moore, B. 187–9
Moore, J. R. 417–18

Moore, T. S. 319
Morgan, T. H. 282
Morgan, W. C. 423
Morison, C. G. T. 76, 115–16, 119, 122–3
Morrell, J. H. 386
Morris Motors, Cowley, Oxford 106–7, 444
Morrison, W. 114
Morrison, W. S. 125
Moseley, H. G. J. 6, 383
Moseley, H. N. 260
Mott, N. F. 360, 390
Moullin, E. B. 27, 48, 76, 97, 102–6, 379, 427, 448
Moy-Thomas, J. A. 268
Müller, F. 287, 289
Mumford, E. P. 291
Munich University 226, 330, 335, 338, 380
Murray, G. G. A. 372
Murray, K. A. H. 126, 129
Museum Delegacy 78–9, 227
Museum House 34, 40
Museum of the History of Science 76, 202, 218–19, 255, 258, 264–7, 443
Myres, J. L. 84, 200

Nagel, D. H. 319, 330
Naples Zoological Research Station 273, 277, 284
National Birth Control Association 278
National Committee for Astronomy 249
National Institute for Medical Research 189, 214
National Physical Laboratory 135, 137
National Union of Scientific Workers 312
Natural Sciences Faculty Board 23, 83, 265, 321, 324
Nernst, H. W. 384, 387, 393, 397–8, 401–2, 406–7, 418
New College 55–7, 221, 223, 271, 279–80, 285, 291, 383, 392
New York Zoological Society 219, 294, 444, 448

Newall, H. F. 256
Newboult, H. O. 56, 308
Newman, J. E. 137–8, 140
Nicholson, E. M. 218–19, 280–1, 298–304
Nicholson, J. W. 231, 309
Nobel Prizes:
  Oxonian winners 26
  to Chain 50, 205, 211
  Chandrasekhar 317
  Eccles 164, 175, 449
  Florey 50, 205, 211, 449
  Granit 164, 176
  Hinshelwood 328, 332
  Hodgkin 218, 226
  Krebs 193, 375–6
  Medawar 51
  Ochoa 375–6
  Robinson 305, 346, 348
  Schrödinger 409
  Sherrington 166, 175, 177, 449
  Soddy 353
  Todd 349
Noether, E. 308, 378–9
Norrish, R. G. W. 306, 333
Norway, N. S. 97
Nottingham University College 345, 350, 433
Nuffield, Lord (Morris, W. R.) 45, 82, 106–11, 161–2, 244, 249, 267, 338, 430, 436, 443–4, 447
Nuffield College 82, 107–11, 244
Nuffield Institute for Medical Research 161, 165, 212–13, 378
Nuffield medical benefaction 161–3, 165, 193, 203, 208, 215, 375–8, 436, 441

O'Brien, J. R. P. 192
Observatory, University 31, 34–5, 219, 244–9, 253–4, 256–8, 267, 317, 426
Ochoa, S. 375–6, 378
Odgers, P. N. B. 76, 200
Odling, W. 318, 339–40
O'Gorman, M. J. P. 385
Ogston, A. G. 62, 165, 192–3

Oliphant, J. N. 152–5, 158
Oliphant, M. 414
Olmsted, J. M. D. 176, 180
Oppel, A. 223
Ord, M. G. 165
Ormsby-Gore, W. G. A. 158, 447
ornithology 273–5, 284, 298–304
Orr, J. 129
Orr-Ewing, J. 210–11, 230, 450
Orwin, Charles Stewart 13, 124–30, 136
Orwin, Christabel Susan 129
Osborn, H. F. 274
Osborn, T. G. B. 234, 242–3, 375
Osler, W. 220, 261–3
Owen, B. J. 71–3, 112, 125, 130–5, 147, 244, 302, 437
Owen, J. B. B. 101
Owen-Smith, H. G. O. 449
Oxburgh, E. R. 225
Oxford Bird Census 219, 299–301, 303
Oxford by-election (1937) 426–7
Oxford, City of 30, 40, 69, 107
Oxford Instrument Company 392
Oxford Ornithological Society 219, 260, 284, 299
Oxford Preservation Trust 252
Oxford University Press 25, 27, 104, 107, 177, 264–7, 313, 333, 363, 395

Page, T. L. 257
Paget-Wilkes, A. H. 275–6
Palmer, G. 286
Pannekoek, A. 254
Papworth Village Settlement 376
Paris, Laboratoire de L'Électroaimant de Bellevue 415–16
Parker, A. P. D. 200
Parkes, G. D. 57, 326
Parks (University) 23–5, 29–41, 437
patents 134–5
pathology 31, 164–6, 182–6, 204–11, 216, 278, 297, 375–6, 441–3, 446–7, 449
Paton, W. D. M. 165
Pauling, L. 336, 352
Peel, W. R. 116

Pembroke College 108
Penfield, W. G. 175, 178, 449
penicillin 208–11, 352, 367
pensions 21, 79–80
Penston, N. L. 239
Perkin, W. H., jnr. 338–52
  and college laboratories 321, 324, 326
  commercial research 354
  as connoisseur 438
  as consultant 366
  supported by Department of Scientific and Industrial Research 442
  dies in post 79
  as discipline builder 179, 305, 318, 389, 439
  as D.Phil. supervisor 28, 75
  lobbies government 17
  headhunted from Manchester 167
  encourages Hume-Rothery 359, 361
  promotes new degrees 14–15
  suspects physical chemistry 362
  supported by Royal Commission for 1851 Exhibition 445, 448
  war work 8
Perrin, M. 365, 367–8
Perutz, M. F. 230
Peskett, G. L. 193
Peters, R. A. 151, 186, 189–94, 216, 375–6, 442, 444, 446–7
Pharmaceutical Society 214, 375
pharmacology 165–6, 184, 211–16, 278, 375, 440
Phillips, C. G. 165
Phillips, E. G. 307
philosophy 82–94, 179–80, 312, 315
Philpot, J. St L. 165, 192–3, 444
Physical Sciences Faculty Board 251, 324, 337, 392, 410, 412
physics 312–14, 381–432, 439, 448, 449
physics, Clarendon Laboratory 41–2, 45, 225, 247–8, 257, 370, 381–3, 385–432, 439, 444, 448–9
physics, Electrical Laboratory 40, 370, 373, 381, 383, 386, 392, 427, 431, 448–9

physiology 161–81, 186–8, 191–3, 199, 203–4, 205–6, 212–16, 277, 376, 442, 446–7, 449
Pickard, G. L. 419, 421, 426
Pickard-Cambridge, A. W. 289
Pidduck, F. B. 11, 76, 309
Pilgrim Trust 240, 301
Pilkington, C. 238, 450
Pitt Rivers, A. H. L. F. 259–61
Pitt Rivers Museum 201, 217, 258–61, 264, 428
Planck, M. C. E. L. 334
Plant, S. G. P. 345, 350
Plaskett, H. H. 35, 244, 253–8, 267, 305, 317, 374, 440
Plaskett, J. S. 254
Plummer, H. C. 252–3
Polanyi, M. 423
Politics, Philosophy, and Economics degree (PPE) 65, 82–4, 121, 125, 141
Pollak, E. 377
Pollock, J. D. 102, 443, 447
Polunin, N. 281
polythene 367–8
Pontius, R. B. 417–18
Poole, E. G. C. 27, 307–8, 314
Pope, W. J. 306
Porritt, A. E. 164, 175, 449
Porter, M. W. 228
Potsdam Observatory 246, 253–4, 316, 374
Poulton, Edward Bagnall 12, 15, 26–7, 79, 268–9, 283–4, 286–91, 339, 440
Poulton, Emily 286
Poulton Fund 269, 283, 287
Powell, H. M. 75, 226, 228–30, 360–1
Pregl, F. 192
Prewett, F. J. 129
Prichard, C. R. 367
Prichard, H. A. 92
Priestley, J. G. 10, 76, 167, 169–73, 175, 376
Princeton University 309, 313–14, 375, 440
professors 21, 77–80, 438, 451

Prothero, R. E. 126
psychology, experimental 85–92, 170, 373, 443, 447
Psychology, Institute of Experimental 85, 90–2
Pullinger, B. 207
Pusey, H. K. 268
Pye, D. R. 12, 27, 48, 96, 103

Queen's College 346, 373, 375, 412
Queen's College chemistry laboratory 319–24, 326–7, 349

Radcliffe Camera 46, 219
Radcliffe Infirmary 161–3, 182, 208, 244, 249, 377
Radcliffe Observatory 162, 219, 244–5, 248–52, 254–5, 267, 317, 401
Radcliffe Science Library 25, 45–6, 78, 305, 314
Radcliffe Trustees 45–6, 219, 245, 248–55, 401
radiology 184
Radium Corporation, Czech (Imperial and Foreign Corporation of London) 184, 353–4
Rado, R. 308, 413
Raikes, H. R. 8, 49, 322, 328, 331
Raiment, P. C. 193
Rambaut, A. A. 245, 248
Ramsay, W. 339, 397
Ramsden, W. 48, 167, 187
Rankine, W. J. M. 94
Ranson, R. M. 293, 298
Rawcliffe, G. H. 105
Rayleigh, Lord (Strutt, J. W.) 382, 397
Raynor, G. V. 360
readers 76–7, 218, 375
Reading University 112–16, 118, 123, 141–2, 236, 357, 434–5, 447
Rees, M. G. 55
refugees 308, 335, 369–82, 395, 401–32, 439
registrars 68–9
retirement, compulsory 21, 79–80
Rew, H. 126

Rhodes scholars:
 appear in 1903 259
 in astronomy 257
 in Bureau of Animal Population 298
 in chemistry 332, 354
 in pathology 185–6, 204
 in physics 404, 427, 432
 in physiology 174–5, 205, 449
 and Radcliffe Observatory 250
Rhodes Trustees 45, 102, 212, 432, 444–5, 448
Rice, D. T. 196
Richards, O. W. 28, 269, 285–6
Richards, T. W. 330, 440
Richardson, H. G. 132
Riley, D. P. 230, 232
Rio-Hortega, P. del 376–8
Ritchie, J. 182
Roaf, D. 424, 432
Robbins, L. C. 55
Roberts, E. A. H. 209
Robinson, G. M. 348
Robinson, Robert 345–52
 as consultant 366
 as D.Phil. supervisor 75
 inspires and helps Hodgkin 228–30
 honours 26, 305
 aids penicillin project 211
 as Perkin's chemical son 338
 suspicion of physical chemistry 361
 and polythene 367
 aids refugees 372, 380–1, 439
 supported by Royal Commission for 1851 Exhibition 445, 448
 attacked by Sidgwick 364
 helps Sutton 336
 dismisses Wrinch's cyclol theory 232
Robinson, Roy Lister 146–7, 157–8, 447
Robson, G. C. 285–6
Rockefeller Foundation:
 supports anatomy, human 203
 benefactor, third biggest 443, 447–8
 supports biochemistry 31, 34, 166, 186–7, 189–92, 443
 supports Bodleian Library 42–3, 314, 443

rejects botany 243
supports Cambridge University 45, 141, 443, 447-8
supports chemistry 336, 349
rejects craniology 197
supports experimental psychology 90
rejects Museum of History of Science 267
supports pathology 206-8, 210
supports pharmacology 215
and refugees 379-80, 391, 403, 414-15
supports Wrinch's research 231-2
supports zoology 277
Rollin, B. V. 417, 419, 431
Rose, J. D. 365
Rosenhain, W. 360
Rothamsted Experimental Station 123-4
Rouse Ball, W. W. 315, 443, 447
Royal Army Medical Corps 7, 10
Royal Astronomical Society 245, 247, 252, 317, 391
Royal Commission for the Exhibition of 1851: 345, 350, 445, 448
Royal Flying Corps 7-8, 11-12, 14, 87
Royal Geographical Society 223, 281
Royal Greenwich Observatory 391
Royal Holloway College, London 47, 319
Royal Society of London:
 supports biochemistry 192, 444, 448
 supports botany 239
 supports Bureau of Animal Population 219, 295, 444, 448
 supports chemistry 334-5
 and Florey 206
 supports metallurgy 360, 444, 448
 supports Oxford University Exploration Club 281
 supports physics 394-6, 400, 404, 415-16, 431
 presidency 26, 166
 reform of 356
 and Sidgwick 363
Russell, A. S. 22, 321, 327, 353, 422

Russell, B. 312
Rutherford, E. 6, 320, 370, 372, 383, 389, 396, 400, 404, 414, 430
Rydon, H. N. 349
Ryle, G. 92
Ryle, M. 424-5

Sachs, H. 376
Sachs, I. 378
Sadler, M. E. 372
St Bartholomew's Hospital 198
St Edmund Hall 64, 225
St Hilda's College 378, 449-50
St Hugh's College 449-50
St John's College 40, 113-14, 120, 139, 143, 302, 358, 367, 412, 419
St Peter's College 108
Sale, C. V. 292-3
Salters' Company/Institute of Industrial Chemistry 423, 432, 444
Sanders, A. G. 209-11
Sandford, K. S. 221-3, 277, 280
Schiebold, E. 228
Schlesinger, B. 377-8
Schlesinger, F. 249
Schlich, W. 13-14, 142-4, 146, 159
Schoental, R. 204
Schrödinger, E. 316, 390, 409-10, 422
Schwarz, H. 376
Science Area 29-46
 agriculture excluded 123
 and Asquith Commission 22-5
 and astronomy 254
 and botany 234, 240, 242
 and chemistry 347
 engineering science excluded 94-5
 experimental psychology excluded 91
 and forestry site 112, 146, 151
 and geology 223
 and mathematics institute 314
 and physics 428
 and physiology 177
 as Veale's monument 69, 437
Science Greats 82-5, 401
Science Masters' Association 330-1
Scott, S. G. 7, 167

Seward, A. C. 235
Sheffield University 206–7, 217, 345, 433–6, 438, 447
Shelford, V. E. 291, 293
Shell-Mex Oil Company 45, 221, 225, 444, 448
Sherrington, C. S. 164–81
  supports biochemistry 186–91
  admired by Clark 199
  as connoisseur 438
  and Dunn Trustees 184
  and experimental psychology 86–8
  supports Florey 205–6
  befriends Franklin 213
  supports Goodrich 272
  headhunted from Liverpool 438
  honours 26
  supported by Medical Research Council 442, 446
  admired by Peters 193
  and refugees 372
  research orientated 216, 389
  resigns aged seventy-eight 79
  and Rhodes Scholars 449
  and Science Area 40
  war work 9–10
Shotover Hill, Oxford 395
Shull, A. F. 290
Sidgwick, N. V. 360–4
  and Chemical Society 26
  as college fellow 60–1, 78, 331, 335–7, 357
  as connoisseur 58, 438
  and Department of Scientific and Industrial Research 446
  and forestry site 151
  lectures at Magdalen 319
  dismisses Perkin's research 344
  readership 76
  provokes Robinson 350
  and Royal Society reform 356
  and Science Greats 84–5
  denigrates Soddy 321, 353
  as systematizer 305, 340
  love of USA 440

Simon, F. E. 76, 374, 381, 384, 397, 400–4, 406–7, 412–23, 432
Sinclair, H. M. 165, 180, 193
Singer, C. J. 220, 261–3
Singer, D. W. 262
Skilbeck, D. 120–1
Skinner, H. W. B. 396
Sladen Fund 281
Smith, G. E. 198
Smith, G. W. 6, 270, 274
Smith, J. C. 350
Smith, M. 87
Snow, G. R. S. 233, 374, 450
Snowden, P. 118
Society for Freedom in Science 280
Society for the Protection of Science and Learning 369, 371–3, 376–7, 379, 423
Soddy, F. 353–61
  hostility to colleges and college laboratories 21, 306, 320–2, 324–6, 328
  relations with demonstrators 324, 451
  ejects Egerton 397–8
  honours 26
  ejects Merton 396
  and physical chemistry laboratory 320–2
  and Radium Corporation 184
  opposes redefinition of Dr Lee's chair 336–7
  discourages H. W. Thompson 334
Sollas, W. J. 12, 80, 221–4, 373
Solvay Physics Conference 384, 390
Somerville, W. 13, 113–16, 127, 133
Somerville College 228–30, 233, 378–9, 449–50
Sommerfeld, A. 409
Southern, H. N. 298
Southwell, R. V. 27, 40–2, 46–7, 97–106, 136, 141, 151, 379–80, 399, 428
Spencer, W. B. 260
Speyer, E. 271
Spiller, R. C. 228, 367
Spooner, W. A. 55
sport 270, 341

Index

Springall, H. D. 350
Squire, C. F. 414–15
Squire, H. B. 103
Stallybrass, W. T. S. 19, 65
Stamp, J. C. 331
Staveley, L. A. K. 331, 335–6
Steel, J. L. S. 366–7
Steiner, G. 377
Stephenson, W. 90–2
Stevens, T. S. 346
Stewart, M. J. 206
Stiles, W. 242
Stout, G. F. 86
Stratton, F. J. M. 256
Strauss, F. 380
Sucksmith, W. 360
Sudhoff, K. F. J. 262
sugar beet 128, 133–4, 136
Sugar Beet and Crop Driers Ltd 71–3, 134, 136
Sutton, L. E. 61, 331, 335–6, 350, 352, 363, 442
Svedberg, T. 192, 444
Swallow, J. C. 368
Sydney University 242, 438
Szilard, L. 407, 422–4

Tallents, S. G. 293
Tansley, A. G. 27, 34–5, 152, 234–42, 280, 293, 296, 298, 374
Taylor, F. S. 267
Taylor, T. W. J. 350, 363
Thomas, H. H. 242
Thompson, C. H. 308
Thompson, H. W. 331, 334–6, 353, 358, 366, 442
Thompson, J. H. C. 307, 309–10
Thompson, R. H. S. 165, 193
Thomson, A. 16, 21, 80, 194–7, 202–3
Thomson, J. J. 65, 382, 437
Thorpe, J. F. 340
Tinbergen, N. 274
Titchmarsh, E. C. 314–15
Tizard, H. T. 11, 26, 47, 65, 105–6, 360–1, 383–6, 397, 400, 426, 446
Todd, A. R. 26, 349, 352

Topley, B. 365
Toronto University 387
Touch, A. G. 424
Townsend, J. S. E. 11, 80, 84, 105, 151, 339, 370, 383, 386–7, 392, 427–8, 431
tractors 134, 139–40
Trapnell, C. 280–1
Trikojus, V. M. 346
Trinity College 64, 95, 310
  see also Balliol–Trinity chemistry laboratory
Troup, R. S. 31, 34, 142–56, 159
Tuck, J. L. 423–5, 432
Tucker, B. W. 218, 268–9, 273, 277, 284, 298–301, 303–4
Turner, H. H. 31, 79, 244–8, 250–1, 253, 257, 447
Tylor, E. B. 259–60
Tyndall, A. 390

Ubbelohde, A. R. J. P. 399–400
universities, British, comparisons 433–6
University College, London 47, 90, 198, 201, 204, 220, 263, 339, 397, 421, 433–6, 449
University College, Oxford 64
University Grants Committee 108, 159, 295, 428, 432, 436–7, 441, 445, 447
Upavon, Central Flying School 8, 11, 385, 394
Uppsala University 192

Van de Graaff, R. J. 427
Vatican Observatory 246, 374
Vaughan, J. M. 165
Veale, D. 68–73
  and agriculture 120, 141
  and botany 239–40, 243–4
  and biology teaching 270
  supports Bureau of Animal Population 295–6
  on entomology 291
  as facilitator 219, 303
  and forestry 38, 151–2, 155, 158–9
  alliance with Furse 147

Veale, D. (*cont.*)
  as modernizing bureaucrat 437
  and Museum of History of
    Science 266
  and Nuffield College 109–10
  on physiology laboratory 181
  on professoriate 78
  and Radcliffe Observatory 254
  and refugees 371
  and Science Area 38
  on vocational training 159
Veblen, O. 309, 313, 440
Verdon-Smith, W. R. 58
Vernon, H. M. 10
Vevers, H. G. 281
Vice-chancellors 65, 68
Vienna University 373, 376–7, 395
Vines, S. H. 234–5
vocational training 93–4, 96–9, 115, 159, 161–3, 445
Vollum, R. L. 185, 204, 442

Wadham College 310, 392–3
Walden, A. F. 56–7, 331, 356–7
Walker, E. 191
Walker, J. J. 27, 289, 291
Walls, E. 365
Walshe, F. M. R. 175
War, First World 6–17, 128, 169–71, 183, 189, 262, 341–2, 355–6, 383, 385, 437
War, Second World 204, 234, 296, 327, 424–5, 428, 432, 451
Warrell, J. 440
Waterhouse, P. 343
Waters, E. G. R. 288–9
Watson, J. A. S. 116, 119–21, 123, 128, 136
Watts, Mrs H. 90–1, 443, 447
Webb, C. C. J. 84
Weissberger, A. 380
Weiler, Gerhard 380
Weiler, Grita 378
Welch scholarship 269
Welch Trust 295
Weldon, T. D. 62

Wells, A. F. 229
Wells, A. Q. 297
Wells, J. 265
Whitehead, J. H. C. 63, 308–9, 379, 440
Whitley, E. 182, 186–9, 191, 443, 447
Whitrow, G. J. 74
Whitteridge, D. 164
Wilde, H. 85–7
Wilkins, W. H. 237–40, 242–3, 442
Williamson, A. T. 332
Willmer, E. N. 275
Wilson, E. B. 358
Wilstätter, R. 330, 380–1
Wimperis, H. E. 426
Winternitz, A. 378–9
Witts, L. J. 162, 208, 378
Witherby, H. F. 301, 304
Witwatersrand University 49
Wohl, K. 374–5
Wolfenden, J. H. 328, 331, 333–4
women:
  in agriculture 129
  in anatomy, human 197, 200–1
  in botany 235, 238
  in chemistry 354–5
  and college fellowships 310, 449–51
  and college headships 165
  in engineering science 101
  and entomology 286
  and experimental psychology 90–1
  in mathematics 313, 450
  become full members of university 16
  in mineralogy 228–33
  in pathology 204, 207, 210–11
  in pharmacology 214–15
  in physics 417–18
  in physiology 174–6, 179
  refugees 375, 377–9
  in zoology 268
Woodforde, F. C. 289
Woodward, L. A. 327, 331, 335–6
Woodward, R. B. 352
Woodward, R. C. 119
Woollard, H. 198–9, 204
Worcester College 108
Wright, E. M. 313

Wright, S. J. 135–6, 140
Wrinch, D. M. 231–2, 450
Wynne-Edwards, R. M. D. 97
Wynne-Edwards, V. C. 268, 284–5, 299
Wynne-Jones, W. F. K. 334

X-ray crystallography 225–33

Yale University 175, 178–80
Yemm, E. W. 238

Yerbury, J. W. 289
Young, J. Z. 28, 62, 268–9, 273, 277–8, 443

Zanstra, H. 255, 374
zoology 217–19, 268–304, 373, 428, 441, 449
zoology tutorials 62
Zuckerman, S. 50–1, 201–4, 224–5, 442, 447